1st International Symposium on Advances and Innovations in Mechanical Engineering (ISAIME 2019)

AA007635

Advancing in Mechanical Engineering Innovation for Sustainable Development

IOP Conference Series: Materials Science and Engineering
Voulme 694

Jakarta, Indonesia
9 - 10 October 2019

ISBN: 978-1-7138-0986-9
ISSN: 1757-8981

Printed from e-media with permission by:

Curran Associates, Inc.
57 Morehouse Lane
Red Hook, NY 12571

Some format issues inherent in the e-media version may also appear in this print version.

Additional copies of this publication are available from:

Curran Associates, Inc.
57 Morehouse Lane
Red Hook, NY 12571 USA
Phone: 845-758-0400
Fax: 845-758-2633
Email: curran@proceedings.com
Web: www.proceedings.com

TABLE OF CONTENTS

IOP Conf. Series: Materials Science and Engineering **694** (2019) 011001 doi:10.1088/1757-899X/694/1/011001

Preface

International Symposium on Advances and Innovations in Mechanical Engineering (ISAIME) is the first international symposium initiated by Badan Kerja Sama Teknik Mesin (BKSTM), Indonesian national organization on mechanical engineering. The symposium was hosted by the Department of Mechanical Engineering, Universitas Trisakti, Indonesia and held on 9 – 10 October 2019 in Aston Grogol Hotel Jakarta and Trisakti campus, Jakarta, Indonesia. This symposium is to accommodate the eagerness of researchers, lecturers, and students to publish the results of their research at a reputable international seminar. BKSTM holds an annual national seminar called Seminar Nasional Tahunan Teknik Mesin (SNTTM), an annual national seminar on mechanical engineering which takes place at different BKSTM member universities. ISAIME was held in conjunction with SNTTM XVIII 2019. The theme for ISAIME is "*Advancing in Mechanical Engineering Innovations for Sustainable Development*". The event is bringing together researchers, scientists, and engineers from academia and industry to present and exchange ideas with a focus on the advancement of mechanical engineering and technologies.

The Editorial Committee members have been working very hard and patient to meet the deadline of review and to prepare the conference proceedings. The final ISAIME Program consists of 50 papers mainly on the topic of mechatronics and automation, design and optimization in mechanical and thermal system, energy conversion, advanced manufacturing, and materials engineering. The proceedings would be published in a volume by IOP Conference Series: Materials Science and Engineering.

We would like to express our sincere gratitude to all the committee members for their hard work, precious time and endeavor preparing for the conference. Our deepest thanks also go to the sponsors, staffs, and volunteers for their generosity and long-hours work they have given to make the conference possible. The last but not least, we would like to thank our keynote speakers and all of the authors, speakers and participants for their great contributions to the success of ISAIME 2019.

Chairman of ISAIME 2019

Daisman P.B. Aji, Ph.D

IOP Conf. Series: Materials Science and Engineering **694** (2019) 011002 doi:10.1088/1757-899X/694/1/011002

List of Committee

Advisory Board:
1. Prof. dr. Ali Ghufron Mukti, MSc, PhD (Rector, Universitas Trisakti)
2. Prof. Dr. Indra Surjati (Dean of Faculty of Industrial Technology, Universitas Trisakti)

Steering Committee:
1. Dr. Ario Sunar Baskoro, ST, MT, M.Eng (Universitas Indonesia & BKSTM)
2. Ir. Tono Sukarnoto, MT (Universitas Trisakti)
3. Dr. Rianti Dewi Sulamet Ariobimo, ST, M.Eng (Universitas Trisakti)

International Advisory Board:
1. Professor Dr.Eng. Teuku M. Indra Mahlia (Universitas of Technology Sydney)
2. Associate Professor Dr. M. Akbar Rhamdani (Swinburne University)
3. Associate Professor Dr. Denni Kurniawan (Universitas Teknologi Brunei)

Chairman: Daisman P.B. Aji, ST, Ph.D

Editorial:
1. Dr. Sally Cahyati (Universitas Trisakti, Indonesia)
2. Dr. Triyono (Universitas Trisakti, Indonesia)
3. Professor Dr. Yulfian Aminanda (Universiti Teknologi Brunei)
4. Dr. Fethma M. Nor (Curtin University, Malaysia)
5. Professor Dr. Guojie Huang (GRIMAT Engineering Institute, Beijing, China)

Secretariat: Renny, SH

Treasurer: Tumini, SH

Event : 1. Dr. Dorina Hetharia
2. Dr. Supriyadi

Publicity : 1. Fajar, ST
2. Abigunto, ST

Proceeding : Dr. Sjahrul Annas

Sponsorship: Ir. Noor Eddy, MT

Equipments & Transportation: 1. A. Gozhali
2. Ir. Yusep Mujalis, MT

IOP Conf. Series: Materials Science and Engineering **694** (2019) 011003 doi:10.1088/1757-899X/694/1/011003

Photos

Photo Session

Welcome Speech by the Chairman

IOP Conf. Series: Materials Science and Engineering **694** (2019) 011003 doi:10.1088/1757-899X/694/1/011003

Plenary Session

Keynote Speech

Peer review statement

All papers published in this volume of *IOP Conference Series: Materials Science and Engineering* have been peer reviewed through processes administered by the proceedings Editors. Reviews were conducted by expert referees to the professional and scientific standards expected of a proceedings journal published by IOP Publishing.

IOP Conf. Series: Materials Science and Engineering **694** (2019) 012001 doi:10.1088/1757-899X/694/1/012001

Intelligent System for Gearbox Fault Detection & Diagnosis Based on Vibration Analysis using Bayesian Networks

Dedik Romahadi[1], Hui Xiong[2] and Hadi Pranoto[1]

[1] Mechanical Engineering, Mercu Buana University, Indonesia
[2] Mechanical Engineering, Beijing Institute of Technology, China

E-mail: dedik.romahadi@mercubuana.ac.id

Abstract. Moving energy from one machine to another and functioning to reduce speed while increasing torque is the ability of the gearbox. Due to many components and the structure between the components is fairly complex, thus to be able to detect the initial damage, sophisticated methods is needed. Vibration analysis is a method that has been effective in detecting the initial damage that occurs in machine. But it takes time and costs are not small to implement. The purpose of this study is to create an intelligent system capable of detecting gearbox damage based on data obtained from vibration measurements. Merging of two methods of vibration analysis and Bayesian networks is done to be able to design the system with the expected results. A series of multistate nodes are applied to the network and a system review is performed. Results are given and compared with results provided by the manual analysis. The results indicate that the system is feasible and reasonable which can assist inidentifying gearbox damages. This study definitively answers the problem of how to design an expert system capable of replacing the work of an expert on vibration analysis services.

1. Introduction

Almost all industries use gearboxes to carry out a series of production, gearboxes may operate under varying load conditions. In industrial applications, gearboxes are a very important component. Gearbox must work optimally to be able to complete the task as planned. Termination of the production process and other undesirable things can occur due to gearbox damage, even human casualties. Therefore, damage to the gearbox needs to be known optimally as early as possible to ensure the operation of an efficient mechanical transmission system and prevent accidents [1,2].

One condition monitoring technique that are reliable, easy to do and is applied in real life is vibration analysis. Vibration frequency characteristics are able to show most types of damage that occurs on a rotating machine and hence, analysis of vibration characteristics useful to identify most types of fault. Vibration analysis is an attempt to minimize damage to the engine components propagate to other engine components. Vibration analysis also helps to know the structure of the machine, whether it is still in stable condition, in accordance with the standard or vice versa. The latest technique commonly used to diagnose gearbox is by analysing vibration signals taken from the gearbox casing. The aim is to detect the location and type of damage at the initial stage of development and monitor the condition, to estimate the lifetime of the machine and prepare appropriate treatment. Gear Mesh Frequency (GMF) and its repetitions or harmonics and modulation phenomena that produce sidebands are the main components in the gearbox domain frequency (spectrum). Disturbance conditions can be observed from increases in the number and amplitude of sidebands. In addition, the source of damage related to the sidebands distance [3].

Not all engineers know how to analyse vibration on industrial machines, especially on the gearbox. To analyse the vibrations of the gearbox requires a long learning time and experience in analysing using standard software from vibration equipment manufacturers. Usually in this case the company must pay dearly for expert staff services to measure and analyse the condition of the gearbox using vibration data. Even companies that already have a vibration analyser still have to call and pay for expensive because they do not understand how to analyse vibration data taken from gearbox. In addition, an expert staff in vibration analysis also still needs time to analyse spectrum data. Therefore, in solving these problems the authors propose to create a system that can read the spectrum of vibration data specifically to diagnose gearbox. A system that requires only input spectrum data and specifications from the gearbox. Furthermore, the system can provide important information about the gearbox condition, covering where the point of damage, what actions are needed and how much damage level.

The medical domain has inspired the concept of diagnosis. At present, the types of machine failures can be identified as root causes with sophisticated diagnostic methodologies. However, the diagnosis is made after the damage has occurred, including in the reactive action. In fact, observing certain machine conditions has a problem fault diagnosis in detecting the characteristics of damage that might be hidden in the dataset. Besides, detecting various damage to the rotating machine is suitable using monitoring vibration conditions. Signal processing, data acquisition, extraction of other features and features to reduce steps if required and the last correct classification step that can be obtained from historical data, expert insight and physical formation, all of which are a common framework for damage diagnosis systems [4, 5]. BN is able to make the possibilities that occur in the future very well, BN is able to handle uncertain causal relationships, update probabilities, multi-state variables, carry out two-way interpretations and handle the lack of data [6, 7, 8, 9]. In various scientific fields, BNs are widely used. BN's application practices in life are very widely used. BN has long been an interesting topic and continues to this day, supported by the rapid development of computer science, especially in data mining and machine learning [10]. Uncertainty arises and become the main component when designing knowledge-based reasoning and decision makers methods [11]. Related to expert system in manufacturing, Kobbacy and McNaught et al. emphasize that BNs are most useful when dealing with uncertainty [12]. BN is a reliable tool for developing expert systems in the field of artificial intelligence. It is advantageous in reflecting and diagnosing intricate systems with incomplete, uncertain and even conflicting information. The probabilistic graphical model of BN shows a probabilistic dependency relationship in a variable connected via an acyclic graph. Pearl introduced in the early 1980s [13, 14], in probabilistic inference and the realm of knowledge discovery has been applied successfully [15].

Previous techniques of artificial intelligence systems for damage detection use fault tree analysis (FTA), It is an analytical tool that graphically translates combinations of errors that cause system failures. This technique is useful for describing and assessing events in the system. This FTA method is effective in finding the core of the problem because it ensures that an unwanted event or the loss caused does not originate at one point of failure. FTA identifies the relationship between causal factors and is displayed in the form of a fault tree involving a simple logic gate [16, 17]. The advantage of using the FTA method because it has a clear, logical, intuitive, and usable system compared to some other systems. Nevertheless, the computation of the probability of failure becomes intricate when the structure of a Fault Tree is large, therefore an uncertain causal relationship and a multi-state variable cannot be done with an FTA [18, 19, 20], the component fault and reasoning generally do not pay attention to frequency failures and relationships for different components of the system. Calculation and analysis assuming that failures for independent components or subsystems can be subject to major errors. Regarding intelligent fault inference system for rotating machine and the methods used in this study was adopted from his paper, Bin Gang Xu used system fault assessment is structured by the BN concept, makes several failure sequences and the relationship between component failure with the BN model and validated fault assessment estimates. The proposed network applications in a flexible rotor's vibration fault diagnosis showed strong inferential abilities. Various error inferences can be continued, related to a single error, a combination of maximum errors and maximum network status, including with reasonable probabilistic information [21]. Developing BNs applications for rotating machines, this study uses spectrum frequencies as input data to identify symptoms of damage that occurs in the gearbox.

2. Methodology

2.1. Bayesian networks

Bayesian Inference and Bayesian network (BN) are decompositions of the Bayes' theorem. The development of the Bayes theorem began in the 1760s, which new information to update possibilities. Some previous statisticians, for instance Pierre-Simon La-Place, decision-making methods and systematic statistical inference are developed from the Bayes formula [22, 23]. A BN was proposed by Pearl in 1988, which is the method that continues to be discussed today [24]. A special method called Bayesian Analysis that uses previous statistical information and success in many practical implementations.

Bayes' Theorem:

$$P(N \mid M) = \frac{P(M \mid N) \times P(N)}{P(N)} \tag{1}$$

Building the BN, a system is constructed with a statistical technique known as the Bayes theorem namely Conditional Probability. Calculation of the probability of an event N if known event M has occurred is called Conditional Probability, represented by P(M), as shown in equation 1. This theorem is used to calculate the odds of a data set to enter into a particular class based on inference existing data. The combined probability factoring can be obtained by the regulation rule of probability theory, as shown in equation 2.

$$P(x_1, x_2, x_3, \ldots, x_n) = \prod_i P\left[x_i \mid Parents(x_i)\right] \tag{2}$$

A BN is a Probabilistic one simple graphical model constructed from theory probability and graph theory. Probabilistic theory is related directly with data, while graph theory is related immediately with the representation you want to get. As for example, a BN can represent relationships between disease and symptoms. BN is able to calculate the probability of various symptoms of the disease. Building BN, an expert is needed to apply a concept of knowledge to the network. Knowing how a domain and its uncertainty can be applied in the BN and see how to use BN for domain needs. When observing the values of several variables and condition them in new information. This conditioning process is called probability inference that is carried out through the network. The information flow on the BN is not limited to the direction of the arc. In the probabilistic network, this becomes the task of calculating the posterior probability distribution for a set of request nodes, some evidence nodes provide values [24, 26].

Make probabilistic inference is the ability of the BN. Probabilistic inference is predicting the grade of an element that can't be known immediately by using the values of other known variables. An example of probabilistic inference is determining the conditional probability of the type of gearbox damage based on the symptoms of damage if it is known that the frequency spectrum characteristics of damage occur. Probabilistic inference can be done if first obtained Joint Probability Distribution (JPD) of all the modelled variables. JPD is the probability of all variable events occurring simultaneously. Probabilistic inference can be done if the BN structure has been built. In diagnosing gearbox damage, the relationship between variables and the probability of variable values is unknown. The construction of BN from the data consists of structural construction or also called the qualitative stage, which is looking for connectivity between the modelled variables and parameter estimation or also called the quantitative stage, which is calculating probability values.

2.2. Vibration signature analysis

Analysis of rotating machine damage usually uses vibration signals in the form of spectrum frequencies obtained from Fast Fourier Transform (FFT). This method is a reliable tool for analyzing the location

IOP Conf. Series: Materials Science and Engineering **694** (2019) 012001 doi:10.1088/1757-899X/694/1/012001

of damage. FFT combined with vibration level values provides information about the condition of the machine, location and cause of damage. Based on the measurement history can reveal the remaining life of the machine or until the damage becomes critical. The spectrum frequency also presents the amplitude of the vibration in each source of damage occurring. Therefore, we can track where the location of the damaged component is based on its frequency. Every damage to the machine creates a frequency that has its own characteristics so that it can be used for diagnosis [27]. Vibration characteristics or vibration signatures are obtained when the machine is operating. Actual vibration signals derived from transducers can be considered as frequency characteristics, but usually refer to spectrum frequencies. Vibration signal measurement with standard transducer assisted acceleration produces accurate vibration analysis. The FFT is guided by the principle that all signals (not ideal signals) can be generated by the sum of sine waves. Conversely any signal can be broken down on its components in the form of a sine wave, it is important to say that the frequency spectrum fully represents the vibration signal. No information is lost because of the conversion from time domain to spectrum frequency, if the phase difference between components is also included [28, 29].

Spectrum characteristics provide valuable information about the damage location. The frequency and magnitude in the spectrum are related to machine rotation. For example, an amplitude line appears on 1X, 2X, 3X, ... 7X or something else. The relationship that belongs to each frequency line, for example the line that appears at 1X is greater than 2X or appears harmonic or there are sidebands. Knowing the source of vibration amplitude originates, for example 1X is an unbalance fan, 5.34 is bearing damage, etc.

2.3. Vibration assessment

Figure 1 shows type of ISO provides a standard for measurements carried out on site regarding the level of vibration [30]. The standard set applies to machines that have a capacity more than 15 kW and operating speeds in the middle of 120 RPM and 15000 RPM.

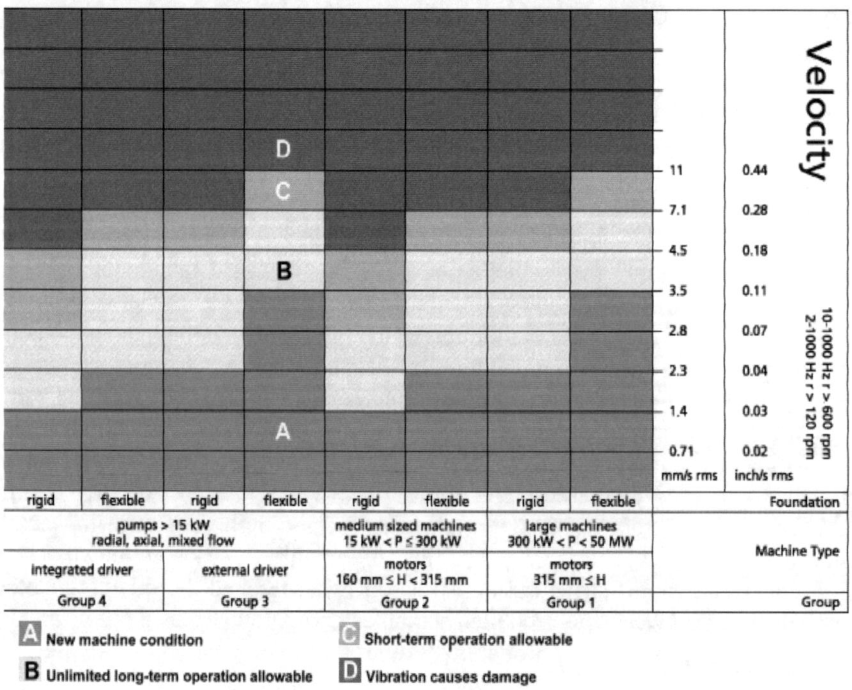

Figure 1. 10816-3 ISO Standard

IOP Conf. Series: Materials Science and Engineering **694** (2019) 012001 doi:10.1088/1757-899X/694/1/012001

3. Implementation

3.1. System analysis

Figure 2 shows the process of intelligent system. The input data in diagnosing the gearbox damage is in the form of gearbox specification data including gearbox RPM input, gear RPM, number of teeth on gear and pinion, number of ball bearings, vibration value and also input spectrum data. The data already entered will be processed to find out the type of frequency that appears in the data spectrum. Furthermore, the identification will be used as evidence to indicate whether or not there is a spectrum line related to the damage to the gearbox. Identifying the frequency using predefined formulas. Suppose the obtained frequency value of GMF is 275 Hz, then the system will see whether there is a line that appears at that frequency or not. Existing evidence will be processed by calculating its probability value using BNs method.

Based on the input data obtained will be calculated probability value through several stages using BN method, which starts from the determination of parameter values for each symptoms of damage, then determine the value of conditional probability, after the two values obtained then the system will be calculated the value of joint and posterior probability of each type of gearbox damage that is adjusted to the BNs structure and the posterior probability value is used as the probability inference of the type of gearbox fault. BN produces relational information and conditional probabilities through two-way propagation between input and output nodes. Also, in the application of real cases generally use multistate nodes. So, from the consultation conducted by the user will get the type of fault that occurs on the gearbox and the percentage of the fault. Users will know what action takes precedence.

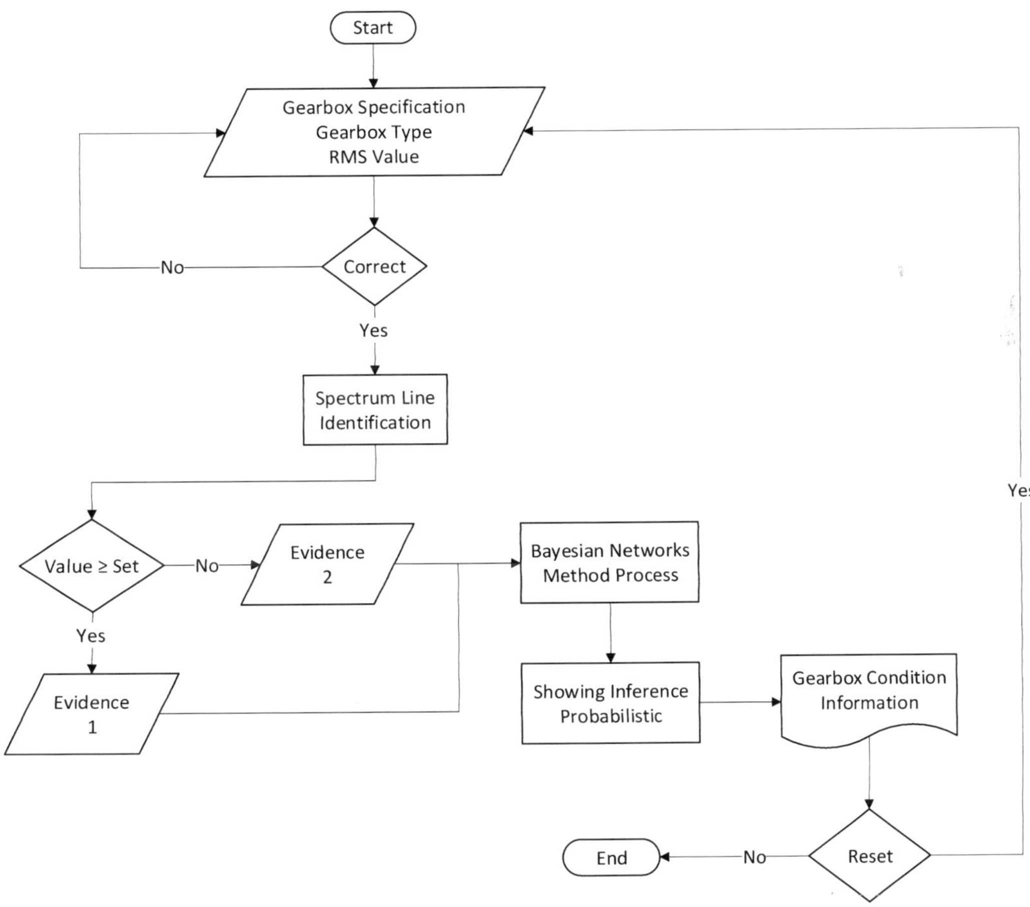

Figure 2. Intelligent system flowchart

3.2. Bayesian Networks approach

Designing expert systems in accordance with the concept map that you want to address in solving problems. This concept supports the decomposition of knowledge that explains each component and work procedure as discussed in figure 2, then created an expert system flowchart to illustrate in detail the process flow of the system works.

According to the stage of the vibration analysis in gearbox fault detection, we can build the structure of BN. In the first process, we assume the cause of the damage that occurs in the gearbox based on the emerging spectrum line. Then built a BN in such a way as to be able to show what kind of damage and what actions should performed by the user. The structure BNs model is shown in Fig. 3. Corresponding to Pearl, relational probability distribution is obtained from expert knowledge [24]. Information on object history is useful in determining prior probability values. This type of information is usually available in every corporation for instance machine damage report. So here BN acts as a knowledge base while the part of the control system is Bayesian inference rules.

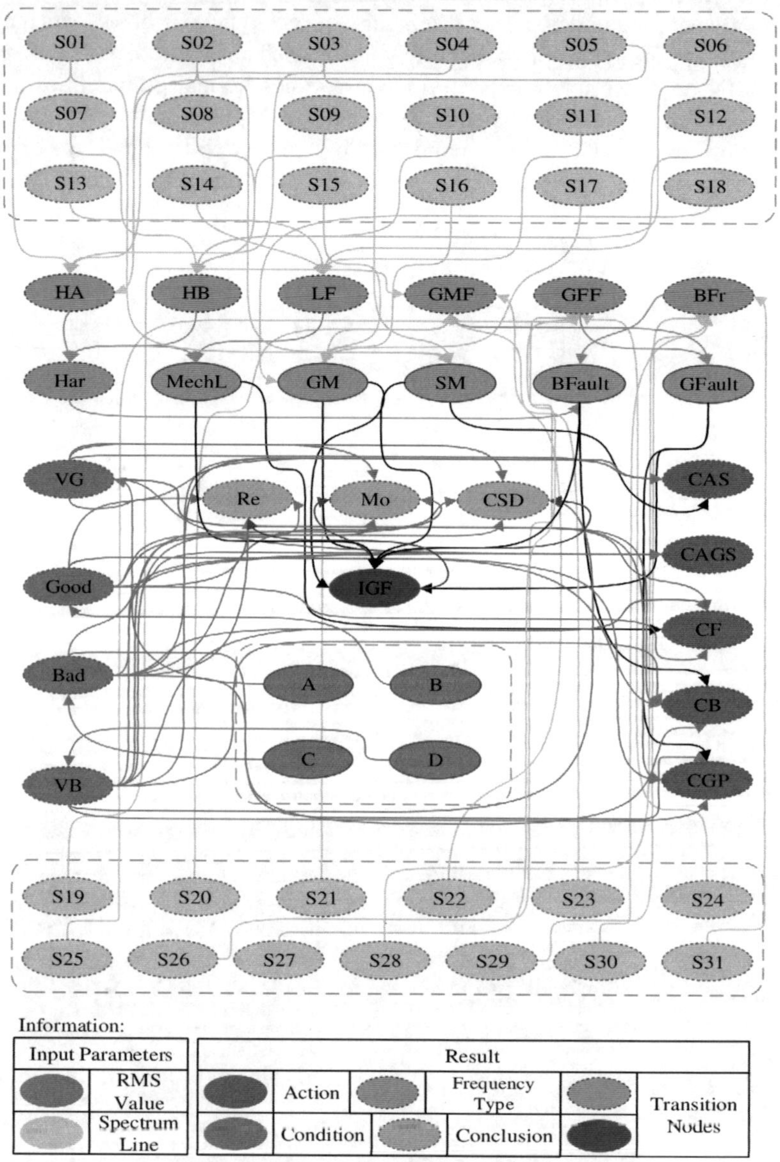

Figure 3. Bayesian Networks system

There are 60 nodes in BNs structure, as shown in figure 3. In solving the problem of a large number entering conditional probability value and to further ease the computer work, some nodes are added to the network such as IGF, VG, Good, Bad, VB, HA, HB, etc. For instance, at node HA, there are five parent nodes which are interconnected. The HA node will generate 64 CPT values, so time and memory will be more needed if the root node has more parent nodes. Associated CPT held every root node, its slump, only a row of representative prior probabilities. So, when a network has many parent nodes or parent nodes containing a lot of classification values, CPT will also be very large. The CPT criterion decided by the number of parent nodes. CPT for Boolean variables with x parent node having a 2x + 1 probability formula. Nodes such as Monitoring, Repair and Check Data Spectrum in Detail (CSD) are added to serve to infer what actions should be performed by the user. Previous probabilities are needed to make casual dependencies.

The CPT table shows the level of damage to the gears. An expert is required to fill each table with a percentage of 0% - 100%. Suppose that the chance of occurrence of GMF if obtained evidence of symptoms of S19 and S20 is 90%. Determination of the value of 90% is obtained from information obtained by an expert. In analysing the damage of the gearbox using vibration data, the vibration value generated on the velocity spectrum is very small because generally gear mesh frequency (GMF) occurs at high frequency, and this will affect the small value of the resulting vibration. However, although the value is small does not mean there is no damage to the gears. While one of the main advantages of the BN when additional evidence is given, the BN system can determine conditional probability values in each node and calculate more accurate probability propagation. In the diagnosis of gearbox damage, conditional probabilities provide important information.

3.3. Software implementation

Material testing uses vibration data of multi stage gearbox, as shown in figure 4, that measurement was taken from a synthetic yarn factory where is the data contains several kinds of measurement such as overall velocity and spectrum velocity data.

Figure 4. Multi stage gearbox

gearbox has a rigid type foundation. There are 10 gears that rely on each shaft and bearings. Measurement of vibration is taken at 4 points of shaft and each point is taken 3 axis point covering horizontal, vertical and axial. Each of the 3 points is taken which has the greatest vibration value. Figure 5 shows the spectrum obtained from gearbox measurements at each accessible bearing location. There are 4 spectrums chosen because it has the largest overall vibration value. Figure 6 describes the gearbox structure in detail including the type of bearing and number of teeth that are used. According to the complete specification information of the employee staff, the gearbox is grouped into ISO group 2 rigid standard.

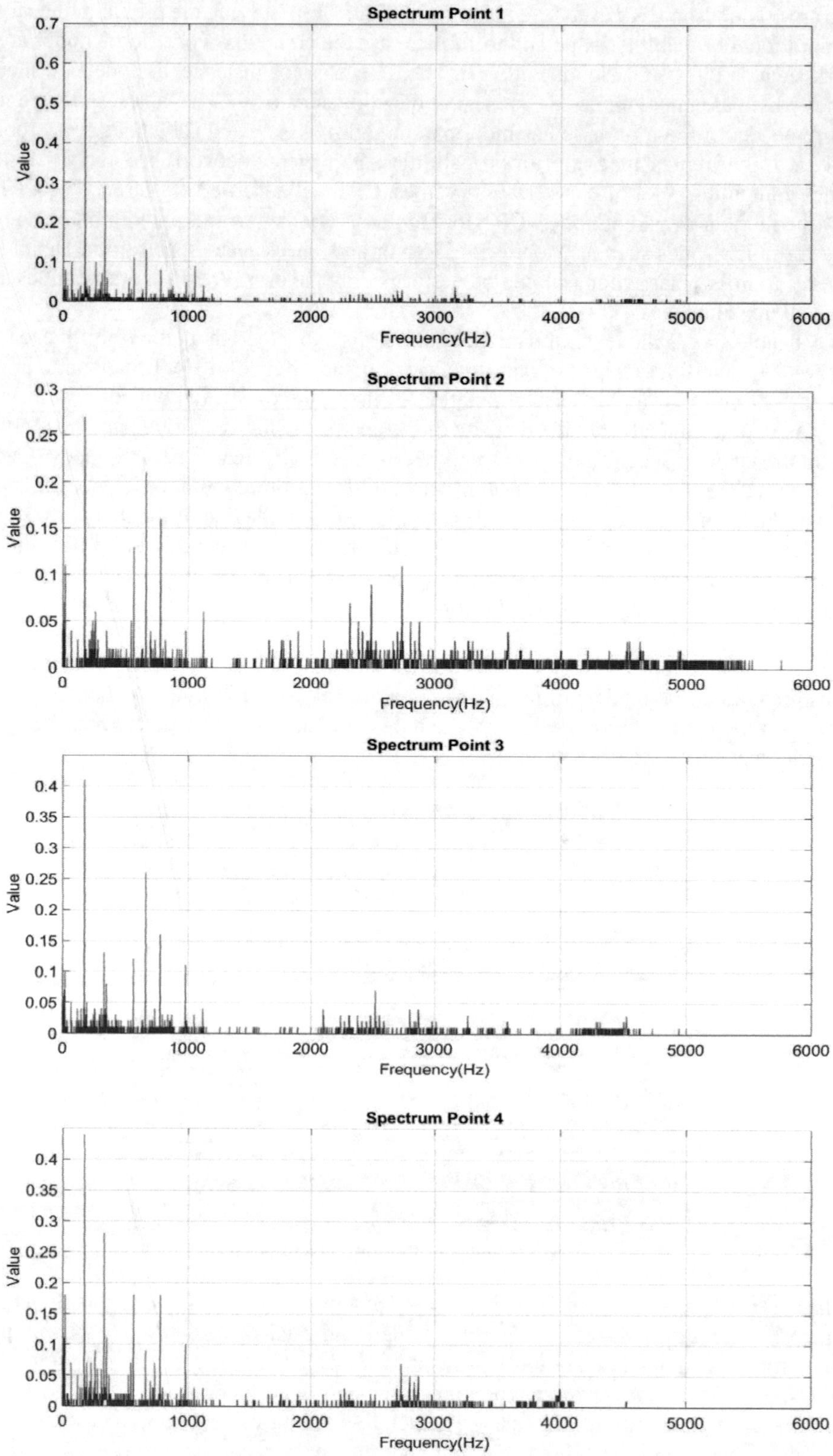

Figure 5. Spectrum of gearbox

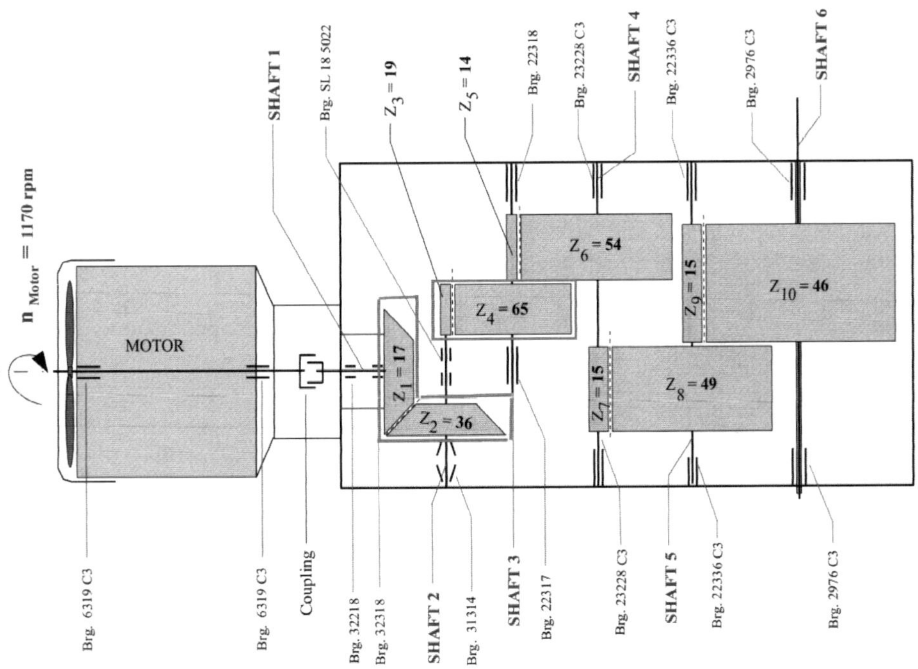

Figure 6. Gearbox structure

3.4. Diagnosis result

Figure 7 shows the calculation results of BNs that have been built, there are symptoms of possible indications of damage to the gears and bearings.

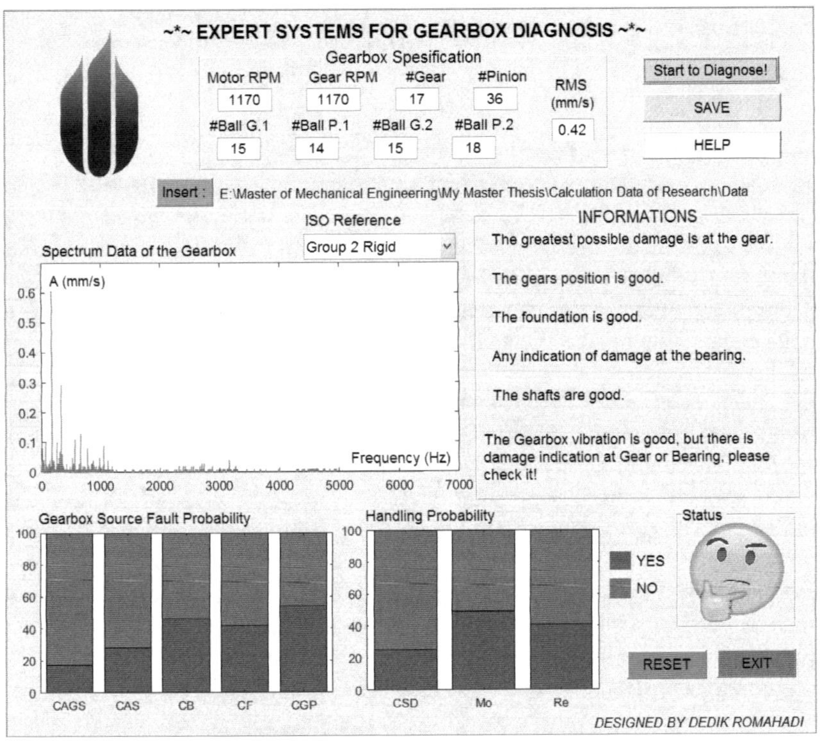

Figure 7. Diagnosis result of point 1

On the handling probability side, the highest percentage value is monitoring, this is because the vibration value on the gearbox is still in good condition. As noted earlier, in particular, gears and bearing problems, the small vibration values cannot be used as a benchmark to determine that the gears and bearings are still in good condition or vice versa. Users should still check the state of the gears and bearings according to experience or history of the gearbox.

Table 1. Fault detection and diagnosis result

Measurement Point		1A	1B	2	3	4
Fault Probability (%)	Gear	0.54	0.55	0.52	0.55	0.2
	Gear Mis.	0.17	0.15	0.15	0.59	0.15
	Foundation	0.42	0.42	0.42	0.42	0.42
	Bearing	0.46	0.46	0.46	0.48	0.47
	Shaft Mis.	0.28	0.28	0.28	0.28	0.29
Action Prob. (%)	Monitoring	0.49	0.49	0.49	0.49	0.5
	Repairing	0.41	0.41	0.41	0.41	0.41
	Checking Spectrum	0.25	0.25	0.25	0.25	0.25

Table 2. Comparison of Diagnosis result

No	Diag. Point	Symptoms at Spectrum	Manual Diagnosis	Intelligent System
1	1A	0.3X-0.7X, 3.5X, Gear, GMF, GAPF, FHT, GFN and BPFI	Gear Mesh, Hunting Tooth, Gear Phase Assembly and Inner Race Bearing Problems	Gear and Bearing Problems
2	1B	0.3X-0.7X, 3.5X, GMF, 2GMF, 5GMF, GAPF, 2GAPF, FHT, GFN, BPFO and 2BSF	Gear Mesh, Hunting Tooth, Gear Phase Assembly, Outer Race Bearing and Ball Bearing Problems	Gear and Bearing Problems
3	2	0.3X - 0.7X, 5GMF, FHT, BPFI and FTF	Gear Mesh, Hunting Tooth, Inner Race Bearing and Cage Bearing Problems	Gear and Bearing Problems
4	3	0.3X-0.7X, Gear, 2Gear, Pinion, 2Pinion, GMF, 5GMF, GAPF, 2GAPF, FHT, GFN, BPFI, BPFO, 2BSF and FTF	Gear Misalignment, Gear Mesh, Hunting Tooth, Gear Phase Assembly, Inner Race Bearing, Outer Race Bearing, Ball Bearing and Cage Bearing Problems	Gear, Mis. Gear and Bearing Problems
5	4	1X, 0.3X - 0.7X, BPFO, 2BSF and FTF	Outer Race Bearing, Ball Bearing and Cage Bearing Problems	Bearing Problems

According to a series of diagnostic tests that have been done on the gearbox, we can know easily what damage happens to the gearbox. The results of each calculation using different BNs methods in each trial, this is due to the difference in evidence obtained from each spectrum data against the gearbox specification data. We can clearly examine the evidence data that is entered on the BNs method. Based on evidence of symptoms obtained, we can conclude whether the results of the diagnosis are appropriate or not. The results of each diagnosis indicate the suitability of the design with the intended expectations.

4. Conclusion

Designing an intelligent system, which considers both vibration analysis and BN system with multiply states, using discrete failure probability distribution. BN based on the vibration frequency characteristic analysis procedure applied in the implementation system to better present BN applications in expert systems so as to produce accurate information from the gearbox condition. The advantages of BN are indicated by two-way inference for conditional probabilities. Damage to the components of the gearbox can be seen after proof of the spectrum added to the BN. A more accurate and faster decision can be made to ensure the performance of the expert system and provide valuable information regarding damage detection and gearbox maintenance. Results of diagnostic testing obtained matching results between the tests conducted manually and by the expert system. BNs method have been effectively used to the diagnostic expert system for the source of gearbox damage, so it can provide a fast diagnosis and the probability of occurrence of the location of the damage.

5. Acknowledgments
This work is financially supported by the Research Project of Mechanical Engineering in Beijing Institute of Technology and Mercu Buana University. The research data supported by Putranata Adi Mandiri, Inc.

References
[1] G. Diwakar, M. Satyanarayana and P. Kavi Kumar, "Detection of Gear fault using vibration analysis," International Journal of Emerging Technology and Advanced Engineering, vol. 2, no. 9, pp. 122-125, 2012.
[2] L. Jing, M. Zhao, P. Li and X. Xu, "A convolutional neural network based feature learning and fault diagnosis method for the condition monitoring of gearbox," Measurement, vol. 111, pp. 1-10, 2017.
[3] A. Aherwar and M. S. Khalid, "Vibration Analysis Techniques for Gearbox Diagnostic: A Review," International Journal of Advanced Engineering Technology, vol. 3, no. 2, pp. 1-7, 2012.
[4] J. Lee, F. Wu, W. Zhao, M. Ghaffari, L. Liao and D. Siegel, "Prognostics and health management design for rotary machinery systems—Reviews, methodology and applications," Mechanical Systems and Signal Processing, vol. 42, no. 1-2, pp. 314-334, 2014.
[5] M. Y. Asr, M. M. Ettefagh, R. Hassannejad and S. N. Razavi, "Diagnosis of combined faults in Rotary Machinery by Non-Naive Bayesian approach," Mechanical Systems and Signal Processing, vol. 85, pp. 56-70, 2017.
[6] D. N. Dongiovanni and T. Iesmantas, "Failure rate modeling using fault tree analysis and Bayesian network: DEMO pulsed operation turbine study case," Fusion Engineering and Design, vol. 109, pp. 613-617, 2016.
[7] W.-S. Wu, C.-F. Yang, J.-C. Chang, P.-A. Château and Y.-C. Chang, "Risk assessment by integrating interpretive structural modeling and Bayesian network, case of offshore pipeline project," Reliability Engineering & System Safety, vol. 142, pp. 515-524, 2015.
[8] X. Li, G. Chen and H. Zhui, "Quantitative risk analysis on leakage failure of submarine oil and gas pipelines using Bayesian network," Process Safety and Environmental Protection, vol. 103, pp. 163-173, 2016.
[9] W. Wang, K. Shen, B. Wang, C. Dong, F. Khan and Q. Wang, "Failure probability analysis of the urban buried gas pipelines using Bayesian networks," Process Safety and Environmental Protection, vol. III, pp. 678-686, 2017.
[10] H. Langseth and L. Portinale, "Bayesian networks in reliability," Reliability Engineering & System Safety, vol. 92, no. 1, pp. 92-108, 2007.
[11] I. Boersch, J. Heinsohn and R. Socher, Wissensverarbeitung: Eine Einführung in die künstliche Intelligenz für Informatiker und Ingenieure, vol. 2, Elsevier, 2007, p. 397.
[12] K. A. Kobbacy, K. McNaught and A. Chan, "Bayesian networks in manufacturing," Jnl of Manu Tech Mnagmnt, vol. 22, no. 6, pp. 734-747, 2011.
[13] J. Pearl, Bayesian Networks: a model of self-activated: memory for evidential reasoning, Los Angeles: UCLA, Computer Science Department, 1985.
[14] J. Pearl, "Fusion, propagation, and structuring in Belief Networks," Artificial Intelligence, vol. 29, pp. 241-288, 1986.
[15] L. Yang and J. Lee, "Bayesian Belief Network-based approach for diagnostics and prognostics of semiconductor manufacturing systems," Robotics and Computer-Integrated Manufacturing, vol. 28, no. 1, pp. 66-74, 2012.
[16] Z. Xuejun and C. Yu, "The research of intelligent fault diagnosing methods based on FTA," Microcomputer Information, vol. 21, no. 6, pp. 123-124, 2005.
[17] Y. Zhi-Ling, W. Bin, D. Xing-Hui and L. Hao, "Expert System of Fault Diagnosis for Gear Box in Wind Turbine," The 2nd International Conference on Complexity Science & Information Engineering, vol. 4, pp. 189-195, 2011.

[18] N. Khakzad, F. Khan and P. Amyotte, "Safety analysis in process facilities: comparison of fault tree and Bayesian network approaches," Reliability Engineering & System Safety, vol. 96, no. 8, pp. 925-932, 2011.
[19] N. Khakzad, F. Khan and P. Amyotte, "Dynamic safety analysis of process systems by mapping bow-tie into Bayesian network," Process Safety and Environmental Protection, vol. 91, no. 1-2, pp. 46-53, 2013.
[20] N. Khakzad, F. Khan and P. Amyotte, "Quantitative risk analysis of offshore drilling operations: a Bayesian approach," Safety Science, vol. 57, pp. 108-117, 2013.
[21] B. G. Xu, "Intelligent fault inference for rotating flexible rotors using Bayesian belief network," Expert Systems with Applications, vol. 39, pp. 816-822, 2012.
[22] D. Zhou, "The Application of Bayesian Networks in System Reliability," Master Thesis in Arizona State University, pp. 1-13, 2014.
[23] J. Grover, Strategic economic decision-making: Using Bayesian belief networks to solve complex problems, vol. 9, Springer Science & Business Media, 2012.
[24] J. Pearl, Probabilistic Reasoning in Intelligent Systems: Networks of Plausible Inference (Morgan Kaufmann Series in Representation and Reasoning), San Mateo, California, 1988.
[25] K. B. Korb and A. E. Nicholson, Bayesian Artificial Intelligence (Second Edition), London: CRC Press, 2011.
[26] R. E. Neapolitan, Learning Bayesian Networks, Chicago, Illinois: Northeastern Illinois University, 2003.
[27] J. Mais, "Spectrum Analysis: The key features of analyzing spectra," SKF USA Inc., pp. 3-28, 2002.
[28] S. K. Sar and R. Kumar, "Techniques of Vibration Signature Analysis," International Journal of Advanced Research in Computer and Communication Engineering, vol. 4, no. 3, pp. 240-243, 2015.
[29] S. R. W. Mills, Vibration Monitoring and Analysis Handbook, vol. 1, Northampton: The British Institute of Non-Destructive Testing, 2010.
[30] NEN, "Mechanical Vibration - Evaluation of machine vibration by measurements on non-rotating parts," International Standard, vol. II, pp. 1-10, 2009.

Mode Shape Analysis of EV-Bus Chassis with Reverse Engineering Method

N Nazaruddin[1,2,a], A Syehan[1], G Heryana[1,3], M Adhitya[1] and D A Sumarsono[1]

[1]Research Center for Advanced Vehicle (RCAVe), Universitas Indonesia, 16424, Indonesia
[2]Department of Mechanical Engineering, Universitas Riau, 28293, Indonesia
[3]Department of Mechanical Engineering, Universitas Wastukancana, 41151, Indonesia

[a]Corresponding author: nazaruddin@eng.unri.ac.id

Abstract. Universitas Indonesia has developed several electric vehicles, one of which is an electric bus. In order to fulfill the local content of this vehicle, reverse engineering of the R260 ladder frame type chassis is carried out. In this paper, to fulfill the local content, we used material of the ladder frame is type SS400 from PT. Krakatau Steel. After the model was created successfully with finite element software, a dynamic analysis was carried out by taking the 5 lowest frequencies from the simulation results. Loading is carried out evenly over the two main beam ladder frames totaling 14,200 kg. The support placed in the mounting position of the front and rear wheel leaf springs. The distance of the front overhang is 2380 mm, the wheelbase is 6000 mm, and the rear overhang is 3290 mm. The resulting approach is carried out assuming a three-beam and two-overhang beam model. The five lowest frequencies are 11,798 Hz, 21,009 Hz, 23,876 Hz, 30.84 Hz, and 36,473 Hz, respectively.

1. Background

Development of electric cars (Electric Vehicle, EV) in Indonesia, accordance with Presidential Decree No. 55 of 2019 concerning the Acceleration of the Battery-Based Electric Motor Vehicle Program for Transportation, which is in the context of energy efficiency and energy security and to realize clean air quality. The use of electric cars can increase the efficiency of energy use from batteries to wheels above 77% compared to hybrid cars 38% and cars with internal combustion (ICE) only 25%[1-4]. The high level of efficiency in electric cars because the distribution of energy from the power source to the wheels has a relatively low level of energy loss.

Universitas Indonesia, UI, launched an electric car based on the 52nd anniversary of the Faculty of Engineering UI on Monday (7/18/2017) at the Faculty of Engineering, UI Campus Depok. There were four electric cars displayed at this event, namely Bus Electric Vehicle (EV), Makara Electric Vehicle (MEV) 01, City Car MEV 02, and City Car MEV 03. These electric cars are innovations by FTUI lecturers and students who are members of the UI Molina Team. The EV bus is a vehicle with a capacity of 60 passengers with motor power of 120 kW and 300 Ah. In the future this bus is expected to gradually be able to replace the overall function of the UI yellow bus as an internal transportation tool for the UI campus[5].

Car chassis is a significant component and is very important, where the chassis is supporting the axle, steering to adjust the direction of the vehicle, wheels, tires, and brakes to stop the vehicle when walking. The system on the chassis is very directly influential on comfort when driving, stability and others [6,7]. Static analysis on the chassis and dynamic ladder frame chassis has been done [8-15].

In this paper, the author wants to explain the mode shape analysis for the structure of the University Indonesia's electric bus chassis by reverse-engineering the ladder frame that was installed on this bus with type R260. The goal is to find the first five vibrating modes by simulating chassis material, which is the SS400 type produced by PT. Krakatau Indonesia. This modal analysis is essential to avoid resonance due to some interference from the engine and other rotating components mounted on the chassis having the same frequency as the frequency of the mode shapes.

2. Basic Theory: Vibrate Mode of Continuous Beam System with One or Two Overhangs

The characteristic equation for a Bernoulli overhang beam with density per unit length ρ(x), with stiffness elasticity, $\boldsymbol{r(x) = EJ(x)}$ for a range of 0 to l can be written in the equation: [16]

$$[r(x)u^{n}(x)]^{n} = \omega^{2}\rho u(x) \tag{1}$$

where ω is a personal frequency, u (x) is a displacement mode that corresponds superscript is a derivative of x. Footing conditions with three stretches and with two overhangs like the following figure 1.

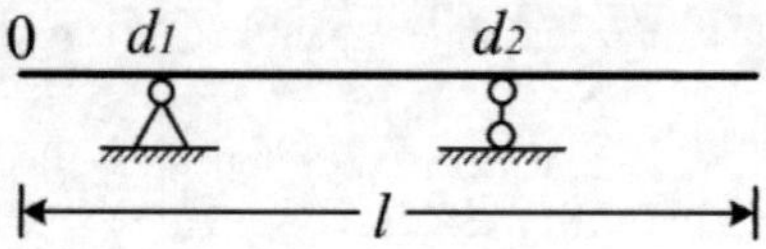

Figure 1. Beam with two supports and three spans [16]

The boundary conditions for this beam are : [16]

$$r(0)u^{"}(0) = 0, \ \ [r(x)u^{"}(x)]'_{x=0} = 0; \tag{2}$$

$$r(l)u^{"}(l) = 0, \ \ [r(x)u^{"}(x)]'_{x=l} = 0; \tag{3}$$

and

$$u(d_i^{+}) = u(d_i^{-}) = 0 \tag{4}$$

$$u'(d_i^{+}) = u'(d_i^{-}), \tag{5}$$

$$r(d_i^{+})u^{"}(d_i^{+}) = r(d_i^{-})u^{"}(d_i^{-}), \ \ (\boldsymbol{i = 1,2;\ 0 < d_1 < d_2 < l}) \tag{6}$$

Here are the various vibration mode shapes of one bar with two overhangs.

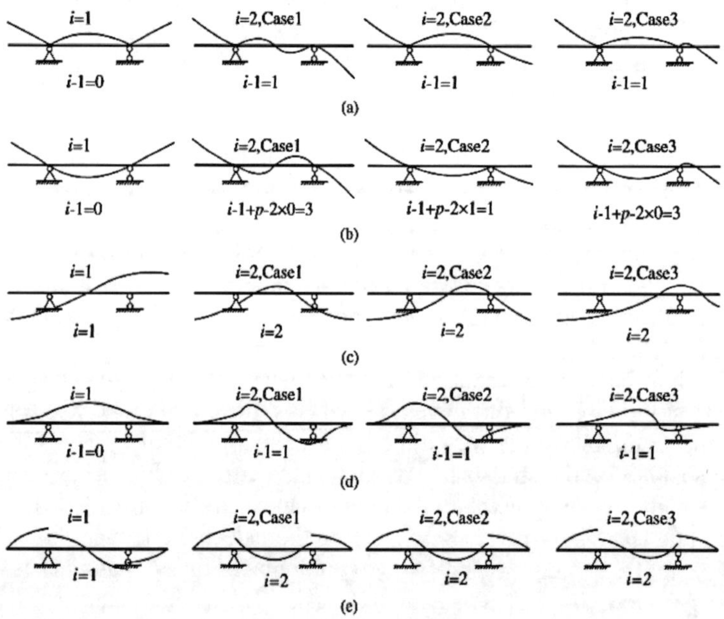

Figure 2. First and second mode shapes for the beam with three spans and two overhangs; (a) the first and second modes of mathematical models, (b). First and second displacement modes, (c) first and second rotation modes, (d). first and second bending moment modes, (e) first and second shear force modes [16]

3. Methodology

3.1 Reverse Engineering

The University Indonesia's electric bus, like Figure 3 uses R260 chassis (figure 4), using the reverse engineering method to get the dimensions of this chassis with the main dimensions, as in figure 5. The type of the R260 chassis is a ladder frame with two main beams and four cross member beam and two cross pipe beam.

Figure 3. Universitas Indonesia's Electric Bus

Figure 4. *Rolling Chassis* Model of Electric Bus

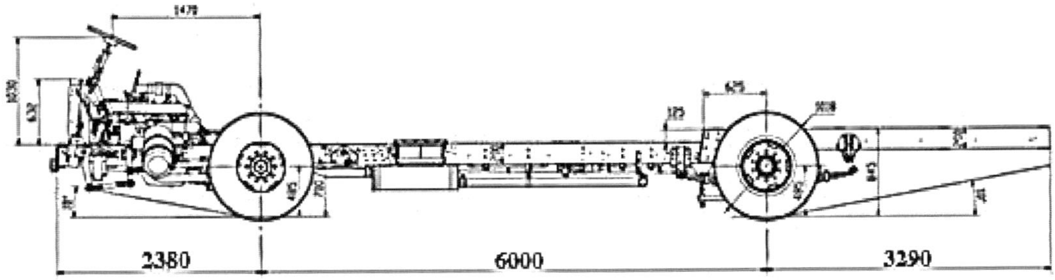

Figure 5. Primary Dimension of Rolling Chassis Electric Bus.

From Figure 5, it can show that dimensions of the front overhangs are 2380 mm, the wheelbase is 6000 mm, and the rear overhang is 3290 mm.

Figure 6. The rolling chassis parts are measured

The reverse engineering process is carried out by measuring the main parts of this electric bus chassis as shown in figure 6.

3.2 Modeling with CAD software

After all, dimensions have obtained, a redrawing using CAD software is carried out with the following results, as can be seen in figure 7.

Figure 7. Rolling Chassis Model with CAD software

3.3 Material Selection

As a commitment to increase the local content of Indonesian electric cars, the material selection is carried out in such a way by using the production of PT. Krakatau Steel with SS400 code. Following are SS400 material properties from PT. KS and from several other sources.

IOP Conf. Series: Materials Science and Engineering **694** (2019) 012002 doi:10.1088/1757-899X/694/1/012002

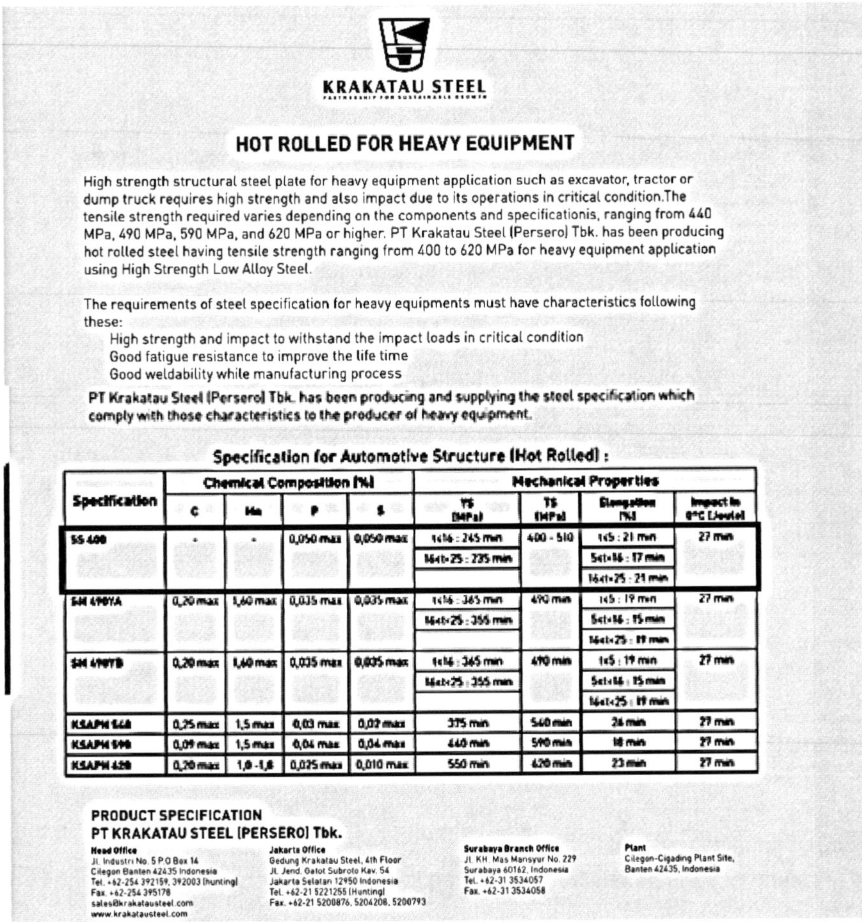

KRAKATAU STEEL

HOT ROLLED FOR HEAVY EQUIPMENT

High strength structural steel plate for heavy equipment application such as excavator, tractor or dump truck requires high strength and also impact due to its operations in critical condition.The tensile strength required varies depending on the components and specificationis, ranging from 440 MPa, 490 MPa, 590 MPa, and 620 MPa or higher. PT Krakatau Steel (Persero) Tbk. has been producing hot rolled steel having tensile strength ranging from 400 to 620 MPa for heavy equipment application using High Strength Low Alloy Steel.

The requirements of steel specification for heavy equipments must have characteristics following these:

High strength and impact to withstand the impact loads in critical condition
Good fatigue resistance to improve the life time
Good weldability while manufacturing process

PT Krakatau Steel (Persero) Tbk. has been producing and supplying the steel specification which comply with those characteristics to the producer of heavy equipment.

Specification for Automotive Structure (Hot Rolled) :

Specification	Chemical Composition (%)				Mechanical Properties			
	C	Mn	P	S	YS (MPa)	TS (MPa)	Elongation (%)	Impact in 0°C (Joule)
SS 400	-	-	0,050 max	0,050 max	t≤16 : 245 min 16<t≤25 : 235 min	400 - 510	t≤5 : 21 min 5<t≤16 : 17 min 16<t≤25 : 21 min	27 min
SM 490YA	0,20 max	1,60 max	0,035 max	0,035 max	t≤16 : 365 min 16<t≤25 : 355 min	490 min	t≤5 : 19 min 5<t≤16 : 15 min 16<t≤25 : 19 min	27 min
SM 490YB	0,20 max	1,60 max	0,035 max	0,035 max	t≤16 : 365 min 16<t≤25 : 355 min	490 min	t≤5 : 19 min 5<t≤16 : 15 min 16<t≤25 : 19 min	27 min
KSAPH 540	0,25 max	1,5 max	0,03 max	0,02 max	375 min	540 min	26 min	27 min
KSAPH 590	0,09 max	1,5 max	0,04 max	0,04 max	440 min	590 min	18 min	27 min
KSAPH 620	0,20 max	1,0 -1,6	0,025 max	0,010 max	550 min	620 min	23 min	27 min

PRODUCT SPECIFICATION
PT KRAKATAU STEEL (PERSERO) Tbk.

Head Office
Jl. Industri No. 5 P.O Box 14
Cilegon Banten 42435 Indonesia
Tel. +62-254 392159, 392003 (hunting)
Fax. +62-254 395178
sales@krakatausteel.com
www.krakatausteel.com

Jakarta Office
Gedung Krakatau Steel, 4th Floor
Jl. Jend. Gatot Subroto Kav. 54
Jakarta Selatan 12950 Indonesia
Tel. +62-21 5221255 (Hunting)
Fax. +62-21 5200876, 5204208, 5200793

Surabaya Branch Office
Jl. KH. Mas Mansyur No. 229
Surabaya 60162, Indonesia
Tel. +62-31 3534057
Fax. +62-31 3534058

Plant
Cilegon-Cigading Plant Site,
Banten 42435, Indonesia

Figure 8. Material SS400 from PT. Krakakatau Steel

Mechanic Properties

- Elastic (Young's, Tensile) Modulus : 190 *GPa* (27 x 10^6 *psi*)
- Elongation at Break : 22 %
- Fatigue Strength : 200 *MPa* (29 x 10^3 *psi*)
- Poisson's Ratio : 0.29
- Shear Modulus : 73 *GPa* (11 x 10^6 *psi*)
- Shear Strength : 300 *MPa* (44 x 10^3 *psi*
- Tensile Strength: Ultimate (UTS) : 480 *MPa* (69 x 10^3 *psi*)
- Tensile Strength: Yield (Proof) : 290 *MPa* (41 x 10^3 *psi*)
- Brinell Hardness : 140

Thermal Properties

- Latent Heat of Fusion : 250 *J/g*
- Maximum Temperature: Mechanical : 400 °*C* (750 °*F*)
- Melting Completion (Liquidus) : 1460 °*C* (2670 °*F*)
- Melting Onset (Solidus) : 1420 °*C* (2590 °*F*)
- Specific Heat Capacity : 470 *J/kg-K* (0.11 *BTU/lb-°F*)
- Thermal Conductivity : 50 *W/m-K* (29 *BTU/h-ft-°F*)
- Thermal Expansion : 11 $\mu m/m$-*K*

IOP Conf. Series: Materials Science and Engineering **694** (2019) 012002 doi:10.1088/1757-899X/694/1/012002

Process of inputting material properties into the material menu in the CAD software can be seen in figure 9.

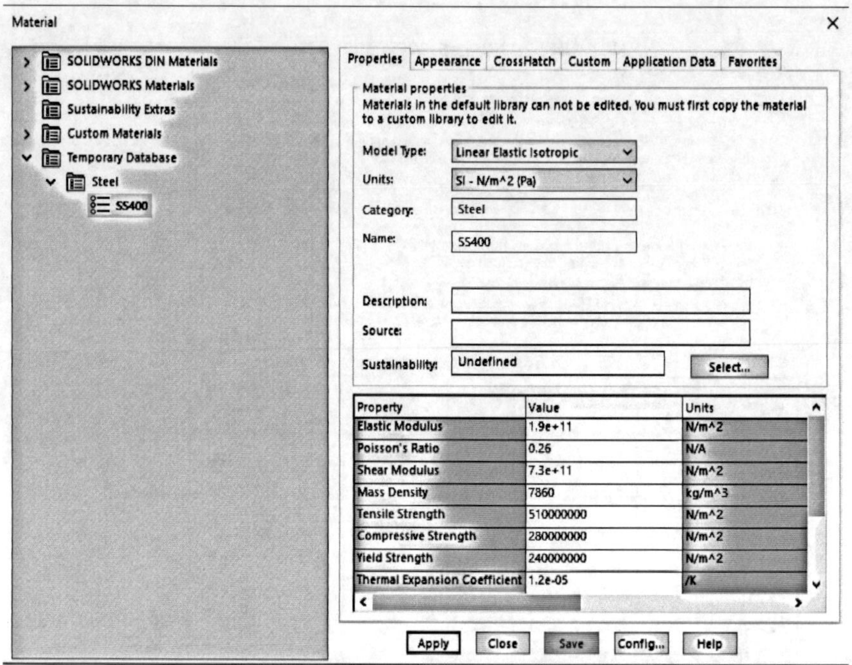

Figure 9. Material Input of CAD Software

3.4 Chassis Model of Ladder Frame

From the CAD rolling chassis model in Figure 5, an analysis is carried out on the ladder frame chassis with the shape as shown in figure 10.

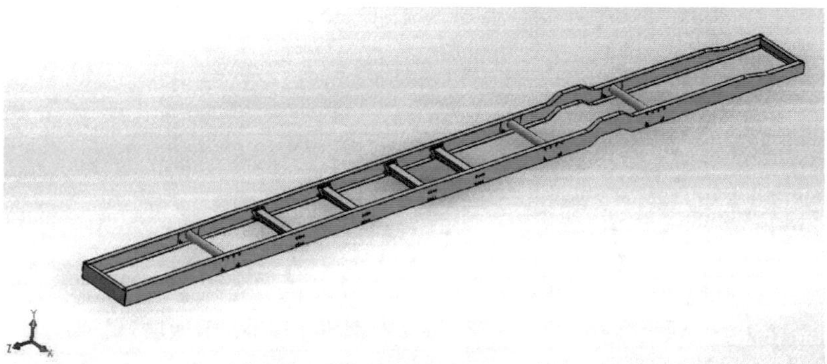

Figure 10. Model Ladder Frame Chassis Electric car Universitas Indonesia

4. Result and Discussion

4.1 Determination of Load and Support in Chassis

In Figure 5, where the support was placed on the mounting front-wheel springs and rear-wheel springs. While the load was given in the form of an even distribution along the two main beams of the chassis for a total Gross Vehicle Weight (GVW) = 14200 kg, which is in accordance with the specifications of the R260 rolling chassis. The following form of the support and the load on CAD modeling.

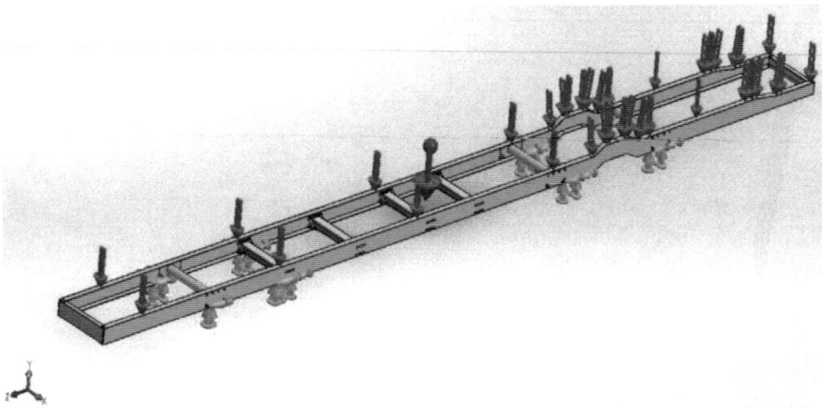

Figure 11. Support and Load Mode of Universitas Indonesia's Electric Bus

4.2 Meshing on Chassis Ladder Frame

After modeling of support and loading and material selection in the previous stage, before finite element analysis is carried out to determine the vibrational mode of Universitas Indonesia's electric car chassis meshing is performed on the CAD software. The meshing results can show in the following figure.

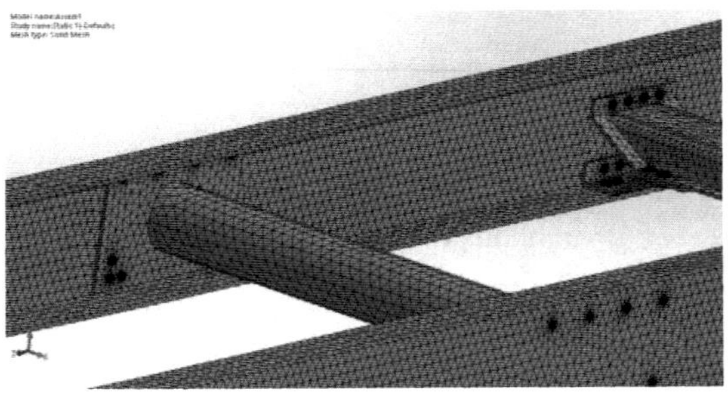

Figure 12. The meshing of Chassis Frame (Front Section)

Figure 13. The meshing of Chassis Frame (Middle Section)

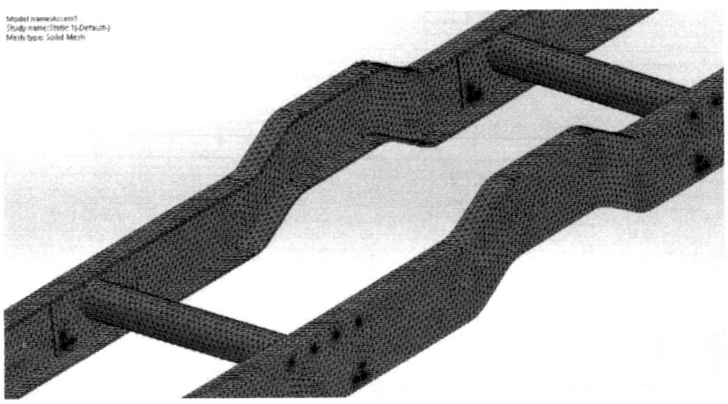

Figure 14. The meshing of Chassis Frame (Rear Section)

4.3 Mode Shape Analysis

The next step can be performed mode shape analysis from dynamic analysis on CAD software. The following 5 first vibrate mode are displayed.

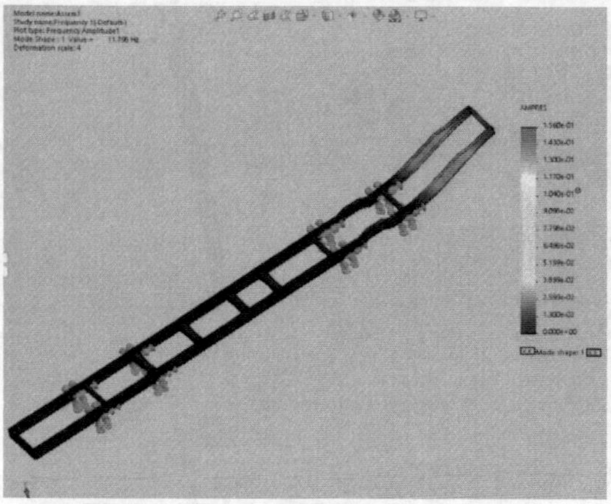

Figure 15. Mode Shape 1 : 11.798 Hz

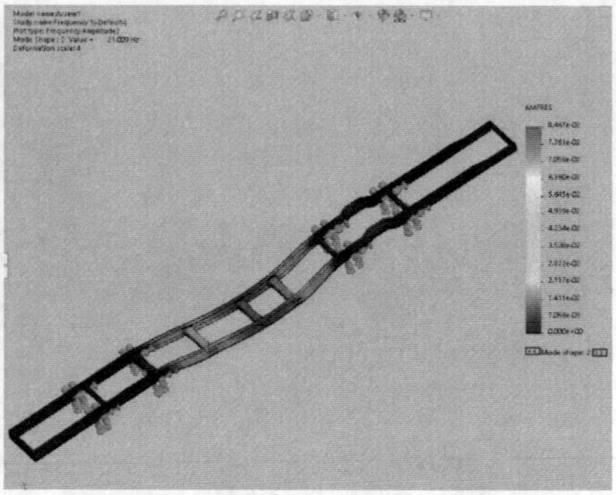

Figure 16. Mode shape 2: 21.009 Hz

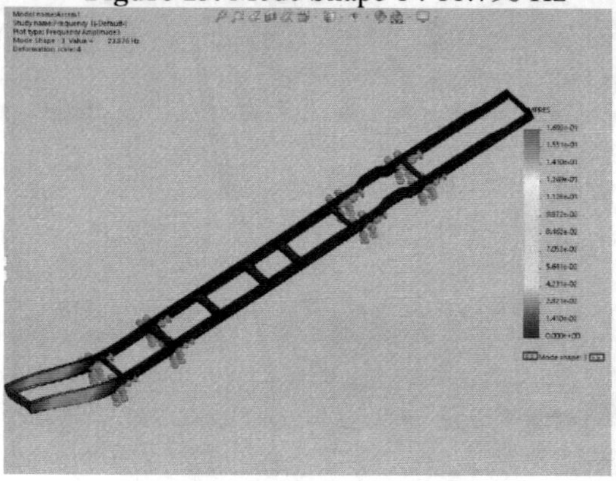

Figure 17. Mode shape 3: 23.876 Hz

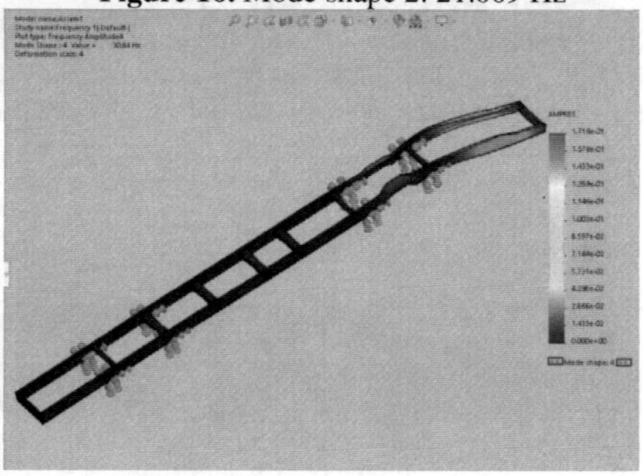

Figure 18. Mode shape 4: 30.84 Hz

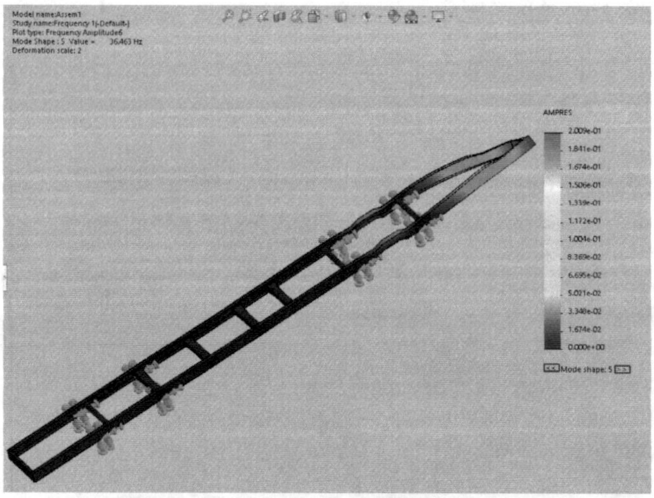

Figure 19. Mode shape 5: 36.463 Hz

The shape mode as shown in Figure 15, up to Figure 19 has been equivalent to the theoretical mode shapes as shown in Figure 2. In the first mode seen in Figure 14, the rear overhang experience deflection first, because the rear overhang distance is 3290 mm longer than front overhang distance is 2380, where there is no cross member beam as a stiffener between that distance. The second module that experiences more significant deflection is the distance between the two axes (wheelbase) with a distance of 6000 mm, although there is already a cross member beam in between. The third vibrate mode occurs in the front overhang, with the most significant deflection in the forward direction upwards. The fourth vibration mode occurs again in the rear overhang with deflection downward direction, while the fifth vibration mode occurs in the rear overhang in the form of a twist between the two main beams of the ladder frame. Further research can be done by adding a cross member beam at the rear overhang because, in this section, mounting will be made to the holder of the electric motor as the rear-drive of this electric bus.

5. Conclusion

The conclusion that can be obtained from the mode shape analysis of the Universitas Indonesia's electric ladder frame bus chassis are:

1. The material used is Krakatau Steel SS400 type steel with ladder frame structure model
2. The total workload distributed to the chassis structure with a total of 14,200 kg
3. The model approach used is a beam with two overhangs.
4. Front overhang distance is 2380 mm, wheelbase 6000 mm, and rear overhang 3290 mm.
5. The mode shapes from 1 to 5 with the lowest frequency of 11.798 Hz, 21.009 Hz, 23.876 Hz, 30.84 Hz, 36.463 Hz.

Acknowledgments

The authors would like to thank DRPM Universitas Indonesia through "**TADOK 2019**" scheme with contract number **NKB-0172/UN2.R3.1/HKP.05.00/2019** and **Indonesian Endowment Fund for Education** (**LPDP**) for funding this research.

References

[1] Thomas, J., *Drive cycle powertrain efficiencies and trends derived from EPA vehicle dynamometer results.* SAE International Journal of Passenger Cars-Mechanical Systems, 2014. **7**(2014-01-2562): p. 1374-1384.

[2] Lohse-Busch, H., et al., *Ambient temperature (20 F, 72 F and 95 F) impact on fuel and energy consumption for several conventional vehicles, hybrid and plug-in hybrid electric vehicles and battery electric vehicle.* 2013, SAE Technical Paper.

[3] Carlson, R.B., J. Wishart, and K. Stutenberg, *On-road and dynamometer evaluation of vehicle auxiliary loads.* SAE International Journal of Fuels and Lubricants, 2016. **9**(1): p. 260-268.

[4] Rhodes, K., et al., *Estimation of the Effects of Auxiliary Electrical Loads on Hybrid Electric Vehicle Fuel Economy.* 2017, SAE Technical Paper.

[5] Ayu, W., *UI Perkenalkan Mobil Listrik di Dies Natalis ke-52 FTUI.* 2016.

[6] Milliken, W.F., D.L. Milliken, and M. Olley, *Chassis design: principles and analysis.* Vol. 400. 2002: Society of Automotive Engineers Warrendale.

[7] Ribbens, W., et al., *Automotive engineering: powertrain, chassis system and vehicle body.* 2009: Butterworth-Heinemann.

[8] Mahmoodi-k, M., et al., *Stress and dynamic analysis of optimized trailer chassis.* Tehnički vjesnik, 2014. **21**(3): p. 599-608.

[9] Rajappan, R. and M. Vivekanandhan. *Static and modal analysis of chassis by using FEA.* in *Proceedings of the "National Conference on Emerging Trends In Mechanical Engineering.* 2013.

[10] Renuke, P.A., *Dynamic analysis of a car chassis.* International Journal of Engineering Research and Applications (IJERA), 2012. **2**(6)

[11] Patel, V.V. and R. Patel, *Structural analysis of a ladder chassis frame.* World Journal of science and Technology, 2012. **2**(4): p. 05-08.

[12] Singh, A., V. Soni, and A. Singh, *Structural analysis of ladder chassis for higher strength.* International Journal of Emerging Technology and Advanced Engineering, 2014. **4**(2): p. 253-259.

[13] Francis, V., et al., *Structural Analysis of Ladder Chassis Frame for Jeep Using Ansys.* International Journal of Modern Engineering Research, ISSN, 2014: p. 2249-6645.

[14] Patil, H.B., S.D. Kachave, and E.R. Deore, *Stress analysis of automotive chassis with various thicknesses.* IOSR Journal of Mechanical and Civil Engineering, 2013. **6**(1): p. 44-49.

[15] Chandra, M.R., S. Sreenivasulu, and S.A. Hussain, *Modeling and Structural analysis of heavy vehicle chassis made of polymeric composite material by three different cross sections.* International Journal of Modern Engineering Research.

[16] Wang, Q., et al., *Some qualitative properties of the vibration modes of the continuous system of a beam with one or two overhangs.* Journal of Engineering Mechanics, 2011. **138**(8): p. 945-952.

IOP Conf. Series: Materials Science and Engineering **694** (2019) 012003 doi:10.1088/1757-899X/694/1/012003

Design and Prototyping of Weigh-In-Motion and Overload Detection System for Freight Vehicle

A H Masyhur[1], I P Nurprasetio[1], B A Budiman[1,2] and T P B Utomo[1]

[1] Faculty of Mechanical and Aerospace Engineering, Institut Teknologi Bandung, Bandung, Indonesia
[2] National Center for Sustainable Transportation Technology, Institut Teknologi Bandung, Bandung, Indonesia

E-mail: hakim@ftmd.itb.ac.id, ipn@ftmd.itb.ac.id

Abstract. The high number of freight vehicles that exercised overloading in Indonesia has led to a significant reduction in the quality of road pavement. The obsolete overload detection system using weigh bridges have been ineffective, since it requires vehicles to maneuver and enter the weighing station. In the above system, after manual checking, overloaded vehicles will be penalized. The Weigh-in-Motion (WIM) device that is developed in this research utilized a bending plate sensor with strain gage. As overloaded vehicle runs over the WIM, axle weight will be automatically recorded, and the data is processed further to assess the compliance to maximum freight weight as specified for the vehicle. In this research work, the WIM prototype is developed and tested using motorcycle under normal operation and overloaded conditions at different speeds. Arduino based signal processing and analysis system has been developed such that the overload detection may be performed automatically. When the system senses an overloaded vehicle, it will automatically trigger the alarm and a camera will capture the vehicle. The error of the prototype is within acceptable range and the system has shown the potential to detect whether the vehicle, i.e., the motorcycle, is overloaded or not. As a continuation, the development of a reliable microcontroller-based WIM and a robust automatic overloaded freight vehicle will be our future work.

1. Background

Maintaining road pavement quality in several areas in Indonesia is essential since logistic distribution is mainly carried using land transportation. Based on Indonesian Minister of Public Works Regulation, the design service lifetime for main road in Indonesia is at least ten years, while for local road is about five years [1]. However, the quality of road pavement of most main road in Indonesia have been deteriorated before it reaches the design lifetime. This condition also causes the road maintenance cost to increase significantly [2]. Huge traffic of overloaded vehicle on the road is the main cause why the road pavement quality decrease faster than the expected lifetime. Wang et.al findings show that every 1% of overload in trucks would cause 1.8% reduction of road lifetime [3].

Weight measurement especially for truck becomes an important activity in order to prevent road damage and reduce the risk of road accident occurred due to overloaded vehicle [4]. Weighbridge is a common method that has been applied in Indonesia for measuring the total weight carried by a freight vehicle. However, this method has a drawback since the vehicle weight is measured in static condition. It is very difficult to weigh all commercial vehicles while it should stop by the weighing station, since it could create congestion at the station. Based on this issue, a method called Weigh-in-Motion (WIM) is developed. The main strategy of WIM is to predict total weight of the vehicles by measuring the dynamic tire forces exerted by the wheels to the sensors on the ground. This method enables the weight measurement to be carried out while the vehicle is still moving.

In this research, a simple WIM device with overload detection system is developed. As a case study, the prototype is implemented to two-wheeler or motorcycle. The WIM prototype is based on bending plate – strain gage sensors, while the overload detection system is developed using Arduino microcontroller with the addition of alarm and camera connection.

2. Methodology

2.1. Bending Plate Design

There are three types of weight sensors that is commonly used for WIM application, i.e., piezoelectric, bending plate, and load cell. In this research, bending plate device using strain gage sensor is chosen after considering several aspects. Bending plate device is relatively easier to be designed and manufactured. In addition, obtaining strain gage in the market is also relatively easy as compared with the other two sensors.

The ideal position for the bending plate sensor should be in the road pavement to ensure level surface between the sensor and the road. However, due to location constraint, the bending plate will be placed over the road pavement. Wood panels with similar thickness and width with the bending plate is attached in order to compensate the effect of level difference between the plate and road surface that could produce sudden change in vertical acceleration while the vehicle passes through the plate. The layout of the testing rig is depicted in Figure 1.

The plate is made from low carbon steel. Coating treatment is applied to the plate surface in order to avoid corrosion.

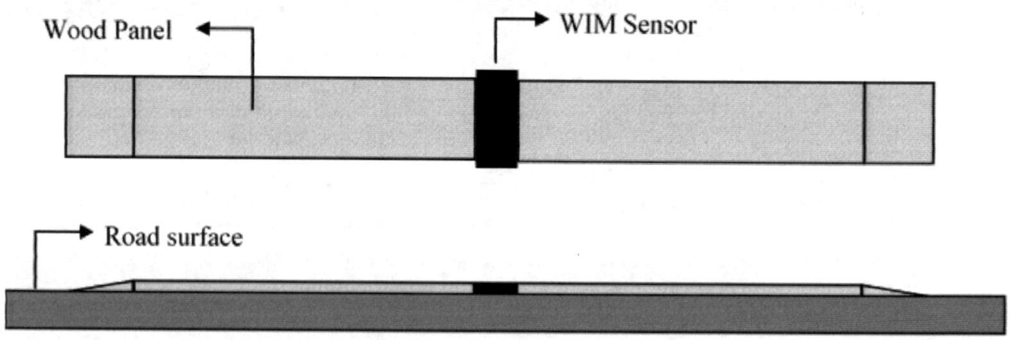

Figure 1. Testing rig layout

Strain gage is attached at the bottom surface of the bending plate. The location of the strain gage should be in the middle of the plate and its orientation should be in parallel with vehicle direction of movement as illustrated in Figure 2. Since only one strain gage is used, the vehicle's rider should ensure the vehicle tires make a contact with the bending plate at the point over where the strain gage is placed, such that it could reduce reading error by the WIM system.

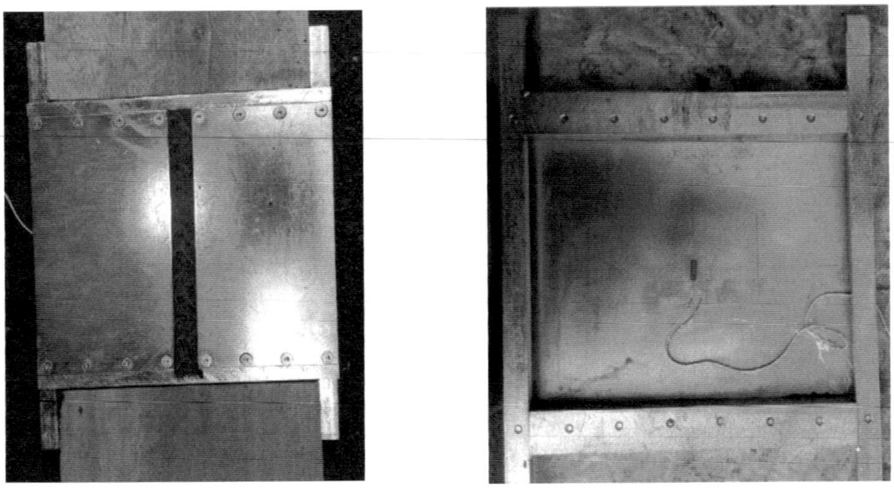

Figure 2. Strain gage location on the bending plate.

In this bending plate system, the strain which is occurred at the plate by tire's contact is converted into electrical voltage signal. Then, this signal is processed until the total weight of the vehicle is known and could be saved in the computer. The block diagram for WIM signal processing unit is shown in Figure 3.

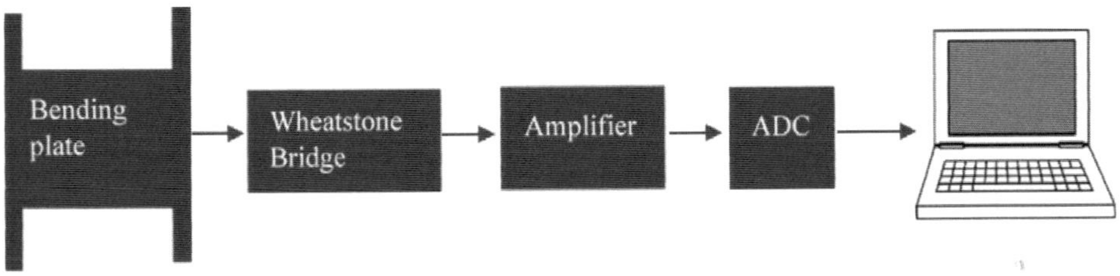

Figure 3. Block diagram for WIM signal processing unit

2.2. *Overload Detection System*

The overload detection system is also developed in this research, such that the overloaded vehicle could be identified within a few seconds. This system utilizes the signal data that comes from the bending plate which is combined with the application of alarm and camera using Arduino Uno as a microprocessor. The microcontroller receives signal from the strain gage that has been ampilifed by the HX711 strain amplifier module. The alarm will be triggered once the system detects an overloaded vehicle and the camera will capture the picture of the vehicle including its identity. Figure 4 illustrates the block diagram for the overload detection system, while Figure 5 shows the position layout for each component of the system. In this case, the overloaded vehicle is defined as a motorcycle occupied by three passengers.

IOP Conf. Series: Materials Science and Engineering **694** (2019) 012003 doi:10.1088/1757-899X/694/1/012003

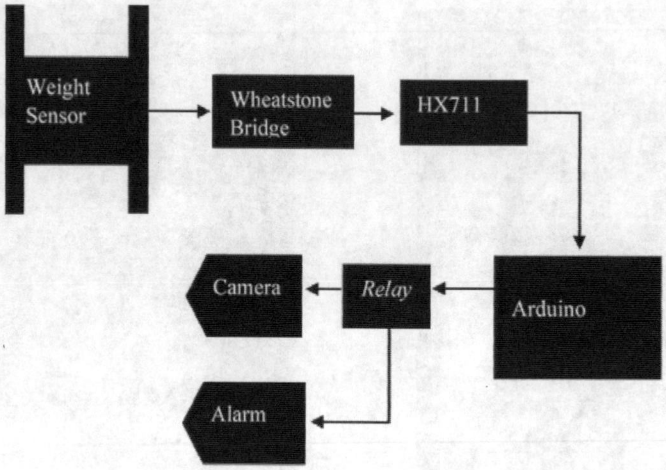

Figure 4. Block diagram for overload detection system

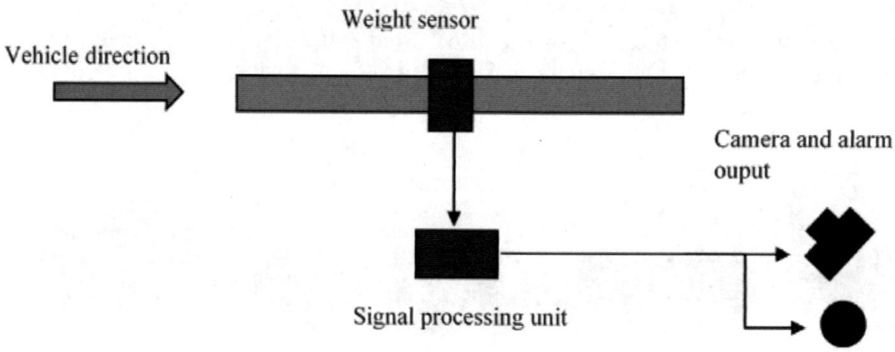

Figure 5. Overload detection system layout

3. Analysis and Discussion

3.1. Static calibration

In order to obtain relation between exerted force to the bending plate as an input and output voltage, static calibration should be conducted first. The calibration process is carried out by measuring the weight distributed at the motorcycle's front and rear tires in static condition. Three types of motorcycle with different curb weights are used for benchmarking. Variation of the data is also obtained by adding one passenger for each vehicle. Table 1 shows the weight data during static calibration.

Table 1. Vehicle's weight data for static calibration

Motorcycle type	Without Passenger [kg]		With Passenger [kg]	
	Front	Rear	Front	Rear
Type A	47	67	70	103
Type B	37	55	63	88
Type C	63	71	79	114

From the static calibration process, it is discovered that the relation between vehicle weight (m_k) and output signal from the bending plate (V_S) is linear with positive slope. Equation (1) as shown below can then be determined using simple regression method. The quality of the curve fitting is depicted in Figure 6.

$$V_S = 0.0053 m_k \quad (1)$$

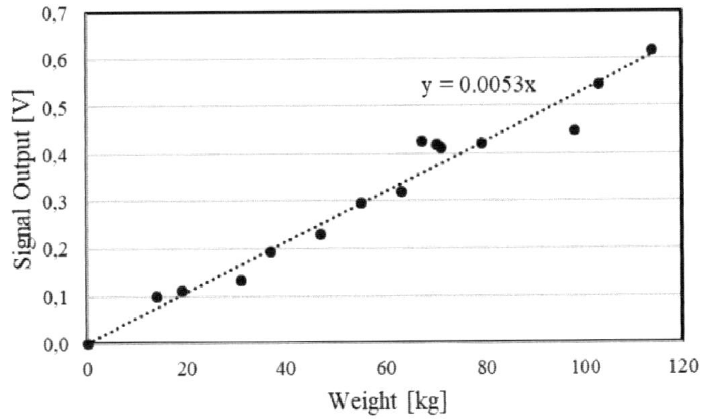

Figure 6. Static calibration result of the bending plate

3.2. Dynamic testing experiment
A straight path with distance of 30 m is set to be the lane for dynamic testing experiment. The WIM prototype is placed 20 m beyond the start line. The condition of road pavement for the testing lane is relatively flat and the testing is carried out by using single type of motorcycle with one passenger.

Three different speeds are chosen for reference, which are 10 km/h, 20 km/h, and 30 km/h. Each speed is repeated 30 times to satisfy statistic requirement. Speed variation occurred during the tests since the rider tends to maintain the speed manually. Figure 7 shows the box plot of the vehicle speed during the dynamic test. It can be seen that the variation of the speed is not significant which can be neglected in the next analysis.

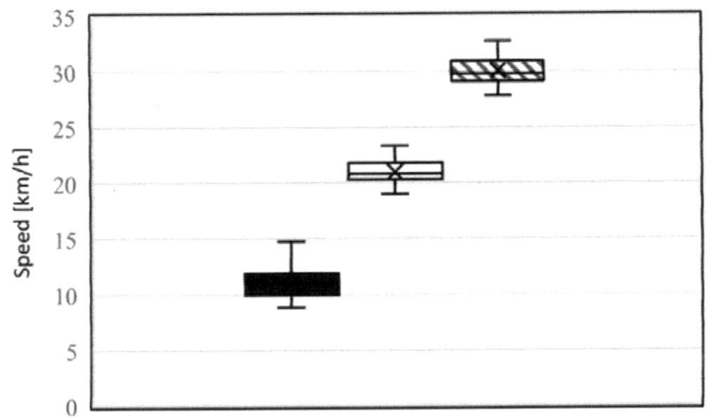

Figure 7. Box plot of speed variation during dynamic testing

3.3. Noise reduction

Figure 8 illustrates a sample of the signal obtained during dynamic test for theWIM system. The signal is in the form of time series of the voltage coming from the HX-711 strain amplifier. It may be observed that there are two peaks with different values. The peak on the left, with lower maximum value than other peak, indicates the contact signal from the front tire, and the other peak appears as the rear tire make a contact with bending plate.

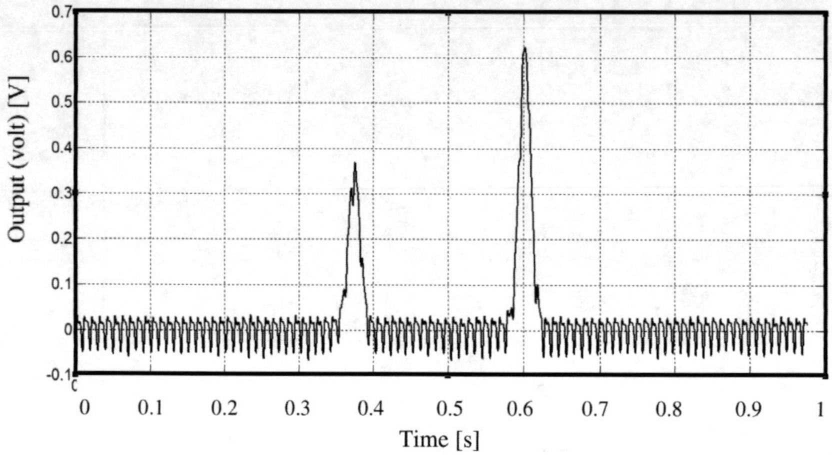

Figure 8. Output signal from WIM dynamic testing

The signal that is obtained from the data acquisition comes with high frequency noise with stable frequency. Averaging method is selected to reduce the noise level, considering its simplicity and effectiveness to remove harmonic effect in the signal [5]. The method is done by accumulating and divide the number of signals used in order to get the value of arithmetic average that can be written as in equation (2) .

$$\mathrm{A}M = \frac{1}{n}\sum_{i=1}^{n} a_i \tag{2}$$

In this research, adjustment should be included in the averaging method in order to limit the amount of data that is used in this experiment. Sample signal that consist of noise only is separated with output signal. Then, the noise signal is translated in the time domain until the phase difference between noise signal and output signal is close to zero. Final result of the new output signal is obtained by subtracting the original output signal with noise signal after averaging, which is shown in Figure 9.

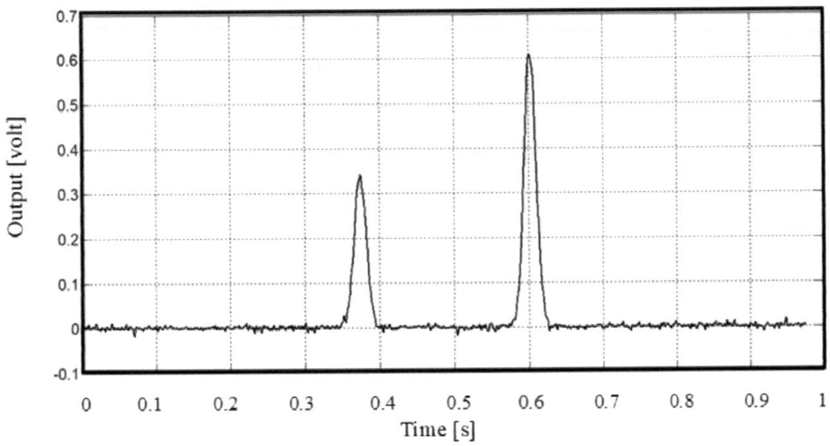

Figure 9. New output signal after noise reduction

3.4. Vehicle speed correction factor

Vehicle speed data is required during testing and analysis. In testing process, group of data is decided based on vehicle speed, and vehicle speed correction factor should be determined in analysis process. The true vehicle speed is calculated by dividing vehicle wheelbase (L_{base}) with time difference (Δt) between the starting time of two peak signal coming from front and rear tire.

Measurement result from the bending plate for similar vehicle in dynamic testing are different from static test measurement. The condition of uneven road surface creates sudden acceleration in vertical direction. Change in acceleration and deceleration during dynamic testing contributes to the change of weight distribution at the axle for each wheel.

Different speed input also gives different output response both for front and rear tire as illustrated in Figure 10. Higher speed will have smaller signal amplitude that is probably caused by greater time constant in the bending plate and strain gage. The signal has not reach maximum amplitude or yet to be in steady state condition after the vehicle makes a contact with the bending plate in higher speed.

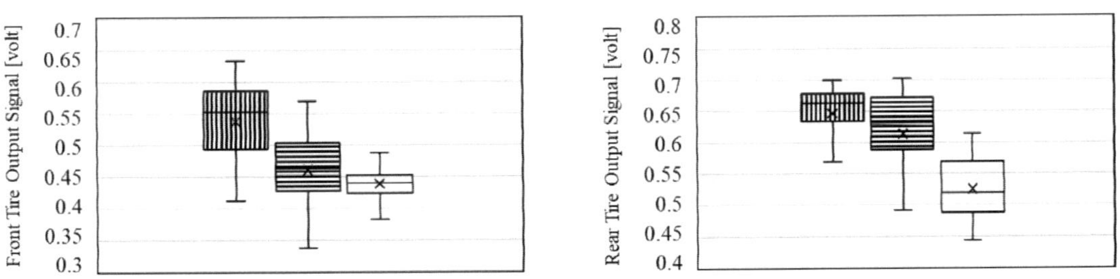

Figure 10. Box plot trend of signal amplitude according to dynamic testing for front tire (left) and rear tire (right)

Equation (1) could be used to convert voltage signal output into weight value. However, additional variable is needed as correction factor due to effect given by vehicle speed to the output signal. Vehicle speed correction factor (K_v) is a function of vehicle speed (v) that is used to correct the value of output signal when the vehicle is moving in different speed. By adding this correction factor, the weight of the vehicle estimated by WIM system (M_w) can be determined by equation (3). The value of K_v is obtained

IOP Conf. Series: Materials Science and Engineering **694** (2019) 012003 doi:10.1088/1757-899X/694/1/012003

using equation (4) with V_s' is the average of output signal amplitude from each tire at three different reference speed values.

$$M_w = m_k(V_s) \times K_v(v) \tag{3}$$

$$K_v = \frac{0.0053 m_k}{V_s'} \tag{4}$$

Value of K_v for each reference speed is shown in Figure 11. It can be observed that the value of K_v tend to increase when the speed becomes higher. This trend corresponds with our previous claim that time constant in the WIM system is causing the signal amplitude to be lower for the same load at the higher speed. Therefore, speed correction factor is required. Since the value of K_v for each reference speed seems to have linear relation, by using linear regression, K_v can be calculated using equation (5). By combining equation (1), (3), and (5), the weight of the vehicle can be predicted based on data measured from WIM system (signal amplitude V_s and time difference Δt) using equation (6). This equation is developed based on the apriori knowledge about the wheelbase of the vehicle.

$$K_v = 0.0009v + 0.6443 \tag{5}$$

$$M_w = \left(\frac{V_s}{0.0053}\right) \times \left(0.0009\frac{L_{base}}{\Delta t} + 0.6443\right) \tag{6}$$

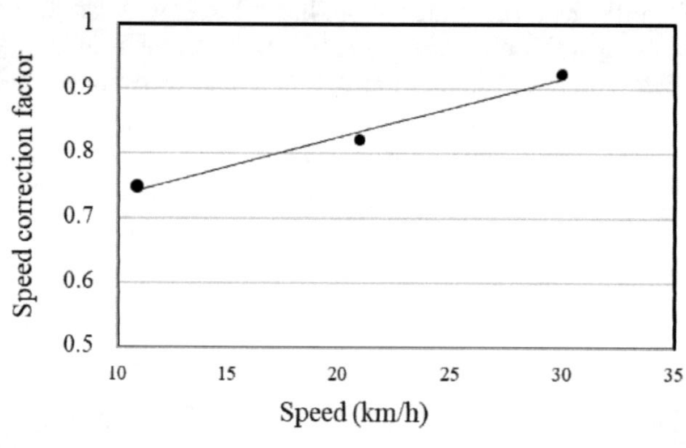

Figure 11. Relation between speed correction factor with vehicle speed

3.5. Error in measurement results for WIM system

Figure 12 gives an example of weight calculation results for motorcycle moving with reference speed of 10 km/h and 30 times repeated test. The upper and lower limit in the graph respectively indicates a value of twice standard deviation above and below the average value comes from WIM system measurement which is indicated by the straight line. By statistic, 95% of this measurement results will be within that range. Meanwhile, the dotted line indicates weight measured in static condition. It can be seen in the Figure 11 that the average WIM measurement for front tire always greater than static measurement. On the other hand, the measurement in rear tire gives a result in opposite way. These phenomena also occur for WIM measurement with reference speed of 20 and 30 km/h.

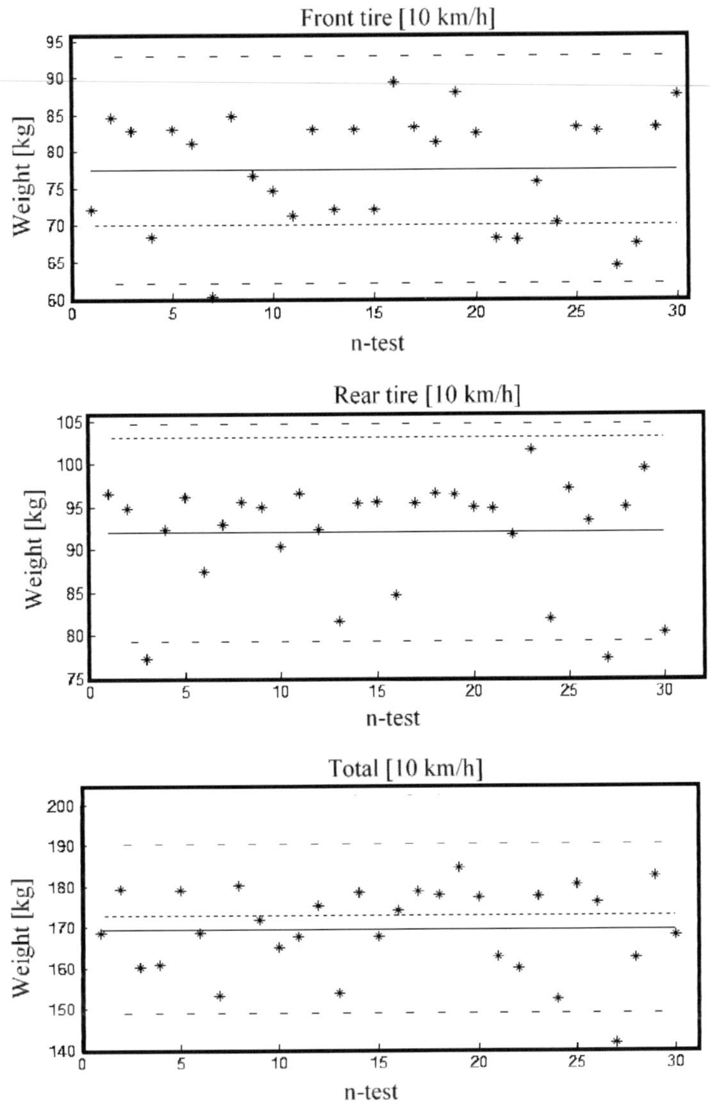

Figure 12. Results from WIM measurement with reference speed of 10 km/h.

Sudden change in vehicle acceleration due to condition of uneven road surface is an unavoidable circumstance that produce random results for WIM measurement in this experiment. Based on best practice, WIM device should be placed in flat road surface with no obstacles within distance of 100 m. Another limitation that contributes to this result is the application of single strain gage. The ideal condition for the measurement is the contact between tires and bending plate should be exactly over the strain gage location. However, the contact does not always happen at the desired position and it leads to produce signal output with lower amplitude.

Table 2 shows comparison between static and WIM measurement for three reference speed values. It is obvious that errors in terms of total weight are relatively smaller since the drop of weight distribution in one axle will be compensated with increasing weight in another axle.

Table 2. Vehicle's weight data for static calibration

Speed	WIM Measurement [kg]			Measurement Error		
	frront	rear	total	front	rear	total
10 km/h	62–93	79–105	149–190	-11% – 33%	-23% – 2%	-14% – 9%
20 km/h	67–88	76–110	152–189	-5% –26%	-26% – 7%	-12% – 9%
30 km/h	68–88	77–105	152–186	-3% – 26%	-25% — 2%	-12% – 8%
Front wheel load	70 kg					
Rear wheel load	103 kg					
Vehicle's total weight	173 kg					

3.6. Experiment Testing for Overload Detection System

The Instruments, which is used in previous experiment, has different gain value from HX711 module that is used for detection system. New calibration factor should be detemined using an identical procedure from previous experiments and the result is shown in Figure 13.

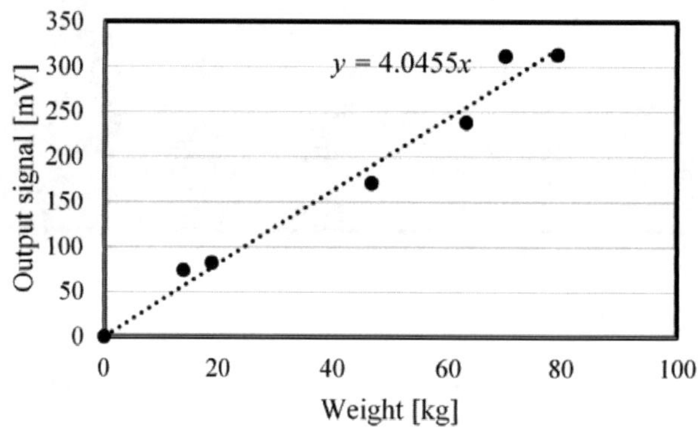

Figure 13. New calibration factor for overload detection system.

By replacing calibration equation from previous equation, it is obtained new equation for determining total weight of the vehicle. In equation (7), the value of vehicle speed correction factor is the same as from previous experiment since it came from the same sensor with same characteristic. Therefore, different instrument would not affect the result significantly.

$$M_w = \left(\frac{m_k}{4.0455}\right) \times \left(0.0009\frac{L_{base}}{\Delta t} + 0.6443\right) \tag{7}$$

In this experiment, an overloaded vehicle is defined as a motorcycle with three passengers, while only one passenger occupying the motorcycle to indicate normal vehicle. A motorcycle with three adult passengers has total weight within range of 360-450 kg. Whereas single occupant motorcycle has a total weight between 235-260 kg. Therefore, weight value of 320 kg is set as a minimum threshold for overloaded vehicle. If the system detects the total weight of vehicle greater than the threshold value, then the alarm and camera is activated.

This experiment proves that the overload detection system is able to differ whether the vehicle which passes through the bending plate is overload or not. The system is also able to trigger the alarm and activate the camera for taking picture of the vehicle when overloaded vehicle is detected and giving no responses when normal vehicle is passing by. The camera should be located at the proper position, so that the identity of the vehicle will always be captured when it is indicated to be overloaded. The illustration when the overloaded vehicle is captured by camera is shown in Figure 14.

Figure 14. Illustration for overloaded vehicle identity captured by camera

4. Conclusion

In this research work, Weigh-in-Motion (WIM) prototype based on bending plate technology using single active strain gage with overload detection system has been developed. Static and dynamic testing was carried out in order to check whether the WIM sensor could produce an output signal and perform the intended purpose in obtaining vehicle weight as it passes the sensor. Noise that appears in the output signal is alleviated by averaging method. Random WIM measurement results occur due to uneven road surface and inconsistent location of tire contact to the bending plate.

Over all, the overload detection system based on Arduino microcontroller and HX711 strain amplifier performed really well to trigger the alarm and camera when overloaded vehicle is detected. The above outcome gives an idea to develop a reliable microcontroller-based WIM and a robust automatic overloaded freight vehicle detection in our future work.

References

[1] Public Works Office 2011 *Minister of Public Works Regulation No. 19 Year 2011 on Road Technical Requirements and Design Criteria* (Jakarta - PUPR) – in Indonesian

[2] Budget Review Center, Indonesian People's Representative Council 2017 *Road Preservation Study 2015 and 2016* (Jakarta: DPR RI) – in Indonesian

[3] Wang H, Zhao J and Wang Z 2015 *Impact of Overweight Traffic on Pavement Life Using Weigh-In-Motion Data and Mechanistic-Empirical Pavement Analysis* (Department of Civil and Environmental Engineering, The State University of New Jersey)

[4] Jacob B and Beaumelle V 2010 Improving Truck Safety. Potential of Weigh-in-Motion Technology *IATSS Research* **34(1)** 9

[5] Xiong L, Zhuo F, Liu X, Wang F and Chen Y 2015 Optimal Design of Moving Average Filter and Its Application in Distorted Grid Synchronization, *2015 IEEE Energy Convers. Congr. Expo.* pp 3449-3454

IOP Conf. Series: Materials Science and Engineering **694** (2019) 012004 doi:10.1088/1757-899X/694/1/012004

Design and Performance Analysis of Brushless Direct Current (BLDC) Motor Controller for Electric Scooter

M N Yuniarto[1], Estiko Rijanto[2] and Agus Mukhlisin[1]

[1]Center of Excellence for Automotive Control & System, Institut Teknologi Sepuluh Nopember Surabaya 60111 Indonesia
[2]Research Center for Electrical Power and Mechatronics, Indonesian Institute of Sciences, LIPI, Bandung, Indonesia

E-mail: muhammad.nur3006@gmail.com

Abstract. Electric vehicle is an alternative to substitute internal combustion engine when fuel availability is difficult to obtain. Electric energy can be obtained from renewable energy and non-renewable energy. Electric vehicles like electric scooters consist of many parts. The parts used in an electric scooter have requirements such as reliability, safety, good performance, and high efficiency. Brushless direct current (BLDC) motor controller is one of the main components in the electric scooter. This paper presents a design and performance test of the BLDC motor controller. The controller uses PID current control with six step commutation in a trapezoidal control. The motor current ripple is eliminated by using a digital isolator. Tests were carried out using an Eddy Current Dynamometer. The results show that the current ripple is sucessfuly reduced which makes the controller suitable for electric scooter applications.

1. Introduction

A feed back controller of a brushless direct current (BLDC) motor usually incorporates hall effect sensors, back electromotive force (BEMF) signals, and current sensors. The quality of the controller is affected by the accuracy in measuring the feedback signals. Ripple current may occur, but how much the ripple current can be tolerated. This ripple current issue is important so that the motor can be rotated properly. Ripple currents can cause a rough motor response. Factors causing the current ripple are the error reading of the current sensor, load variations in the motor, switching frequency and inductive interference.

A current measurement method compensated the offset and scaling errors separately [1]. This study used two current measurements in phases A and phase C. Current measurement error consists of noise on hall sensors, filter circuits, and analog to digital converters [2]. Errors occur periodically so that they can be removed by a high pass filter. Ripple torque can be reduced by using a low switching frequency or constant switching frequency [3,4]. The optimal switching is calculated at each switching cycle to satisfy the minimum ripple based on the torque slope equations. Three-phase inverters use switching frequency limited around 0.5–1 kHz to minimize torque ripple [4]. This depends on the design of electric motor. There are many other methods to reduce the current ripple [5-9].

In this paper, a new current control is proposed which minimizes ripple current measurement using digital isolator. This method is effective to reduce the inductive effects of a power circuit. The high voltage ground is separated from the ground of the microcontroller so that the high voltage does not affect the performance of the microcontroller. ADUM 1200 can be used to separate the ground. This paper is organized as follows. The issues arising in a proportional, integral, direvative (PID) current control technique are described in section II. Design of a BLDC motor controller in an electric scooter application is presented in Section III. The implementation and evaluation of this control strategy are presented in Sections IV. Finally, the paper conclusion is in Section V.

2. PID Current Control Method

The current control technique consists of two controls (cascade control) as illustrated in figure 1. The first level control is speed control which uses the hall effect sensor as speed feedback. A speed reference is obtained from the throttle position sensor. Speed error is the difference between the speed reference and the present speed which is then compensated with the PID control technique so that it generates a reference current (I_{s_ref}). The second level control is the current control technique that uses the current sensor in phase as feedback. Current error that occurs will be compensated with the PID control. The PID current control will produce duty cycle and frequency values that will be used in Pulse Width Modulation (PWM). Next go to the commutation logic with the hall sensor feedback. The Hall sensor will also provide the position information of the rotor angle so that the commutation can be determined. Commutation logic can be realized in a PWM signal to activate the gate driver on the 3-phase inverter so that each phase becomes in an appropriate supply.

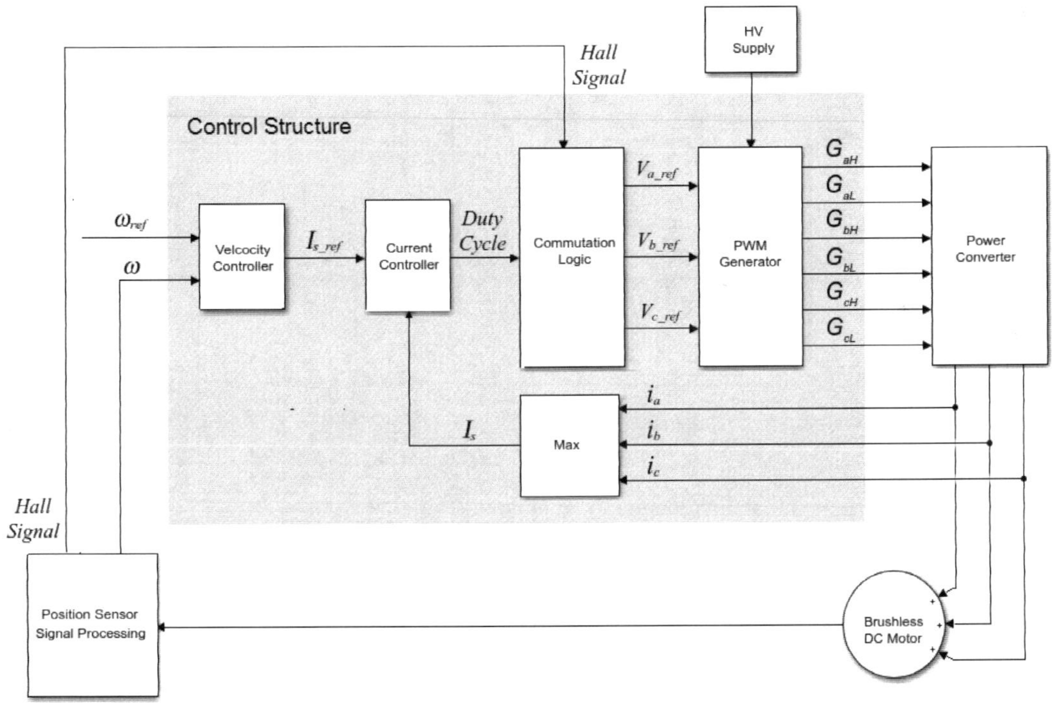

Figure 1. Current Control of BLDC Motor [10]

A model of a BLDC motor fed by a three phase inverter can be shown in Figure 2. The inverter consists of 3 half bridge MOSFETs as a switching phase A, B and C. Each switch Q1, Q2, Q3, Q4, Q5 and Q6 will be activated by the gate driver. Commutation logic of each switch is shown in Figure 3.

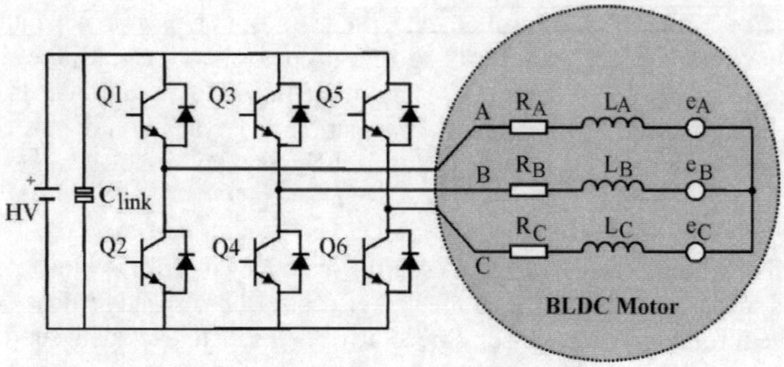

Figure 2. Model of BLDC Motor fed by 3 phase inverter

One electrical cycle consists of six-step commutations while one mechanical cycle will be composed of several electrical cycles. The amount depends on the configuration of pole pairs magnet mounted on the motor. Figure 3 shows a example of six-step commutation obtained from experiments. The hall sensor signals H1, H2, and H3 are ploted at the upper side. The commuation logic can be observed from the switching signals of Q1, Q2, Q3, Q4, and Q6. The soft PWM scheme is used where the high leg of a certain positive phase is switched in PWM mode while the lower leg of the companion negative phase is kept ON. Notice that the low leg of the same positive phase is switched in the opposite PWM mode using the PWM complementer. We name it as PWM mode of type (Hp-PWM, Ln-ON, Lp-inverted PWM), where subscripts p and n denote positive and negative phases of the motor.

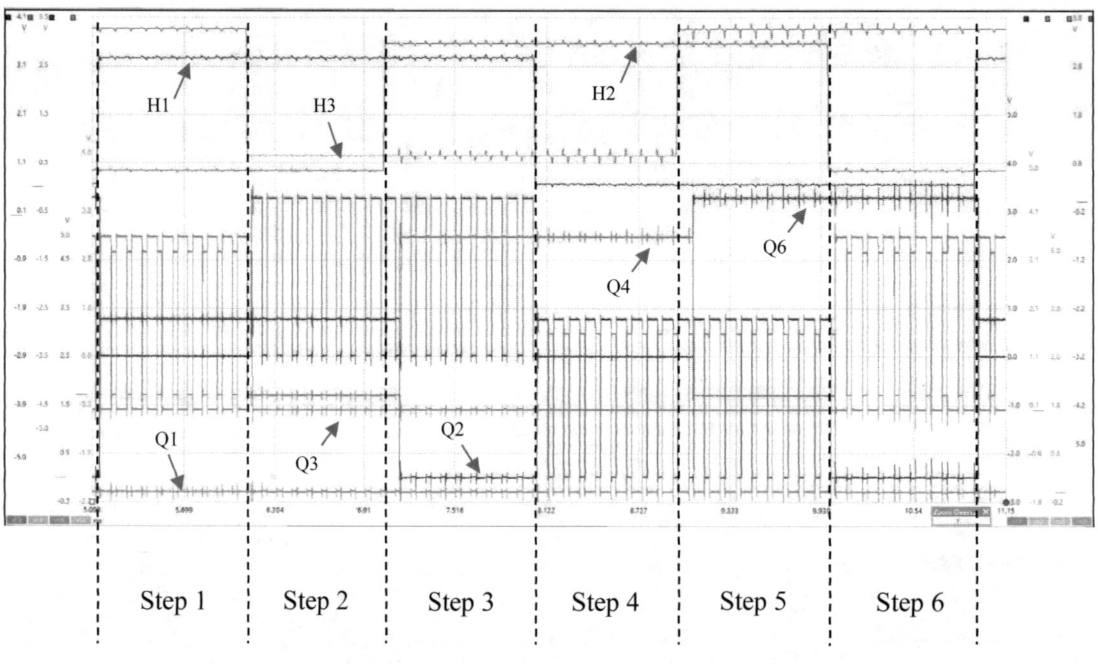

Figure 3. Six Step Commutation Using PWM Type of (Hp-PWM, Ln-ON, Lp-inverted PWM).

This type of commutation, which uses PWM type of (Hp-PWM, Ln-ON, Lp-inverted PWM), has a smaller torque ripple compared to (High-ON, Low-PWM) or (High-PWM, Low-PWM) [11]. The results in figure 3 can be summarized as six-step commutation logic according to table 1.

Table 1. Truth Table of Six Step Commutation of a BLDC Motor Controller

Step	Signal Hall Effect Sensor			Phase A		Phase B		Phase C	
	H1	H2	H3	Q1	Q2	Q3	Q4	Q5	Q6
1	1	0	1	Open	Closed	PWM	PWM inverted	Open	Open
				Negative		Positive		NC/Off	
2	1	0	0	Open	Closed	Open	Open	PWM	PWM inverted
				Negative		NC/Off		Positive	
3	1	1	0	Open	Open	Open	Closed	PWM	PWM inverted
				NC/Off		Negative		Positive	
4	0	1	0	PWM	PWM inverted	Open	Closed	Open	Open
				Positive		Negative		NC/Off	
5	0	1	1	PWM	PWM inverted	Open	Open	Open	Closed
				Positive		NC/Off		Negative	
6	0	0	1	Open	Open	PWM	PWM inverted	Open	Closed
				NC/Off		Positive		Negative	

This paper uses a 5 KW BLDC Motor operating at 100VDC. The parameter values of the BLDC motor are listed in table 2.

Table 2. Parameter values of the BLDC motor [12]

Parameter	Unit	Value
Resistance (R)	mΩ	8.005
Inductance (L)	μH	158.485
BEMF Constanta (k_e)	Volt/krpm	16.915
Torque Constanta (k_t)	N.m/A	0.1616
Time Constant (τ_m)	s	0.1791
Friction Constanta (B)	N.m.s	0.0280
Moment of Inertia (J)	Kg.m^2	0.0356

3. Design of BLDC Motor Controller

The BLDC motor controller in this paper is illustated by a block diagram in figure 4. An important point to consider in designing a 3-phase inverter is that HV must be isolated with a microcontroller supply. This is done to avoid spike interference in the HV which can cause a microcontroller reset. Digital isolator ADUM1200 is used. HV sensing uses optical operational amplifiers to measure battery voltage and BEMF. Other sensors such as hall effect sensors and current sensors in phase A and C also use galvanic isolation.

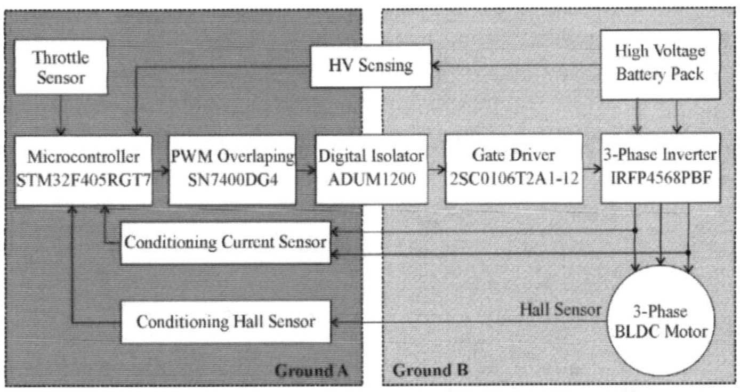

Figure 4. Diagram of the BLDC Motor Controller

IOP Conf. Series: Materials Science and Engineering **694** (2019) 012004 doi:10.1088/1757-899X/694/1/012004

Overlapping PWM is needed to ensure that PWM channels high and low sides do not meet. To really protect, certain value of dead time is added. Gate drivers use modules from the manufacturer so that gate driver optimization has been done well by the manufacturer. It is hoped that the failure mode on MOSFETs can be avoided. Isolation can also be done by separating between boards, both the controller board, gate driver board, and the 3-phase inverter board. The designed controller is shown in figure 5.

Figure 5. BLDC Motor Controller

4. Result and Analysis

The BLDC motor controller performance was tested using an Eddy Current Dynamometer as shown in figure 6. The dynamometer can be used to determine power, torque, and speed (rpm). The signal measurement is done using an oscilloscope.

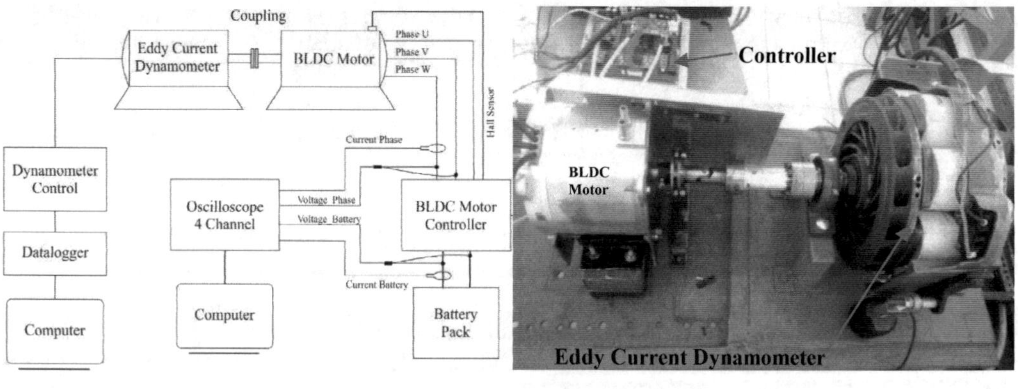

Figure 6. Experiment Set-up

4.1 Signal Measurement

Figure 7 shows the hall effect sensor and BEMF signals. BEMF signals are trapezoidal so that this control technique is called trapezoidal control. A signal was found in the fault hall sensor, which was on the red circle. The signal does not directly affect the commutation logic because the signal does not reach the digital input level to be low. But of course the problem needs to be formulated a solution. BEMF signals on blue circle marking can cause switching losses that need to be eliminated. In general the signal is in accordance with the BLDC motor drive theory.

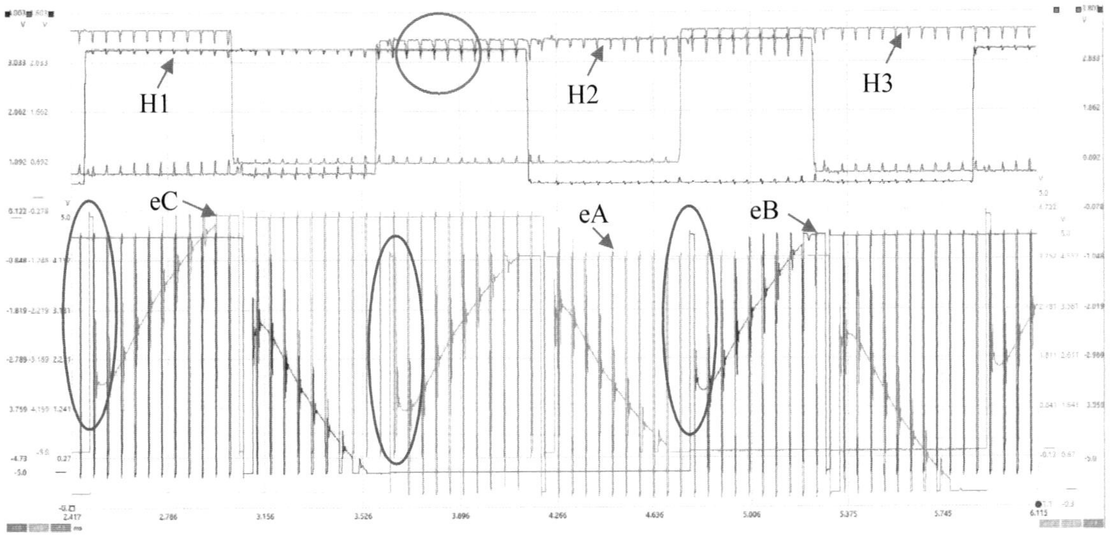

Figure 7. Measurement of Hall Sensor and Back Electro Motive Force (BEMF)

Figure 8 shows the comparison of current signals of the motor phase A using a digital isolator and not using a digital isolator. Design without a digital isolator can cause ripple current which can lead to compensation calculation errors on the PID controller. The ripple also affects the ripple torque so that the motor rotation is not smooth. Whereas current using digital isolator has less current ripple compared to not using digital isolator. The author suggests that the BLDC motor controller should be completely isolated so that external interference from the motor does not directly affect the system control.

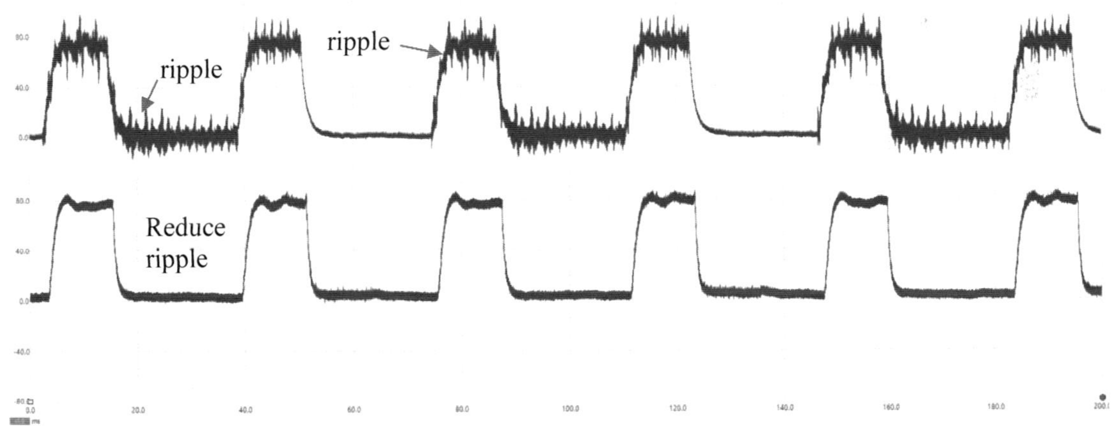

Figure 8. Ripple Current Elimination Using Digital Isolator

Figure 9 shows current measurements results of each phase. Part a is the phase A current at a full speed condition while part b measures the current during start-up. By observing these results, we confirm that the shape of the current signal is in accordance with the BLDC motor drive theory. An interesting thing that needs to be discussed is that the phase current at start-up is very large. This requires that the MOSFET 3-phase inverter can handle large currents, further increasing production costs.

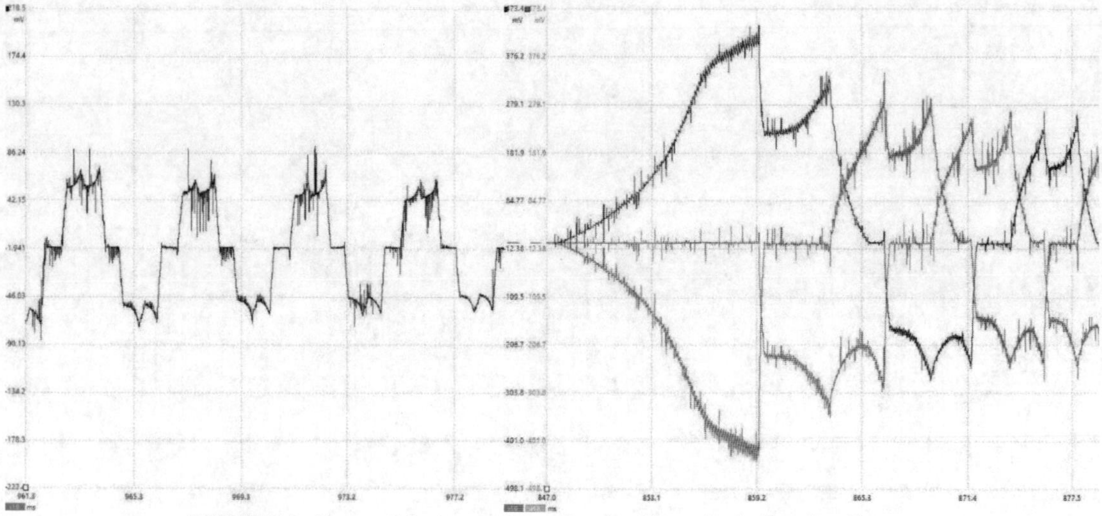

FIGURE 9. Current Measurement: a. Current Phase A , b. Starting Current

4.2 Performance Test Results

Figure 10 shows the BLDC motor controller performance testing results. The testing was done by rotating the motor until it reached the maximum rotation speed without any load, then new load was added in stages until the motor rotation was unable to resist the dynamometer load. The results indicate that the controller can operate at 5000 Watt instantenously with the torque reaching 60 N.m at the speed of 1000 rpm. This BLDC motor controller can be applied to electric scooters.

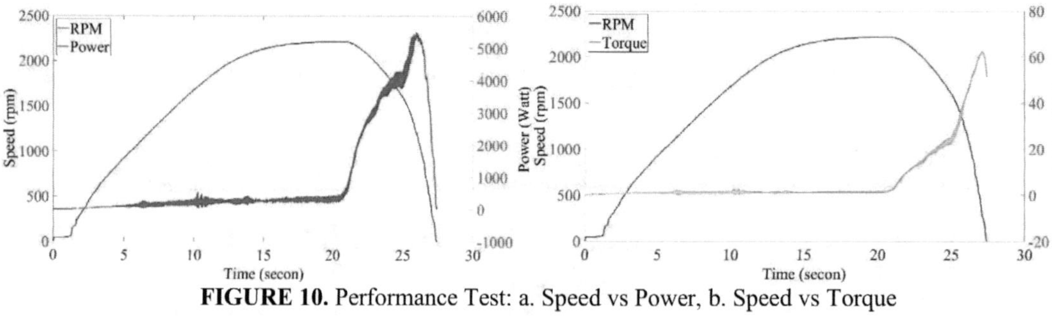

FIGURE 10. Performance Test: a. Speed vs Power, b. Speed vs Torque

5. Conclusion

We have designed and tested the 5 kW BLDC motor controller using a PID current control with trapezoidal commutation technique. The ripple current in the BLDC motor was reduced by a digital isolator. Performance test results indicate that the controller is suitable for electric scooter applications.

References

[1] M. Kim, S. Sul and J. Lee, "Compensation of Current Measurement Error for Current-Controlled PMSM Drives," in IEEE Transactions on Industry Applications, vol. 50, no. 5, pp. 3365-3373, Sept.-Oct. 2014.

[2] S.-H. Song, J.-W. Choi, and S.-K. Sul, "Digitally controlled AC drives," IEEE Ind. Appl. Mag., vol. 6, pp. 51–62, 2000.

[3] Jun-Koo Kang and Seung-Ki Sul, "New direct torque control of induction motor for minimum

torque ripple and constant switching frequency," in IEEE Transactions on Industry Applications, vol. 35, no. 5, pp. 1076-1082, Sept.-Oct. 1999.

[4] Kyo-Beum Lee, Joong-Ho Song, Ick Choy and Ji-Yoon Yoo, "Torque ripple reduction in DTC of induction motor driven by three-level inverter with low switching frequency," in IEEE Transactions on Power Electronics, vol. 17, no. 2, pp. 255-264, March 2002.

[5] Y. Cho, K. Lee, J. Song and Y. I. Lee, "Torque-Ripple Minimization and Fast Dynamic Scheme for Torque Predictive Control of Permanent-Magnet Synchronous Motors," in IEEE Transactions on Power Electronics, vol. 30, no. 4, pp. 2182-2190, April 2015.

[6] Q. Liu and K. Hameyer, "Torque Ripple Minimization for Direct Torque Control of PMSM With Modified FCSMPC," in IEEE Transactions on Industry Applications, vol. 52, no. 6, pp. 4855-4864, Nov.-Dec. 2016.

[7] Kyo-Beum Lee, Joong-Ho Song, I. Choy and Ji-Yoon Yoo, "Improvement of low-speed operation performance of DTC for three-level inverter-fed induction motors," in IEEE Transactions on Industrial Electronics, vol. 48, no. 5, pp. 1006-1014, Oct. 2001.

[8] G. H. B. Foo and M. F. Rahman, "Direct Torque Control of an IPM-Synchronous Motor Drive at Very Low Speed Using a Sliding-Mode Stator Flux Observer," in IEEE Transactions on Power Electronics, vol. 25, no. 4, pp. 933-942, April 2010.

[9] Y. Zhang and J. Zhu, "A Novel Duty Cycle Control Strategy to Reduce Both Torque and Flux Ripples for DTC of Permanent Magnet Synchronous Motor Drives With Switching Frequency Reduction," in IEEE Transactions on Power Electronics, vol. 26, no. 10, pp. 3055-3067, Oct. 2011.

[10] Mathwork team, "BLDC Current Controller with PWM Generation". https://www.mathworks.com/help/physmod/sps/ref/bldccurrentcontrollerwithpwmgeneration.html : Data Access : 1 October 2019.

[11] Niapour, S.A.KH & Garjan, GH & Shafiei, Mehdi & Feyzi, M. Reza & Danyali, Saeed & Bahrami Kouhshahi, Mojtaba. (2014). Review of Permanent-Magnet Brushless DC Motor Basic Drives Based on Analysis and Simulation Study. International Review of Electrical Engineering (IREE). 9. 930. 10.15866/iree.v9i5.827.

[12] A. Mukhlisin, Estiko. R, I.N.Sutantra, M.N. Yuniarto and A.S.Pramono, "Towards an Intelligent Reconfigurable BLDC Motor Controller: Commutation Configuration and Parameters Detection," in The 4th International Conference on Mechanical Engineering (ICOME), August 2019.

IOP Conf. Series: Materials Science and Engineering **694** (2019) 012005 doi:10.1088/1757-899X/694/1/012005

Simulation of float of 19 passenger aircraft during landing on water surface

Riki Ardiansyah and Mohammad Adhitya

Research Center for Advanced Vehicle Universitas Indonesia (RCAVe)

E-mail: ardiansyah_riki@yahoo.com

Abstract. The 19-passenger plane was modified to land in the water by making a landing gear with a float. The floats used are imported products that have previously been used by similar aircraft. Float must be able to support the aircraft and can accept impact loads when landing. This study aims to determine the effect that occurs in float when making water surface landings. Stress and displacement are simulation outputs that are the focus of research. These parameters are obtained by numerical simulation methods based on Civil Aviation Safety Regulation (CASR) by varying input parameters such as: landing speed. From the results of the study it was found that the faster the velocity then the output of the stress and displacement response from the float becomes larger.

1. Introduction

The 19-passenger aircraft Amphibious aircraft that will be produced is a modified aircraft from the 19-passenger aircraft which is given a float to be able to take off and land in the waters. The floats used are imported products that have previously been used by similar aircraft. The use of floats is very beneficial to be applied to aircraft because it doesn't change the configuration of the airframe much, only the weight and design of the float shape must be adjusted so that the performance of the aircraft is not reduced. [1]

The impact load is one of the important factors to know because at the time of landing a large energy absorption process occurs from the surface of the water to the surface of the float. The impact between the water and the basic structure of the ship can cause critical loads or even result in structural damage [2].

Von Karman [3] was the first person to put forward a theory related to the analysis of amphibious aircraft landings. After that, many studies have been conducted regarding the effects of water impact. Numerical methods and experimental tests by dropping objects vertically have been done several times. The models that have been studied include: horizontal cylinders [4], wedges [5], cone 20°, 45° dan hemisphere [6], plate [7], boat [8], deformable sphere [9].

This 19-passenger aircraft float impact study was carried out using a static simulation method. Checking is needed to find out the strength of the imported float used on the 19-passenger aircraft. The float must be able to support the aircraft and can accept impact loads when landing in accordance with CASR (Civil Aviation Safety Regulation) Subpart C Water Load that issued by the Ministry of Transportation.

2. Methods

2.1 Float Specification

19 passenger aircraft Amphibian is a twin seaplane float. This float is located on both sides of the fuselage. The first series Amphibious 19 passenger aircraft float is an imported product made by a company from the United States specializing in float making for floatplane and amphibian, namely: WIPAIRE, Inc. This float is made of aluminium material, can float with a maximum load of 6,5 Tons each side and already used by Twin Otter 300 series.

Table 1. Float Dimension [10]

Float	Dimension
Length	9900 mm
Height Hull	1140 mm
Width Hull	1535 mm
Dead Rise	23°

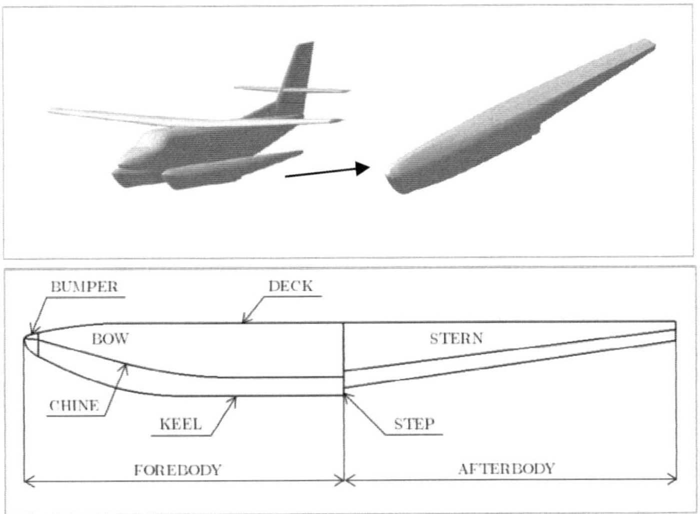

Figure 1. Illustration of Float 19 passenger aircraft Amphibian & Part of Float

Table 2. Sheet metal thickness Wipline13000 [10]

Sheet metal 6061-T6		Thickness (inch)
Top Skins	Forward & Aft	0.04
	Centre	0.063
Side Skins	Forward & Aft	0.04
	Centre	0.063
Bottom Skins	Bow to Wheel Well	0.063
	Afterbody	0.032
	All Remaining	0.05
Bulkhead	Nose	0.1
	Aft Wheel Well	0.063
	All Remaining	0.04

Float is made of 6061-T6 series aluminium material. (Mg and Si alloy). This type of aluminium has very good corrosion resistance and excellent seepage despite reducing the strength in the welding zone [11]. To restore its strength can be done by reheat treatment and artificially aging [12]. This type of aluminium has medium fatigue strength. In addition, this type of aluminium has excellent joining characteristics and acceptance of good coating applications.

Table 3. Al 6061-T6 Mechanical Properties [13]

Properties	
Density	2,7 g/cm3
Ultimate tensile strength (24°C)	310 MPa
Modulus of Elasticity	68,9 GPa
Poisson's Ratio	0,33
Shear Modulus	26 GPa
Shear Strength	207 MPa

2.2 Water Loads

In NAS 807 [14] (National Aerospace Standard) which is the standard specification for twin seaplane float, it is stated that two seaplane floats must have 80% more buoyancy needed to support the maximum weight of the aircraft in water. Strength requirements are specified in terms of limit load and ultimate load. Limit load is the maximum load that occurs during operation. Ultimate load is the same as the limit load multiplied by the safety factor. Safety factor used 1.5.

Based on CASR 23.525 Application of Load (b) In applying the loads may be distributed over the hull or main float bottom using pressures not less than those prescribed in 23.533(c). [15]

The following pressure distributions shall be used for the design of the frames, keel and chine structure. These pressures shall be applied simultaneously over the entire float bottom and their distributions shall be uniform.

The magnitudes of the symmetrically distributed pressures in psi shall be given by.

$$P = \frac{C_4 K_2 V_{SO}^2}{\tan\beta} \tag{1}$$

Where:

P : Pressure (psi)
C_4 : $0{,}078C_1$ ($C1 = 0.012$)
K_2 : Float station weighing factor (figure 2.)
V_{so} : Stalling speed (knots) (Vs = 59 knots, Vso = 45.4 knots)
β : Angle of dead rise. (β = 23°)

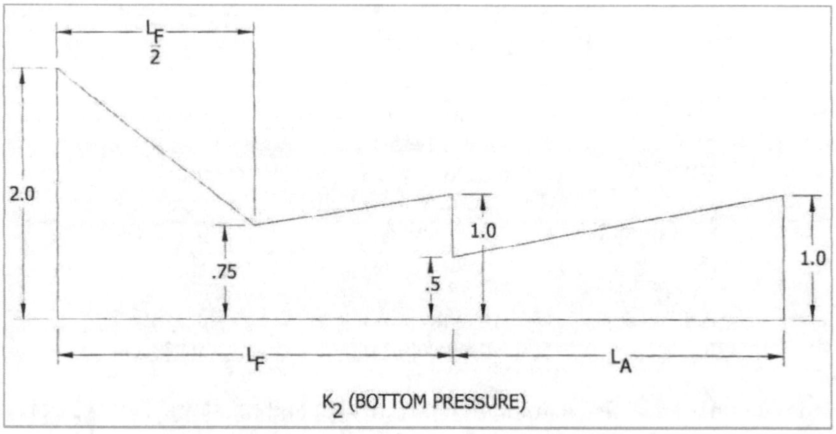

Figure 2. Float Station Weighing Factor

IOP Conf. Series: Materials Science and Engineering **694** (2019) 012005 doi:10.1088/1757-899X/694/1/012005

2.3 Numerical Modelling

The float model is designed in CATIA-V5 Software. The designed model is shown below.

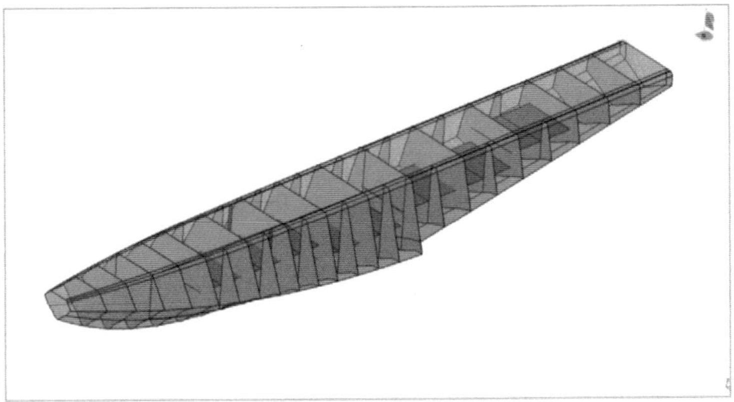

Figure 3. Float Model

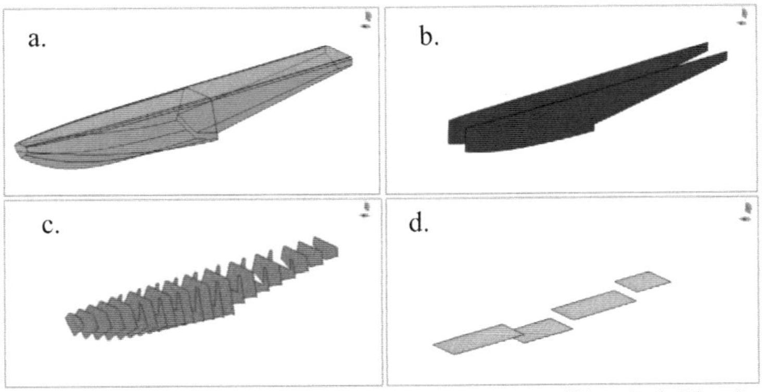

Figure 4. Parts of Float (a. Skin; b. Frame; c. Bulkhead; d. Floor)

The results of the Catia software image are then imported for further analysis. Design analysis is done using Abaqus software.

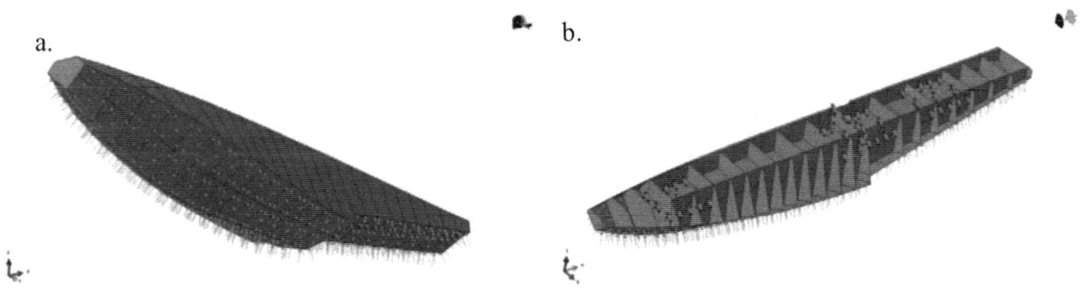

Figure 5. a. Load Distribution; b. Boundary Condition

Load distribution is given to all the lower surfaces of the float. Boundary conditions are placed on the floor inside the float, this is because in that part the landing gear strut is attached. The mesh used in the simulation is a quad type.

IOP Conf. Series: Materials Science and Engineering **694** (2019) 012005 doi:10.1088/1757-899X/694/1/012005

3. Results

3.1 Loads

By using equations (1) and divide the float into sections based on the position of the bulkhead, then the load obtained in the form of pressure at the bottom of the float as follows:

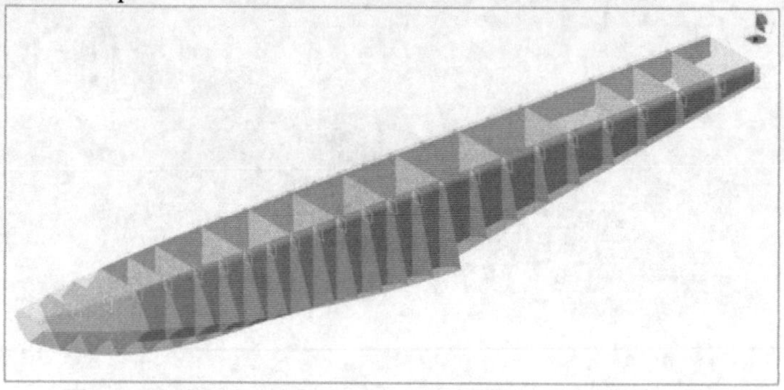

Figure 6. Section of Float Based on Bulkhead Position

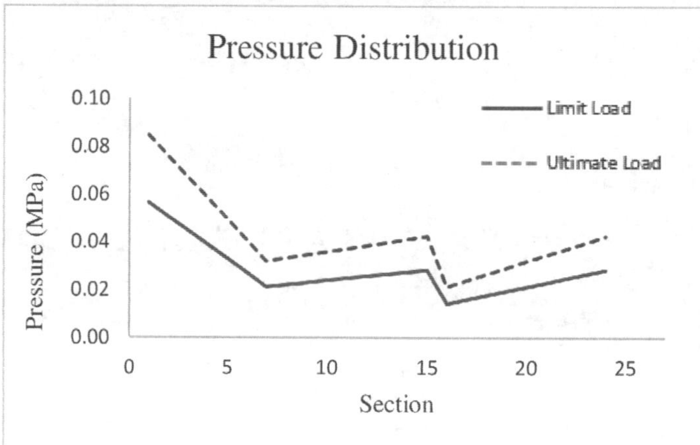

Figure 7. Pressure Distribution Under the Float

3.2 Mesh Convergence

Six mesh sizes were selected (70, 60, 50, 40, 30 and 20) for the float simulation. These meshes are simulated to obtain convergent points of the simulated model mesh size.

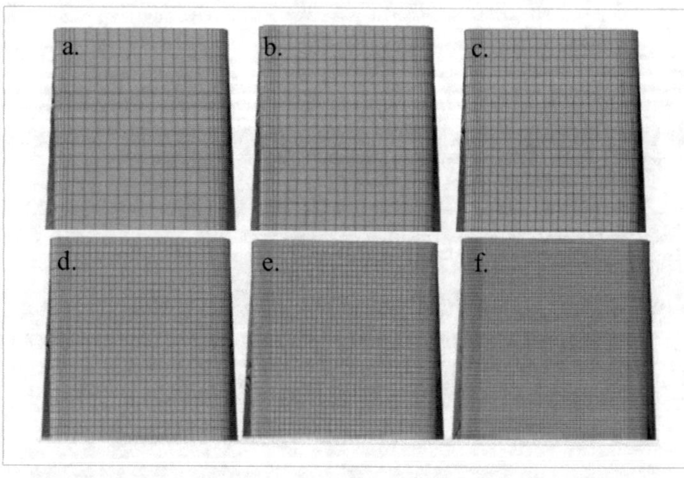

Figure 8. Mesh size (a. 70; b. 60; c. 50; d. 40; e. 30 and f. 20)

IOP Conf. Series: Materials Science and Engineering **694** (2019) 012005 doi:10.1088/1757-899X/694/1/012005

Based on the simulation results, it is found that with reduced mesh size, the stress produced is getting smaller. Convergent points occur when entering mesh size 40 and stable up to size 20. With this result, for the next modeling, a mesh with a size of 40 will be used.

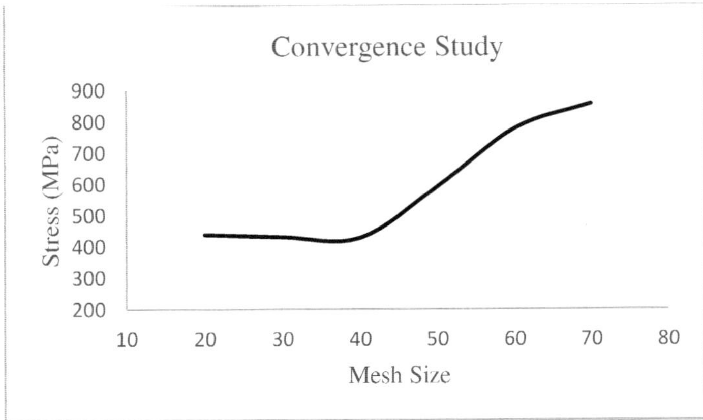

Figure 9. Mesh Convergence Study

3.3 Numerical results

The results of the simulation with the distribution pressure limit load on the float with aluminum float thickness of 2.5 mm on the entire surface produce von misses stress of 236 MPa. Whereas in the loading of the ultimate load (1.5 times the limit load) von misses stress that occurs on the float is 353 MPa. If the float only gets a limit load, the float structure can still withstand the pressure received because the stress received is smaller than the allowable stress Al 6016-T6 but when the float receives the ultimate load, the float structure will fail.

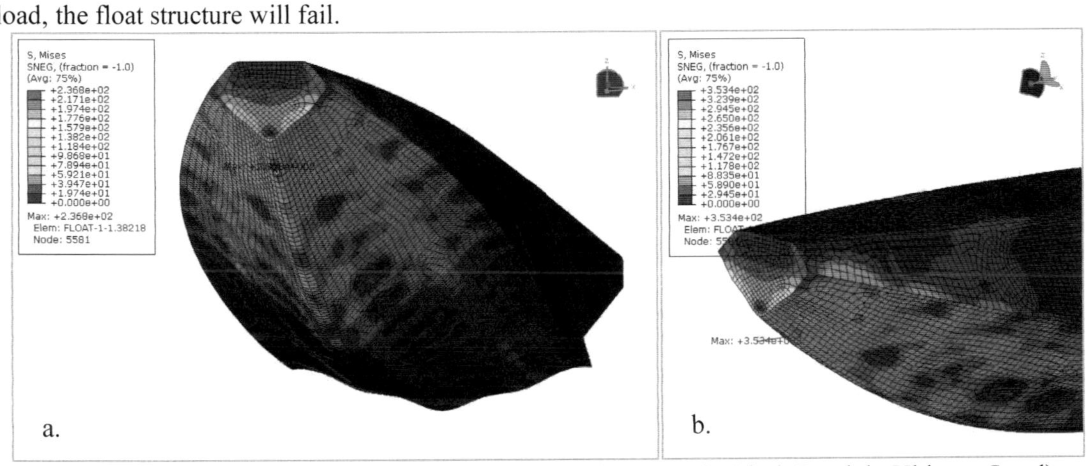

Figure 10. Simulation Results with A Thickness Of 2.5 mm (a. Limit Load; b. Ultimate Load)

From Figure 10 it can be seen that max stress occurs in the first section of the bottom skin float. It is understandable because in that part according to Figure 2, float station weighting factor in that position has the biggest K factor. other than that, in that section there are only skins supported by bulkhead without additional support from the frame and floor. See figure 11.

IOP Conf. Series: Materials Science and Engineering **694** (2019) 012005 doi:10.1088/1757-899X/694/1/012005

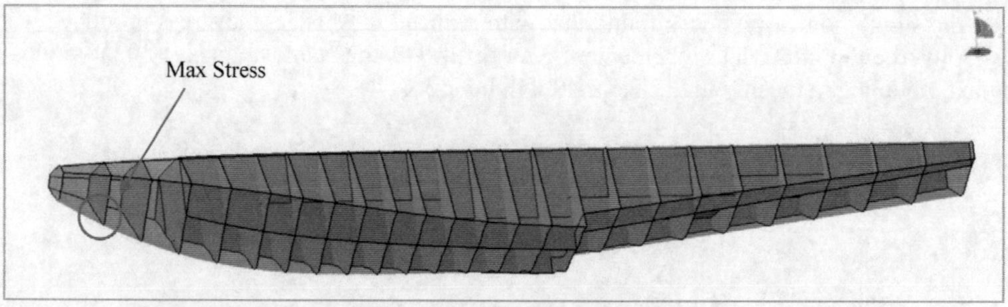

Figure 11. Max Stress Location Without Frame and Floor Support

To resolve this problem can be done by extending the frame to the front or adding an extra floor between the front bulkhead to bulkhead behind and one more solution that can be done is to increase the thickness of the bottom surface that receives max stress.

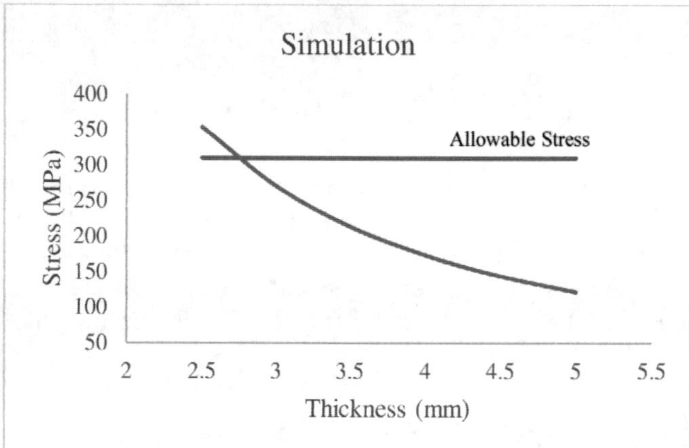

Figure 12. Simulation Result After Increasing The Thickness

Figure 12 shows that with an increase in float surface thickness the stress will be reduced. At 3mm thickness the stress on the float surface becomes 207.3 MPa so that the float will be safe when receiving pressure.

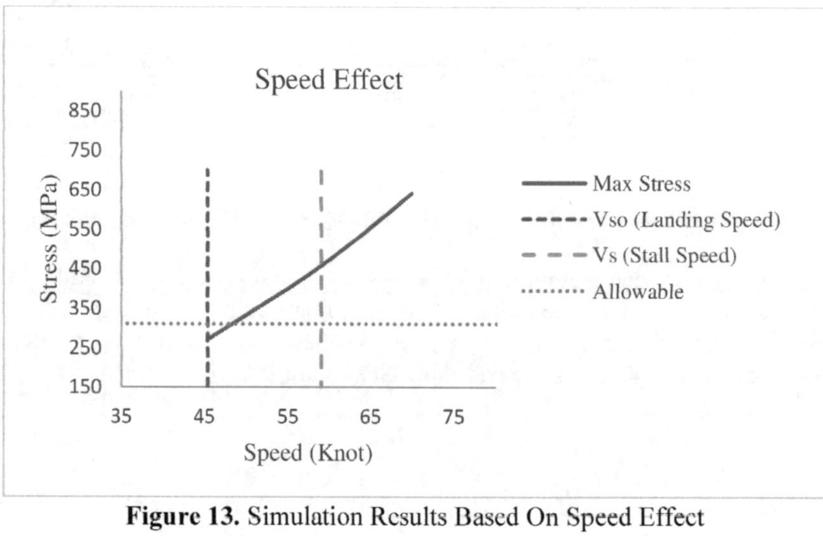

Figure 13. Simulation Results Based On Speed Effect

The effect of increasing speed on stress received by float can be seen in Figure 13. At landing speed (V_{SO}) the stress received was 207.3 MPa, with increasing speed becoming the stall speed (V_S) stress received to 457.1 MPa. and when it reaches a speed of 70 knots the stress it receives becomes even greater, which is 641.1 MPa. Based on equation 1 it can be seen that the speed variable has a large influence on pressure. Velocity has a quadratic function in the pressure distribution equation. so rising speed has a very large effect on increasing the pressure and stress received by float.

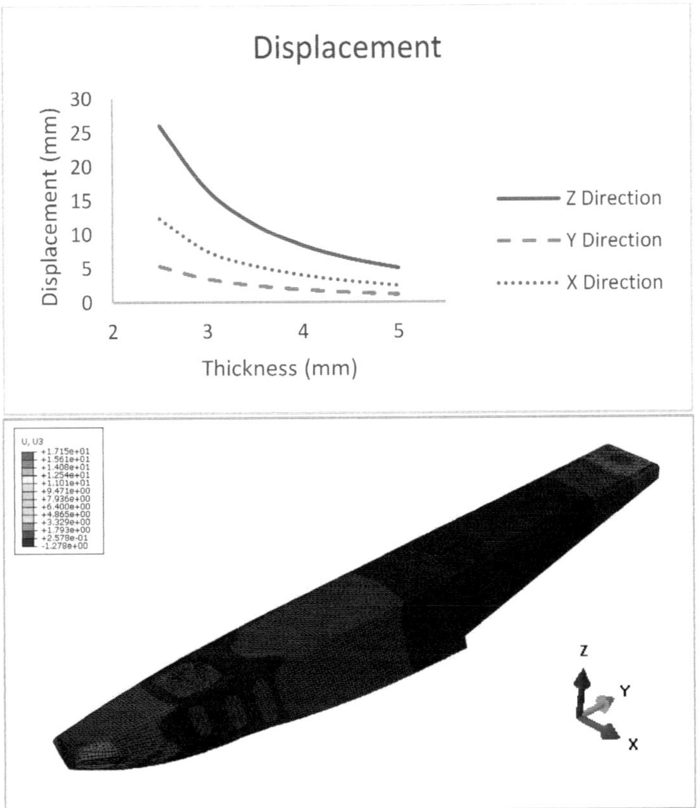

Figure 14. Displacement on Float & Axis Information

When the float receives pressure on the lower surface, the float will be displaced along with the stress received. From the simulation results in Figure 14. it can be seen that the greatest displacement occurs towards the z direction, because in this direction it is in the direction of the pressure received by the float. greatest displacement is on the front and rear of the float. this is because this point is the farthest point from the boundary so that the moment obtained is greater and the displacement is even greater.

4. Conlusions

In this paper we have discussed float design and analysis. The float is analyzed using simulation software for stresses and displacements. It was observed that the stress experienced by the float decreased with increasing surface thickness and stress increases with increasing aircraft speed when landing. The results show that with a thickness of 3mm, the float can accept impact loads based on CASR 23. For future work is to do dynamic simulations on float to get closer to the actual conditions and get more accurate results.

Acknowledgement
The authors gratefully acknowledge the financial support of the PITTA Grant 2019 Universitas Indonesia for this publication and Pustekbang LAPAN for the useful data, helpful discussions and suggestions. Many thanks and gratitude also go to Saintek Ristekdikti scholarship for the financial support of this study.

References

[1] redaksi. *2022, Pesawat N219 Amphibi Uji Coba Terbang*. 2018; Available from: https://www.lapan.go.id/index.php/subblog/read/2018/5389/2022-Pesawat-N219-Amfibi-Uji-Coba-Terbang/1349.
[2] Lv, J. and J.L. Grenestedt, *Analytical study of the responses of bottom panels to slamming loads.* Ocean Engineering, 2015. **94**: p. 116-125.
[3] Karman, V., *The Impact on Seaplane Floats During Landing.* National Advisory Committee for Aeronatics, 1929. **Technical Note no.321**: p. 309-313.
[4] Sun, H. and O.M. Faltinsen, *Water impact of horizontal circular cylinders and cylindrical shells.* Applied Ocean Research, 2006. **28**(5): p. 299-311.
[5] Luo, H., H. Wang, and C. Guedes Soares, *Numerical and experimental study of hydrodynamic impact and elastic response of one free-drop wedge with stiffened panels.* Ocean Engineering, 2012. **40**: p. 1-14.
[6] Wang, S. and C. Guedes Soares, *Numerical study on the water impact of 3D bodies by an explicit finite element method.* Ocean Engineering, 2014. **78**: p. 73-88.
[7] Wang, S., et al., *Dynamic structural response of perforated plates subjected to water impact load.* Engineering Structures, 2016. **125**: p. 179-190.
[8] Facci, A.L., M. Porfiri, and S. Ubertini, *Three-dimensional water entry of a solid body: A computational study.* Journal of Fluids and Structures, 2016. **66**: p. 36-53.
[9] Hurd, R.C., et al., *Water entry of deformable spheres.* Journal of Fluid Mechanics, 2017. **824**: p. 912-930.
[10] Wipaire, i., *Twin-Otter-Wipiare-Product & Modification Guide*. 2019.
[11] Aalco. *Aluminium Alloy 6061-T6 Extrusions*. 2018; Available from: http://www.aalco.co.uk/datasheets/Aluminium-Alloy-6061-T6-Extrusions_145.ashx.
[12] Hydro. *Aluminum I Beam*. Available from: https://www.hydro.com/en-US/profiles/6061-t6-aluminum-properties/.
[13] Aerospace_Specification_Metals_Inc. *Aluminum 6061-T6; 6061-T651*. 2001; Available from: http://asm.matweb.com/search/SpecificMaterial.asp?bassnum=ma6061t6.
[14] Standard, N.A., *Twin Seaplane Floats*. 2017, Aerospace Industries Association of America, Inc.
[15] Indonesia, M.o.T.R.o., *Civi Aviation Safety Regulation (CASR) Part 23*. 2001.

IOP Conf. Series: Materials Science and Engineering **694** (2019) 012006 doi:10.1088/1757-899X/694/1/012006

Review of the Topology and Energy Management Hybrid Energy Storage on Electric Vehicle

Ayuning Fitri Desanti[1], Yoga Uta Nugraha[1], M Nur Yuniarto[2] and Alief Wikarta[2]

[1]Department of Electrical Engineering, Center of Excellence for Automotive Control & System, Institut Teknologi Sepuluh Nopember, Surabaya, Indonesia
[2]Department of Mechanical Engineering, Center of Excellence for Automotive Control & System, Institut Teknologi Sepuluh Nopember, Surabaya, Indonesia

E-mail: ayuningfitridesanti@gmail.com, uta.utu@gmail.com, muhammad.nur3006@gmail.com, aliefwikarta@gmail.com

Abstract. There are several ways to improve the mileage of the electric vehicle (EV). High-efficiency drivetrain and high energy density are the main factors. Energy storage source for EV obtained from the battery, ultra-capacitor (UC) and fuel cells (FC). Each of that has own characteristics and capacity. The battery has a good energy density, UC and FC are superiors on power density. Combining some of the energy storage will improve performance and mileage. On this paper, the topology of combining energy storage are reviewed. In addition, also review the power management strategy of hybrid energy storage. By using a convenient strategy, hybrid energy storage becomes very worth.

1. Background

Energy storage in the electric vehicle must comply for power density and energy density. High power density energy storage such as battery consists of some types. The common use in this era is Lithium Ion (Li-Ion). Although there are some other types of the battery such as Lead–Acid, Nickel–Metal Hydride (NiMH), Nickel–Zinc (Ni–Zn) and Nickel–Cadmium (Ni-Cd) [1]. Lead-Acid is popular in automotive use, but not suitable for EV because of the low energy density and power density. NiMH has twice energy density than Lead-Acid and also can be recycled. But if repeatable use in high load power demands the cycle count will reduce 200-300 cycle. For applying on EV need advance power management because of dynamic power demands. Ni-Cd has economically priced and long lifetime. However if Ni-Cd if it is not fully charged it can cause problems with capacity because there is a momory effect [1][2]. The upcoming energy storage is Ultra-Capacitor (UC). UC basicly have similar construction like existing capacitors. Using adavance materials can make UC with better performance than the capacitor by the same dimension. Appropriate to capacitors characteristics, UC can charge and discharge instantaneously.

Purpose of research on an electric vehicle is improving the milage range. Using the best performance energy storage is the solution. But every energy storage has primacy and limitation. Combining energy storage can fullfil the limitation. This paper [1] explaining energy sources and hybrid energy storage. And also detailed comparise on the aplication and the price. In addition, to get a better result, some research did optimization. There are two way of optimaization, standart clustering and using aritifical intelegent. Paper [3] combining batteery, UC and fuel cells. Using the SOC parameter for clustering the capacity of each source. Paper [4] proposed a new energy management strategy using Pontryagin's Minimum Principle (PMP) topology.

This paper reviews topology and energy management of hybrid energy storage on an electric vehicle. Section II presents the topology of hybrid energy storage. Section III explain about management system on hybrid and section IV presents the summary and conclusion

2. Review of Hybrid Topology

There are three basic topologies that use on hybrid energy storage. Series, parallel and series-parallel configurations [3]. These topologies developed on energy storage for electric vehicle based on the capacity of electric vehicle's drive train. And also, all of the topologies can be connected as a passive or active configuration. Advantage of active configuration is can be a potentially smaller dimension that passive configuration as the same load.

2.1. Basic Passive Parallel

In a simple way, the passive configuration doesn't need any kind of power electronics/converter on the hybrid system[5]. By that's condition there is a major problem for utilized the hybrid energy storage.

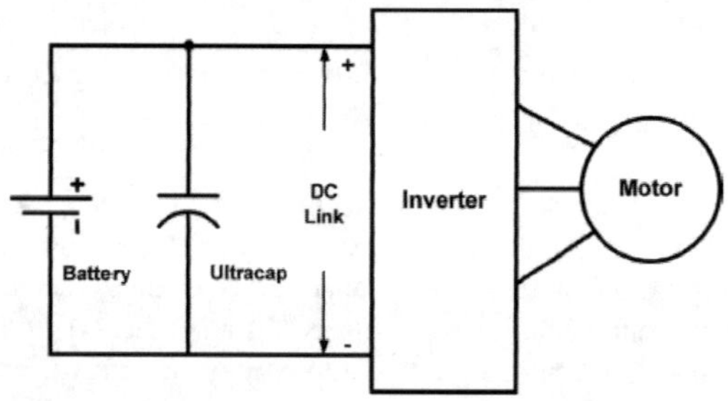

Figure 1. Basic passive parallel configuration

2.2. Ultracapacitor (UC)-Battery Configuration

This configuration is the most popular because of a simple system. Also, because using a bidirectional dc / dc converter makes this system applicable in many conditions. The development of hybrid energy storage systems starts with this system. So this system becomes the initial gate of research in terms of hybrid energy storage.

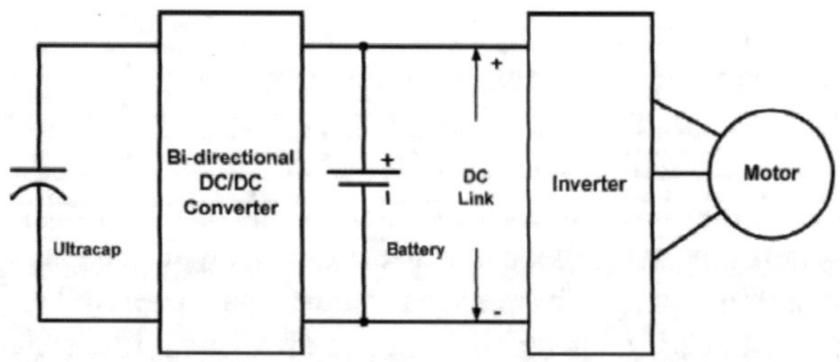

Figure 2. Ultracapacitor (UC)-Battery Configuration

Figure 2 shows the diagram of UC - Battery configuration. The bi-directional DC/DC converter is located between UC and battery, so it needs larger size converter to handle the power of UC. Furthermore, the voltage of UC can be lower from converter specifications and the battery directly connected to the DC link that causes the DC link voltage cannot be changeable.

IOP Conf. Series: Materials Science and Engineering **694** (2019) 012006 doi:10.1088/1757-899X/694/1/012006

2.3. Battery-UC configuration

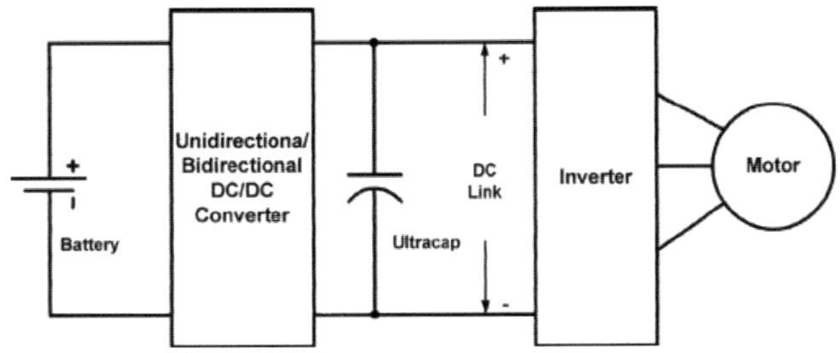

Figure 3. Battery-UC configuration

Another conjuration is changed the position between UC and battery. Battery located before the converter. By this condition, the battery can be hold onto lower or higher than the UC voltage [5]. Shows on figure 3, UC located after the converter and connected to DC link. UC also working as low pass filter because of this condition. The advantage of this configuration is the control of DC link voltage can be adjusted within a range so that the UC energy can be more effective.

2.4. Cascade Configuration

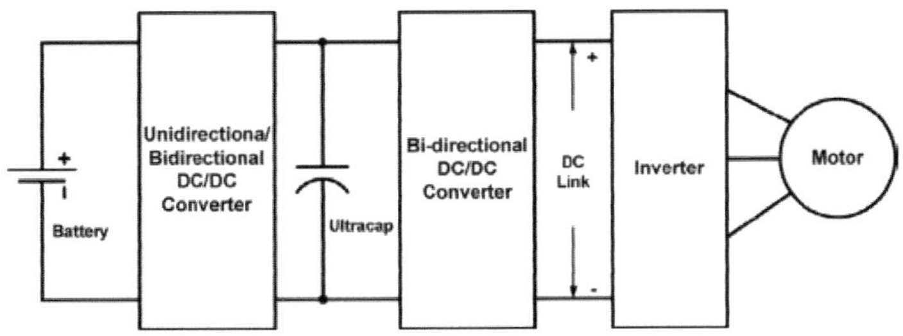

Figure 4. Cascade configuration

Diagram of the cascade configuration shows in figure 4. Using two converters, unidirectional/bidirectional converter located between battery and UC, other bidirectional converter located between UC and DC link [5]. The development of these configurations explained below.

2.5. Multiple converter configuration

From the cascade, the configuration can develop become parallel connection the inverter output that connected to the DC link. The diagram shows below in figure 5.

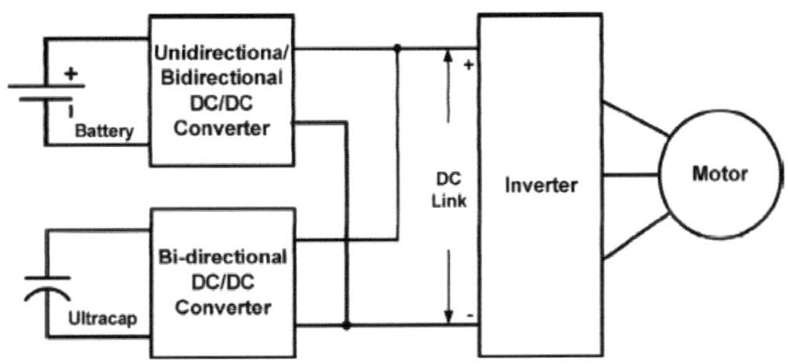

Figure 5. Multiple configuration

The advantage is output voltage from both converter can be hold onto lower than the dc-link voltage, so the voltage on DC link can be balanced [5]. In addition, the UC using can be maximized because of the wide range of voltage. But this configuration needs two powerful converters that will affect price and dimension.

2.6. Multiple input converter configuration

To reduce the problem on multiple converter configuration, the new configuration is researched [5]. Named multiple input converter configuration. The diagram shows in figure 6 below.

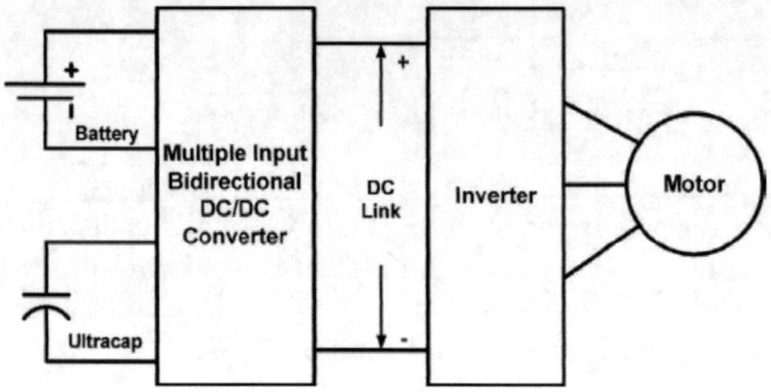

Figure 6. Multiple input converter configuration

The use of hybrid energy storage systems can use a DC / DC converter with less power as charging a UC bank. Even with a relatively constant load profile will make battery life longer because there is no need for a fluctuating load. So using this system has the potential for efficiency if sizing and adjusting is done accurately.

2.7. Threeway Configurations

Three way systems consist of battery units (BU), fuel cells (FC) and ultra capacitors (UC). This system is quite complicated to apply because each energy source has different characteristics. Sizing between components must be absolutely right because by using many different energy sources the risk of failure will also increase. In addition, high installation costs are still a challenge in this system.

A bidirectional step up/step-down dc/dc converter, as shown in Fig. 7, connects each power source to the dc link. From each source to be able to supply the load using a step-up dc / dc converter so that the resulting voltage will be stable. Conversely, when it functions to store energy regenerative braking, a step-down dc/dc converter is used to charge the battery and the UC. In this condition the system needs to be added to the safety system so that the current is flowed according to the current charging capacity of each source.

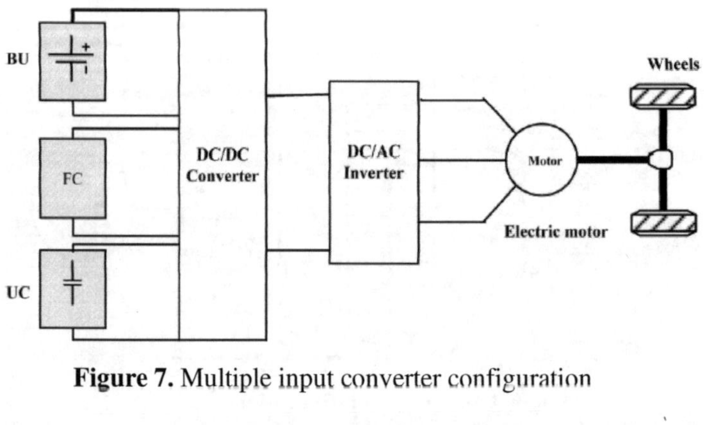

Figure 7. Multiple input converter configuration

IOP Conf. Series: Materials Science and Engineering **694** (2019) 012006 doi:10.1088/1757-899X/694/1/012006

3. Power Management

Supply energy to the loads needs to manage the power on each energy source, such as the battery, UC and RFC. Purposes of power management extend the range, improve vehicle performance and extend the battery life [6][1]. In general there are two power management strategies in HESS, the most widely used system today is the rule base method with certain restrictions. In addition, fuzzy methods based on artificial intelligence are widely studied. [4]. Another optimization is developed from those strategies. The main target of optimal sizing to get the best combination between each energy source which can minimize the weight and maximize the power output.[7].

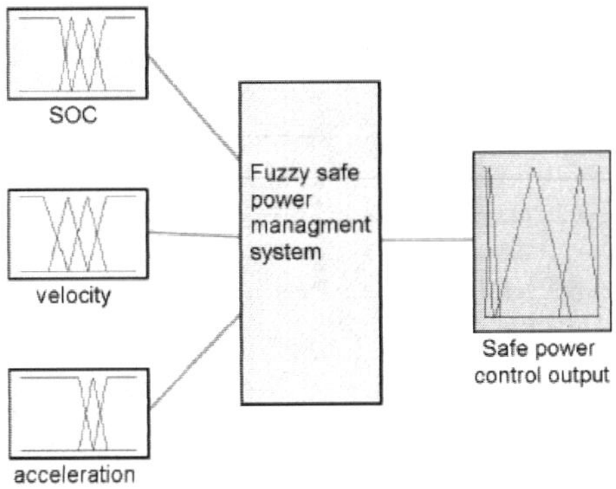

Figure 8. Fuzzy on power management system

In the research [8], author focus research on power management using fuzzy logic based on the safe condition. The safe condition means that the electric vehicle is on the low power state, so every energy source needs to keep the performance and condition above minimal value and characteristic. On fig 8. Diagram of fuzzy for power management using three data input, there are SOC, velocity and acceleration. For energy storage that uses RFC, research [7] use clustering on frequency to get the best optimal sizing of RFC.

To determine the power consumtion, research [5] separated as four driving modes. On each mode has own characteristics so need to define own management system. There are low constant speed mode, high constant speed mode, accelerations mode and decelerations mode. Low consant speed mode when power of dc-dc converter (Pconv) can handle the power demand (Pdmd). Otherwise, if Pdmd higher than Pconv it means on high constant speed mode. In low speed mode, the voltage of UC can be preserved higher than the battery current directly supply to the DC link and dc/dc converter not operated.

In acceleration mode, fig 9 shows first condition that UC voltage higher than the battery voltage. When Pconv < Pdmd, UC voltage will decrease. UC and the dc/dc converter are supplying power for the vehicle acceleration. After a while, fig 10 shows UC voltage will be the same as the battery voltage then dc/dc converter will energize the vehicle demand and UC.

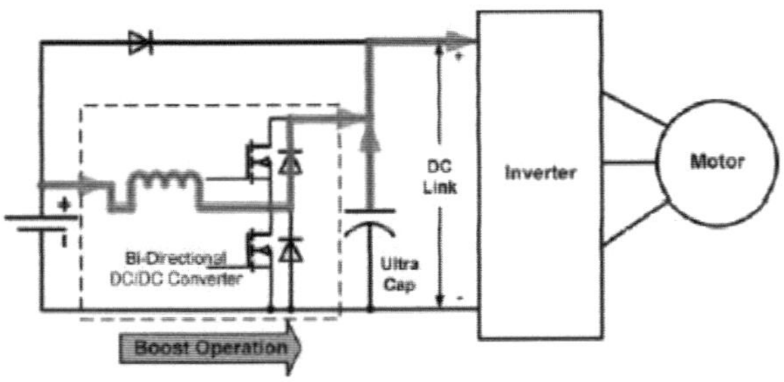

Figure 9. Acceleration mode 1

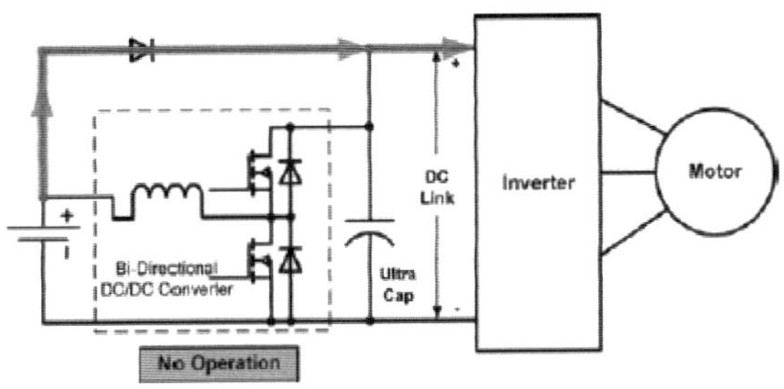

Figure 10. Acceleration mode 2

IOP Conf. Series: Materials Science and Engineering **694** (2019) 012006 doi:10.1088/1757-899X/694/1/012006

There is two steps on deceleration mode. First, on fig 11 and 12 the regenerative power will energize UC. The dc/dc converter on boost operation or not depand on the UC voltage. Then on fig 11, when regenerative power is in continusly, dc/dc converter will act like a buck converter. So the battery will absorb the energy from the vehicle. This condition will safe the UC from over voltage condition.

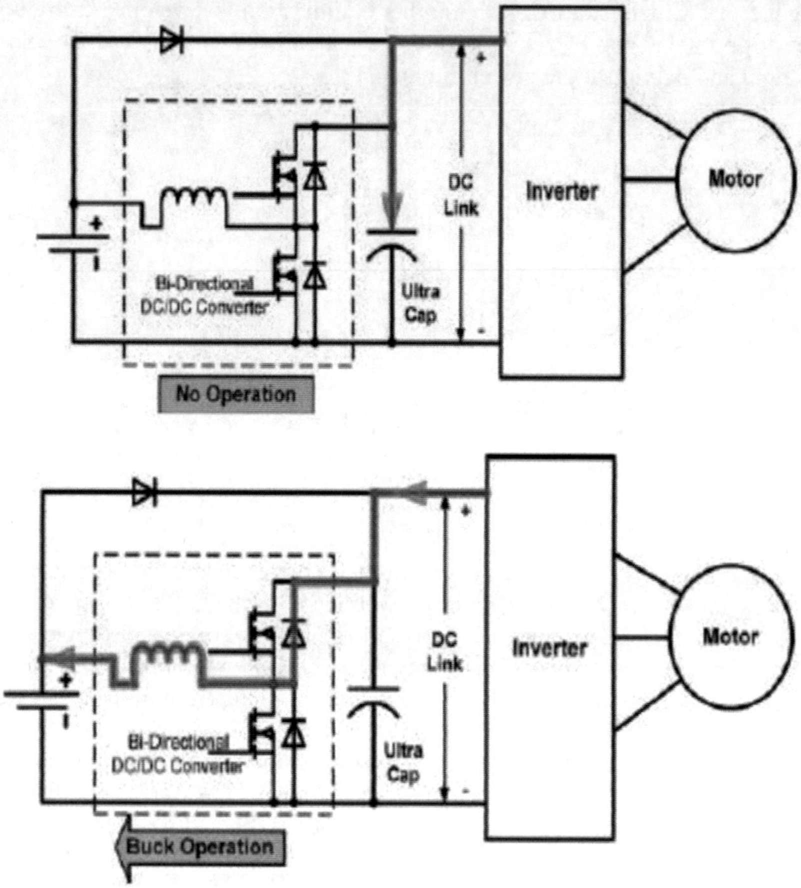

Figure 11. Deceleration mode 1 & 2

Another way to optimizing is by using multiobjective optimization algorithm [9]. By using wavelet-based algorithms, the battery state of health (SOH) is highly concerned with dimensions and battery life. The algorithm performs smart settings related to power delivery between the battery and the UC. UC handles very high transient power requirements. These conditions cannot be done with batteries. Conversely, for stable conditions, a battery source is used. Until now there has been no comparison between methods with fuzzy and rules in certain applications.

4. Conclusion

In this paper, topologies of hybrid energy storage for electric vehicle present on some type. Each topology has own advantages. DC/DC converter has an important rule because of every topologies strategy use its device as the load management. There are two power management strategies in HESS, the most widely used system today is the rule base method with certain restrictions. In addition, fuzzy methods based on artificial intelligence are widely studied. By using power management, application on hybrid energy vehicle can be optimized.

5. References

[1] A. Khaligh and Z. Li, "Battery, ultracapacitor, fuel cell, and hybrid energy storage systems for electric, hybrid electric, fuel cell, and plug-in hybrid electric vehicles: State of the art," *IEEE Trans. Veh. Technol.*, vol. 59, no. 6, pp. 2806–2814, 2010.

[2] "Nickel-based Batteries Information – Battery University." [Online]. Available: https://batteryuniversity.com/learn/article/nickel_based_batteries. [Accessed: 23-May-2019].

[3] Z. Amjadi and S. S. Williamson, "Power-electronics-based solutions for plug-in hybrid electric vehicle energy storage and management systems," *IEEE Trans. Ind. Electron.*, vol. 57, no. 2, pp. 608–616, 2010.

[4] C. Zheng, W. Li, and Q. Liang, "An Energy Management Strategy of Hybrid Energy Storage Systems for Electric Vehicle Applications," *IEEE Trans. Sustain. Energy*, vol. 9, no. 4, pp. 1880–1888, 2018.

[5] J. Cao and A. Emadi, "A new battery/ultracapacitor hybrid energy storage system for electric, hybrid, and plug-in hybrid electric vehicles," *IEEE Trans. Power Electron.*, vol. 27, no. 1, pp. 122–132, 2012.

[6] X. D. Xue, K. W. E. Cheng, S. R. Raman, Y. C. Fong, and X. L. Wang, "Investigation of energy distribution and power split of hybrid energy storage systems in electric vehicles," *2016 Int. Symp. Electr. Eng. ISEE 2016*, vol. 1, no. 1, 2017.

[7] J. Snoussi, S. Ben Elghali, M. Benbouzid, and M. F. Mimouni, "Optimal sizing of energy storage systems using frequency-separation-based energy management for fuel cell hybrid electric vehicles," *IEEE Trans. Veh. Technol.*, vol. 67, no. 10, pp. 9337–9346, 2018.

[8] V. Galdi, A. Piccolo, and P. Siano, "A fuzzy based safe power management algorithm for energy storage systems in electric vehicles," *2006 IEEE Veh. Power Propuls. Conf. VPPC 2006*, 2006.

[9] L. Zhang, X. Hu, Z. Wang, F. Sun, J. Deng, and D. G. Dorrell, "Multiobjective Optimal Sizing of Hybrid Energy Storage System for Electric Vehicles," *IEEE Trans. Veh. Technol.*, vol. 67, no. 2, pp. 1027–1035, 2018.

IOP Conf. Series: Materials Science and Engineering **694** (2019) 012007 doi:10.1088/1757-899X/694/1/012007

Design and Performance Analysis of Brushless DC Motor Cooling System for Low-Cost Manufacturing Process

Yoga Uta Nugraha[1], Arief Cahyadi[1], Juniono Raharjo[2], M Nur Yuniarto[3] and Alief Wikarta[3]

[1]Department of Electrical Engineering, Center of Excellence for Automotive Control & System, Institut Teknologi Sepuluh Nopember, Surabaya, Indonesia
[2]Department of Naval Architechture, Center of Excellence for Automotive Control & System, Institut Teknologi Sepuluh Nopember, Surabaya, Indonesia
[3]Department of Mechanical Engineering, Center of Excellence for Automotive Control & System, Institut Teknologi Sepuluh Nopember, Surabaya, Indonesia

E-mail: uta.utu@gmail.com, ariefc72@gmail,com, juniono10@gmail.com muhammad.nur3006@gmail.com, aliefwikarta@gmail.com

Abstract. Thermal analysis on BLDC Motors is very important to be learned in order to obtain a low-cost motor design and have a high level of efficiency. The design of the existing motor cooling system uses a complex design that must use sophisticated machining tools and spend a lot of money in the manufacturing process. The purpose of this paper is to get the geometry of the cooling system on BLDC motor with simple manufacturing process thus saving the cost of production. The expected cooling system design is a simple geometry so that manufacturing can be done using 3-axis CNC machining. Design of cooling system using water-jacket with an additional cover. On this study, Author modeled the motor with Ansys Maxwell to obtain the heat source parameters of the motor geometry. Software Fluent is involved to perform the temperature simulation on the motor using the designed cooling system then validated with an experiment. Thermal analysis of this study compares motor performance with the use of water and oil. The flow rate compared to the radiator power efficiency.

1. Background

Currently the Brushless DC motor, or often referred to as BLDC, is a brushless motor that is controlled by a 3 Phase Inverter. This type of motor has been chosen as the prime mover of electric vehicles because it has high efficiency, high power density, and low noise. The motor consists of main parts such as stator, rotor, shaft, coils, housing, and bearings. Sometimes for large motors require additional cooling to accelerate the heat propagation from the inside of the motor to the surface of the motor.

In this paper we will discuss thermal analysis of cooling systems that can be applied to existing motors. The experimental method is carried out by simulation using the Solidworks Flow Simulation Software. The fluid cooling model that is often used is circular geometry. This requires a fairly complex design and manufacturing process. Therefore it is necessary to conduct research on making cooling systems that are cheap but efficient.

Chang et al. [1] in his research proves that differences in motor design will have an impact on the amount of motor losses that will be a source of heat. Baggu et al [2] in his research modeling thermal analysis using an electric equivalent circuit. Yoon et al [3] in his research proves that motor performance can be improved by improving the cooling system.

2. Motor Loss Source Model

2.1. Winding Loss

Winding the motor uses enamelled wire with the main material being copper. Copper flowing current in an electric motor also produces copper losses can be formulated into:

$$P_{cu} = 3I^2R \tag{1}$$

Where P_{cu} represents copper losses that are affected by current value I and internal resistant R. Value of R affected by the length of the conductor ℓ, measured in metres (m), the cross-sectional area of the conductor A measured in square metres (m²), and ρ (rho) is the electrical resistivity (also called specific electrical resistance) of the material. The mathematical equation can be written as:

$$R = \rho\frac{l}{A} \tag{2}$$

2.2. Core Loss

Stator and rotor are made of electrical steel material which is both a conductor and ferromagnetic. Therefore Stator and Rotor have losses called Hysteresis Loss and Eddy Current Loss. Hysteresis Loss is caused by the process of magnetization and demagnetization of the stator when an electric current is flowed in a forward and reverse direction. When the current is directed forward, the flux is positive. With the same energy rate, there is still a flux value so it needs reverse energy to make zero flux. This difference becomes one of the energy losses in the core.

While eddy current loss refers to Faraday's Law where a conductor that moves to cut the flow of magnetic flux will cause an induced voltage. Core construction is shaped like a closed loop circuit so that the current flows individually to the core. This current is called the Eddy Current. To reduce eddy current effect, cores are usually made with layers of conductor with silicon laminates. Core loss values are obtained from simulation results using ANSYS MAXWELL electromagnetic simulation software.

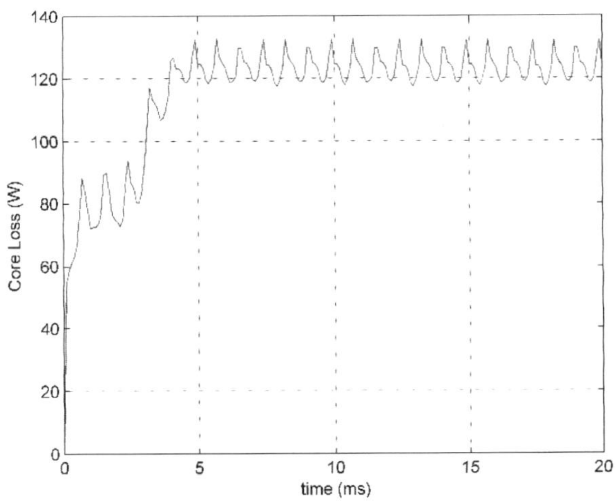

Figure 1. Coreloss Value

IOP Conf. Series: Materials Science and Engineering **694** (2019) 012007 doi:10.1088/1757-899X/694/1/012007

2.3. Bearing Friction Loss
All bearings have friction loss. Factors that affect magnitude friction loss are influenced by type of bearing, bearing size, bearing design, load, oil viscosity, oil flow, and speed [4]. In this paper, friction loss is neglected.

3. Physical Model
The types of motors studied in this paper are Radial Brushless DC (BLDC) motors with axial way cooling and internal fans. This motor has a power rating of 20 kW with a voltage of 300 V. This motor is designed with 12 stator slots and 8 magnetic poles. The combination of the number of slots and poles has a winding factor of 0.866 [5]. The motor part consists of a stator and rotor made of electrical steel material, housing made of aluminum, shafts made of stainless steel, permanent magnet Neodymium N52, and coil winding.

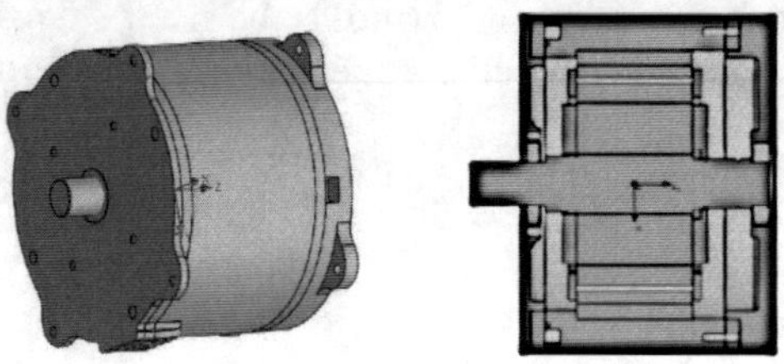

(a) (b)

Figure 2. 2 CAD model of BLDC motor: (a) outside view, (b) cross section view

4. Simulation Result
Thermal modeling is done using Solidworks Flow Simulation software. This software is a Computational Fluid Dynamics (CFD) embedded in Solidworks 3D software. Initially the motor operated full load at a speed of 3600 rpm, with a room temperature of 280C, the results are shown in Fig.3.

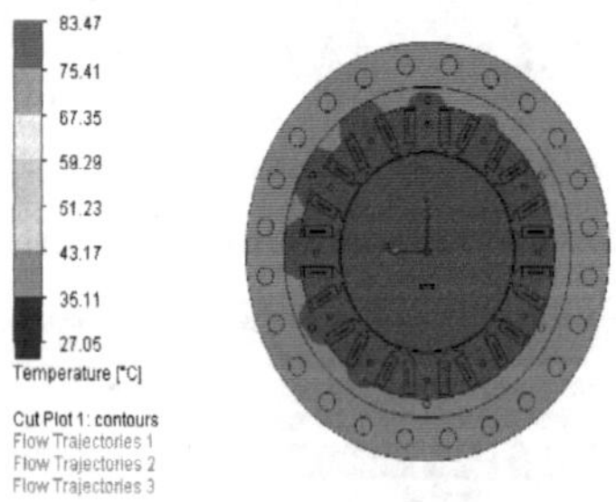

(a)

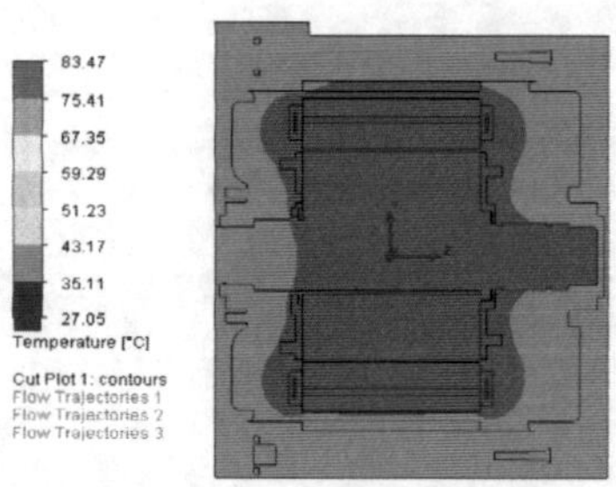

(b)

Figure 3. Cross section of thermal simulation without cooling system, (a) : Front View, (b): Side View

From the simulation results it can be seen that when the motor is operated full load, the average temperature of the motor is 800C. Therefore, this motor design requires additional cooling systems such as fans or water flow.

a) Cooling system with water flow

Water is a type of cooling fluid that is often used by motor manufacturers with medium to large capacity [6]. Besides being economical, the advantages of using water as a cooling medium will not cause crust on the channel. The cooling channel is designed with geometry as in Fig.4 to simplify and reduce production costs. Circular inlet and outlet with a surface area of 0.000785 m2. Water flow rate is set at 8L / m.

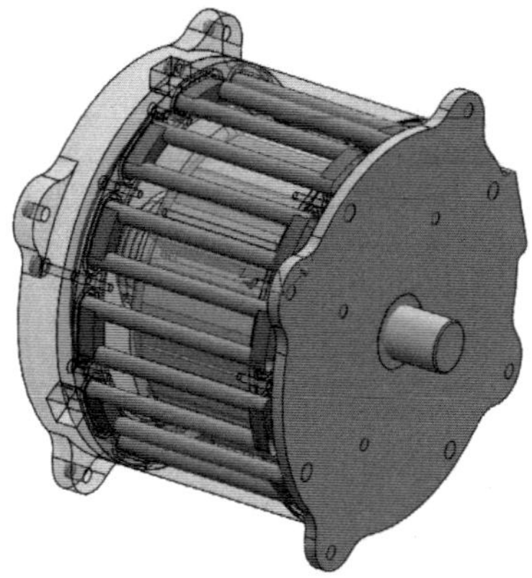

Figure 4. Water cooling line design

The manufacturing process is carried out by utilizing a cylindrical ring shape on the housing then forming 20 axial holes. After that the main housing is zigzagged so that water can flow on all sides of the motor.

Figure 5. Cross section of thermal simulation with watercooling system, (a): Front View, (b): Side View

From the simulation results it can be seen that the temperature of the motor when operated full load by cooling water has better results than without cooling. The inside of the motor has an average temperature of 65^{0}C while the housing has an average temperature of 33^{0}C.

b) Cooling system with fan

The second experiment is using two fans on the inside. The shape of the fan is shown as in Fig.6. The fan is paired with a shaft, so the fan has the same rotational speed as the motor. The fan is made of PLA material made with a 3D Printing machine. The results of thermal simulation using two fans and without cooling water are shown in Fig. 7.

Figure 6. Water cooling line design

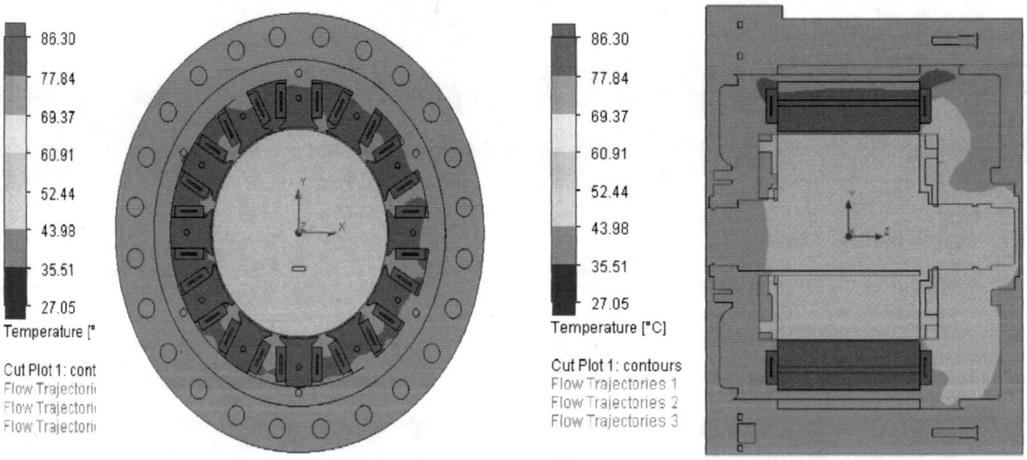

(a) (b)

Figure 7. Cross section of thermal simulation with Fan cooling system, (a): Front View, (b): Side View

Cooling design with the fan is only able to reduce the temperature in the rotor area. From the simulation results, the rotor temperature obtained from 60°C-70°C, at the stator section 70°C to 85°C. While the housing has an average temperature of 75°C.

c) Cooling System with Fan and Waterflow

The third cooling option is to use both water flow and fan . When the load and speed parameters are set to the same as before, the simulation results are shown as in Fig.8.

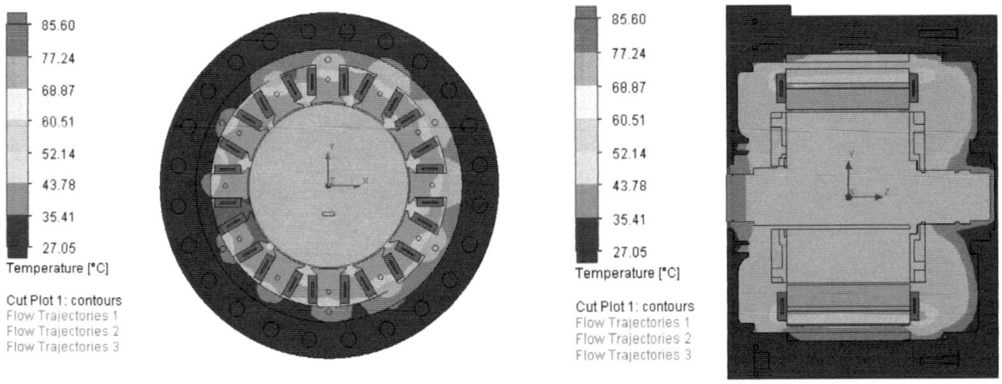

(a) (b)

Figure 8. Cross section of thermal simulation with Fan and water cooling system, (a) : Front View, (b): Side View

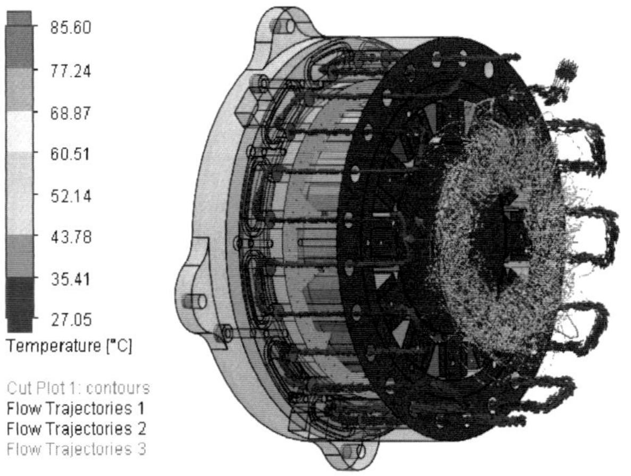

Figure 9. Fluid flow vector simulation

With the fan inside the motor, the air will circulate quickly so that heat rapidly travels through the main housing. As a result, the temperature in the motor and housing dropped dramatically.

5. Simulation Result

The motor generates thermal energy when operated due to mechanical or electrical factors. For medium to large capacity motors an additional cooling system needs to be provided to reduce the heat generated by the motor parts. If the motor is very hot will cause insulation failure on the copper wire and reduce motor efficiency.

Some ways that can be done to reduce the heat from the motor by using a water flow or fan. From the simulation results in this paper, the motor without cooling has an average temperature of 85^0C on the inside of the motor, while the main housing is 75^0C. While using water and fan, the average temperature of 65^0C is obtained and the main housing section is similar to room temperature.

IOP Conf. Series: Materials Science and Engineering **694** (2019) 012007 doi:10.1088/1757-899X/694/1/012007

6. References

[1] C. Chang, Y. Kuo, J. Wang, and S. Chen, "Air cooling for a large-scale motor," *Appl. Therm. Eng.*, vol. 30, no. 11–12, pp. 1360–1368, 2010.

[2] M. M. Baggu, H. L. Hess, and K. Rink, "Thermal Modeling of ' Direct Lamination Cooling (DLC)' Induction Motor for Hybrid Electric Vehicle Applications .," no. Dlc, pp. 242–246, 2005.

[3] M. K. Yoon, C. S. Jeon, and S. K. Kauh, "Efficiency Increase of an Induction Motor by Improving Cooling Performance," vol. 17, no. 1, pp. 1–6, 2002.

[4] W. J. Anderson, "Rolling-Element Bearings Rolling-Element Bearings," 2019.

[5] S. E. Skaar, O. Krovel, and R. Nilssen, "Distribution, coil-span and winding factors for PM machines with concentrated windings," *XVII Int. Conf. Electr. Mach. ICEM 2006*, p. 346, 2006.

[6] R. Pechánek and L. Bouzek, "Analyzing of Two Types Water Cooling Electric Motors Using Computational Fluid Dynamics," no. September, 2012.

IOP Conf. Series: Materials Science and Engineering **694** (2019) 012008 doi:10.1088/1757-899X/694/1/012008

Integrated CAD customization system for fused deposition models in additive manufacture with 3D printing machine

Achdianto[1], Sally Cahyati[1], Triyono[1] and Saifudin[2]

[1]Department of Mechanical Engineering, Faculty of Industrial Technology, Universitas Trisakti, Jl.Kyai Tapa No.1 Grogol, Jakarta, Indonesia
[2] Department of Information Technology Engineering, Faculty of Industrial Technology, Universitas Trisakti, Jl.Kyai Tapa No.1 Grogol, Jakarta, Indonesia

E-mail: sally@trisakti.ac.id

Abstract: A computerized addictive aided design based on manufacturing system that is digitally integrated, has been modified, built, and designed. The system refers to the smart manufacturing sectoral (smart manufacturing) which to answer the challenges of industrial revolution 4.0, namely the purpose for efficiency of the distance, cost efficiency, time efficiency, quality improvement and increased flexibility. The concept of manufacturing systems with fused deposition modelling 3D printers integrated with the internet or intranet has been modified, for the purpose to connected each other under one platform being namely as category as a Computer Aided Manufacture (CAM) system. A synthesis between the production room and the customer, real-time depiction of 3D CAD users can be presented and connected directly to the manufacturing locations where the desired product is located. Seeing this as an opportunity growing for the industrial society, the author conducted research to modified and collaborate the system. The method used in this study was carried out with rapid application design or better known as Rapid Application Development / RAD aimed at simplifying and accelerating the design of the system for 3D print web application system. CAD drawings can be processed from the user's device to the main webserver and then forwarded into the operator's production system space, resulting in a final product / finished goods, with no restrictions on distance, area and allows low production costs (low production costs). The results of this study represent the manufacturing industrial that is being supported by the application of information technology, where are manufacturing is connected with IoT (Internet of Things). Presenting the benefits of the Industry 4.0 concept. The customization system to be able to add value added for the competitiveness small and medium companies in Indonesia on manufacturing process and can support micro economic growth.

1. Introduction

The development of manufacturing sectoral lines is increasingly showing its latest innovations with each line by continuous improvement, taking various best steps. All improvements in existing processes are according to the kaizen principle has already been made by previous founder and researcher in Mechanical Engineering. An industry concept 4.0 is synonymous with the transformation of manufacturing activities into smart manufacturing, designed to meet and exceed current challenges. The application is carried out with a shorter product manufacturing cycle, an adjustment to the manufacturing of the product through competitive measures, one of the applications being to eliminate borderless and boundaries.

IOP Conf. Series: Materials Science and Engineering **694** (2019) 012008 doi:10.1088/1757-899X/694/1/012008

Several previous studies present and discuss the use of the manufacture Industrial area which support by Internet of things (IoT), related topic in this case using 3D Printing connected to the internet will be discuss. Among of them are Z.Bi, L.Da Xu, and C. Wang, discussion presents a strength of the internet in the modern manufacturing world [1] but does not discuss in detail the relationship of the component-components related to the modern manufacture. Other researchers, namely from Barbosa and Aroca, discussed the direct use of a 3D printer with a controlling monitor directly to the device, where the user enters directly into the machine's performance framework, in therefore the author examines and sees that a filter is needed to develop and maintain that users cannot enter directly into the manufacturing system, namely by using a website / world wide website, so that it can be used online, as an interface unit platform or user filtering [2] and Mai, Jingeng in their research discussed the approach to integration of 3D Printing in cloud [3] in the discussion there was not much written detail about the mechanism of performance or configuration of hardware or software and tools that will be used to create a prototype of the synthesis of 3D Printing integration via internet or intranet. All of these studies primary considered the perception and access of traditional machine resources from the perspective of manufacturing capacity, but they did not analyze and discuss solution further for integration for user into manufacture for 3D printers and 3D printing services.

Regarding based on previous researcher before, the author doing some research which related to customization and the configuration of 3D printer. The specification of hardware, local servers, external servers and software being used, that were being connecting between others. Mainly the *3D printer Fused Deposition Model* are in different locations, can collaborate as one integrated connectivity as one manufacturing platform area, which can be connected by internet or intranet by entering the main component system framework which is very much needed [4] for the purpose from the user in different area by accessing the website platform. Thus, it is essential to examine how to get the best integrate and apply 3D printing services in web manufacture.[5]

2. Research Objectives

This research work is presented to design a framework for a production system additive manufacture - fused deposition modeling (FDM) - 3D Printer using an online network (website) as a manufacturing service that is integrated with hardware and software online. [6,7] The purpose and benefits of this research are:

a) With modeling and simulations without distance limitation can help synthesize devices and data as alternatives in manufacturing, evaluate the performance of uploaded CAD drawings whether they are in accordance with the user's intentions, can be a fast alternative choice for the industrial world, and can configure designs that are optimal. It is expected to make a manufacturing solution for operators and small manufacturing industries.[5]

b) Can facilitate the user / user in making a dynamic product, and can facilitate the operator to integration of making 3D Printing products FDM (Fused Deposition Modeling) [8,9]

c) Connecting and shortening the distance between Production and Users. In this case cutting the value chain, supply chain in an integrated manner, with the worldwide Web Cloud-Based Design and Manufacturing as the basis for the development of the manufacturing industry in Indonesia.

d) Producing output CAD readings both dimensions, price, material / material and the processing time of the 3D Printing machine.

The results of this study are highly expected and can be a reference for the world of industrial production design and other similar research. The main contribution of this research relates to the following two aspects:

1. Integrating and building a service or platform from CAD (Computer aided design) users to a 3D cloud web printer which then translates the CAD image to be able to support more efficient production and can reach in a variety of different locations / regions. By using online tools and integrated with 3D CAD drawing model construction in purpose to be integrated as computerize aided manufacture.

2. Development and design of a manufacturing system for 3D Printer machine production services, including a dynamic 3D Product evaluation facility with monitoring and assembly via the web and cloud and a combination of production services, as well as scheduling the work of the 3D Printer machine

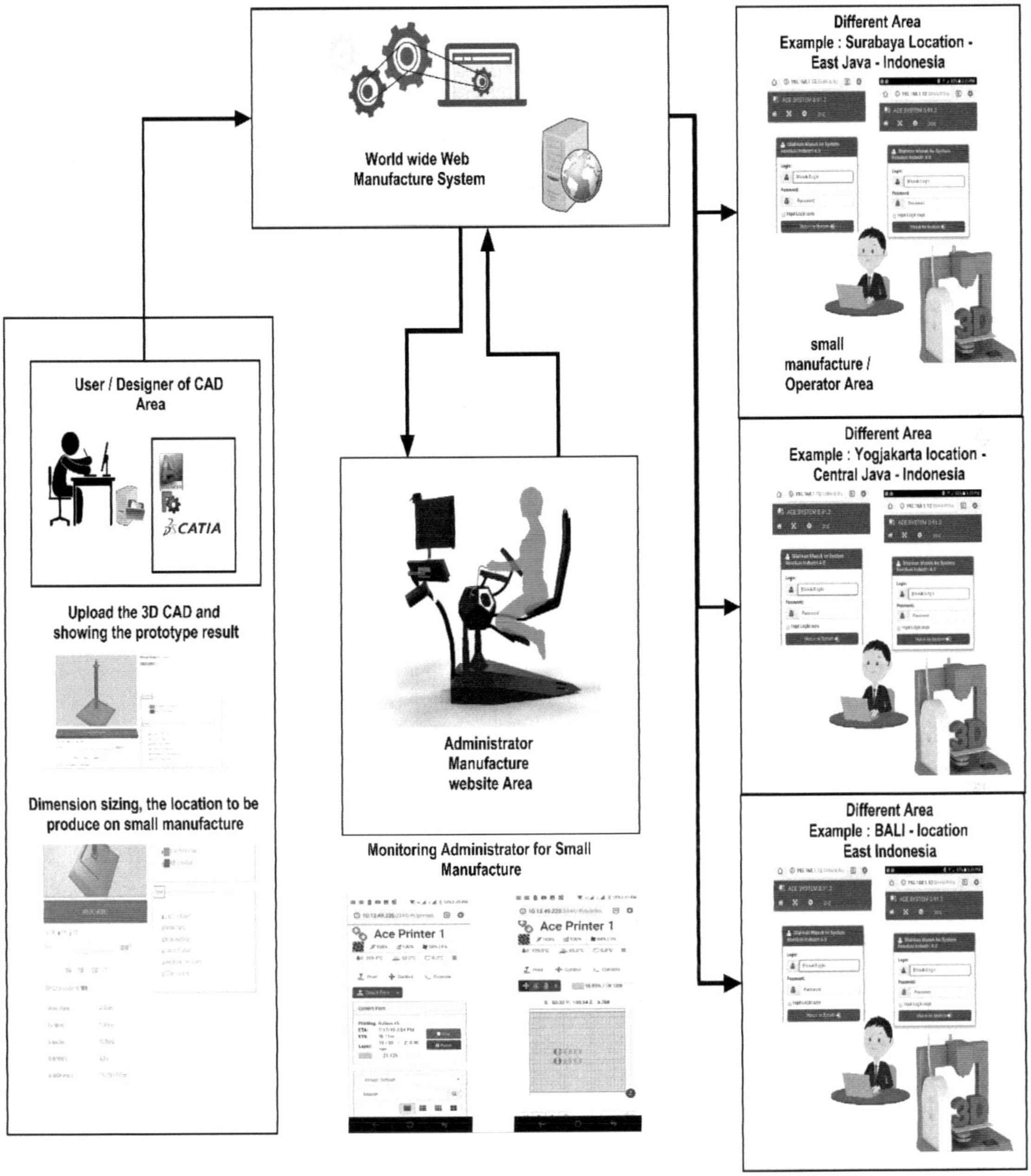

Figure 1. Manufacture world wide web computerize aided manufacture integration connected to operational area with monitoring system.

IOP Conf. Series: Materials Science and Engineering **694** (2019) 012008 doi:10.1088/1757-899X/694/1/012008

3. Significance of Implemented Techniques

RAD (Rapid Application Development) is an object-oriented approach to system development which includes a development method and software. RAD aims to shorten the time that is usually required in the life cycle of traditional system development between the design and implementation of a system. In the end, RAD also tried to meet the conditions of development that were changing rapidly [10].

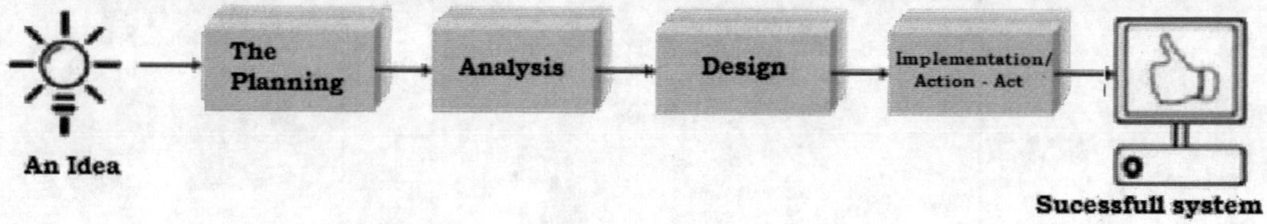

Figure 2. The basic phase of making a system

The methodology used in this study is the RAD prototype method carried out in Figure 2 which consists of several stages, namely:

1. Planning: in this phase, the user and the analyst meet with each other by identifying the objectives of the application or system, and to identify the information requirements arising from those objectives. The orientation in this phase is to solve the problems that will exist or exist, the focus will always remain on the efforts to achieve the goal of making the system [10].Without perfect planning, counting the strengths and weaknesses of the system, software development is meaningless without a plan. Planning starts the project perfectly and can positively influence its progress.
2. Analysis: This step is about analysing the system performance of both software and hardware at various stages and making notes about additional requirements. Analysis is a very important step to go further into the next step.
3. Design: After the analysis is complete, the design step is to take over, which basically is to build and develop architectural projects with various development innovations. This step helps eliminate the possibility of weaknesses by setting some standards and trying to track them.
4. Implementation: In this implementation phase, the analyst works intensively with users during the workshop and designs the business and non-technical aspects of the company. As soon as these aspects are agreed upon and systems are built and filtered, new systems or parts of the system are tested and then introduced to the organization [10].

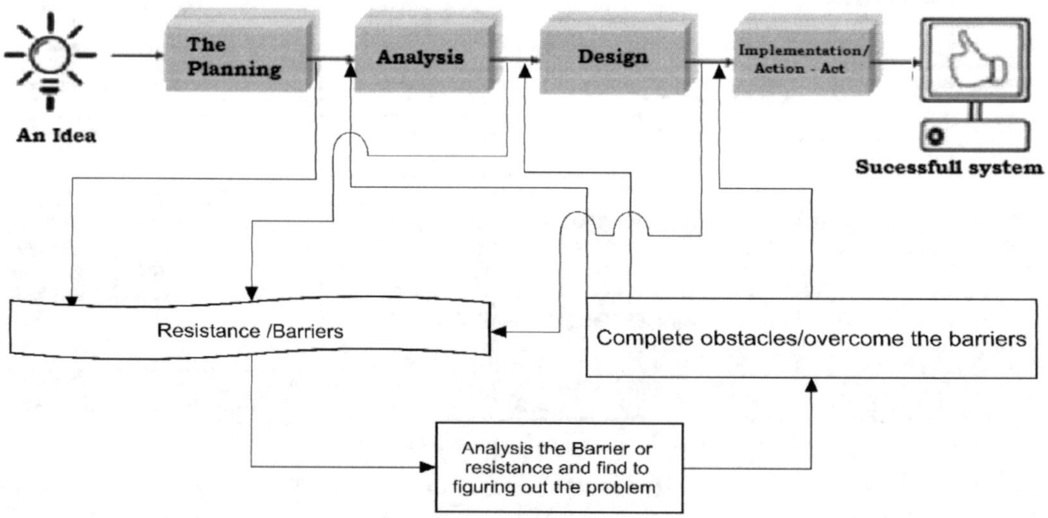

Figure 3. The analogy process implementation making the system

Manufacturing via website [11] which provides services for users with various objectives framework are surely needed for the flexibility and reducing the cost for small manufacture. Which can be modified to the needs of users themselves, it can be directly print in user demand exactly location. Web manufacturing production technology using CAD (Computerize aided design) to a 3D Printer can provide an adjustment for a production system, especially when 3D printer technology can be shared (sharing) through networks in online manufacturing [12]. Compared to traditional manufacturing methods, 3D printers can significantly reduce some of the dependencies at very complex stages of the manufacturing process

Framework purpose are to combine the software to make it easier for users, to create an application for a website or a system in which there are various functions related to them, including using a plugin, and is a concept for forming a particular system to be more structured and neatly structured.[10]

The framework contains following functions are being needed to construct and modified the system (1) webserver for user interface, (2) cpanel access to controlling the system of worldwide web manufacture (2) advance wordpress within integrated with customized wp3dprint and others plug in into the system, (3) customized repetier server within integrated with (4) repetier host as local network area between operator computer and 3D Print fused deposition model.

Table 1. Description for actor use case diagram

No	Actor	Description
1.	Admin/Administrator	Admin or administrator website is the main actor who performs documentation data, delete documentation data, edit documentation data, add admin, delete admin, and edit admin. Admin communicates with the user, as well as the operator to be addressed, adds and configures the user's wishes (user) in accordance with user expectations. Forward the image in the form of format stl (stereolithography) and obj (Object) to the operator, and can configure the 3d printing machine (Fused deposition model) to be addressed for the production process. Can see the web virtual machine condition fused deposition model (3D Printing) with the help of API (Application programmable Interchange) connection from the repetier server installed by the destination operator.
2	Operator	Receiving orders from the Admin as the main actor to carry out execution of the 3D Printing -Fused deposition machine The operator also monitors the machining periodically and maintains the machine's performance. The operator also checks the availability of materials that will be used periodically and is informed to the Admin.
3	User	The user is an actor who can see the documentation data that will be uploaded in the obj / stl format. The user (user) performs work instructions (standing instruction) through the web filtered by the main actor (Admin). Users (users) can edit the profile (personal data) and include the wishes of the material to be used, as well as the desired destination location.

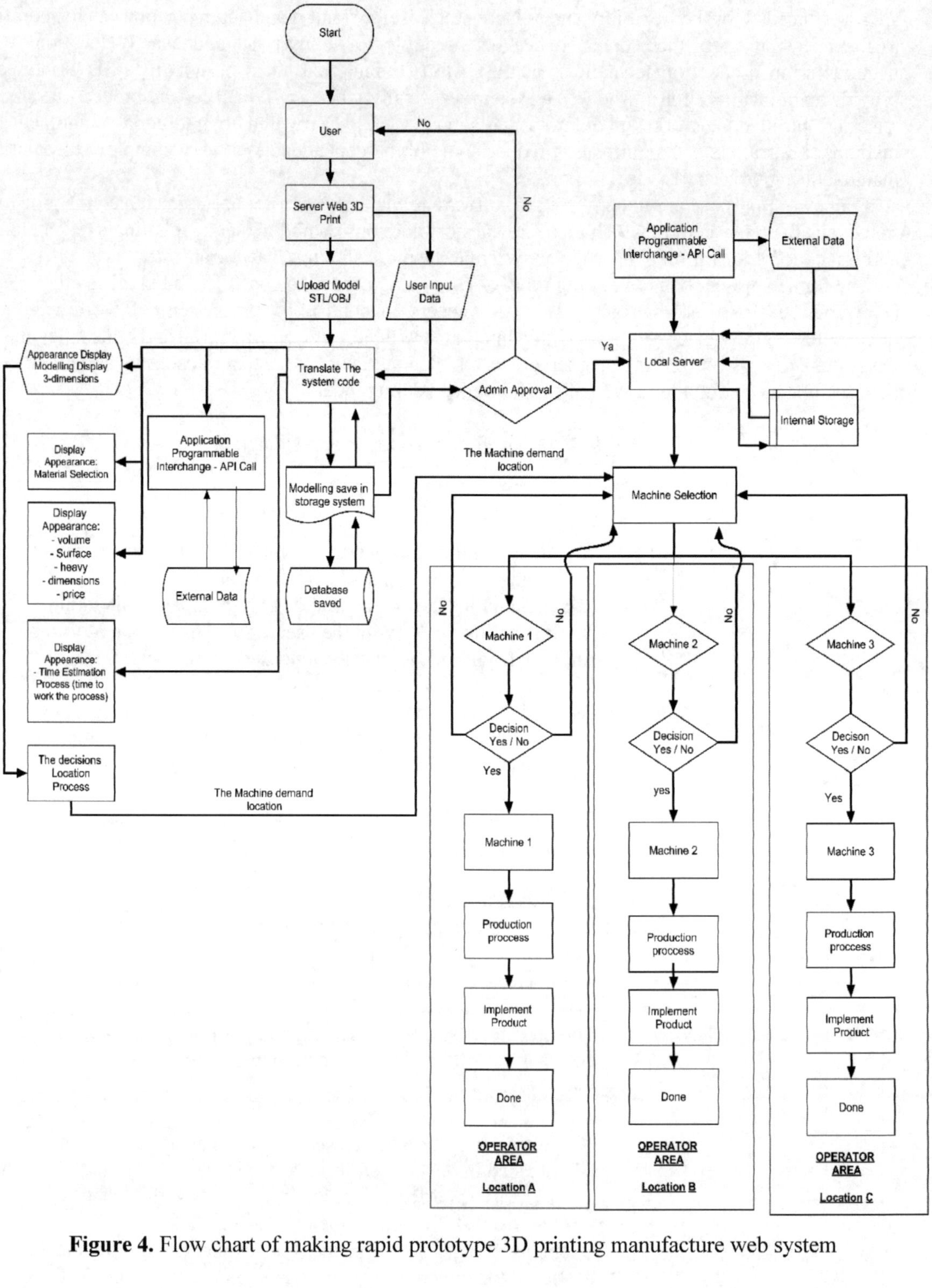

Figure 4. Flow chart of making rapid prototype 3D printing manufacture web system

4. Processes in the System Design

Figure 3 describes a process design will be made in this research and customized implementation. The user will be directed to open a site, where the user (user) can upload a model / image that you want to print on the expected 3D Printing, the image is in the format stl / obj (stereolithographic / Object file), the user completes the order data and provides notes regarding images that are expected to become a finished goods process. After the order is confirmed, the admin takes action on the approval confirmation process. In the next process the admin forwards the order confirmation of the goods to the intended operator according to the user's wishes, after the data is received by the operator then the operator executes execution to the 3D printer machine to do the work so as to produce the finished product.

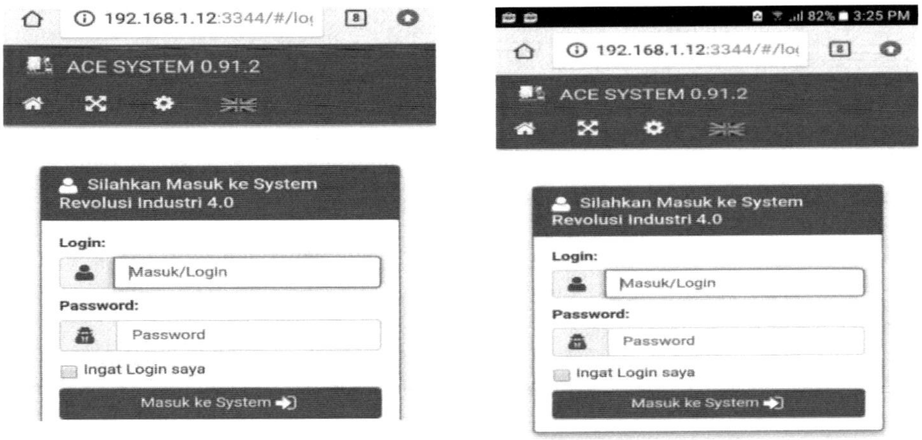

Figure 5. Administrator and Operator manufacture web login system

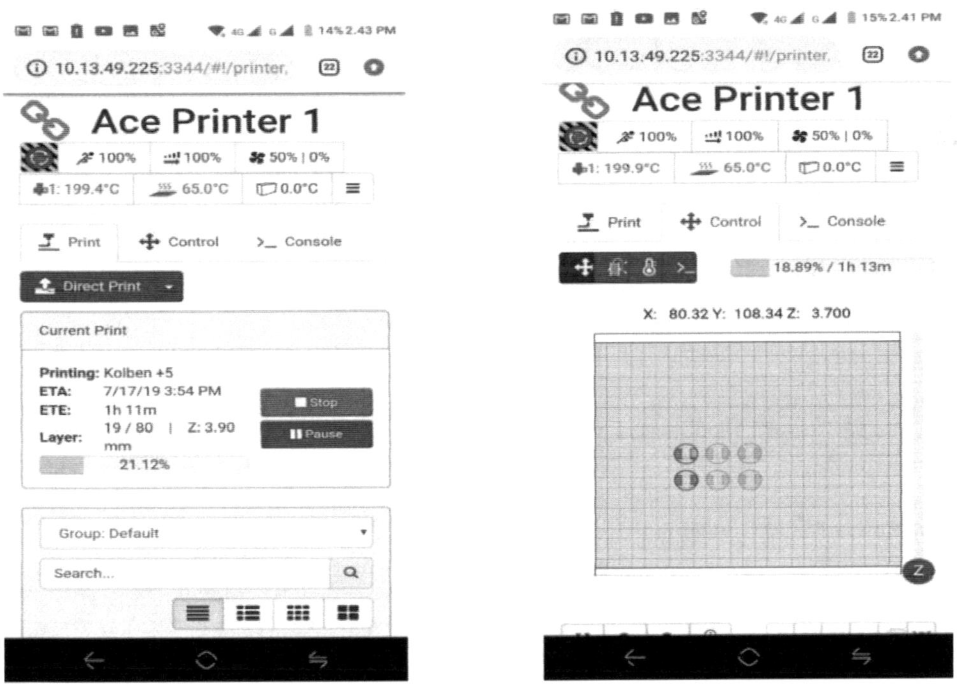

Figure 6. Inside monitoring system for administrator website and operator small manufacture

5. Result and Discussion

In our research, there's are three area include, in production area, the operator run the 3D print machine which integrated with the repetier software [13] which in this sector are the customization repetier firmware and being configured with the repetier host. separate system placement software in the local application and it will be connected to customized system of server repetier to be acknowledge as one each other's on local area.

The system server will get command from the operator itself or by website administrator to run the process.as shown in Figure 4 and 5.

In user area sectors, user can directly upload the 3D data (STL, Obj) to front interface website as fig.6, and also they can exactly choose what material to be used, location for the 3D Print will be process, the dimension for they need, and they can customized the size and also the user can get information for the price regarding to the dimension and material they want to use.

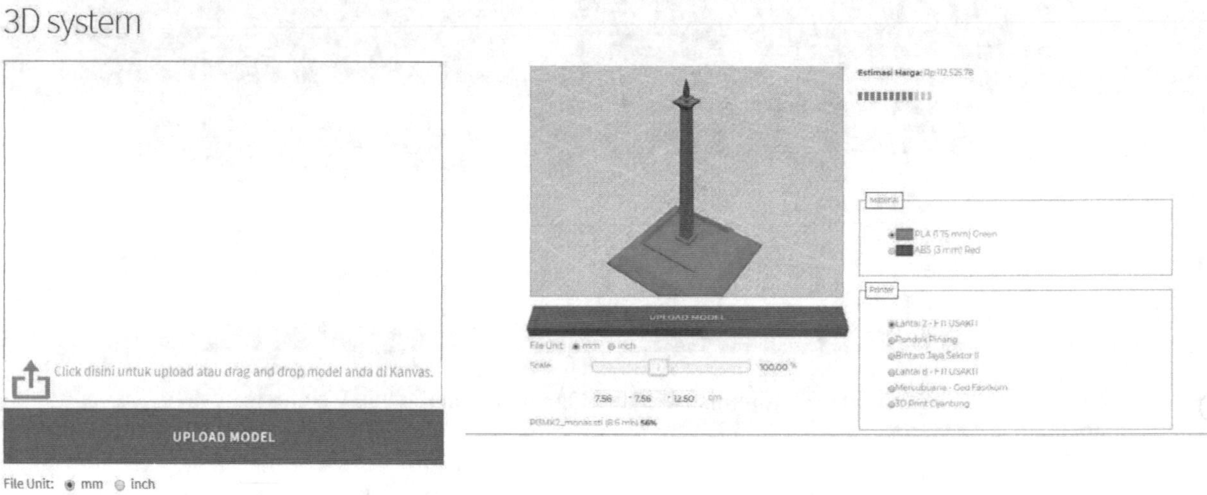

Figure 7. Manufacture web upload area

With the 3D CAD output with STL or Obj file as input it has a function to describe the dimensions of the box (length, width, height), and can be used to make a decision, to produce or not to produce; to tell which customer which is appropriate to their product.

In the web manufacturing system that we customized and developed based on repetier server and advance php wordpress 3d system, the interaction will be providing directly in front of user (rotating, zooming in, zooming out) the 3D model is used interactively in the decision making process.

6. Conclusion

Regarding to the merge some of the existing system, the aim for 3D CAD data communication as our path research, a useful collaboration among others system can be applied as one synthetic system, repetier advanced server on local server and php wordpress 3D [14] attach it on the server website, and others detail component include, will be an advantage to better realization for the process. The system can collaborate with each other and can meet the shortcomings of each other. The ease of interaction between the user and the operator and administrators will be closer and easier to understand.

Collaboration system can be observed by each other and can be monitored between users, admin and operator. This became the foundation of its scope and the purpose of advance manufacturing which cannot be separated from a flexibility, distance efficiency, costs efficiency, and monitor with ease.

With this system which customize the 3D CAD transfer data, hopefully it can facilitate the controlling on ongoing projects and is desired by 3D Printing users to make virtual goods manufactured without knowing the distance and space restrictions.

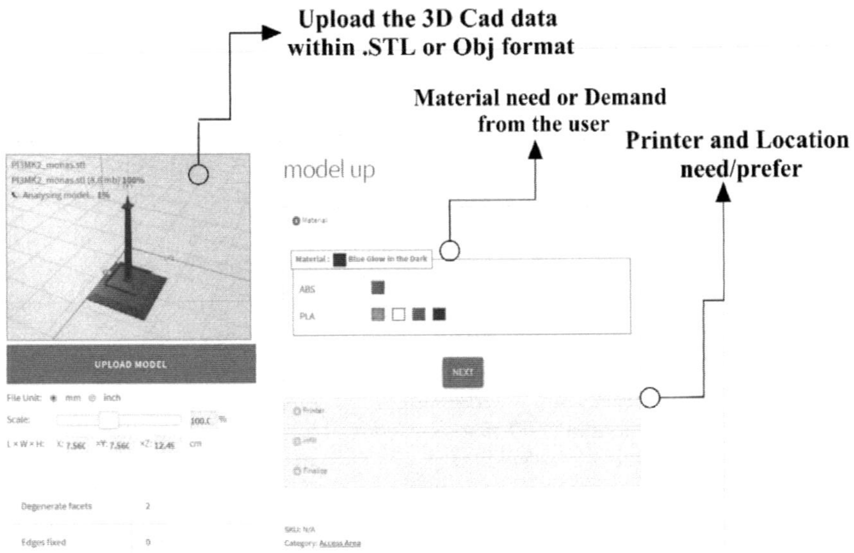

Figure 8. The user interface with location and material need

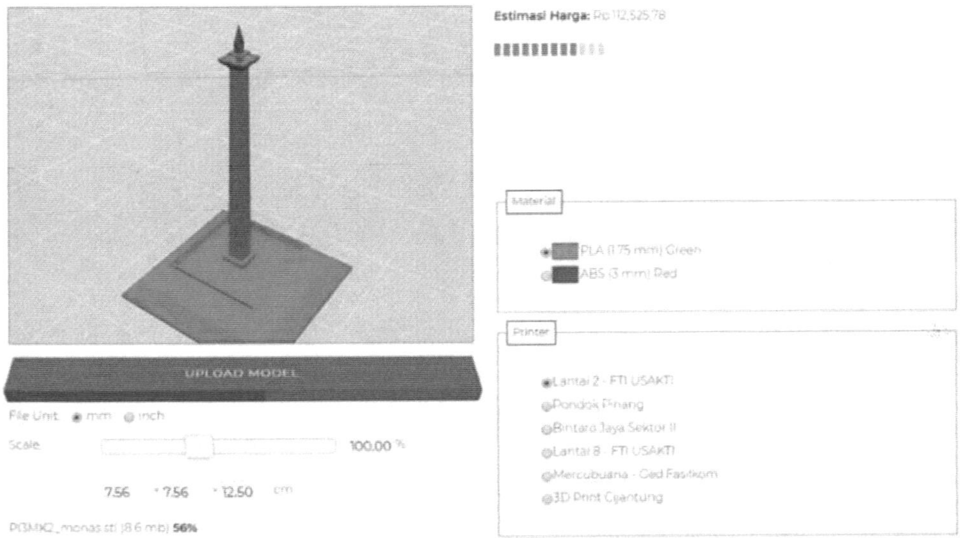

Figure 9. The 3D CAD upload model within price, material and location printer as user demand and friendly use.

References

[1] Z. Bi, L. Da Xu, and C. Wang, "Internet of things for enterprise systems of modern manufacturing," *IEEE Trans. Ind. Informatics*, vol. 10, no. 2, pp. 1537–1546, 2014.

[2] B. GF and A. RV, "An IoT-Based Solution for Control and Monitoring of Additive Manufacturing Processes," *J. Powder Metall. Min.*, 2017.

[3] J. Mai, L. Zhang, F. Tao, and L. Ren, "Customized production based on distributed 3D printing services in cloud manufacturing," *Int. J. Adv. Manuf. Technol.*, 2016.

IOP Conf. Series: Materials Science and Engineering **694** (2019) 012008 doi:10.1088/1757-899X/694/1/012008

[4] P. Nyamsuren, S. H. Lee, H. T. Hwang, and T. J. Kim, "A web-based collaborative framework for facilitating decision making on a 3D design developing process," *J. Comput. Des. Eng.*, vol. 2, no. 3, pp. 148–156, Jul. 2015.
[5] P. Siltanen and S. Valli, "Web-based 3D mediated communication in manufacturing industry," in *Advanced Concurrent Engineering*, 2013, pp. 1181–1192.
[6] F. W. Baumann and D. Roller, "Closed-loop control of 3D printers via web services," in *CEUR Workshop Proceedings*, 2017, vol. 1826, pp. 44–50.
[7] I. Gibson, D. Rosen, and B. Stucker, *Additive manufacturing technologies: 3D printing, rapid prototyping, and direct digital manufacturing, second edition*. 2015.
[8] I. Gibson, D. W. Rosen, and B. Stucker, *Additive manufacturing technologies: Rapid prototyping to direct digital manufacturing*. 2010.
[9] A. Alafaghani, A. Qattawi, and M. A. Ablat, "Design Consideration for Additive Manufacturing: Fused Deposition Modelling," *Open J. Appl. Sci.*, 2017.
[10] Kenneth E. Kendall and Julie E. Kendall, *Analysis Y Diseno de Sistemas*, 2001
[11] P. Nyamsuren, S. H. Lee, H. T. Hwang, and T. J. Kim, "A web-based collaborative framework for facilitating decision making on a 3D design developing process," *J. Comput. Des. Eng.*, 2015.
[12] M. Livesu, S. Ellero, J. Martínez, S. Lefebvre, and M. Attene, "From 3D models to 3D prints: an overview of the processing pipeline," *Comput. Graph. Forum*, 2017.
[13] Hot-World GmbH & Co. KG.2011-2019, "Repetier Server." [Online]. Available: https://www.repetier-server.com/.
[14] Burkov, Sergey.,"Wp3DPrinting,"2019.[Online].Available: https://www.wp3dprinting.com.

IOP Conf. Series: Materials Science and Engineering **694** (2019) 012009 doi:10.1088/1757-899X/694/1/012009

Design optimization for torque density in brushless DC motor with IPM V-type using PSO Method

Yoga Uta Nugraha[1], Arief Cahyadi[1], M Nur Yuniarto[2] and Indra Sidharta[2]

[1] Department of Electrical Engineering, Center of Excellence for Automotive Control & System, Institut Teknologi Sepuluh Nopember, Surabaya, Indonesia
[2] Department of Mechanical Engineering, Center of Excellence for Automotive Control & System, Institut Teknologi Sepuluh Nopember, Surabaya, Indonesia

E-mail: uta.utu@gmail.com, ayuningfitridesanti@gmail,com, muhammad.nur3006@gmail.com, indra.sidharta@gmail.com

Abstract. The analysis model is the first stage used to determine the design parameters of the motor. It is coupled with an optimization method to find a combination of design parameters. The purpose of this paper is optimizing the torque density of the Brushless DC (BLDC) Motor with Interior Permanent Magnet (IPM) V-type. The reference design is 3-phase and 12/8 pole BLDC motor with a concentrated winding. To obtain this optimal design, this is done by eliminating the rotor area that is not covered by magnetic flux without reducing the main performance of the motor. Elimination area prediction refers to initial motor design simulation which is operated on rating condition. Mathematical modeling is based on the 2-D finite element analysis (FEA) using software Ansys Maxwell. Particle Swarm Optimization (PSO) algorithm with multi-objective function has been applied to determine the area of the rotor structure to be eliminated, resulting in an optimal torque-density design. Matlab code is also included to execute the optimization algorithm. Excitation uses the ideal sinusoidal signal by ignoring harmonic distortion. This paper features a new optimal design for IPM by considering torque density, cogging torque, harmonic distortion in back-electromagnetic force (back-EMF), and material demagnetization constraints using PSO. Quality of PSO optimization can be proven by increasing the torque density without compromising the main performance of the motor compared with FEA. With this method also provide disadvantage, such as the cost of production is more expensive because it should add a profile of the rotor mold.

1. Background

Brushless DC (BLDC) motor has the advantages of efficiency, torque, and ease of control. But the geometry of the motor needs to be considered more serious so that it is more efficient. One of them is the torque density value, which is the ratio between torque and the overall mass of the motor. Modifying certain parts of the motor needs to be done but the performance of the motor does not decrease significantly.BLDC motor has the main construction of stator, rotor, magnet and coil. Stator is a fixed motor part, while the rotor is a rotating motor part. A permanent magnet is placed on the rotor.

To increase the torque density of the motor, one of them is by modifying the rotor shape as has been done by [1]–[4]. Multi Objective can use a variety of algorithms such as Bat-Inspired Algorithm [5], Genetic Algorithm and Particle Swarm Optimization [6].

IOP Conf. Series: Materials Science and Engineering **694** (2019) 012009 doi:10.1088/1757-899X/694/1/012009

In this paper the motor that will be used is a type of radial BLDC motor with a configuration of 12 stator slots and 8 magnetic poles. The combination of slots and stator affects the magnitude of the winding factor. The 12 slot combination and 8 pole magnet have a windung factor of 0.866 [7] which has the advantage of easy winding and high efficiency.

Figure 1. Physical model of motor: (a): Stator, (b): Rotor and shaft, (c): Chasing, and (d): Motor Assembly.

Table 1. Mass of each part

Part	Material	Volume (cm^3)	Mass Density (g/cm^3)	Mass (g)
Stator	Electrical Steel	812.27	7	5697.96
Enamled Wire	Copper	-	8.96	1477.05
Shaft	Stainless Steel	201.48	8	1611.88
Magnet	NdFeB	-	-	800.00
Casing	Alumunium	-	2.7	8566.44
Bearing		-	-	200.00
Rotor (existing)	Electrical Steel	520.59	7	3644.16
Total				**21997.52**

2. Modelling of Motor Torque

To get a calculation of the motor torque value is done by using the concept of conservation of energy. This concept is carried out by modeling the energy flow from electrical energy to magnetic energy and mechanical energy. Rules that must be fulfilled can be formulated into $dW_e = dW + dW_m$, where dW_e is a differential value of electrical energy, dW is a differential value of magnetic energy, and dW_m is a differential value of magnetic energy[8].The Torque can be influenced by differential of Inductance L, Reluctance R, magnetic flux Φ to the rotational angel of rotor θ. Mathematically it can be formulated into[8]:

$$T = \frac{1}{2} i^2 \frac{dL}{d\theta} - \frac{1}{2} \Phi^2 \frac{dR}{d\theta} + Ni \frac{d\Phi}{d\theta} \tag{1}$$

Finite Element Method is used for modeling. In this study the software used was 2-D Ansys Maxwell simulation. Simulation using 2-D reduces time rather than 3-D simulation. There are three basic topologies that use on hybrid energy storage. Series, parallel and series-parallel configurations [3]. These topologies developed on energy storage for electric vehicle based on the capacity of electric vehicle's drive train. And also, all of the topologies can be connected as a passive or active configuration. Advantage of active configuration is can be a potentially smaller dimension that passive configuration as the same load.

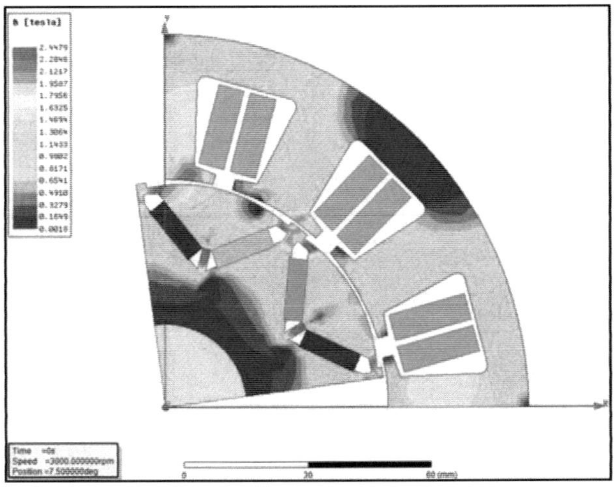

Figure 2. Magnetic flux distribution of intial model

In existing conditions, the rotor has an area that is not aligned by magnetic fluxes. This zero area of magnetic flux must be eliminated but must still have a sturdy structure. The elimination is extended towards an area that has a larger flux, but the expected final result is that the torque does not decrease significantly. The average torque produced is 34.41 Nm with a total mass of 21.977 kg. That means the existing motor has a torque density of 1.56 Nm / kg.

3. Optimization Strategy and Methodology

Particle swarm optimization (PSO) is an algorithm used to determine the optimal parameters that adopt the habits of birds in foraging. PSO was first initiated by Kennedy and Eberhart [9]. Some particles are scattered randomly in a certain range. Each particle is calculated the solution value. The best solution of all particles is called the global best (gbest). The other particle moves to gbest with a certain speed. The position of the new particle is also a potential solution. The best value of each solution in each particle is called particle best (pbest). The value of the partilce transfer speed depends on the value of the gbest and pbest. These steps are repeated to get the steady state gbest value.

To find particle speed at any time can be calculated by formula [9].

$$v_i(t) = Wv_i(t-1) + C_1 r_1 \left(x_{pbest\,i} - x_i(t)\right) + C_2 r_2 (x_{gbest} - x_i(t)) \tag{2}$$

IOP Conf. Series: Materials Science and Engineering **694** (2019) 012009 doi:10.1088/1757-899X/694/1/012009

where v_i is particle speed, W is inertia weight, C_1, C_2 are cognitive and social parameter, $x_{pbest\ i}$ is position that has best solution for each particle, x_{gbest} is particle position taht has best solution for all particle, and r_1, r_2 are a random value betwen 0 and 1.

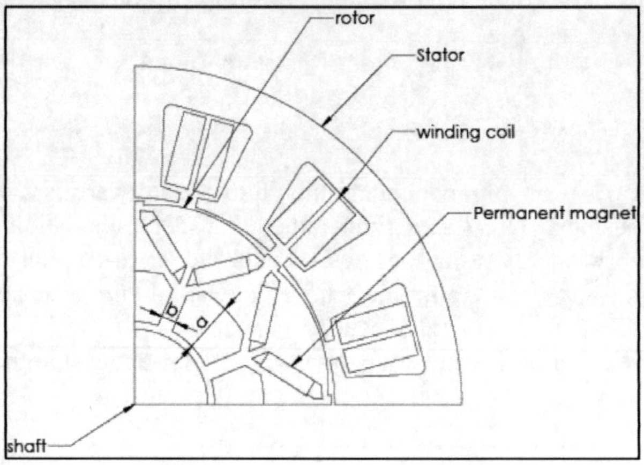

Figure 3. Cross section of ¼ purpose model

Table 2. PSO Parameter

Parameter	Symbol	Value	Unit
-	a	[3,23]	mm
-	b	[2,15]	mm
Weight Inertia	W	1.2	-
Cognitive and Social Parameter	C_1	1.2	-
	C_2	0.8	-
Number of Particle	-	100	-
Number of Iteration	-	20	-

4. Optimization result

The first step in carrying out optimization is to determine the particle with dimensions a and b randomly with a range according to table II. Each particle was evaluated for the value of torque density using Ansys Maxwell software. The largest density of all random particles is called gbest. Determine the next position of each particle using equation (2). Matlab software is used to speed up calculations. The particle that initially spreads in a large range, then moves closer to the position that has the best solution value (See Fig. 4 and Fig. 5).

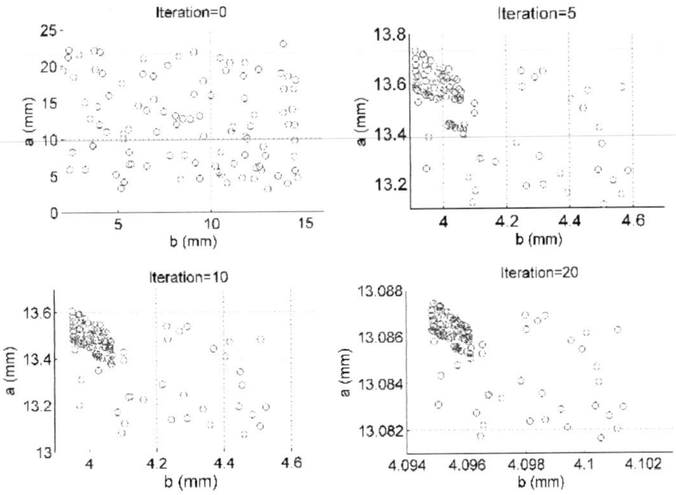

Figure 4. Particle position for 0, 5, 10, and 20 iteration.

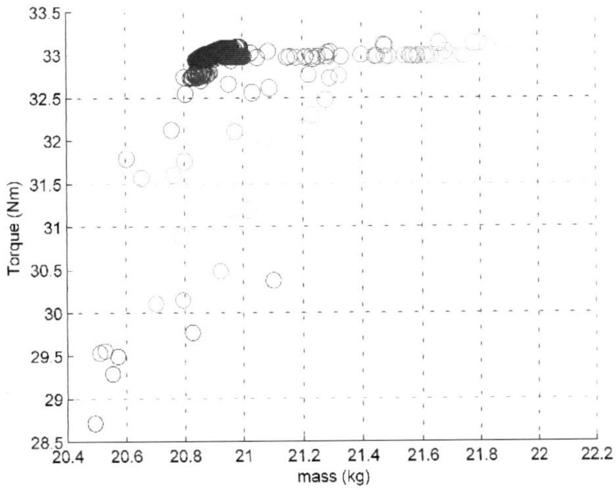

Figure 5. Torque vs mass of all particle.

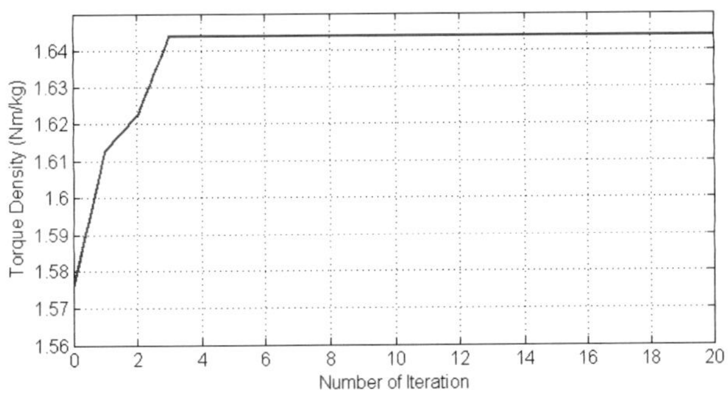

Figure 6. Value of best torque density each iteration.

IOP Conf. Series: Materials Science and Engineering **694** (2019) 012009 doi:10.1088/1757-899X/694/1/012009

The final result of 20 iterations is a design with a value of a = 4.096 mm and b = 15.481 mm. With this design, the average Torque value = 34.34 Nm and the overall mass is 20.89 kg. So that the torque density value of the desired design = 1.6438 Nm / kg.

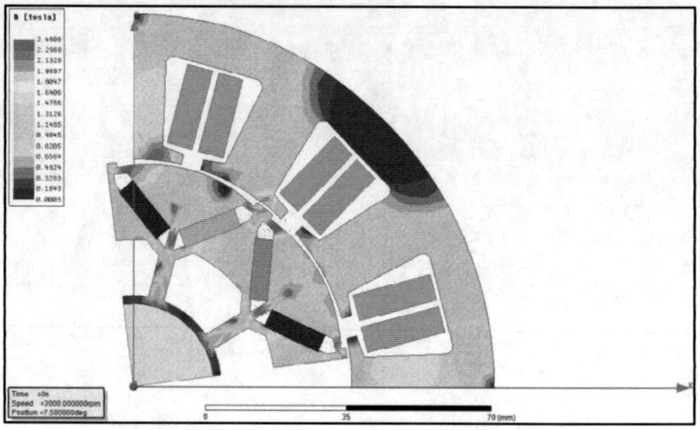

Figure 7. Magnetic flux distribution of purposed model.

Examined from the distribution of magnetic flux, the rotor and stator parts have an average value of 1.5 T. Meanwhile, there are only small areas that exceed the saturation limit of electrical steel material, which is 1.8T. That design is still possible to do prototyping.

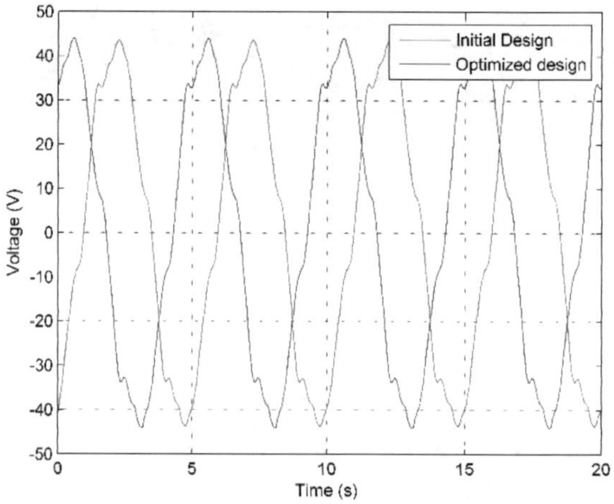

Figure 8. Induced voltage of initial design and purposed design.

The peak-to-peak induced voltage generated by the initial design is 88 V. The peak-to-peak induced voltage produced by the purposed design is 87.2 V. Both the initial design and purposed design have a voltage drop from the source voltage with a value 100 V. This is due to the armature reaction. Initializing the speed in the simulation software is equal to 3000 rpm. The torque produced in the purposed design is 34.34 Nm. 0.2% lower than the initial design which has a value of 34.41 Nm. However, the torque ripple produced by the initial design is higher at 4.36% compared to the purposed design of 3.89%. Greater torque ripple results in increased vibration and acoustic noise on the motor. The maximum cogging torque value for both initial design and purposed design is 572.425 mNm. There is no difference between the two designs because there is no change in shape between the surface of the rotor or the adjacent stator surface.

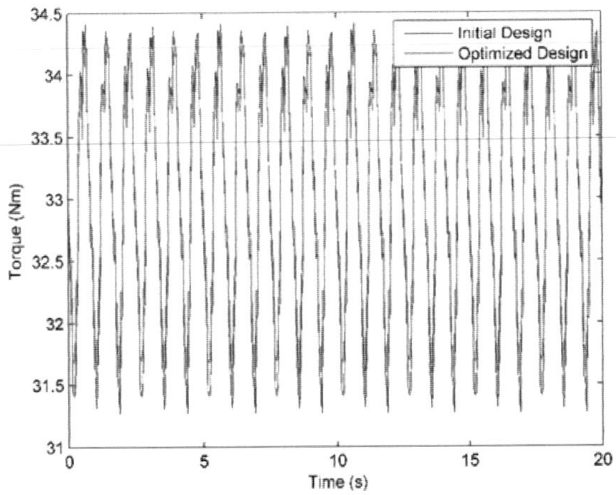

Figure 9. Torque produced by initial design and purposed design

Table 3. Comparison of initial design and purposed design

Parameter	Initial Design	Purposed Design	Improvement
Average Torque	34.41 Nm	34.34 Nm	-0.2 %
Total Mass	21.977 kg	20.89 kg	-4.9 %
Torque density	1.56 Nm/kg	1.6438 Nm/kg	5.37 %
Maximum flux desnity	2.44 T	2.46 T	0.8 %
Ripple Torque factor	4.36 %	3.89 %	12 %
Peak to peak Induced voltage	88 V	87.2 V	0.9 %
Maximum cogging torque	572.425 mNm	572.425 mNm	0 %

5. Conclusion

In this paper, increasing the value of torque density is done by eliminating certain areas of the rotor. The algorithm used is PSO with 100 particles and 20 iterations. The purpose of the design with the PSO algorithm is the dimensions of a = 4,096 mm and b = 15,481 mm.

The maximum torque of the purposed design is lower than the initial design, but the purposed design has a lower torque ripple. The induced purposed design voltage is lower than the initial design when rotated at 3000 rpm. Meanwhile the maximum value of cogging torque does not change because the focus of the research is not to change the stator and rotor surface.

IOP Conf. Series: Materials Science and Engineering **694** (2019) 012009 doi:10.1088/1757-899X/694/1/012009

6. References

[1] K. Y. Hwang, S. B. Rhee, B. Y. Yang, and B. Il Kwon, "Rotor pole design in spoke-type brushless DC motor by response surface method," *IEEE Trans. Magn.*, vol. 43, no. 4, pp. 1833–1836, 2007.

[2] A. Fatemi, D. M. Ionel, M. Popescu, Y. C. Chong, and N. A. O. Demerdash, "Design Optimization of a High Torque Density Spoke-Type PM Motor for a Formula e Race Drive Cycle," *IEEE Trans. Ind. Appl.*, vol. 54, no. 5, pp. 4343–4354, 2018.

[3] S. Nikam, R. Vandana, and B. G. Fernandes, "A high torque density permanent magnet free motor for in-wheel electric vehicle application," no. c, 2011.

[4] M. H. Mohammadi, "Rotor Design Optimization of Permanent Magnet – Assisted Synchronous Reluctance Machines for Traction Applications," no. November, pp. 4–8, 2015.

[5] T. C. Bora, L. D. S. Coelho, and L. Lebensztajn, "Bat-inspired optimization approach for the brushless DC wheel motor problem," *IEEE Trans. Magn.*, vol. 48, no. 2, pp. 947–950, 2012.

[6] A. Sarikhani and O. A. Mohammed, "Multiobjective design optimization of coupled PM synchronous motor-drive using physics-based modeling approach," *IEEE Trans. Magn.*, vol. 47, no. 5, pp. 1266–1269, 2011.

[7] S. E. Skaar, O. Krovel, and R. Nilssen, "Distribution, coil-span and winding factors for PM machines with concentrated windings," *XVII Int. Conf. Electr. Mach. ICEM 2006*, p. 346, 2006.

[8] H. Duane C, Brushless permanent magnet motor design. 2006.

[9] J. Kennedy and R. Eberhart, "Particle swarm optimization," in *Proceedings of ICNN'95 - International Conference on Neural Networks*, 1995, vol. 4, pp. 1942–1948 vol.4.

IOP Conf. Series: Materials Science and Engineering **694** (2019) 012010 doi:10.1088/1757-899X/694/1/012010

Increasing the weight of onion bulbs due to the reduction of *Spodoptera exigua* using a portable light trap

R Wulandari[1], R E Santoso[1], D Prasetiyo[1], A Lestari[1], A Mizar[1] and P Puspitasari[1]

[1]Mechanical Engineering Department, Universitas Negeri Malang, Jalan Semarang 5 Malang, Indonesia

E-mail: retno.wulandari.ft@um.ac.id

Abstract. Red onion is one of the strategic horticultural commodities whose distribution is almost in all regions of Indonesia. The problem of this this commodity is the low productivity resulting from the *Spodoptera exigua* pest attack. Based on the problem above, the purpose of this research is to design a portable pest control lamp. This lamp will be measured its effectiveness in trapping pests. This study also aims to determine the effect of trap effectiveness on increasing the weight of onion bulbs. This study used a randomized block design with 3 treatments. The observed variables were the number of pest catches in each treatment as well as the weight of onion bulbs produced by the land that was placed in each treatment. The results showed that T3, a trap with a constant light, was the best treatment with an average yield of *S. exigua* catch reaching 126.75 *S. exigua*. The second best results were given by T2 and T1 with an average catch reaching 74.75 and 46.75. The best weight is also given by T3 with an average weight reaching 116 g. The second best results were given by T2 and T1 with an average weight of 96 g and 85 g.

1. Introduction

Red onion is one of the strategic horticultural commodities whose distribution is almost in all regions of Indonesia. The majority problem to develope this commodity is the low productivity as a result of *Spodoptera exigua* [1] and [2]. This happens because the onion plant is the main host of *S. exigua*. [3] and [4] state that the *S. exigua* pest attacks onion cultivation from the vegetative phase to the time of harvest and during severe attacks can cause losses of up to 100%.

Facing this problem, what is done by farmers is conventional control using the yellow trap. However, the application of this control is less optimal because at night it is not visible to pests, considering that *S. exigua* pests start to attack at night [5]. In addition, the *S. exigua* is also not too fond of the color yellow. This is in line with the results of research [6] and [7] which state that white lights are more attractive to *S. exigua* than yellow lights.

Knowing such conditions, farmers add another conventional control device in the form of lights [8]. However, its application is also not optimal because there is still a distance between the water tank and the lamp that can cause pest to escape even though it has been trapped before. The relatively short age of the lamp construction and electrical circuit also results in the application of this model to be less economical. The last drawback is that the energy source that uses alternating current often causes electrical short circuit.

Based on the problem above, the purpose of this research is to design a portable pest control lamp. This lamp will be measured its effectiveness in trapping pests. This study also aims to determine the effect of trap effectiveness on increasing the weight of onion bulbs.

2. Methodology

2.1. Preparation of Portable Light Trap

Traps can be made using used bottles, inside the used bottles are mounted with spiral led ribbon lights as in figures 2 and 3. The electrical circuits are installed in a box under the bottles. The outer layer of the bottle is coated with plastic, then coat the outer plastic with glue to trap pests. The traps that have been made are then installed at a flight height of *S. exigua*, 60 cm. The lights are turned on just before dusk, which is around 5.30 p.m until 20.00 p.m. The light energy source uses batteries.

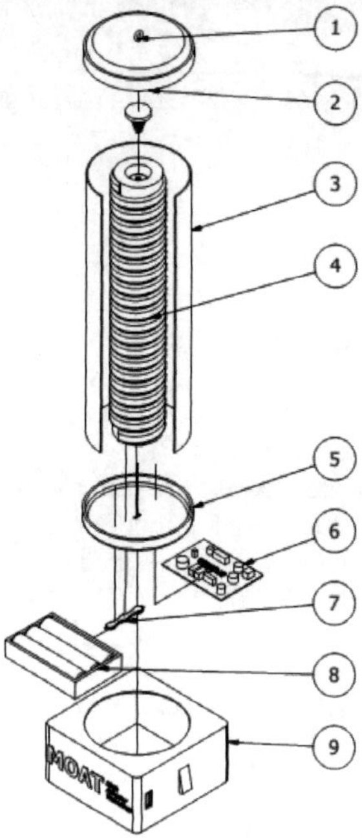

Figure 1. (1) hanger, (2) top side, (3) tube, (4) lamp, (5) bottom side, (6) electronic components, (7) timer, (8) battery, (9) box

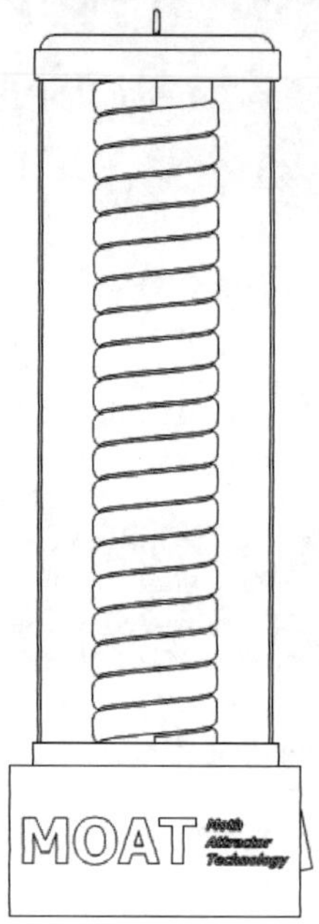

Figure 2. The trap looks sideways

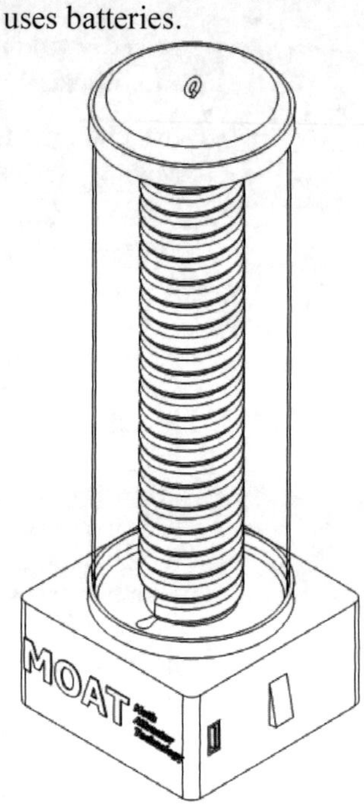

Figure 3. Traps look ahead

2.2. Experimental Design

This study used a randomized block design with 3 treatments. The observed variables were the number of pest catches in each treatment as well as the weight of onion bulbs produced by the land that was placed in each treatment.

Table 1. List of treatments

Treatments	Description
T1	Control tools used by farmers everyday
T2	Treatment of tools based on blinking lights
T3	Treatment of tools based on not blinking lights

2.3. *Treatment Application and Data Collection*

The light trap was applied to 3 red onion fields measuring 180 m^2 located in Gending District, Probolinggo Regency. The three fields are separated from one another by 100 meters. The three fields were also given the same fertilization, which used liquid organic fertilizer. The third application is done when the age of onion enters the second month. Repetition is done 3 times a week, this repetition is carried out for 4 weeks until the age of the onion reaches the age of harvest (2 months).

The captured pests will be observed by adjusting to the *S. exigua* morphology, this is done to ensure that the pests that are caught are actually *S. exigua*. While the weight of onion bulbs will be measured using a digital scale. The catch of pests and onion bulbs weight in the field will be written on research instruments such as table 2.

3. Results and Discussion

3.1. *Effect of Light Trap on S. exigua Population*

Based on observations made, it was found that the best catch was given by T3 with an average number of catches reaching 126.75 *S. exigua*. The second best results were given by T2 and T1 with an average catch reaching 74.75 and 46.75. From this table it is also found that the effectiveness of T3 is almost 3 times compared to the effectiveness of T1 which is a conventional lamp used by farmers. This effectiveness is obtained from the results of the application of the glue-tube-light collaboration system applied to T3. The effectiveness of T3 is also far from leaving T2 with the same lamp but the blazing pattern of the lamp is used blinking.

Table 2. Reduction of *S. exigua* population

Treatments	The average number of *S. exigua* caught per week				Mean
	Week 1	Week 2	Week 3	Week 4	
T1	52	46	48	41	46.75
T2	69	77	74	79	74.75
T3	125	129	122	131	126.75

The above results reinforce research [5] that *S. exigua* uses its vision in adapting to the environment. *S. exigua* is also able to use the moonlight to see and choose which locations they will use to breed [9]. The results of this study also found a finding that the lights were quite effective in suppressing the population of *S. exigua* broodstock. This author found when observing the application of T2, *S. exigua* failed to approach T2 because of the concept of the blinking lights, where at one time the lights were dim and one time bright. This condition actually makes *S. exigua* confused and never approaches the trap.

This finding is in line with the results of the study [10] which states that the strength of the light lamp affects the coming *S. exigua*, a lamp with greater strength will attract the arrival of *S. exigua* more. The results showed that lamp traps with a constant flame were most preferred by *S. exigua* compared to other treatments. This is presumably because the light on the T3 is brighter compared to other treatments.

According to [6], white lights will be more attractive to *S. exigua* moths than yellow lights. The range of wavelengths produced by a 10 watt lamp is around 379.2-640.9 nm [11]. Therefore, *S. exigua* will be more interested in the color spectrum which has a wavelength of 400-600 nm where this range is the wavelength range of its host [12]. Different light colors will emit different wavelengths [13]. Meanwhile the farther or dimmer the light, the less likely the insects to approach [14].

3.2. *Effect of The S. exigua Reduction on Increasing the weight of onion bulbs*

Based on the weighing that had been done, it was found that the best onion bulbs weight was given by T3 with an average weight reaching 116 g. The second best results were given by T2 and T1 with an average weight of 96 g and 85 g. From this table it was also found that the weight of onion bulbs from land given T3 produced the highest weight compared to land given T2 and T1. Increasing the weight of onion bulbs produced will be directly proportional to the high rate of onion production in a field.

Table 3. Increasing the weight of onion bulbs

Treatments	Rata-rata berat umbi basah bawang merah (g)
T1	85
T2	96
T3	116

One of the main factors influencing the yield of onion bulbs weight is the successful control of *S. exigua*. It can be seen in tables 2 and 3 that T3 gives the highest *S. exigua* catch, where these results are directly proportional to the weight of onion bulbs produced by land given T3. In this case, it can be seen that the more insects caught in the light trap, the chance of the insects in breeding will be interrupted. The capture of *S. exigua* will affect the activity of males and females to meet so that the level of attack by *S. exigua* on shallots will also be affected. At the same time, liquid fertilizer given early in the planting period can work optimally to supply nutrient requirements on the leaves where the condition of the leaves of the onion plant is in good condition because the *S. exigua* attack has been reduced.

The high onion bulbs weight produced is thought to be due to the role of the elements Mg, Potassium and Nitrogen contained in the fertilizer used. This is consistent with the results of research [15] which states that Magnesium (Mg) can increase the length and diameter of the tuber better, because magnesium has a role in activating enzymes related to carbohydrate metabolism, respiratory enzymes, and also works as a catalyst. In addition, magnesium also functions as a cofactor in enzymes, especially those that activate the phosphorylase process.

Element potassium is importance because it is able to synthesize proteins to stimulate the formation of tubers more perfect. This is according to the opinion [15], potassium has an important role in the process of photosynthesis directly, able to increase growth and leaf area index in addition to having a function to increase CO_2 assimilation, can also increase the translocation of photosynthetic results out of leaves (to other tissues that need).

The role of Nitrogen is a building material for proteins, nucleic acids, enzymes, and alkaloids in each plant growth, especially plant vegetative growth. This is consistent with studies [16] which reported an increase in nitrogen levels that could significantly increase plant height. Plants that get a lot of nitrogen are thicker green leaves and wider leaves so that the process of photosynthesis has increased. This is consistent with the opinion [16] that the role of nitrogen is very important for vegetative growth, especially the leaves are more green colored, can increase the root shoot ratio, and affect the formation of fruit in seeds.

4. Conclusion

The results showed that T3, a trap with a constant light, was the best treatment with an average yield of *S. exigua* catch reaching 126.75. The second best results were given by T2 and T1 with an average catch reaching 74.75 and 46.75. The best onion bulbs weight is also given by T3 with an average weight reaching 116 g. The second best results were given by T2 and T1 with an average weight of 96 g and 85 g. From this study, a major finding was found that there was a relationship that was directly proportional between the high number of catches of pests and the high weight of onion bulbs. The main factor of this finding is the ability of plants to process nutrient supply from organic liquid fertilizers. Plants can make the maximum process to improve the quality of life if the condition of the leaves and other parts are healthy duc to thc successful control of *S. exigua*.

5. Acknowledgment
The authors would like to acknowledge Laboratory of Mechanical Engineering State University of Malang and Sumber Makmur III Farmers Group at Gending, Probolinggo regency that support this research.

References
[1] Haryati and Agus N 2009 *J. Litbang Pertanian* **28**(2)
[2] Razak 2016 *E-J. Agrotekbisnis* **4**(5) 565
[3] Sumampouw B S 2012 *J. AgroPet* **9**(1)
[4] Rusli, Ameti and Silvia P S 2010 *Manggaro* **11**(1) 25
[5] Prasetyo G W 2016 Pengendalian Hama Ulat Bawang (*Spodoptera exigua*) pada Tanaman Bawang Merah (*Allium cepa*) (Jember: Universitas Jenderal Soedirman)
[6] Rachman A 2005 Hubungan Perangkap Lampu dengan Aktivitas Ngengat Spodoptera exigua Hubn. pada Tanaman Bawang Merah (*Alliumascalonicum* L) (Jember: Universitas Jember)
[7] Rasid M R 2012 Laporan Praktikum Ilmu Hama Tumbuhan Dasar Percobaan Uji Preferensi Warna (Bogor: Institut Peranaian Bogor)
[8] Nugraheni V A 2009 Uji Aktivitas Gel Minyak Atsiri Bunga Kenanga (*Canangium Odoratum* (Lmk.) Hook. & Thoms) sebagai Repelan Terhadap Nyamuk Anopheles Aconitus Betina (Surakarta: Universitas Muhammadiyah Surakarta)
[9] Dent D R and Walton M P 1997 *Methods in Ecological and Agricultural Entomology* p 5
[10] Yunus M, Martono E, Awijonarko, and Soesilohadi R C H 2011 *J. Perlindungan Tanaman Indonesia* **17**(1) 18
[11] Deomedes, Piyoh Y I, Talangas Y A, Sudjito D N, and Rondonuwu F S 2007 *Prosiding Pertemuan Ilmiah* **25** 69
[12] Metcalf R L dan Metcalf E R 1992 *J. Biochem* **24**(12) 1997
[13] Syahri, Thamrin T, and Hutapea Y 2011 Tinjauan Perbaikan Teknologi Pascapanen Padi untuk Menekan Serangan Serangga Hama Gudang (Bandung: Universitas Padjajaran)
[14] Akmal I, Savitri G, and Arimbi N 2007 Simple Garden (Jakarta: Gramedia Pustaka Utama)
[15] Afandie R 2001 Ilmu Kesuburan Tanah (Yogyakarta: UGM Press)
[16] Englestad 1997 Teknologi Penggunaan Pupuk (Yogyakarta: UGM Press)

IOP Conf. Series: Materials Science and Engineering **694** (2019) 012011 doi:10.1088/1757-899X/694/1/012011

Strength Analysis of Metro Kapsul Knuckle Plate and Tie Rod using Finite Element Method

B Budiwantoro[1,2], Y D Raditya[1] and Y A Harjandi[1]

[1]Mechanical Engineering Department, Faculty of Mechanical and Aerospace Engineering, Institut Teknologi Bandung, 40132, Indonesia
[2]National Center for Sustainable Transportation Technology, Ganesha Street number 10, Bandung, West Java, 40132, Indonesia

E-mail: budiwan@edc.ms.itb.ac.id, budiwan56@gmail.com, davianraditya47@gmail.com, yosuaazaryaharjandi@gmail.com

Abstract. Metro Kapsul is a public transportation system developed to solve traffic problems and save travel time in Bandung. It is categorized as a light rail train with a capacity of 70 passengers per capsule. Before it operates, analysis is needed to ensure the vehicle's safety. Two crucial parts to be analysed are the knuckle plate and the tie rod of the train. The knuckle plate and the tie rod must meet the requirements of the standards based on PM 175 2015. Forces on each component are calculated using equations from free body diagrams. Static and fatigue analysis are then conducted using ANSYS with Finite Element Method based on load cases in UIC Code 615-4 standard. The maximum stresses that occurred are 195.64 MPa for the knuckle plate and 118.54 MPa for the tie rod. In terms of fatigue, based on the modified Goodman failure criteria, the knuckle plate and tie rod are proven to be safe.

1. Introduction

Indonesia has a population of 237.6 million and high population growth rate, which is 1.49% every year [1]. High population growth rate leads to an increased number of land transportations especially private cars. With no significant road widening, congestion is inevitable. In order to solve this problem, an efficient form of public mass transportation is needed. Therefore, Metro Kapsul is developed. Metro Kapsul is categorized as a light rail train–its weight makes the load applied on track smaller compared to trains in general. Since the tracks are located above the road, less trestlework and smaller pillars are needed which are suitable for a crowded city.

Further analysis is needed to ensure the vehicle's safety. Two of the most crucial components in Metro Kapsul are the knuckle plate and the tie rod, which are used to transfer forces from the bogie of the train to the wheel and allow the train to travel along corners. In conclusion, the knuckle plate and the tie rod strength analysis is needed.

2. Methodology

The forces acting on each component of Metro Kapsul when traveling along corners were calculated using equations from free body diagrams and UIC Code 615-4. By using the forces calculated, the knuckle plate strength analysis was conducted using ANSYS with finite element method based on UIC Code 615-4 and PM 175 2015.

First, the movement phases of Metro Kapsul to be analysed was defined. Three movement phases were observed in this analysis, as shown in Figure 1. The first being the phase based on UIC Code 615-4 where the train was stationary, the second was when the train was about to enter a corner, and the third was when the train was cornering and there was no movement of components relative to each other. The second and third movement phases are referred to as transient and steady state corners, where the calculations during these two phases are based on the lateral forces distributed along the main wheels and guide wheels of the train bogie.

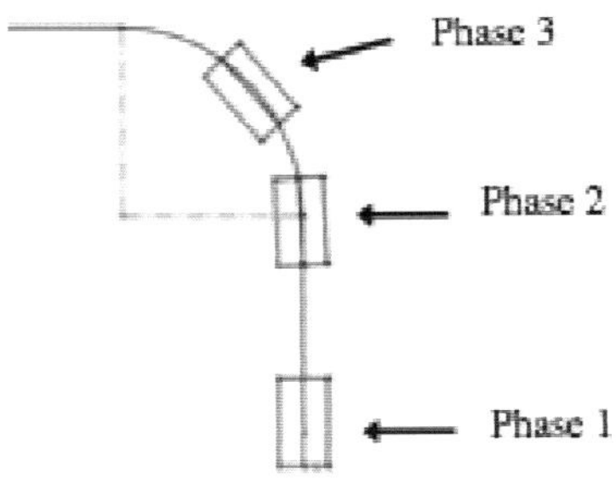

Figure 1. Movement phases of Metro Kapsul.

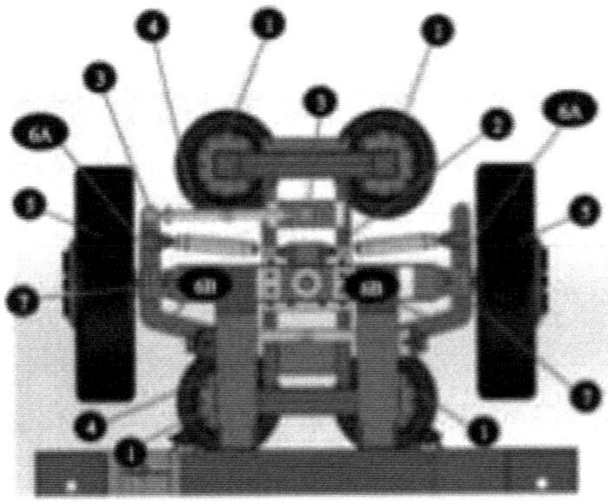

1. Guide wheel
2. Guide wheel frame
3. Tie rod
4. Tie rod pipe
5. Wheel
6. Plate A and B
7. Knuckle

Figure 2. Metro Kapsul bogie components.

The first phase to be analysed was the transient cornering phase. Components included in Metro Kapsul bogie are shown in Figure 2. The first step in calculating the forces during this phase was to draw the free body diagrams of these components, as shown in Figures 3 to 8.

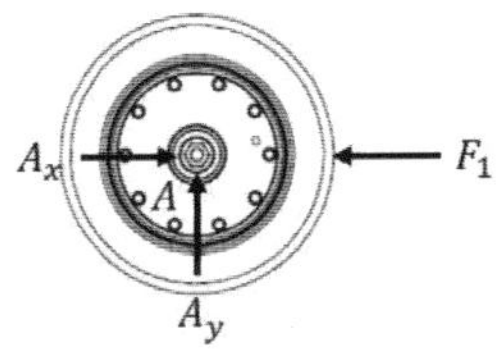

Figure 3. Guide wheel free body diagram.

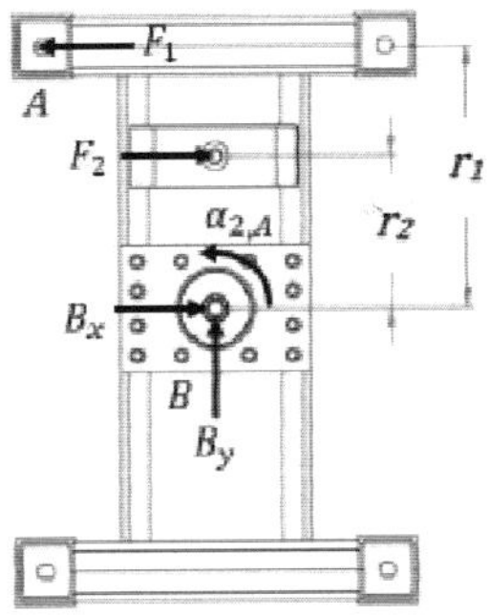

Figure 4. Guide wheel frame free body diagram.

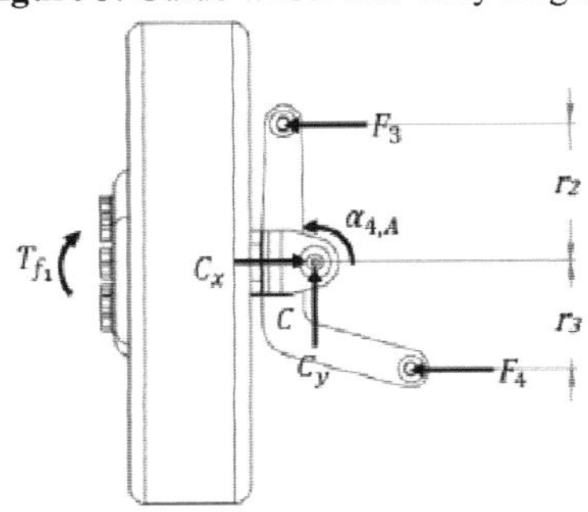

Figure 5. Left wheel and knuckle free body diagram.

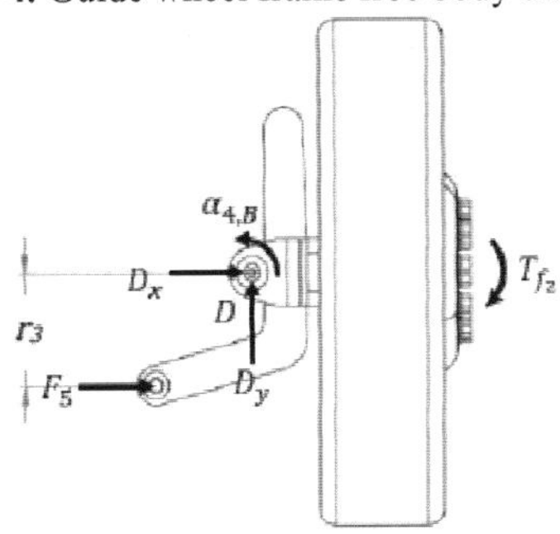

Figure 6. Right wheel and knuckle free body diagram.

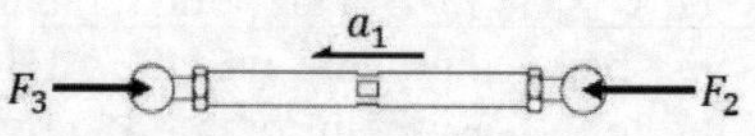

Figure 7. Front tie rod set free body diagram.

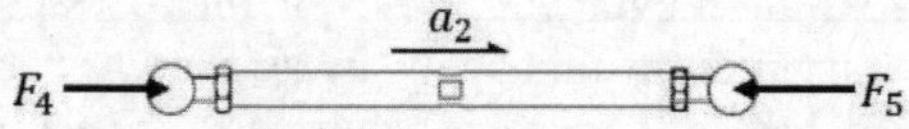

Figure 8. Rear tie rod set free body diagram.

From these free body diagrams, the equations needed for calculating the force on each component were obtained, as shown on Equations 2.1 to 2.7.

$$F_1r_1 - F_2r_2 = I_1\alpha_{2,A} \quad (2.1)$$

$$a_1 = \alpha_{2,A}r_2 \quad (2.2)$$

$$F_2 - F_3 = m_{trd}a_1 \quad (2.3)$$

$$F_3r_3 - F_4r_4 = I_2\alpha_{4,A} + T_{f_1} \quad (2.4)$$

$$a_2 = \alpha_{2,B}r_3 \quad (2.5)$$

$$F_4 - F_5 = m_{trb}a_3 \quad (2.6)$$

$$F_5r_5 = I_3\alpha_{4,B} + T_{f_2} \quad (2.7)$$

To obtain the forces on each component using the equations above, some variables such as angular acceleration, linear acceleration, and torsional friction needed to be calculated first. The angular accelerations were obtained from the kinematic analysis of Metro Kapsul bogie [2] with the results shown in Table 1.

Table 1. Angular acceleration of components [2].

Variable	Value
$\alpha_{2,B}$	$0{,}6245 \frac{\text{rad}}{\text{s}^2}$
$\alpha_{4,B}$	$0{,}4711 \frac{\text{rad}}{\text{s}^2}$

The torsional friction was calculated by determining the contact patch between the tyre and the ground first. Figure 9 shows the vehicle tyre dimension. The contact patch length is a function of vertical deformation purely on the basis of geometry and the contact patch shape is assumed to be perfectly rectangular in shape [3], as shown in Figure 10.

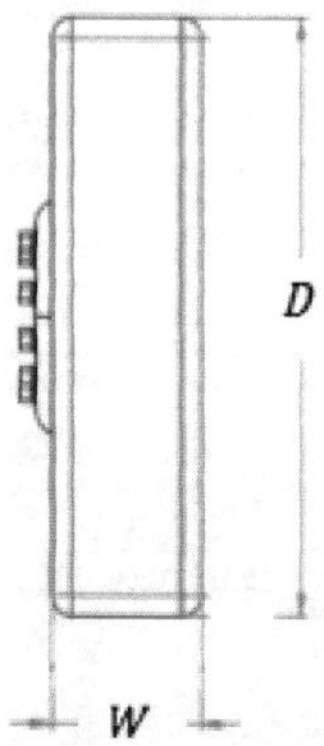

Figure 9. Metro Kapsul tyre dimension.

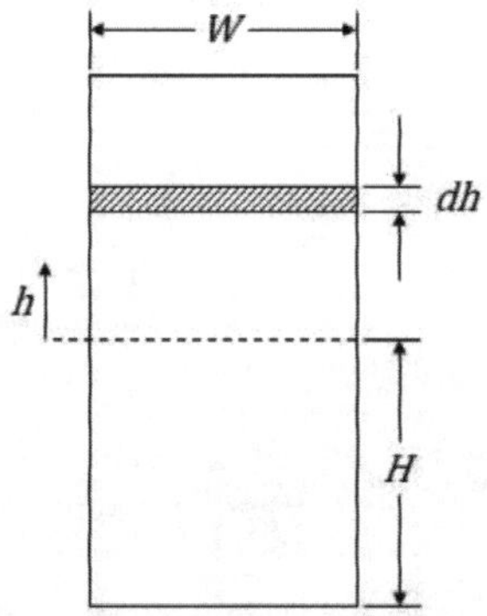

Figure 10. Tyre contact patch.

The vertical stiffness was calculated by using Equation 2.8 [3]. The value of the vertical stiffness and the normal force acting on each wheel were used to calculate the maximum deflection of the tyre using Equation 2.9 [3]. The contact patch length was then calculated using Equation 2.10.

$$K_z = 0{,}0274\, P\sqrt{WD} + 3{,}38 \quad (2.8)$$

$$\Delta z = N/K_z \tag{2.9}$$

$$l_{cp} = \sqrt{R^2 - (R - \Delta z)^2} \tag{2.10}$$

The equation for calculating the torsional friction was obtained by integrating Equation 2.11 derived from Figure 8. Total torsional friction was calculated using Equation 2.12. This torsional friction was later used to calculate the forces on each component using equations obtained from the free body diagrams.

$$dT_f = \mu \cdot dN \cdot h \tag{2.11}$$

$$T_f = \frac{\mu NH}{4} \tag{2.12}$$

The next phase to be analysed was the steady state cornering phase. Two assumptions were used in calculating the forces during this phase, with the first one being the total lateral force during cornering received by the tyres only, and the second one being the total lateral force during cornering received by the guide wheels only. Based on these two assumptions, free body diagrams of the whole vehicle were drawn, as shown in Figures 11 and 12.

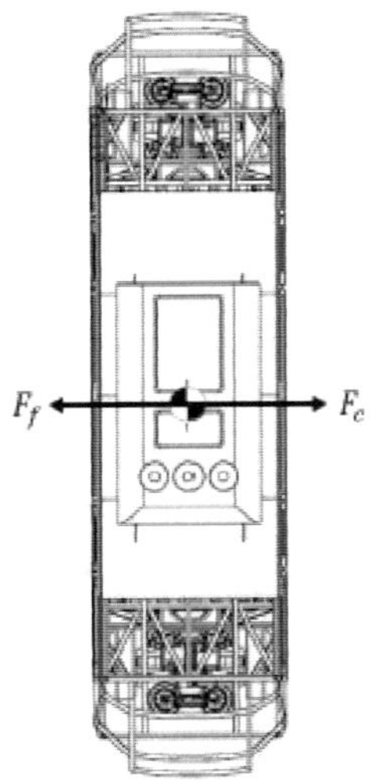

Figure 11. Metro Kapsul free body diagram (top view).

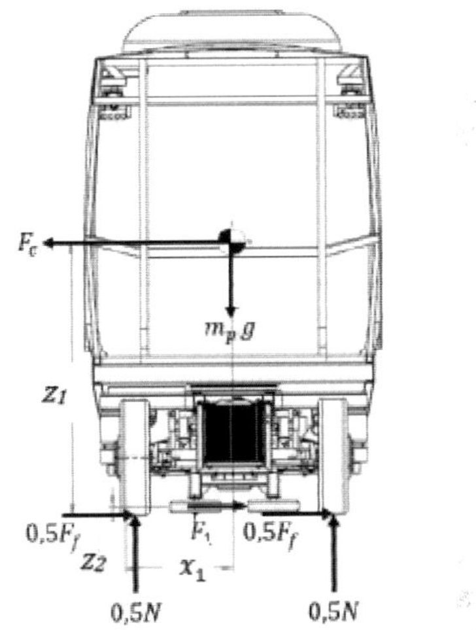

Figure 12. Metro Kapsul free body diagram (front view).

For the first assumption, Figure 11 was used as a reference. The total lateral force received by the tyres was in the form of friction and the maximum possible friction force was first calculated by using Equation 2.13.

$$F_{f\,maks} = \mu N \tag{2.13}$$

For the second assumption, Figure 12 was used as a reference. The total lateral force received by the guide wheels was in the form of normal reaction force and the maximum possible force was calculated at the critical point where the vehicle was about to roll in the opposite direction of the corner. In this condition, the pivot of the vehicle was shifted to point *O* and the free body diagram on Figure 12 changed, becoming the one shown on Figure 13. The forces on all guide wheels were obtained by calculating the centrifugal force on the vehicle and moment of the force about point *O* as shown on Equations 2.14 and 2.15.

$$F_c = \frac{m_p v^2}{R_b} \tag{2.14}$$

$$\Sigma M = F_{1\,maks}\, z_2 + m_p g x_1 - F_c z_1 = 0 \tag{2.15}$$

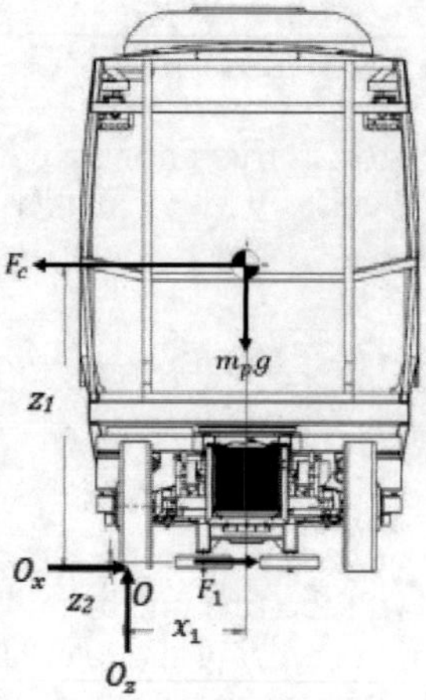

Figure 13. Metro Kapsul free body diagram (front view) the moment before rolling.

The two forces calculated were compared to obtain the percentage distribution of the two forces acting on the Metro Kapsul bogie during cornering. This percentage was validated by using the results from previous simulation conducted using SIMPACK [4], with the results shown in Table 2. The actual forces acting on the wheels and guide wheels were calculated by multiplying the percentage by the value of the centrifugal force.

Table 2. Comparison of theoretical calculation and simulation results.

Component	Percentage force from simulation (%)	Percentage force from theoretical calculation (%)
Wheels	19.20	18.06
Guide wheels	80.80	81.94
Relative error on wheels	5.92	
Relative error on guide wheels	1.41	

The last part of the analysis was the calculation of forces acting on the bogie by using UIC Code 615-4. There are two major loading cases in UIC Code 615-4, exceptional loads, and normal service loads [5]. The load case in exceptional loading is a combination of vertical and lateral forces as illustrated on Figure 14. The exceptional loading case was used as the basis of force calculation to ensure that the bogie received maximum possible forces.

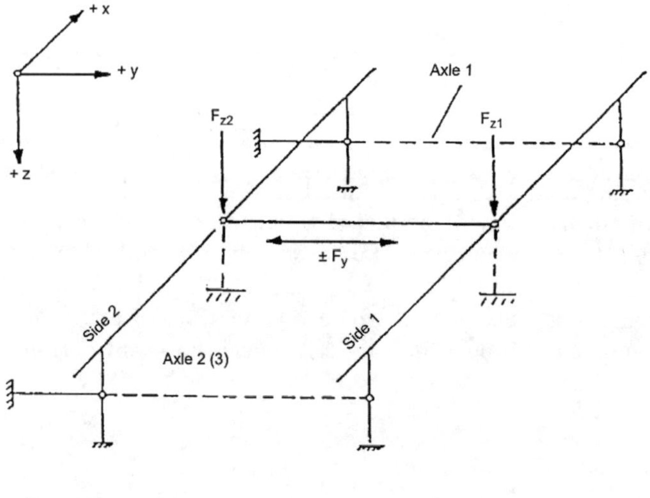

Figure 14. Forces acting on bogie frame based on UIC Code 615-4 standard.

IOP Conf. Series: Materials Science and Engineering **694** (2019) 012011 doi:10.1088/1757-899X/694/1/012011

These forces were calculated by using the equations based on UIC Code 615-4, as shown on Equations 2.16 and 2.17.

$$F_Z(N) = \frac{1{,}4\ g}{2n_b}(m_v + c_1 - n_b m_b) \tag{2.16}$$

$$F_y(N) = 2\left(10^4 + \frac{m_v + c_1}{3n_e n_b}\right) \tag{2.17}$$

The results from the calculations were summarized in the tables below, where the vertical forces and lateral forces are shown in Tables 3 and 4 respectively.

Table 3. Summary of vertical forces.

Component	Vertical force (N)		
	Transient cornering phase	Steady state cornering phase	Exceptional loading on UIC Code 615-4
Knuckle	12,806.96	12,806.96	13,658.46
Total (on bogie)	51,227.82	51,227.82	54,633.85

Table 4. Summary of lateral forces.

Component	Lateral force (N)		
	Transient cornering phase	Steady state cornering phase	Exceptional loading on UIC Code 615-4
Front left guide wheel	1,073.38	4,952.29	-
Rear left guide wheel	0	4,952.29	-
Tie rod (on guide wheel frame)	1,795.87	0	-
Tie rod (on left plate A)	1,794.60	0	-
Left knuckle	2,957.60	1,091.69	5,217.58
Tie rod (on left plate B)	1,162.99	0	-
Tie rod (on right plate B)	1,161.01	0	-
Right knuckle	1,161.01	1,091.69	5,217.58
Total (on Metro Kapsul)	48,351.85	48,351.85	41,740.67

After calculating the forces, the strength analysis on knuckle plate and tie rod was conducted using ANSYS with finite element method based on PM 175 2015 [6] starting from the knuckle. 3D model of the knuckle plate as shown in Figure 15 was made. In this model, the knuckle, the plate, and the housing were assembled for each to be analysed based on UIC Code 615-4 that regulates the method in strength analysis testing of a bogie frame.

The plates were connected to the knuckle using welding connection. Boundary conditions and forces were applied based on UIC Code 615-4. Additional lateral force acting on the plates during the second and the third phase as shown in Figure 16 and the definition of the boundary condition is shown in Table 5.

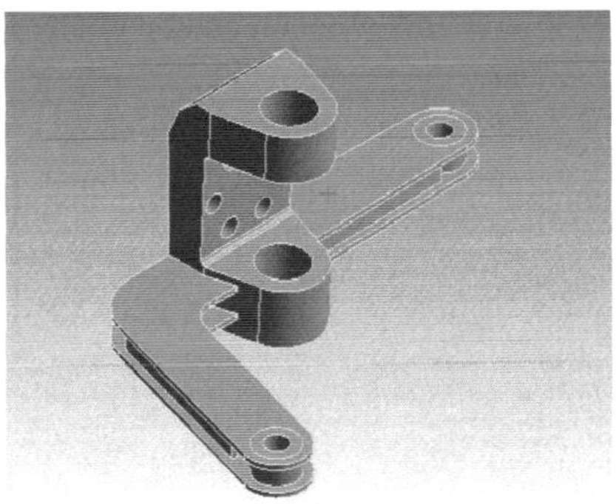

Figure 15. 3D Model of knuckle plate.

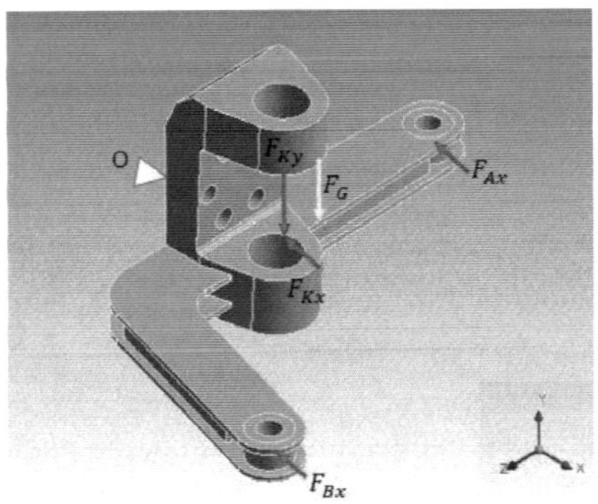

Figure 16. Knuckle plate loading arrangement.

Table 5. Definition of boundary conditions.

	O
dx	0
dy	0
dz	0

Three load cases were used for strength analysis on the knuckle plate, the first being the combination of vertical and lateral forces from the exceptional loading case on UIC Code 615-4, the second one being the combination of vertical forces based on UIC Code 615-4 and lateral forces during the transient cornering phase, and the third one being the combination of vertical forces based on UIC Code 615-4 and lateral forces during the steady state cornering phase. These forces applied on knuckle plate are shown in Table 6.

Table 6. Forces applied on knuckle plate.

No	Parameter	Force (N)		
		Load case 1	Load case 2	Load case 3
1	F_{Ax}	0	1,794.60	0
2	F_{Bx}	0	1,162.99	0
3	F_{Kx}	5,217.58	2,957.60	1,091.69
4	F_{Ky}	13,658.46	12,806.96	12,806.96
5	F_G	319.28	319.28	319.28

After strength analysis on the knuckle plate was carried out, 3D model of the tie rod as shown in Figure 17 was made. The tie rod components were connected using ball joints and they only received lateral forces during the transient cornering phase. Boundary conditions and forces were applied based on the transient cornering phase as shown in Figure 18, and the definition of the boundary condition is shown in Table 7. Forces applied on the tie rod during the transient cornering phase can be referred to Table 8.

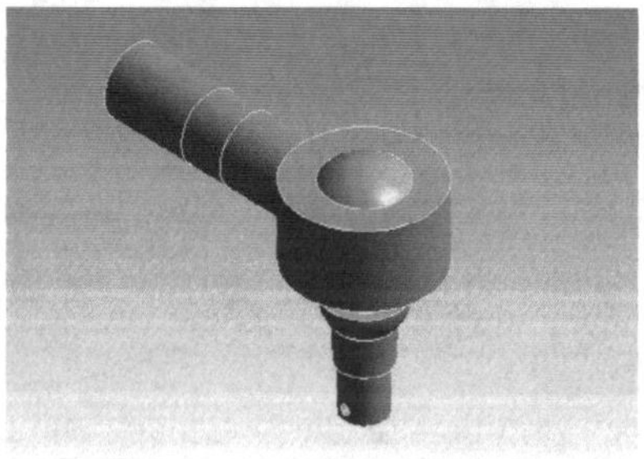

Figure 17. 3D Model of tie rod.

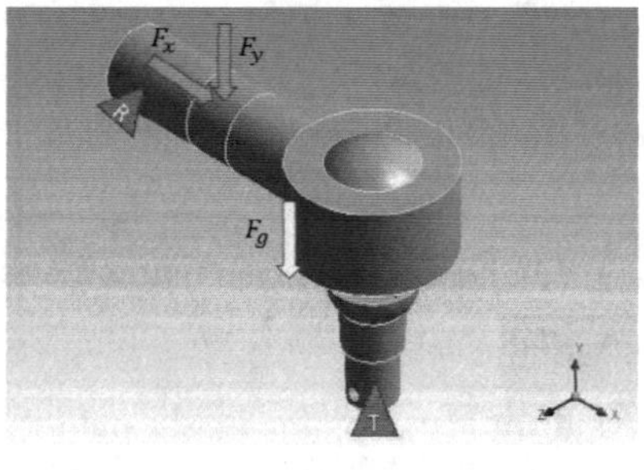

Figure 18. Tie rod loading arrangement.

Table 7. Definition of boundary conditions.

	T	R
dx	0	Free
dy	0	0
dz	0	0

Table 8. Forces applied on tie rod.

No	Parameter	Force (N)
1	F_x	1,795.87
2	F_y	22.74
3	F_g	16.27

Weldox 700 (ASTM A514 Grade B) is the material of knuckle plate and tie rod in which its mechanical properties are shown in Table 9. There should be no permanent deformation after the removal of exceptional loads. While PM 175 2015 only allow stress less than 75% of Yield Strength of the material [3] (Weldox 700) which is 525 MPa. All the stress results of any cases must not exceed 525 MPa. The stress result will also be plotted on modified Goodman Compression and Tension Diagram.

Table 9. Mechanical Properties of Weldox 700.

No	Properties	Symbol	Value	Unit
1	Density	ρ	7,850	$\frac{kg}{m^3}$
2	Yield Strength	S_y	700	MPa
3	Tensile Strength	S_{ut}	780-930	MPa
4	Modulus of Elasticity	E	200	GPa
5	*Poisson Ratio*	v	0.3	

3. Result and Discussion

The results of the strength analysis using finite element method were validated for those results to be reliable. First, meshing was done on each model, and convergence test was performed to determine the correct element size to be used in the simulation. This test was done by increasing the number of elements or decreasing the element size. After that, the results were validated by comparing the normal stress on one of the axis obtained from the simulation to the normal stress on one of the axis calculated theoretically.

3.1 Knuckle Plate

The result of convergence test on the knuckle plate is shown in Table 10 and Figure 19. As the element size decreased from 15 to 6 mm, the relative error started to fall below 10%, and the relative error fell further below 1%. Based on *ANSYS Conference 2008* [7] this result was adequate, therefore the element size of 6 mm was used throughout the knuckle plate strength analysis.

Table 10. Convergence test on knuckle plate.

Element size (mm)	Number of Elements	Equivalent stress (MPa)	Error (%)
15	24,081	82.71	-
10	35,636	93.19	11.25
9	42,073	96.98	3.91
8	52,347	90.46	7.21
6	97,982	96.74	6.50
5	155,051	96.28	0.48
4	283,372	95.96	0.33

The forces on load case 2 were validated by calculating the normal stress on point A located on the knuckle plate where a stress probe was placed using ANSYS, as shown in Figure 20. The normal stress on the *z*-axis calculated from the simulation was -10.83 MPa. For the theoretical calculation, the knuckle plate was cut at point A to calculate the area moment of inertia and the normal stress as shown in Figure 21. The normal stress on point A on the *z*-axis calculated theoretically was -10.49 MPa. The stress calculation is shown in Table 11. Since the error between the simulation result and the theoretical calculation result is below 10%, this simulation can be considered valid based on the "*Concept of Model Verification and Validation*" [8].

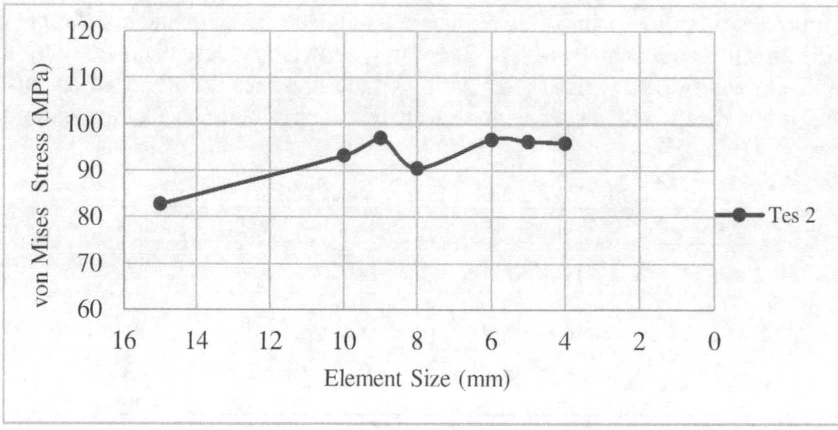

Figure 19. Convergence test on knuckle plate graph.

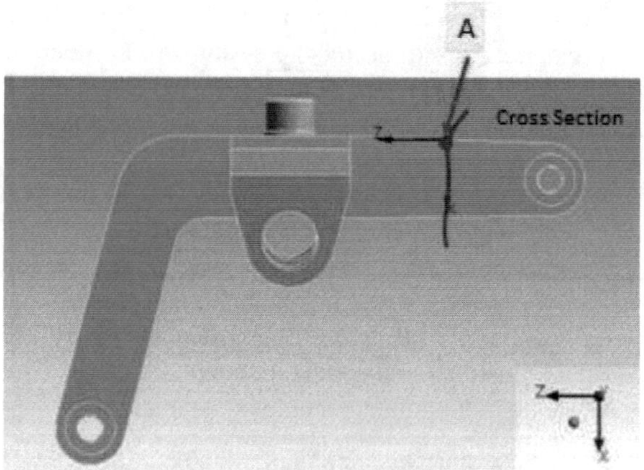

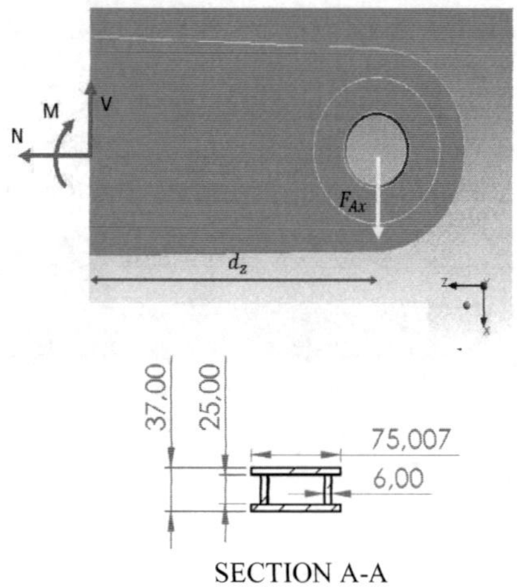

Figure 20. Stress probe on knuckle plate. **Figure 21.** Cross section of knuckle plate.

Table 11 Theoretical calculations on knuckle plate

No	Parameter	Value	Unit
1	F_{Ax}	-1,794.60	N
2	d_z	101.63	mm
3	N	0	N
4	V	-1,794.60	N
5	M	182,376.27	N.mm
6	A	1,200.10	mm^2
7	I_{yy}	633,624.79	mm^4
8	c	36.44	mm
9	σ_x (theoretical)	-10.49	MPa
10	σ_x (simulation)	-10.83	MPa
11	Error	3.26	%

The equivalent stress distribution for load case 2 on knuckle plate is shown in Figure 22. The stress calculated for each load case shown in Table 12, where all of the forces calculated are below 575 MPa.

IOP Conf. Series: Materials Science and Engineering **694** (2019) 012011 doi:10.1088/1757-899X/694/1/012011

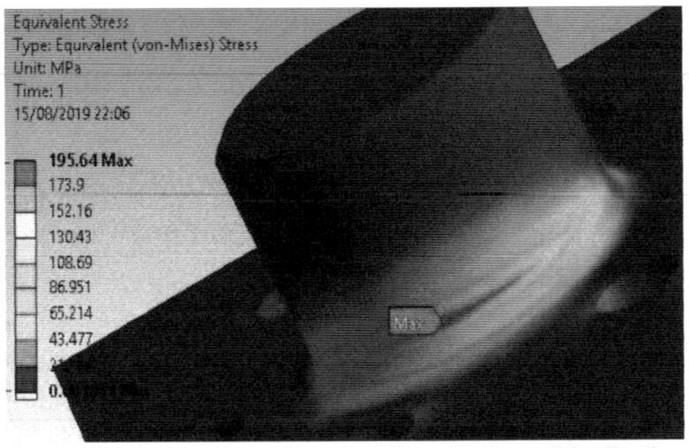

Figure 22. Equivalent stress distribution on knuckle for load case 2.

Table 12. Stress result for all load cases.

Load case	$\sigma_{eq\ max}$ (MPa)	Maximum deformation (mm)	Safety factor
1 – UIC Code 615-4	188.69	0.27	2.78
2 – Transient cornering phase	195.64	0.24	2.68
3 – Steady state cornering phase	164.06	0.22	3.20

From three load cases, there was a critical point on the knuckle plate as a result of each loading, as shown in Figure 23. The mean stress to ultimate tensile strength ratio and the amplitude stress to endurance limit ratio on each of these critical points from each load cases were calculated using their maximum and minimum principal stresses. These values were then plotted on the modified Goodman compression and tension diagram. The mean and amplitude stresses on each of these critical points are first calculated as shown in Table 13.

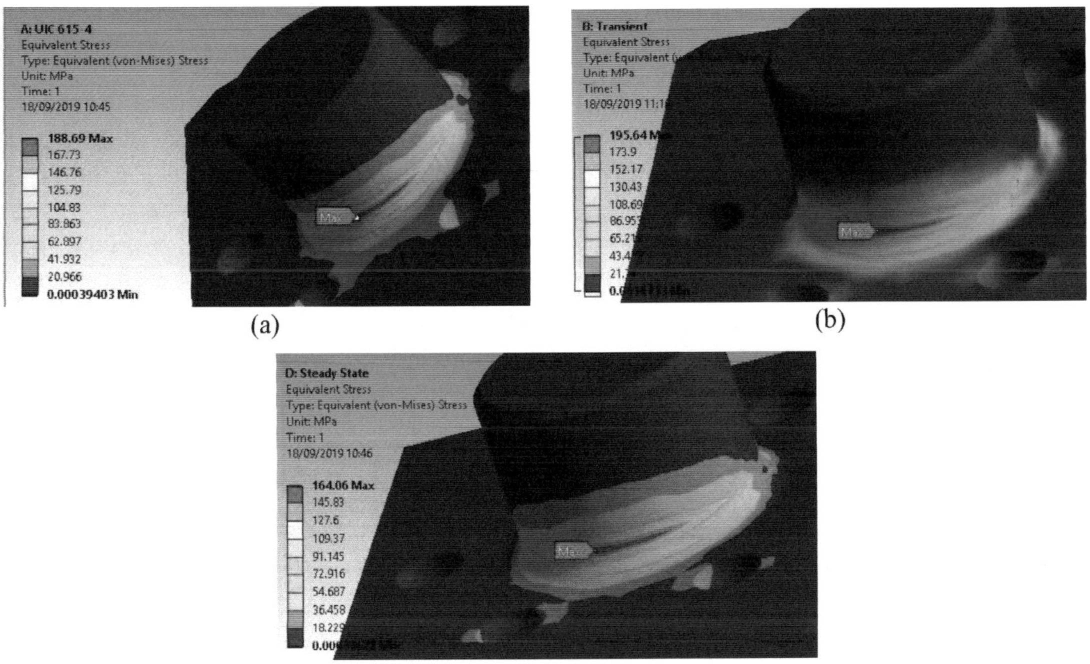

(a) (b)

(c)

Figure 23. Three critical points from (a) load case 1, (b) load case 2, and (c) load case 3.

Table 13. Mean and amplitude stress from each load case.

Load case	Stress (MPa)			
	σ_{pmax}	σ_{pmin}	σ_m	σ_a
1 – UIC Code 615-4	201.99	17.70	92.44	109.85
2 – Transient cornering phase	214.21	12.31	100.94	113.26
3 – Steady state cornering phase	187.36	13.66	86.849	100.51

The value of S_{ut} is 780 MPa, therefore the S_e' is 390 MPa. By applying a correction of 0.75 on the S_e' based on PM 175 2015, the calculated S_e was 292.5 MPa. The mean stress to ultimate tensile strength ratio and the amplitude stress to endurance limit ratio on each critical point are shown in Table 14. These values were plotted on a modified Goodman compression and tension diagram, as shown in Figure 24. From this graph, it can be seen that all of these three points fall below the ABC line, which means that no crack or failure occured on the knuckle plate for each case.

Table 14. Mean stress to ultimate tensile strength and amplitude stress to endurance limit ratio for each load case.

Ratio	Load case 1	Load case 2	Load case 3
$\frac{\sigma_a}{S_e}$	0.3150	0.3451	0.2969
$\frac{\sigma_m}{S_{ut}}$	0.1408	0,1452	0.1288

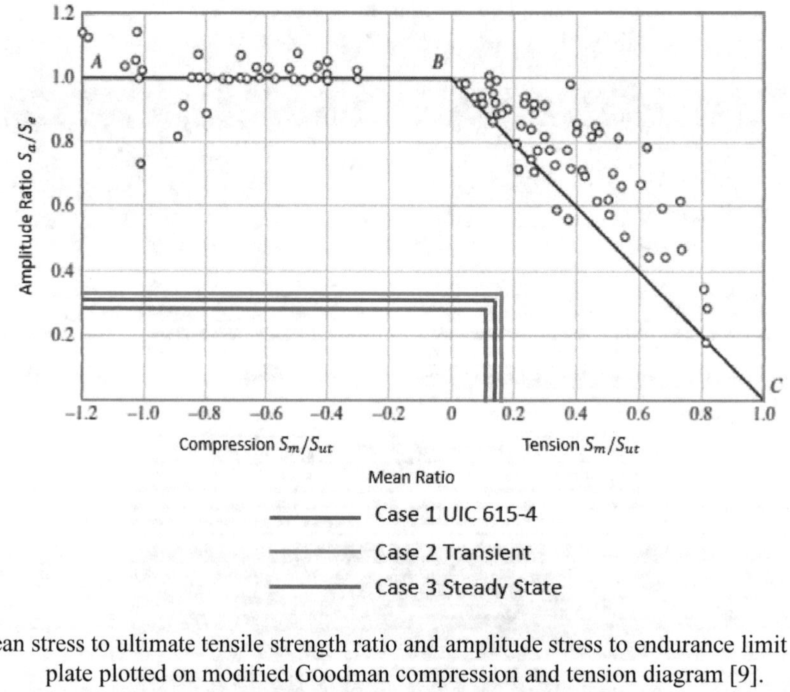

Figure 24. Mean stress to ultimate tensile strength ratio and amplitude stress to endurance limit ratio on knuckle plate plotted on modified Goodman compression and tension diagram [9].

3.2 *Tie Rod*

The result of the convergence test on the tie rod is shown in Table 15 and Figure 25. As the element size decreased from 15 to 3 mm, the relative error started to fall below 5%, and the relative error further fell below 1%. Based on *ANSYS Conference 2008* [7] this result is satisfactory, therefore the element size of 3 mm was used throughout the tie rod strength analysis.

Table 15. Convergence test on tie rod.

Element size (mm)	Number of elements	Equivalent Stress (MPa)	Error (%)
15	849	23.49	-
10	1,950	31.90	26.35
8	3,531	23.75	34.32
6	7,722	19.26	23.32
3	60,785	18.92	1.77
2	203,429	18.77	0.81

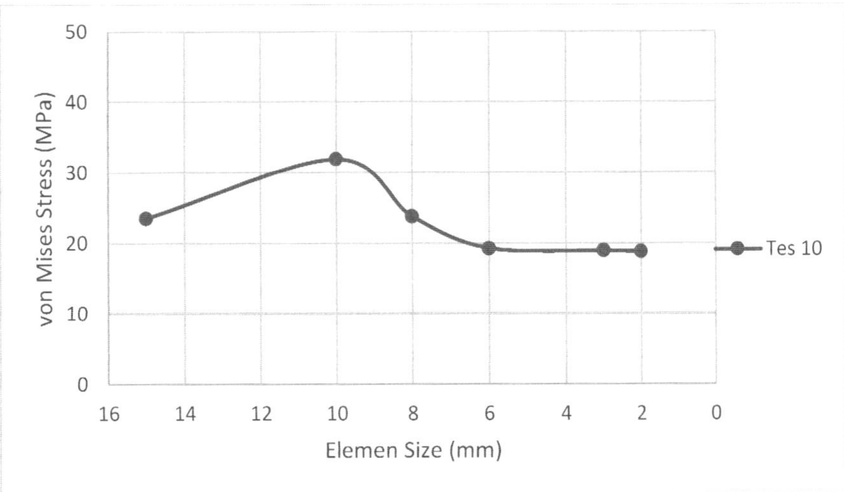

Figure 25. Convergence test on tie rod graph.

The forces on load case 2 were validated by calculating the normal stress on point B located on the tie rod where a stress probe was placed using ANSYS, as shown in Figure 26. The normal stress on the x-axis calculated from simulation was -2.96 MPa. For the theoretical calculation, the tie rod was cut at point B to calculate the area moment of inertia and the normal stress as shown in Figure 27. The normal stress on point B on the x-axis calculated theoretically was -3.05 MPa. The stress calculation is shown in Table 16. Since the error between the simulation and the theoretical calculation result is below 10%, this simulation can be considered valid based on the "*Concept of Model Verification and Validation*" [8].

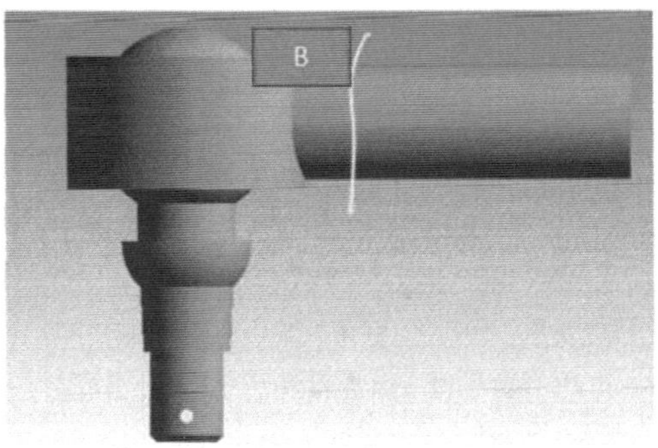

Figure 26. Stress probe on tie rod.

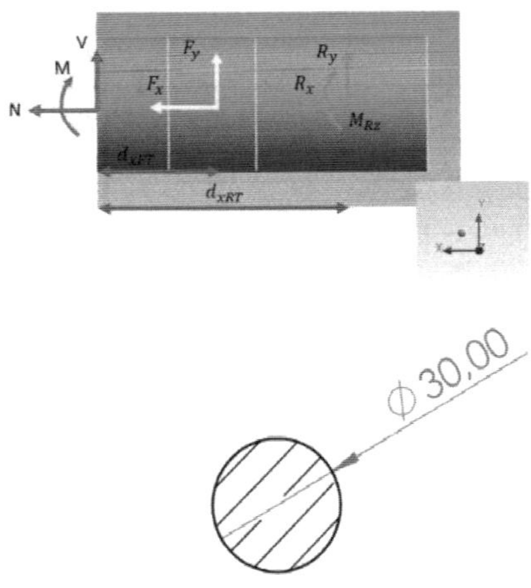

Figure 27. Cross section of tie rod.

IOP Conf. Series: Materials Science and Engineering **694** (2019) 012011 doi:10.1088/1757-899X/694/1/012011

Table 16. Theoretical calculations on tie rod.

No	Parameter	Value	Unit
1	F_x	1,795.87	N
2	F_y	-22.74	N
3	R_x	0	N
4	R_y	75.95	N
5	M_{Rz}	2,250.29	N.mm
6	d_{xFT}	24.95	Mm
7	d_{xRT}	54.95	Mm
8	N	-1,795.87	N
9	V	-53.21	N
10	M	1,355.56	N.mm
11	A	706.86	mm^2
12	I	39,760.78	mm^4
13	c	15	mm
14	σ_x (theoretical)	-3.05	MPa
15	σ_x (simulation)	-2.96	MPa
16	Error	2.98	%

The equivalent stress distribution for load case 2 on the tie rod is shown in Figure 28. The calculated equivalent stress on the tie rod and the safety factor of the tie rod were 118.54 MPa and 4.42 respectively, where the equivalent stress was below 525 MPa. There was a critical point on the tie rod as a result of each loading, as shown in Figure 29. The mean stress to ultimate tensile strength ratio and the amplitude stress to endurance limit ratio on this critical point from load case 2 were calculated using their maximum and minimum principal stresses and then plotted on the modified and tension diagram.

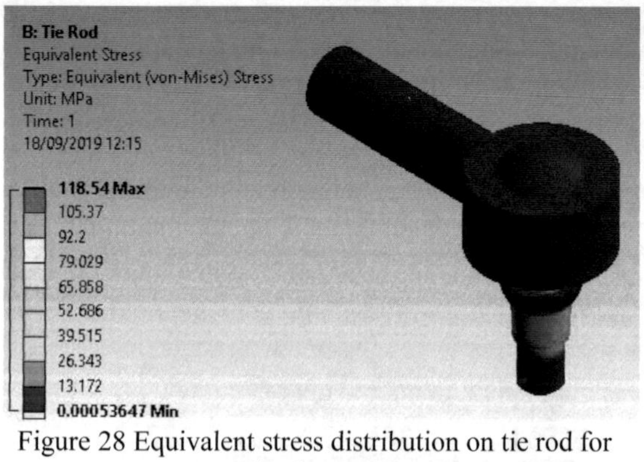

Figure 28 Equivalent stress distribution on tie rod for load case 2

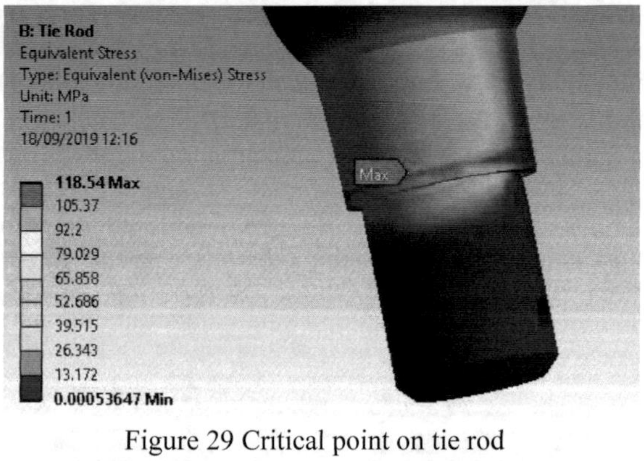

Figure 29 Critical point on tie rod

The value of S_{ut} is 780 MPa therefore the S_e' is 390 MPa. By applying a correction factor of 0.75 on the S_e' base on PM 175 2015, the calculated S_e was 292.5 MPa. The mean stress to ultimate tensile strength ratio and the amplitude stress to endurance limit ratio on the critical point of the tie rod are shown in Table 17. These values were plotted on a modified Goodman compression and tension diagram, as shown in Figure 30. From this graph, it can be seen that all of this point falls below the ABC line, which means that no crack or failure occured on the tie rod.

Table 17. Ratio of Mean Stress and Amplitude Stress on Tie Rod

	σ_{pmax} (MPa)	σ_{pmin} (MPa)	σ_m (MPa)	σ_a (MPa)	$\frac{\sigma_a}{S_e}$	$\frac{\sigma_m}{S_{ut}}$
Tie rod (load case 2)	166.27	59.57	53.35	112.92	0.18	0.14

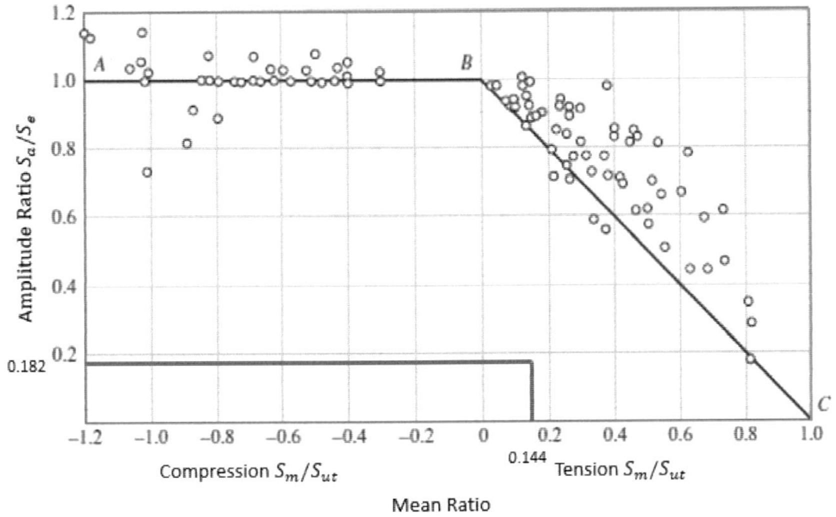

Figure 30. Mean stress to ultimate tensile strength ratio and amplitude stress to endurance limit ratio on tie rod plotted on modified Goodman compression and tension diagram [9].

4. Conclusion

Based on the analysis, the knuckle plate and the tie rod of Metro Kapsul met the requirement on PM 175 2015 since none of the stresses calculated exceeded 525 MPa. Both components passed the failure criteria of modified Goodman for each load case. In the knuckle plate, the highest stress occurred on load case 2 and the highest deformation occurred on load case 1. The safety factors of the knuckle plate calculated using load cases 2 and 1 were 2.68 and 2.78, respectively. The safety factor of the tie rod was 4.42.

5. Acknowledgment

This work was carried out at the Mechanical Engineering Department Engineering Design Center Laboratory, Faculty of Mechanical and Aerospace Engineering, Institut Teknologi Bandung and was financially supported by PT Trekka and USAID under SHERA Program and Institut Teknologi Bandung through P3MI (Program Penelitian Pengabdian Masyarakat dan Inovasi).

References

[1] Statistics Indonesia (BPS) 2018 *Population Growth Rate* (http://sp2010.bps.go.id/.)
[2] Junior I 2019 *Kinematics Analysis of Metro Kapsul Bogie* (Bandung: Intstitut Teknologi Bandung)
[3] Redrouthu B M and Das S 2015 *Tyre Modelling for Rolling Resistance* (Gothenburg: Chalmers University of Technology)
[4] Nugroho H A 2018 *Fatigue Analysis of Metro Kapsul's Guide Wheel Frame* (Bandung: Institut Teknologi Bandung)
[5] Union Internationale des Chemins de Fer 615-4OR 2003 *Motive Power Units – Bogies and Running Gear – Bogie Frame Structure Strength Tests*, ed 2 (Paris: International Union of Railways)
[6] Ministry of Transportation of Indonesia 2015 *Peraturan Menteri Perhubungan Republik Indonesia: Nomor 175 Tahun 2015 Tentang Standar Spesifikasi Teknis Kereta Kecepatan Normal dengan Penggerak Sendiri* (Ministry of Transportation of Indonesia)
[7] Sinclair G B, Beisheim J R and Sezer S 2008 *Practical Convergence Divergence Checks for Stresses from FEA* (International ANSYS Conference)
[8] Los Alamos National Laboratory 2004 *Concepts of Model Verification and Validation* (Los Angeles: Los Alamos National Laboratory)
[9] Budynas R G and Nisbett J K 2011 *Shigley's Mechanical Engineering Design*, ed 9 (New York: McGraw-Hill)

IOP Conf. Series: Materials Science and Engineering **694** (2019) 012012 doi:10.1088/1757-899X/694/1/012012

A damage sensitive feature: Beam case

Fergyanto E Gunawan[1] and Nobumasa Sekishita[2]

[1]Industrial Engineering Department, BINUS Graduate Program - Master of Industrial Engineering, Bina Nusantara University, Jakarta, Indonesia 11480
[2]Department of Mechanical Engineering, Toyohashi University of Technology, Toyohashi, Aichi 441-8580, Japan

E-mail: fgunawan@binus.edu

Abstract. With the adoption of damage tolerance design principle, the health monitoring system has become an integral part of the engineering structure in operation. For the system to work, damage indicator that describes the structural integrity level should be established and monitored. Damage indicator is usually derived from structural responses. Many quantities have been proposed for damage indicator. Some popular damage indicators are natural frequency, mode shape, curvature, strain energy, and t-, F-, and z-statistics. In this paper, we propose a new damage indicator derived from the theory of the strength of materials. We evaluate the proposal for a case involving damage on a simply-supported beam. The required data are established numerically by the finite element method for structure in healthy and damaged conditions. The results suggest that the current proposal is more sensitive to damage than the natural frequency.

1. Introduction

The system of Structural Health Monitoring (SHM) has been implemented in various engineering structures, such as bridges [1–3] and airplanes [4–6]. Damage indicator is required in order to build an SHM system. Natural frequency and mode shape are two most widely used damage indicators [7–13]. The power spectral density was deployed by Ref. [14]. Lately, the SHM system is also built by deploying machine learning techniques [15–17, 4].

The machine learning approach generally requires a large amount of data and intensive computational resources. They often are unavailable on the remote sites where SHM is required.

This paper intends to propose a new indicator that is easy to compute and is better in detecting damage. The indicator is derived by applying the theory of the strength of materials, which is well understood.

2. Research method

This paper proposes a damage indicator derived from the theory of the strength of materials. Notably, it focuses on the Euler-Bernoulli beam theory. Thus, the derived indicator is useful for the SHM of beam-like structures.

The Euler-Bernoulli beam theory is accurate when the beam deflection is small such that the beam cross-section remains flat upon deformation. Also, the section is still perpendicular to the beam neutral plane.

With those assumptions, the theory links the applied distributed loads onto the beam $q(x)$ to deformation $w(t, x)$ with the second-order differential equation of

$$EI\frac{\partial^4 w(t,x)}{\partial x^4} = -\rho\frac{\partial^2 w(t,x)}{dt^2} + q(x) \tag{1}$$

The symbol E denotes the beam elastic modulus, I denotes the second moment of area of the beam cross-section, and ρ is the material density.

The dynamic equilibrium equation is applicable for any point at the beam at any time. We hypothesize that a deviation from the condition may signify a change in either the material properties or the beam geometry or both. Corrosion or crack may alter beam cross-section. Thus, the change may reflect the deterioration in material integrity.

In this work, we use the Euler-Bernoulli theory not to predict a beam deformation, nor to predict the exerted force, but to estimate the beam integrity. For the purpose, we propose:

$$d(t,x) = \left|EI\frac{\partial^4 w(t,x)}{\partial x^4} + \rho\frac{\partial^2 w(t,x)}{dt^2} - q(x)\right| \tag{2}$$

We hypothesize when the beam is intact, and E, I, and ρ are assigned with values on the intact condition, the damage indicator $d(t,x)$ should be zero or very small. The indicator may shift from zero when the beam contains damages. We assume that damages alter the beam deformation $w(t,x)$.

For SHM purpose, we have the freedom to choose the observation or measurement point. We may select the point where the external load is absent. Thus, it simplifies the computation of the damage indicator.

In this paper, we evaluate the above proposed indicator with a case of damage on a simply-supported beam. The beam has 300-mm length, 20-mm width, and 20-mm height. The beam and its discretization are shown in Fig. 2. It is made of steel material with the elastic modulus of 207 GPa, the Poisson ratio of 0.3, and the material density of 8050×10^{-9} kg/mm^3. The beam is subjected to a harmonic load $f(t) = 100 \cdot \sin(2\pi f_p t)$ at the location $x = 100$ mm with $f_p = 2$ kHz.

3. Results

In this paper, we propose the formula in Eq. (2) for SHM. Our early work suggested the formula was able to detect damage when the measurement was performed near the location of damage [18]. It was evaluated on simple cases involving lumped-mass systems. In this paper, we evaluate the theory for detecting the damages in a beam.

For the reason, we focus on the six measurement points depicted in Fig. 2. The points position around the damage. The damage spans the segment between $x = 5$ mm and $x = 60$ mm. The six points are located at $x = 51$ mm, 54 mm, ⋯, 63 mm. Thus, two points are at the ends of the damaged segment, two points on the left of the damaged element, and two points on the right.

For the first result, the beam displacements in the lateral direction at the six measurement points are shown in Fig. 1 for healthy and damaged conditions.

The profiles look rather similar because the six measurement points are separated by the small distance of 3 mm only. It is interesting to witness that the profiles for healthy and damaged conditions look rather identical. The displacements are practically not affected by the five-percent stiffness degradation on Element 20. The fact suggests the quantity is not suitable for damage detection.

For the second result, we focus on the natural frequency. The quantity is computed by solving $\det(\mathbf{K} - \omega^2\mathbf{M}) = 0$, where $\mathbf{K}$ is the stiffness matrix, $\mathbf{M}$ is the mass matrix, and ω is the natural frequency. The computed natural frequency is precise. The approach avoids errors that may occur when the quantity is reduced from the frequency analysis. The computed natural frequency is shown in Table 1. Only the first nine modes are provided. The table presents the quantity for healthy and damaged conditions. On the last column, the change of the frequency due to the damage is provided. The results

IOP Conf. Series: Materials Science and Engineering **694** (2019) 012012 doi:10.1088/1757-899X/694/1/012012

suggest that the change of the natural frequency is too small. The damage does not lead to a significant change in the natural frequency.

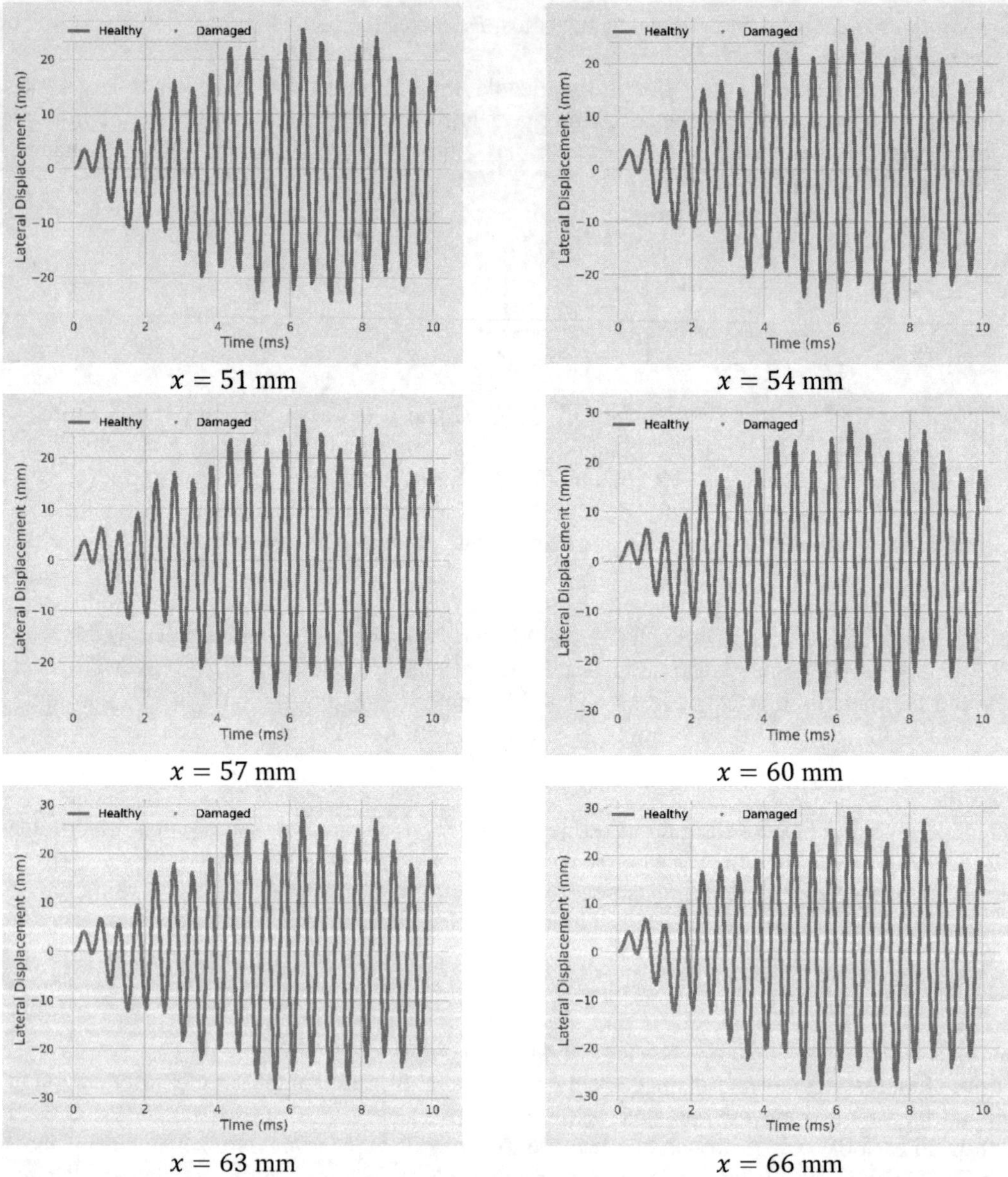

Figure 1. The lateral displacements at the six observation points for the healthy and damaged conditions.

Table 1. The natural frequencies for the healthy and damaged conditions

Modes	Natural Frequency (kHz)		
	Healthy Condition	Damaged Condition	Change (%)
1	0.507	0.507	0.037
2	1.987	1.982	0.095
3	4.226	4.222	0.101
4	4.327	4.323	0.095
5	4.849	4.847	0.037
6	7.382	7.379	0.048
7	9.699	9.690	0.098
8	11.003	11.000	0.022
9	12.678	12.673	0.041

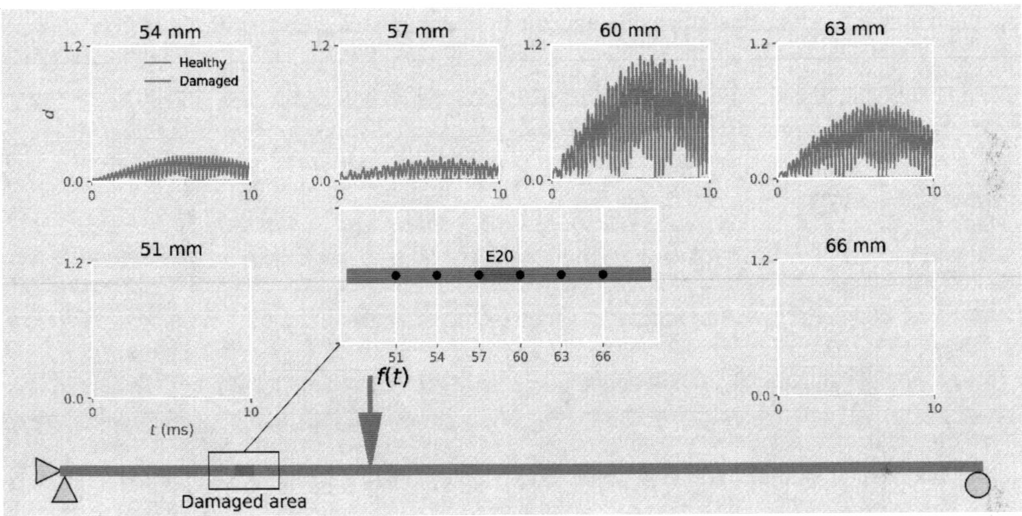

Figure 2. The indicator d computed at the observation points of 51 mm, 54 mm, 57 mm, 60 mm, 63 mm, and 66 mm for the healthy and damaged conditions. The damaged element spans between $x =$ 57 mm and $x =$ 60 mm. The duration of analysis is 10 ms.The lateral displacements at the six observation points for the healthy and damaged conditions.

The natural frequency is well known to be sensitive to damage. Theoretically, damage at any point at the beam should affect the quantity. With this nature, the indicator is said to be global in detecting damage. However, for the present case, it becomes apparent that sufficiently large damage is required to be detectable by the approach. As the natural frequency is proportional to the square root of the stiffness, a 75-percent reduction in the stiffness only changes the natural frequency by half.

For the third result, we present the current proposal of the damage indicator. We compute the indicator at the six observation points. The results are shown in Fig. 2. We observe the indicator closely and obtain the following conclusions. The values of the indicator computed at the six observation points are always zero for the healthy condition. As for the damaged condition, the indicator computed at $x =$ 60 mm is the most sensitive to the damage. The point is on the right end of the damaged segment. The sensitivity to the damage quickly drops by increasing the distance from the damaged element.

IOP Conf. Series: Materials Science and Engineering **694** (2019) 012012 doi:10.1088/1757-899X/694/1/012012

4. Conclusion

In this paper, we propose a damage indicator for assessing the change in structural condition. In general, the indicator is derived from our understanding of mechanics of materials and theory of elasticity. With those theories in mind, we understand and can predict the dynamical behavior of engineering structure upon receiving external loads. The deviation of responses from our prediction may signify alteration of the structural integrity. We evaluate our theory with data collected from a vibration analysis of a simply supported beam. The vibration is simulated numerically by the finite element method. To simplify the issue, we alter the elastic modulus of an assumed damaged element by a small amount. We analyze the structural responses on the healthy and damaged conditions. Then, we compute the proposed damage indicator. We witness the sensitivity of the indicator to the altered element. We also evaluate the change of the natural frequencies due to the damaged element and witness that the change is too small for the frequencies to be affected. With that example, we conclude that the proposed indicator can potentially detect structural changes better than the tradition method based on the natural frequency.

References

[1] You T, Li P, chen Y and Ye H 2012 *Journal of Convergence Information Technology* **7** 28–35
[2] Yu L, hua Zhu J, and Yu L 2013 *Advances in Structural Engineering* **16** 207–2017
[3] Moser P and Moaveni B 2013 *Experimental Techniques* **37** 15–26
[4] Mohanty JR, Verma BB, Parhi DRK and Ray PK 2009 *Archives of computational materials science and surface engineering* **1** 133–138
[5] Gope D, Gope PC, Thakur A and Yadav A 2015 *Applied Soft Computing* **30** 514–528
[6] Alamdari MM, Rakotoarivelo T and Khoa NLD 2017 *Mechanical Systems and Signal Processing* **87** 384–400
[7] Mehrjoo M, Khaji N, Moharrami H and Bahreininejad A 2008 *Expert Systems with Applications* **35** 1122–1131
[8] Yu L, Zhu JH and Yu LL 2013 *Advances in Structural Engineering* **16** 207–217
[9] dos Santos F, Peeters B, der Auweraer HV, Goes L and Desmet W 2016 *Mechanical Systems and Signal Processing* **3** 22–27
[10] de Medeiros R, Sartorato M, Vandepitte D and Tita V 2016 *Journal of Sound and Vibration* **383** 171–190
[11] Quinones MM, Montejo LA and Jang S 2015 *Int J Adv Struct Eng* **7**
[12] Pnevmatikos NG and Hatzigeorgiou GD 2017 *Bulletin of Earthquake Engineering* **15** 227–248
[13] Pedram M, Esfandiari A and Khedmati MR 2017 Journal of Sound and Vibration
[14] Gunawan FE 2019 *International journal of innovative computing, information and control* **15** 1717–1727
[15] Nick W, Shelton J, Asamene K and Albert E 2015 A study of supervised machine learning techniques for structural health monitoring Proceedings of the 26th modern AI and cognitive science conference vol 1353 (Greensboro, NC, USA) pp 133–138
[16] Mohanty J, Verma B, Ray P and Parhi D 2011 Expert Systems with Applications 38 12302–12311
[17] Mohanty J, Verma B, Ray P and Parhi D 2010 Expert Systems with Applications 37 3075–3087 ISSN 0957-4174
[18] Gunawan FE 2019 ICIC Express Letters 13 931–939

IOP Conf. Series: Materials Science and Engineering **694** (2019) 012013 doi:10.1088/1757-899X/694/1/012013

Closed Drain Transfer Pump Failure Analysis using Root Cause Failure Analysis (RCFA) Method

M F Hidayat[1] and Warjito[1]

[1]Mechanical Engineering Department, Faculty of Engineering, Universitas Indonesia, Kampus Baru Depok, Jawa Barat, Indonesia 16424

E-mail: hidayat.farrell@gmail.com

Abstract. Maintenance is a process carried out to maintain the reliability of individual/group assets. Positive displacement type pumps are commonly used in closed drainage systems as transfer pumps in oil and gas processing plants. However, there are times when the pump fails which requires further effort to prevent the failure from recurring. Efforts that can be made to overcome this are using the method of Analysis of the Root Failure (Root Cause Failure Analysis). The RCFA method uses recollection of evidence related to design, procurement, and operation of a tool to find root causes and latent causes of problems. As a result of gaps found between the design process, procurement which leads to the root cause of pump failure, this study recommends pump-operating companies to pay more attention to piping installation to support pump performance and coordination between owner, design, and procurement to prevent this failure from recurring.

1. Introduction

In offshore oil field operations, a drainage system is used which in the context of production operations is tasked with collecting hydrocarbon remnants from the production platform and disposing of them without creating safety risks. In this system, one of the equipment that has an important role is the pump that has the duty to drain fluid and condensate from the flash drum to the liquid burner.

However, in reality, there are still failures at the drainage system pump that can potentially interfere with the performance of the platform or even threaten the safety of workers and the environment in the event of a malfunction of the hydrocarbon waste disposal system. In addition, we know that one of the incidents of upstream oil and gas facilities which caught the public's attention was the Deepwater Horizon Mobile Offshore Drilling Unit (MODU) accident in 2010 [1]. Therefore, as an application of corrective and preventive measures for failure of upstream oil and gas equipment, this study applies the Root Cause Failure Analysis (RCFA) method to be used as a failure analysis method which, in addition to being able to show the root causes of failure, also applies recommendations to prevent such failures from reoccurring, especially in hydrocarbon drainage pump systems.

2. Methods

2.1. Root Cause Failure Analysis (RCFA)

This research employs the Root Cause Failure Analysis (RCFA) method. RCFA is the process of identifying the main root causes of a failure and using that information to determine corrective actions and/or prevent them. Shown below in Figure 1 is a simplified flowchart of a RCFA process.

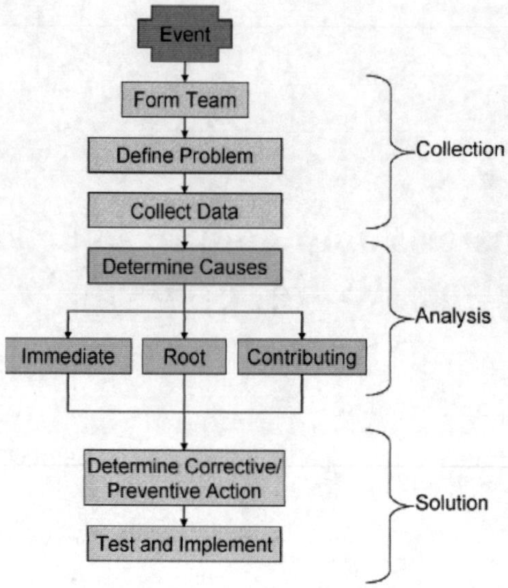

Figure 1. Simple Root Cause Failure Analysis flowchart. [2]

In this study, the system boundary that is the object of research is the progressive cavity positive displacement pump P-6450A and P-6450B closed drain transfer pump unit systems in company P field. This limitation is done because this equipment has a high level of criticality, of which the closed drain transfer pump is assigned to drain the remaining condensate of natural gas production from the KO drum to the liquid burner.

During the data collection process, there might be 3 kinds of data that could be encountered, such as shown in Table 1.

Table 1. Types of data commonly found in RCFA data recollection. [3]

Physical Evidence	Failed components, new (undamaged) components, deposit samples, fluid samples, filter samples
Recorded Evidence	Pictures, operator logs, SCADA logs, field vibration measurements, environmental conditions, inspection reports
Personal Testimony	Operators, maintenance staff, machinists, engineers, managers

The RCFA case investigation, besides referring to the stages proposed by Ransom [2], also uses the concept proposed by Mobley [3]. The first stage carried out in conducting research is to coordinate with company P to conduct an RCFA case review. Furthermore, the case that will be reviewed for RCFA will be the subject of research. Then, the RCFA team for this case was formed with their respective positions and scope of work which included investigations, facilitators, technical inputs, and supervision of RCFA work.

The RCFA process begins with Incident Reporting, where at least incident reporting includes the date, reporter, incident description, specific location and equipment, preliminary causes, and corrective actions taken.

Then, the steps taken are the classification of incidents and from that classification the problem statement of the case under review is determined. According to SINTEF, the severity of failure can be divided into critical failure, degraded failure, incipient failure, and unknown failure [4]. After the problem statement is determined, data is collected that can be in the form of interviews, physical evidence in the form of equipment operating procedures, company policies related to equipment operation, records of operation and maintenance, P&ID of equipment, photo and / or video documentation, evidence of damage, records of the operating conditions of the equipment, as well as reports of incidents that have happened before.

After data recollection is completed within a certain period, existing data is then analyzed to determine the root causes of physical causes, human causes, and latent causes. The tools used in the analysis are Fault Tree Analysis (FTA) and fishbone diagrams. From the root causes that have been found, recommendations and implementation of corrective actions are determined to mitigate the impact of failures and preventive actions to prevent problems with the same root causes from being repeated.

3. Results

3.1. RCFA Problem Definition

The P-6450A / B pump is a screw displacement positive displacement pump located in the closed drain system at the company's Onshore Receiving Facility P pump. The P-6450A / B pump is in charge of flowing the condensate leftover from pig receivers, inlets and blowdown from separator inlets , superheater for fuel gas, as well as sales metering gas package to the liquid burner. This equipment is classified according to the company as having a medium severity level, where the failure of this equipment has the potential to disrupt the productivity of the facility with the opportunity for a not too large security risk.

According to the results of the author's interview with the field technician and related documents, on December 1, 2016 the field technician reported that the pressure transmitters installed in pumps A & B recorded a discharge pressure of 3.0 barg and 2.0 barg respectively, which is a number far below figures for the supplier of the pump unit. According to the pump sheet data from the supplier, the pump unit should be able to produce a discharge pressure of 15.27 barg under normal conditions. Because the pump cannot build pressure, the field team turns off the pump and starts the troubleshooting process.

In the troubleshooting process carried out by the field team, the first step taken is to check the condition of the suction strainer at the pump. After the suction strainer condition is checked, the field team conducts a troubleshooting experiment by closing the VX-6441 and VX-6442 valves of the P-6450A and P-6450B pumps respectively, and draining water from drums placed above the suction strainer of each pump. This is done to ascertain whether this problem is coming from a suction valve that is not optimal.

After the troubleshooting experiment, the data obtained that the discharge pressure generated by P-6450A is 7 barg and P-6450B is 5 barg, which proves that the discharge pressure problem still appears even though the pump is in a by-pass condition. After these steps, the field team decided to dismantle the pump and the results found a kind of debris inside the pump and the condition of the pump stator was in a bad state, as shown in Figures 2 and 3.

As a result of this series of events, the operations engineering team began to initiate the RCFA process and form a team tasked with handling this case.

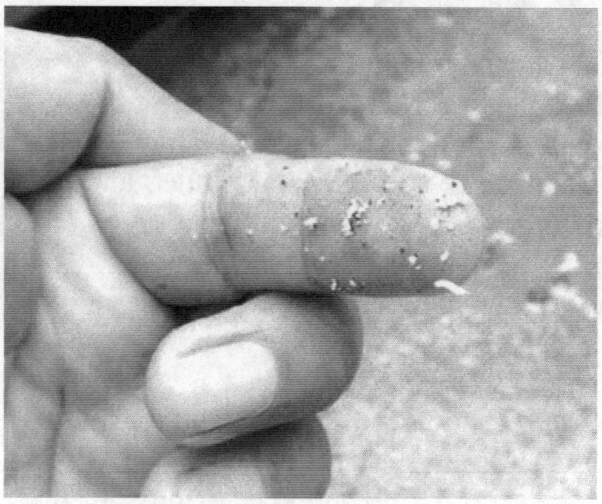

Figure 2. Flakes present after the casing is opened.

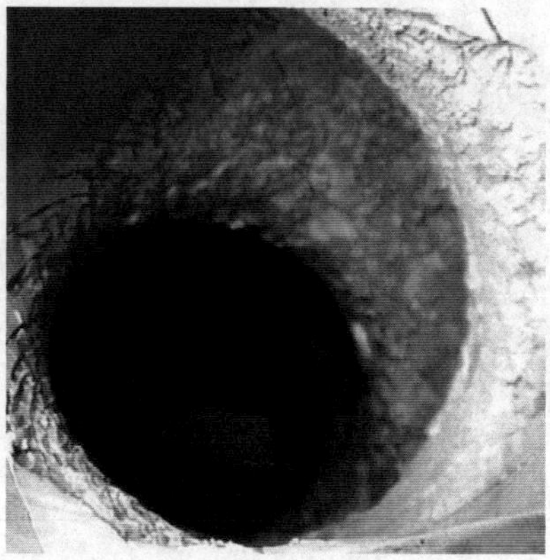

Figure 3. Pump stator conditions after the casing is opened.

3.2. Data Recollection

According to the pump datasheet, the P-6450A / B pump is an AE2N-200ID Allweiler pump capable of producing a discharge pressure of 15.27 barg from a suction pressure of 0.13 barg, with an NPSHr of 2.05 m. This pump is supplied by an electric motor from Siemens with 5.5 kW power and mechanical energy transfer via John Crane gearbox with maximum torque of 315 Nm. Construction and pump material use SS 316 for pump casing, Building for stator and SS 316 TI for rotor and drive shaft from pump. By adapting the ISO 17776 standard [5], this equipment is determined to have a risk category C1.

In the process of data recollection, author found several kinds of things which according to the writer are anomalies that occur in the process of design, procurement, and installation and operation. This will be crucial later in determining the root cause (root cause) of this case. The data collection process is carried out in the following ways:

3.2.1. Interview

The author conducts direct interviews with related crews about this problem and extracts data that can help.

3.2.2. Physical Evidence Recollection

The author recollects physical evidence in the form of photos and documents that can help the writer do the Root Cause Failure Analysis process. During this research period, the authors managed to obtain the following documents:

- P&ID in accordance with ANSI rules [6] for closed drain systems where the system is the functional location of the P-6450A/B pump,
- Pump design calculations and pump datasheet,
- Datasheet for condensate flowing through the pump,
- Factory Acceptance Test from the pump fabricator,
- Isometric picture of piping systems from V-6450 to P-6450A/B, and
- Standard Operating Procedure for manual operation of P-6450A / B pumps.

4. Discussion

The authors used the Fault Tree Diagram and Events and Causal Factor (Ishikawa) Diagrams to find the root cause of problems in the P-6450A / B with the following results, as shown in Figures 4 and 5.

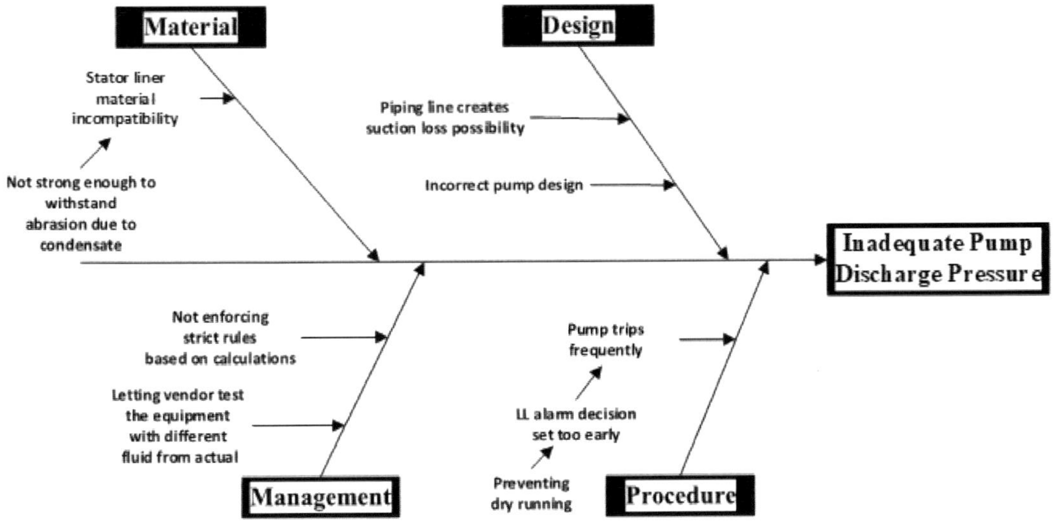

Figure 4. Cause and Effect diagram of Case P-6450A/B.

From the data recollection, the writer found that there were some irregularities during the procurement, installation and operation of the P-6450A/B pump with the following explanation::

4.1. Standard Differences Used

In the Factory Acceptance Test document, it is found that the pump vendors and pump manufacturing plants use the API RP 674 standard as a reference standard in testing P-6450A / B pumps. According to the authors, this is a mistake because API RP 674 is the standard used for reciprocating type positive displacement pumps [7], as shown in Figure 6.

Manufacturer	: [redacted]
Client	[redacted]
Type	: Progressing Cavity
Speed (Rated)	: 268 Rpm
Used Standard	: API-674 Std
Suct./Disch. Press.	: 0.13/15.27 barG (Rated)
NPSH-A	: 4.83 m
NPSH-R	: 2.05 m
Capacity	: 6.1 m^3/h
Differential Pressure	: 15.2 bar
Serial No.	: 13040931
FAT Location	: Allweiler Pump, Kirchhellener Ring 77-79 46244 Bottrop Germany

Figure 5. Incorrect use of standards in P-6450A / B pump testing procedure.

For positive displacement pump types of progressive cavity, the right target standard for use of the API is API Standard 676, which has a special scope in the design, material selection, testing, delivery, installation, and operation of the rotary type positive displacement pump, which also includes positive displacement pump types of progressive cavity [8].

IOP Conf. Series: Materials Science and Engineering **694** (2019) 012013 doi:10.1088/1757-899X/694/1/012013

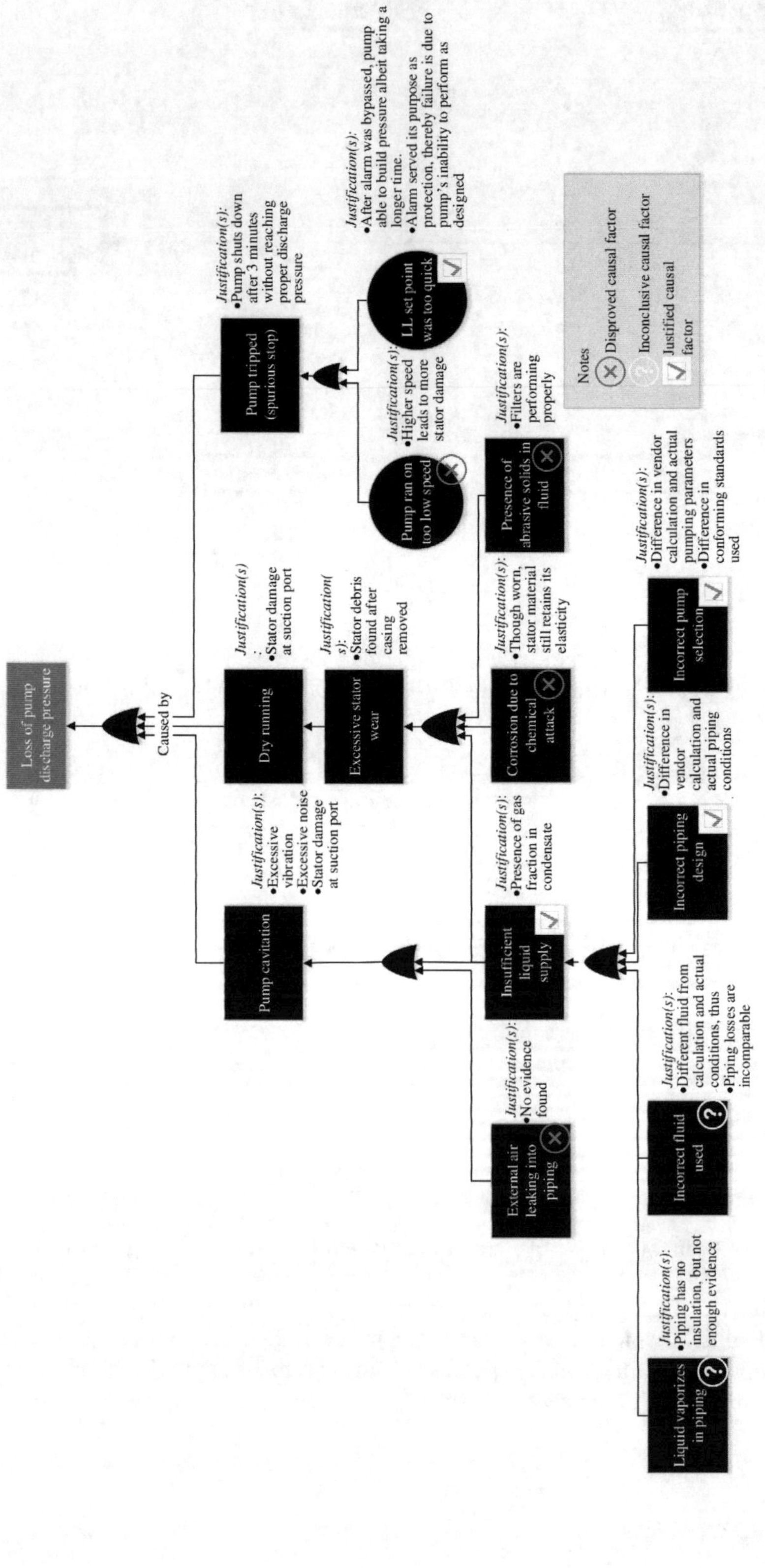

Figure 6. *Fault Tree Analysis* of case P-6450A/B

IOP Conf. Series: Materials Science and Engineering **694** (2019) 012013 doi:10.1088/1757-899X/694/1/012013

4.2. Difference in Working Fluid during Factory Acceptance Test with Routine Operating Conditions

When the vendor performs a Factory Acceptance Test (FAT) at the pump manufacturing plant, the test is carried out using ordinary tap water working fluid which has different parameters from the condensate used in routine operating conditions. This is evidenced by document '11 -ORF-M-REP-06-0004_FAT Report for Closed Drain Transfer Pump_0_AFC_NC' and pump specification document where the FAT document listed the test fluid is water while the routine condensate consists of a mixture of various hydrocarbon chains and almost none. water. This is crucial considering that the two fluids have different specific gravity and have different vapor pressure and friction loss so as to allow a gap for problems when the fluid used in operation is different from the fluid used during testing, as shown in Figure 7.

Prufmedium / test liquid:	Wasser / Water
Mediumtemp. / liquid temp.:	20 °C
Luftdruck / atmospheric pressure:	1011 mbar

Einstellungen und Ergebnisse / settings and results

Druck saugseitig / suction pressure, P_s =	0,1	bar	ΔP =	15,2	bar
Druck drucks. / discharge pressure, P_D =	15,3	bar			
Drehzahl / rpm =	268	min^{-1}			
Fördermenge ohne Unterdruck / flow rate without underpressure =	106	L/Min	6,36	m^3/h	
$NPSH_R$ =	2,05	m			
Fördermenge bei eingestellten Unterdruck / flow rate at set underpressure =	104,5	L/Min	6,27	m^3/h	
Druck saugseitig / suction pressure, P_s =	-0,795	bar	ΔP =	15,2	bar
Druck drucks. / discharge pressure, P_D =	14,405	bar			
Fördermenge min. / min. allowable flow rate =	102,82 (-3%)	L/Min	6,17	m^3/h	

Wir bestätigen hiermit eine Fördermengenreduzierung ≤ 3% und somit einen /
We herewith confirm a reduction of the flow rate by ≤ 3% and therefore a

Pumpen / pump $NPSH_R$ =	2,05	m

Figure 7. Working fluid used when testing.

4.3. Empty Pre-Commissioning Check List

In the document '11 -ORF-M-LST-06-0006_Pre Commissioning Check List for Closed Drain_A_IFR_NC 'does not contain anything that shows the ability of the system pumps to carry out their duties in accordance with the expectations of the owner. Meanwhile, the project implementation process consists of:

- Detailed design
- Fabrication
- Individual Performance Test (IPT)
- Pre-commissioning
- Commissioning and System Performance Test
- Start-up

From this evidence, indications of negligence both from the vendor as the equipment supplier, the EPCC company as the general contractor and the owner company as the user of the tool in ensuring the appropriateness and capability of the equipment before entering operational conditions.

4.4. Differences in Piping Design and Installation

Cavitation and dry-running can occur when the pump does not have enough fluid in the suction section [9]. Due to the lack of incoming liquid fluid, air bubbles will form on the suction part of the pump because the remaining fluid reaches its vapor pressure [9]. The old piping system that was installed allows this to happen, because there is a reverse U-loop in the pipeline without the venting process in the air phase of the condensate as shown in Figure 8. The venting process as stated in the SOP for the pump exists but the process is carried out after pumping process so that it still allows the presence of airlock in the pipeline. thereby reducing the liquid phase discharge entering the suction side of the pump.

Figure 8. The shape of the pipeline that allows airlock.

The presence of air 'stuck' in the piping system (water entrainment) can also explain why the same pump takes longer to build a discharge pressure in accordance with the provisions stipulated in the DCS system, because the presence of air entrainment reduces the effective diameter of the pipe which can be passed through the liquid so that it takes longer to meet the suction side of the pump. This lack of suction flow results in a combined effect of cavitation and dry-running due to the inability of the pipeline to supply sufficient fluid for pump operations to occur at this pump as a result of the existence of the reverse U-loop pipe. In the condition of water entrainment, bubble gas has formed before the process of phase change from liquid to gas. While the cavitation process can still occur even if the positive displacement pump is not as severe as the centrifugal pump in its detection [9]. Dry running can be proved by damage to the stator liner in the form of stator material flakes, but with the drastic change in the composition of the working fluid it can also prove that the stator is not damaged by chemical reactions, but a mechanical reaction to the swelling of the stator due to dry-running.

Figure 9. Pump driveshaft condition of P-6450A/B.

This results in the stator rubbing against the rotor which gradually erodes the stator layer so that debris appears when the P-6450A / B pump is disassembled. Other additional evidence is the condition of the pump suction hole which also shows the presence of a drive shaft that is eroded, supporting the claim that there is a combination of damage between dry-running and cavitation at the P-6450A / B pump, as shown in Figures 9 and 10.

Figure 10. Suction port condition of pump P-6450A/B.

According to Dillon [10], damage to the progressive cavity pump stator can be caused by both cavitation and dry-running, but the difference is in the case of cavitation, the stator damage starts at the suction side and runs to the pump discharge side, while in the case of dry-running, stator damage occurs starting from the middle side of the pump between the suction and discharge sides.

5. Conclusions

5.1. Root Cause Determination

Based on the analysis conducted, the conclusion that can be drawn is that there are 3 types of root causes of the problems that occur at the P-6450A / B pump, namely:

5.1.1. Physical Cause

The physical cause of this problem is the presence of a reverse u-loop line in the pipeline without venting so that the pump has a fluid flow deficiency on the suction side.

5.1.2. Human Cause

The human cause of this problem is the contractor's error in calculating the conformity of pump specifications and designing pump pipeline.

5.1.3. Latent Cause

The latent cause of this problem is management's negligence in applying strict regulations regarding the design and installation of equipment.

5.2. Corrective and Mitigation Actions

As a form of output from the RCFA methodology which requires the results of a solusional investigation, the corrective and mitigative steps taken by the RCFA team are as follows:

- For physical causes, the corrective steps taken are eliminating U-shaped inverted pipelines and replacing them with straight pipe lines to reduce losses and prevent air trapping, as shown in Figure 11.
- Maintains KO drum levels so that the hydrostatic pressure and head are able to resist the head loss of the piping system.

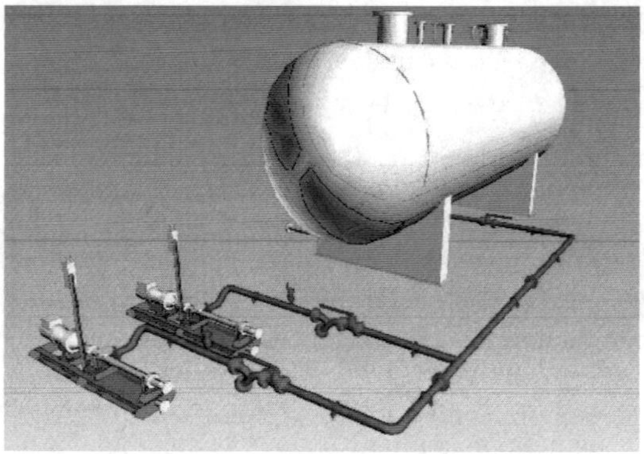

Figure 11. Reverse U-Loop elimination on the pipeline to the P-6450A/B pump.

For human and latent causes, both the owner as the operator and the vendor as the tool provider are responsible for conducting a thorough design review, with important points including:

- The vendor is responsible for recalculating the pump specifications and matching the needs of the owner.
- The owner is responsible for maintaining the operational and maintenance conditions of both the pump and the support system such as piping and KO drum itself, bearing in mind that there are still symptoms of trip on the pump due to setting LL alarms that are too tight.
- Using updated and relevant standards during new equipment design, procurement, and installation processes.

5.3. Evaluation of RCFA Implementation

From the stages of Root Cause Failure Analysis that has been done, the authors propose a number of suggestions which could help the implementation of RCFA in the future at PT. P, which includes:

- Safeguard evidence at every failure occurrence, because of the evidence can help the process of analysis of failures that occur and can be supporting arguments in determining the root of the problem.
- Improve coordination and cooperation between vendors, EPCC contractors, and PT. P as the operator, because it was seen from the design process to the installation there were misunderstandings between each party.

IOP Conf. Series: Materials Science and Engineering **694** (2019) 012013 doi:10.1088/1757-899X/694/1/012013

References

[1] L. Kaufman, «Search Ends for Missing Oil Rig Workers,» 23 April 2010. [En ligne]. Available: https://www.nytimes.com/2010/04/24/us/24spill.html. [Accès le 1 11 2018].

[2] D. L. Ransom, «A Practical Guideline For A Successful Root Cause Failure Analysis,» chez *Turbomachinery and Pump Symposia*, College Station, 2007.

[3] R. K. Mobley, Root Cause Failure Analysis, Oxford: Butterworth-Heinemann, 1999.

[4] SINTEF, OREDA: Offshore Reliability Data Handbook, Høvik: Det Norske Veritas, 2002.

[5] ISO 17776, *Petroleum and natural gas industries -- Offshore production installations -- Guidelines on tools and techniques for hazard identification and risk assessment,* Geneva: International Organization for Standardization, 2000.

[6] ANSI, *ISA-5.1-2009,* Washington, District of Columbia: American National Standard Institution, 2009.

[7] API STANDARD 674, Positive Displacement Pumps--Reciprocating, 3rd éd., Washington, DC: American Petroleum Institute, 2010.

[8] API STANDARD 676, Positive Displacement Pumps--Rotary, 3rd éd., Washington, DC: American Petroleum Institute, 2009.

[9] L. Nelik et J. Brennan, Progressing Cavity Pumps, Downhole Pumps, and Mud Motors, Houston: Gulf Publishing Company, 2005.

[10] M. Dillon et Vüllings, «Applying the NPSHR Standards to Progressing Cavity Pumps,» June 1996. [En ligne]. Available: https://www.rotatingright.com/pdf/seepex/resources/articles/General/NPSHR%20Standard.pdf. [Accès le 26 4 2019].

IOP Conf. Series: Materials Science and Engineering **694** (2019) 012014 doi:10.1088/1757-899X/694/1/012014

Numerical Simulation of Turboexpander on Organic Rankine Cycle with Different Working Fluids

Diki Ismail Permana[1] **and Priyono Sutikno**[2]

[1] Faculty of Industrial Engineering, Institut Teknologi Nasional, Bandung, Indonesia
[2] Faculty of Mechanical Engineering and Aerospace, Institut Teknologi Bandung, Bandung, Indonesia

E-mail: dicky91permana@itenas.ac.id

Abstract. Renewable energy is an alternative to overcome the energy crisis and dependence on fossil-fuel energy that has an environmental impact due to the greenhouse effect. Meanwhile, disadvantage of EBT is depend on weather periodically. Geothermal is a source of EBT that is independent with weather. Geothermal resources dominated by water (brine) can't be directly utilized as a power generated, but it has to separated into water and steam phase. The high enthalpy owned by brine can still be utilized due to development of Organic Rankine Cycle. ORC uses organic refrigerant fluids or hydrocarbons as a working fluid because it has low boiling temperatures. The numerical simulation was analyzed using a CFD ANSYS CFX with turbulence model K-epsilon and Pentane, R245fa, and butane as a working fluids with expansion ratio 5:1 and rotational speed variation at 2000 until 14000 rpm.

1. Introduction

Indonesia is a country that has the greatest geothermal potential in the world around 27 GW. But in application, it only about 1.2 GW and it is divided into 7 regions [1]. One of the areas that has the greatest potential and has been utilized is the Awibengkok area, West Java. Where, the installed plant is around 127 MWe with a reservoir temperature of 250°-290°C and reservoir pressure of 35-85 bar and brine temperature of 160°-180°C and brine pressure of 11 bar [2]. Can be seen the brine temperature needed, then the area of Awibengkok can be utilized as a combined heat generator. One of the combined heat plants is the application of the Organic Rankine Cycle (ORC) which uses organic fluids or hydrocarbons as a working fluids because it has a low boiling temperature point.

At present more than 200 ORC plants have been identified with more than 1800 MW installed worldwide [3]. Factors that causing ORC technology are still not utilized, because the components production is still very lack, especially turboexpander component that require very high costs to be developed and produced [4]. Turboexpander is a substantial component in ORC, where the energy of organic fluid flows into turbine shaft then the mechanical energi from rotating shaft is converted, which moves to the generator to produce electricity. Generally, there are two common turboexpanders, in terms of the direction of fluid passing through the turbine, axial turbine and radial turbine. In this research, the turboexpander type is radial turbine outflow or centrifugal radial turbine is used.

One way to develop ORC technology is to analyze ORC turboexpander. Simulation is a way to get optimal efficiency and is expected to produce maximum power as well. The expander simulation process with the flow that occurs in the turboexpander is done by the CFD method. Spadicini, et al

(2015) have performed numerical analysis with CFD on radial outflow turbines with pentane working fluid with inlet temperature parameters 130°C and 1.3 MPa inlet pressure, where the results of numerical analysis with CFD show the power that can be generated by turbines looking for 2MW at changes of 3000 rpm. Whereas [5] has conducted numerical analysis to discuss turboexpander with radial inflow type for 3.5 kW power by comparing ideal gas and real gas for R245fa fluid.

2. Methods

Awibengkok Geothermal Power Plant in West Java, Indonesia was chosen as a reference for reservoir conditions with a minimum estimated brine injection with a temperature range of 180° – 160°C. In this case, R245fa, Pentane and butane are selected as a working fluid based on ODP (Ozone Depletion Point) characteristics and (Global Warming Potential) GWP that match with ORC characteristics studied.

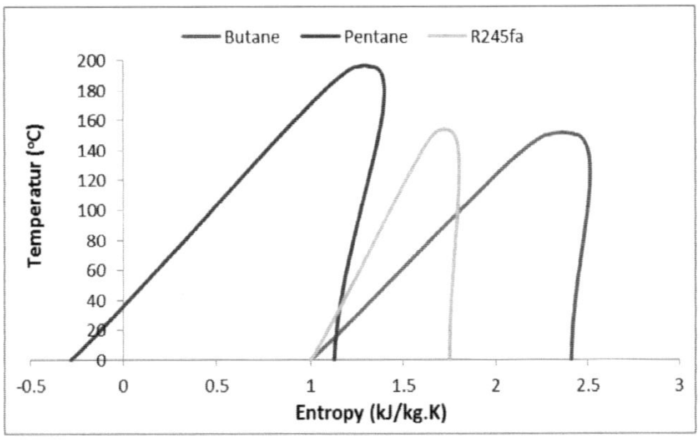

Figure 1. T-s Diagram of working fluids

In the fig 1 above, the three working fluids fall into the dry working fluid category, which is highly recommended for ORC applications to avoid erosion of turboexpander blades due to collision of liquid droplets during the expansion process [6].

2.1 Optimum Condition

Optimum conditions are very important on Organic Rankine Cycle, because in generally ORC has a low efficiency. To optimize the utilization of residual heat from brine PLTP, the cycle must be working at optimum conditions. According to [7] that the maximum output power will be obtained by utilizing the remaining heat as much as possible. Meanwhile according to [8] conducted a study to determine the optimum evaporator temperature in the subcritical cycle. The conditions of the turbine inlet fluid are assumed to be in a saturated vapor condition. The optimum point is determined by optimized the equation that represents the area of the T-s ORC (nett power) curve that is done iteratively. For the utilization of waste heat from brine, the ORC that can be made is limited by two temperature conditions. The brine reinjection temperature at 155°C for upper limit temperature and the cooling water temperature at 25°C as a lower limit temperatur.

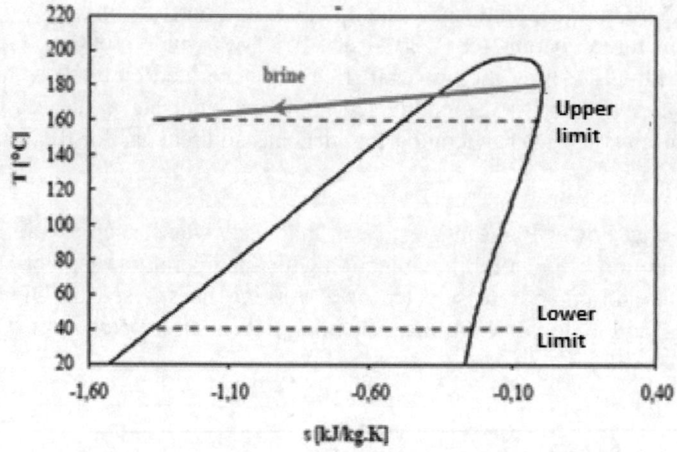

Figure 2. Temperature Limited of optimum condition

From this temperature limit, it can be used to determine the optimal conditions of the three working fluids with only one pressure variation. Furthermore, determined some properties of each fluid that obtained from NIST Refprop and some equations used.

$$\dot{W}_{turbine} = \dot{m}(\Delta h).\eta_{turbine} \quad (1)$$

$$Q_{in} = \dot{m}(\Delta h) \quad (2)$$

$$\dot{W}_{pump} = \frac{\dot{m}(\Delta h)}{\eta_{pump}} \quad (3)$$

The table 1 is the optimum operating state from the calculation of the cycle tempo with the power parameters generated by each working fluid that is set at 5 kW with a thermal efficiency value of each working fluid less than 10%.

Table 1. Optimum conditions of ORC

Parameters	Units	Pentane	R245fa	Butane
Inlet Pressure Brine	MPa	1	1	1
Temperature Brine	°C	180	180	180
$\dot{m}$	Kg/s	0,307	0,307	0,307
Inlet Temperature turboexpander	°C	155	155	155
Outlet Temperature turboxpander	°C	139,8	123,8	132,8
Outlet Temperature condenser	°C	30	30	30
Inlet Pressure turboexpander	MPa	0,4	0,88	1,4
Outlet Pressure turboexpander	MPa	0,082	0,179	0,284
Output Power	kW	5	5	5
Power used by pump	kW	0,28	0,22	1,13
Q_{in}	kW	176,69	91,97	176

2.2 Modeling

The turbexpander that will be simulated in this study is the type of radial outflow turbine produced by infinity turbine. Turboexpander sizes range from 12.7 cm with an output power is around 5 kW. Turboexpander from infinity turbine has the advantage, that is not a gearbox needed [9]. Because the turboexander rotation value is low. Radial outflow turbine has two main parts, namely stator and rotor. The two main components have the shape of impulse blade.

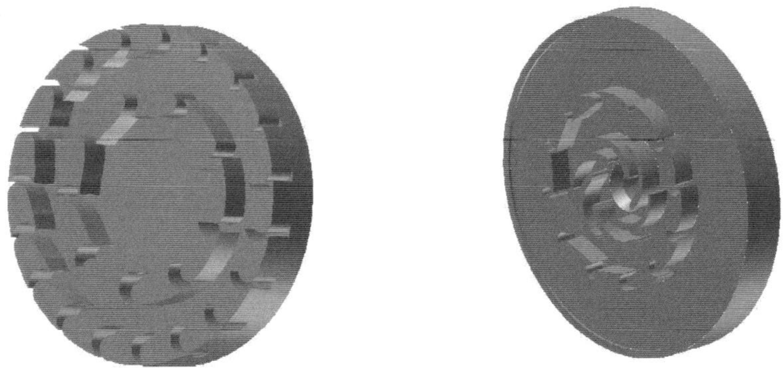

Figure 3. Turboexpander parts : rotor (left), stator (right)

This component has the role of carrying out the process of expansion. The expansion process in the Radial Outflow Turbine occurs entirely due to the flow rate of the working fluid. The flow rate with a certain pressure enters through the inlet section and its expanded to a several levels. At the inlet side, the working fluid is directly header by the first level of stator which also act as a nozzle. Then after exiting from guide blade or stator at the first level, the working fluid flow enters the next level to rotate the moving blade or rotor section. Whereas for numerical analysis using software only the fluid parts are visible with the purpose of simplified numerical calculations of CFD.

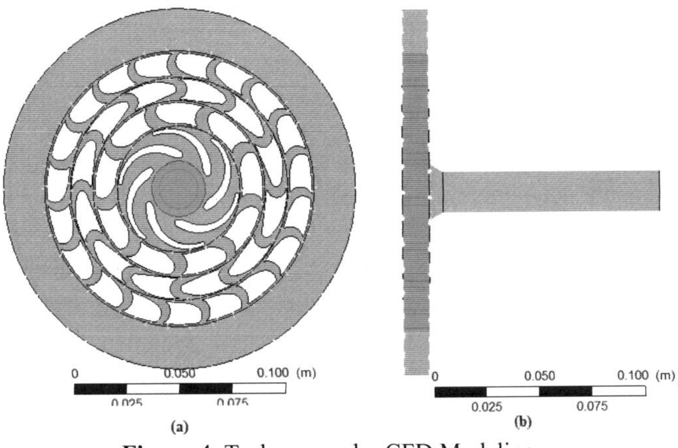

Figure 4. Turboexpander CFD Modeling

Before the simulation process begins, a modeling must be divided into tiny elements or volumes with the purposed of convinient numerical calculations. The number of elements / volumes produced from the meshing results greatly affects to simulation process and results. As for meshing in this turboexpander model, it is in a good category, where the skewness criteria value is around 0.36 and the orthogonal quality value is around 0.78 [10].

In this simulation the software used is ANSYS CFX, where the features in ANSYS CFX are specifically used for the turbomachinery simulation process. But there is disadvantage of working fluid's data especially for R245fa, so that the material tabs need to add the required properties such as molar mass, crtitical temperature, critical pressure, critical volume, acentric factor and boiling temperature. These properties can be found at NIST Refprop. Not only that, because R245fa is in the real gas category, the variables of the fourth order polynomial need to be entered for the correct calculation of the turbulence model of the standard redlich kwong or can be obtained by a linear regression approach. Therefore the coefficient values are.

IOP Conf. Series: Materials Science and Engineering **694** (2019) 012014 doi:10.1088/1757-899X/694/1/012014

$$\frac{Cp}{R} = a_1 + a_2.T + a_3.T^2 + a_4.T^3 + a_5.T^5 \quad (4)$$

$$a_1 = 432{,}29\,;\ a_2 = 1{,}5334\,;\ a_3 = 0\,;\ a_4 = 0\,;\ a_5 = 0$$

3. CFD Results

3.1 Simulation Results

The simulation results from CFD of Turboexpander, give different results when compared with the data of specifications. The operating parameters being compared are turboexpander rotation from 2000 to 14000 rpm with three working fluids of R245fa, Butane and Pentane respectively.

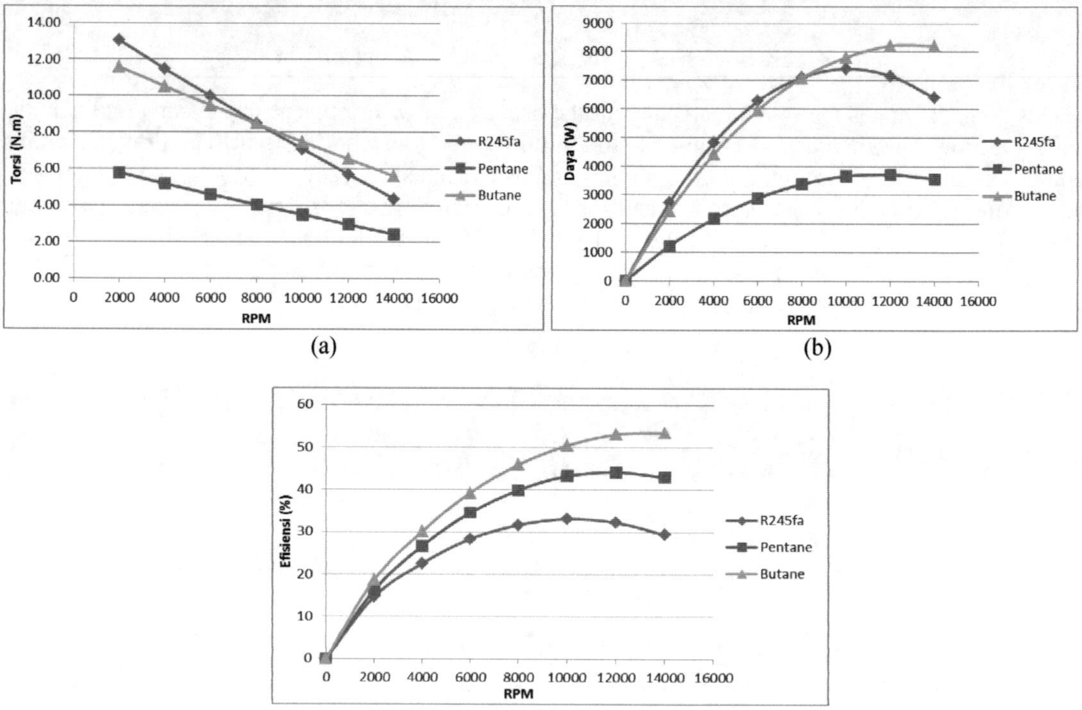

(a) (b) (c)

Figure 5. Graphic results of each working fluid : (a) RPM vs Torque, (b) RPM vs Power, (c) RPM vs Efficiency

From fig 5 we can get the results of each simulation with variations of turboexpander rotating and working fluid variations. Each fluid produces the highest efficiency and power at different rpm, for Pentane produce the power around 3.71 kW and the efficiency of 44.23% at 12000 rpm, whereas for Butane produce power around 8.2 kW and the efficiency is about 32.5% at 14000 rpm, and power produced by R245fa is about of 5.7 kW and 32.25% efficiency at 8000 rpm. One of the indications is that the Butane fluid working area is wider than other fluid working areas.

These results are in accordance with research from [11] which selects the some working fluids for ORC using heat sources with a potential of 1 MWth, where Butane is the most optimum working fluid in the sub-critical cycle with thermal efficiency reaching 15.5% efficiency.

3.2 Analysis of Fluids Contour

The expansion process that occurs in turboexpander involves several parameters when operating. These parameters include pressure, temperature, and velocity. Simulations performed on the ANSYS CFX software provide these parameters.

IOP Conf. Series: Materials Science and Engineering **694** (2019) 012014 doi:10.1088/1757-899X/694/1/012014

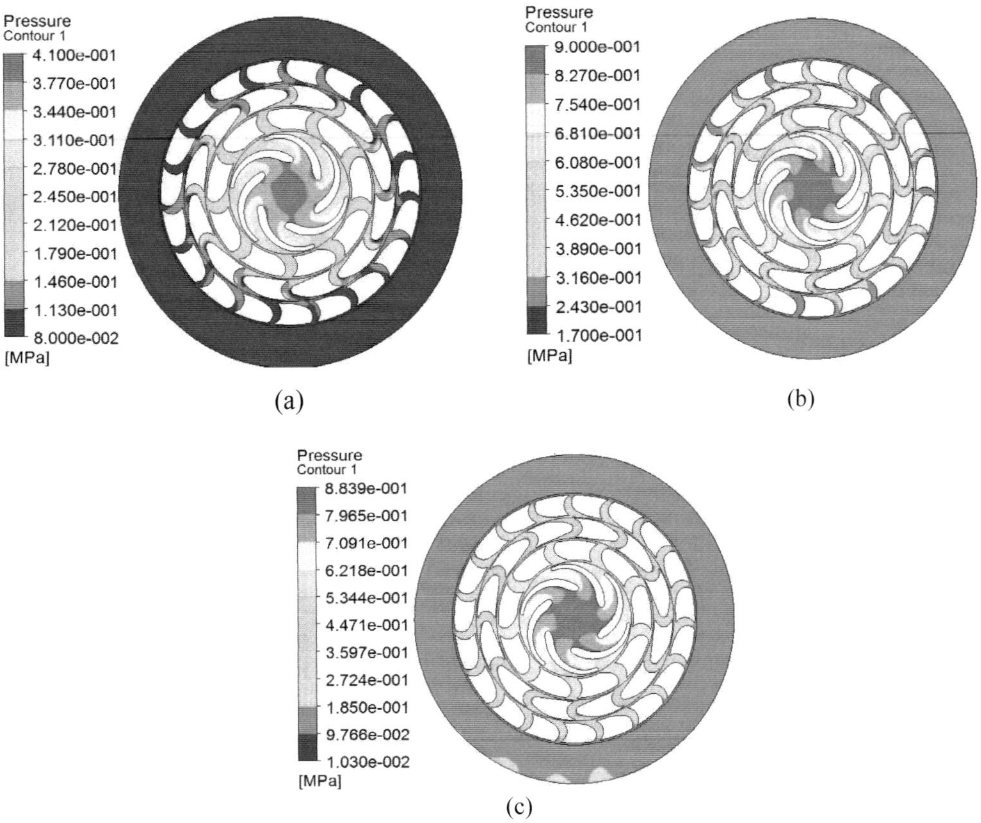

Figure 6. Pressure of fluids static contour: (a) Pentane 12000 rpm, (b) Butane 14000 rpm, (c) R245fa 10000 rpm

From figure 6, the contour of the static pressure of each fluid at turboexpander rotation speed which reaches the highest power and efficiency. The three pressure contour images show the symptoms of expansion in the same place where the fluid is expanded when it enters the leading edge of guide vane and exits through the trailing edge of rotor, the only difference from the expansions is reduction of pressure that occurs. This corresponds to the impulse blade with a zero reaction where a pressure drop occurs at stator or guide vane [12].

From figure 7, the contour of temperature of each fluid where the turboexpander rotation reaches the highest power and efficiency. The three working fluids (Pentane, Butane, R245fa) exhibited the same phenomenon, those are uniform decrease in temperature occurs after the fluid passes through the leading edge of the guide vane with different temperature range values. The small range of temperature reduction that occurs in Pentane compared to other working fluids, is an indication of the low power and efficiency produced.

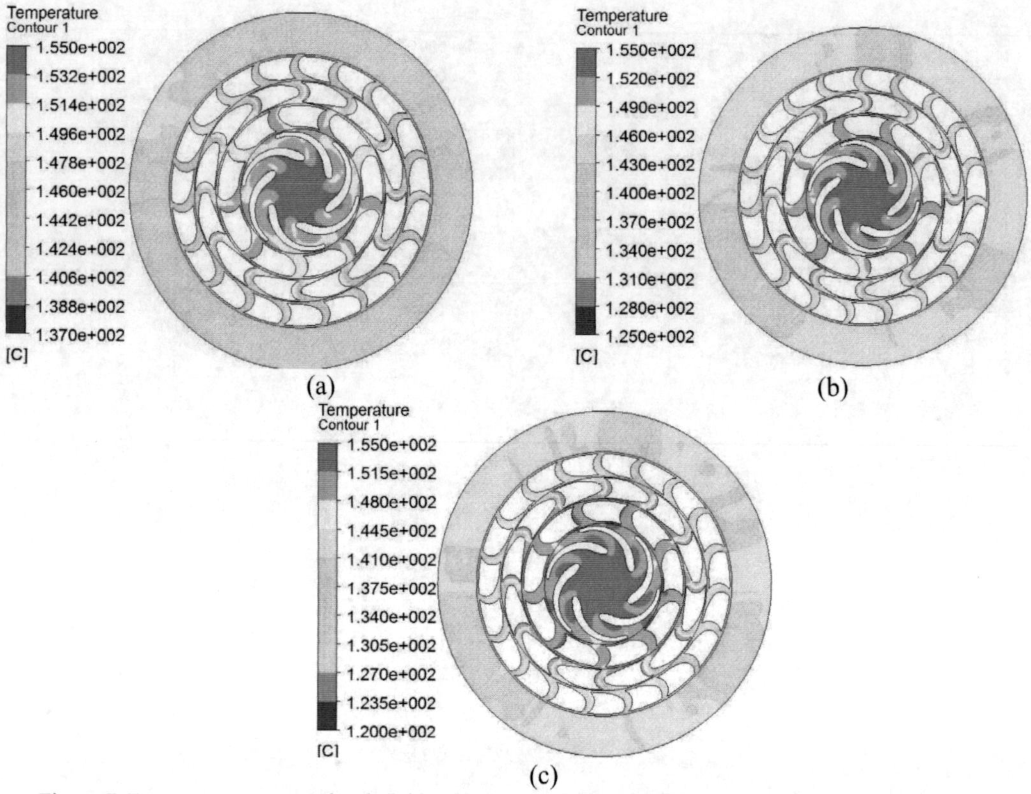

Figure 7. Temperature contour of each fluids : (a) Pentane 12000 rpm, (b) butane 14000 rpm, (c) R245fa 10000 rpm

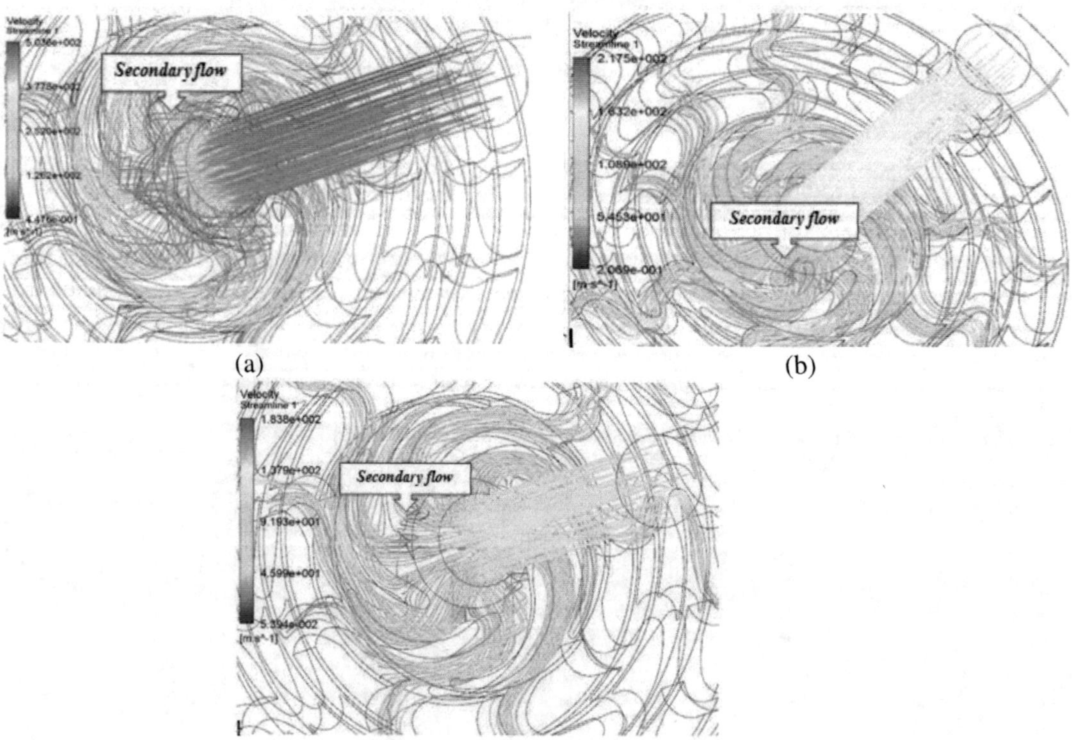

Figure 8. Fluids flow : (a) Pentane 12000 rpm, (b) butane 14000rpm, (c) R245fa 10000rpm

Figure 8 is the direction of flow for each fluid which reaches the highest power and efficiency. It can be seen, there is a secondary flow for each fluid when going through the stator or guide vane. Secondary flow is the losses that often occur in turbines with a percentage of losses of around 30% after leakage losses and friction losses [13]. In this research leak and friction losses are ignored, so secondary flow is the only losses that occurs.

4. Conclussion

The conclusion of this study, there are different results between calculations using the cycle tempo for each working fluid (Pentane, Butane, R254fa) which produces 5 kW of power with CFD simulation results, where Pentane working fluid produces power around 3.71 kW at 12000 rpm, butane produces power around 5.95 kW at 6000 rpm and R245fa produces power around 4.8 kW at 4000 rpm. Whereas Butane obtained the most efficient working fluid at rotational speed, Butane which produces power of 8.24 kW and 53.5% efficiency at 14000 rpm.

References

[1] BPPT.2016. Indonesia Energi Outlook 2016
[2] Irianto, Dani.2016. Thermodynamic Modeling and Simulation of Geothermal Hybrid Flash-Binary Power Plants From the Point of view of Non-Condensable Gas Effect, Master theses, Institut Teknologi Bandung : Bandung.
[3] Jouhara, Hussam et al. 2018. Waste Heat Recovery Technologies and Applications, Elsevier : london, pp. 268-289.
[4] https://orc-world-map.org/biomass-geothermal/, accessed on February 2019
[5] Li, Peng et al. 2018. Analysis of the Effect of Blade Installation Angle and Blade Number on Radial Inflow Turbine Stator Flow Performance, Energies MDPI : china.
[6] Brun, Klau et al. 2012. Analysis of Solid Particle Surface Impact Behavior in Turbomachinery to Assess Blade Erosion and Fouling, Proceeding of Forty First Turbomachinery Simposium : Texas.
[7] Wei, D, et al. 2007. Performance analysis and optimization of organic Rankine cycle (ORC) for waste heat recovery, Energy Conversion and Management, 48(4), pp. 1113–1119.
[8] He, C.,et al. 2012. The optimal evaporation temperature and working fluids for subcritical organic Rankine cycle, Energy, 38(1), 136–143.
[9] https://infinityturbine.com/rot06/ , accessed on July 2019
[10] ANSYS, Inc. 2013. ANSYS CFX-Solver Theory Guide, 15317, pp.724–746.
[11] Qodri, F.2013. Determination Of optimum Temperature And Pressure At Turbine Inlet For Organic Rankine Cycle Based On Brine Temperature And Critical State Of Working Fluid, Master theses, Institut Teknologi Bandung : Bandung.
[12] [Macchi, E. 2016. Organic Rankine Cycle (ORC) Power Systems Technologies and Applications, London: Elsevier Inc.
[13] Cui, J., Rao, V. N., Tucker, P. G. 2017. Numerical investigation of secondary flows in a high-lift low pressure turbine, International Journal of Heat and Fluid Flow, 63, pp 149–157,

IOP Conf. Series: Materials Science and Engineering **694** (2019) 012015 doi:10.1088/1757-899X/694/1/012015

Numerical Simulation of a Small-Scale Cyclone Separator using MP-PIC Method

V J Wargadalam[1], M A Andira[2], C A Putra[2], A I Siswantara[2], G G R Gunadi[2], Hariyotejo[2], C Damis[2] and Aminuddin[1]

[1]PUSLITBANGTEK Ketenagalistrikan, Energi Baru dan Terbarukan, dan Konservasi Energi (P3TKEBTKE), Jakarta, 12230, Indonesia
[2]Faculty of Engineering, University of Indonesia, Depok, 16424, Indonesia

E-mail: verinawargadalam@yahoo.com

Abstract. This paper presents a numerical study of a small-scale cyclone separator performance by means of computational fluid dynamics (CFD). The family of Euler-Lagrange approach named multiphase particle-in-cell (MP-PIC) is employed for modelling gas-solid interaction. The three-dimensional small-scale cyclone separator geometry was discretized with hexahedral-dominant mesh. The gas phase is assumed to be a flue gas which density is 0.363 kg/m^3 and dynamic viscosity is 7.47E-5 Pa s. The solid particles are catalyst used in a small-scale fluid catalytic cracking (FCC) unit which are having 2400 kg/m^3 of density and diameters vary from 90 μm to 130 μm. Large Eddy Simulation (LES) turbulence model was used to accurately predict turbulent flow behaviour of flue gas inside the cyclone. The simulation was performed for 7.82 m/s inlet velocity for both flue gas and catalyst. The pressure drop and catalyst collection efficiency were the variables of interest to be analysed, which were compared to the analytical solutions. It was found that the results generated from numerical simulation using MP-PIC method reflect a good agreement with the analytical solutions.

1. Introduction

Cyclone separator is a device that utilize the centrifugal forces to remove solid particles from flue gases. It has been used in a wide range of industries, such as power generation, oil and gas, incineration plants, iron and steel industry, sand plants, cement plants, coking plants, coal fired boilers, and food industries (Dutta, Gera, Sharma, Singh,Ghosh, & Kushawa, 2009). The simplicity of the design, low-cost manufacturing, easy to maintain, and the ability to adapt in various operating conditions are the reasons behind its extensive use. At the same time, it comes with disadvantage since the cyclone are not capable of collecting the solid particle that is smaller than 10 μm.

The performance of a cyclone separator can be measured by its pressure drop and collection efficiency [19]. These two parameters of cyclone performance are related to each other. Increasing collection efficiency will also increase pressure drop and vice-versa. Hence, the study of a cyclone separator performance is instrumental in cyclone design and analysis.

One of the convenience methods to perform such study is computational fluid dynamics (CFD). CFD became a promising tool to carry out fluid dynamics analysis due to its ability to predict the flow field pattern in a detail manner, by producing the primary flow variables such as pressure and velocity at a given location in a domain. It can be used with some degree of accuracy which depends on several aspects; geometry modelling, mesh configurations, mathematical models, numerical schemes, and algorithms. Under those circumstances, efficient simulation strategy must be employed to obtain the accurate results in a cost-effective means.

There are two multiphase models which can be applied to simulate gas-solid particles flow in a cyclone separator; the Eulerian-Eulerian method and the Eulerian-Lagrangian method. In the Eulerian-Eulerian method, governing equations for both continuous and discrete phase are solved inside a computational cell by weighting the equations terms with continuous phase volume fraction. In the Eulerian-Lagrangian method, computational cell is a placeholder only for continuous phase solution. To determine the trajectory and the other properties of the discrete phase, the statistical calculations are imposed for each particle. It implies that a number of equations that must be solved will be vast. Despite this minor fact, the Euler-Lagrangian method is still a widely used approach to investigate the efficiency of a cyclone separator since it takes into account the wide particle size distribution.

The examination of a cyclone separator performance by making use of computational fluid dynamics (CFD) have been conducted by many researchers. Snider, A. and Johnson, R.A. [18] performed MP-PIC simulations of dense-phase flows in cyclone separators. The CFD results showed excellent agreement with measured experimental data which efficiency value is the order of 85%. Razmi, H. et. al. [19] applied MP-PIC method to investigate the hydrocyclone performance with spiral inlet in Sarcheshmeh copper complex. The variation of the cyclone performance value due to the effect of operating parameters such as particle density, inlet mass flow rate of solid particles, inlet feed velocity, and the effect of design parameters like diameter and height of the vortex finder, the length of the cylinder and conical section, and the type of inlet were analyzed. They came across the conclusion that increasing inlet feed velocity and the height of the vortex finder will improve the cyclone performance.

The present study will carry out the CFD simulation of gas-solid flow with MP-MPIC method inside a small-scale cyclone separator. The cyclone geometry is based on the pilot-scale Fluid Catalytic Cracking (FCC) research project by P3TKEBTKE, which is located after regenerator unit to recollect the solid catalyst particles back to the reactor. This research used CFDSOF-NG® software, a CFD software developed by CCIT Group Indonesia with capability of simulating a wide range of fluid flow and heat transfer phenomena on complex geometry. The pressure drop and particle collection efficiency will be compared to the analytical solutions as an effort to verify the implementation of MP-PIC method in the software.

2. Mathematical model and methodology

2.1. Mathematical model for continuous phase

The governing equations describing the conservation of mass and momentum are solved in each particular computational cell for gas phase. By solving these equations, the pressure and velocity value can be obtained by first defining gas properties such as density and viscosity. Large Eddy Simulation (LES) turbulence model is used in order to accurately capture the velocity field, which is a prominent variable in the cyclone separator case. Henceforth, the governing equations of continuous phase are given in (1) and (2) respectively [19].

$$\frac{\partial(\alpha\rho_g)}{\partial t}+\frac{\partial(\alpha\rho_g u_i)}{\partial x_i}=0 \tag{1}$$

$$\frac{\partial(\alpha\rho_g u_i)}{\partial t}+\frac{\partial(\alpha\rho_g u_i u_j)}{\partial x_i}=-\alpha\nabla p+\frac{\partial}{\partial x_j}\left(\alpha\tau_{\mu ij}+\alpha\tau_{dij}+\alpha\tau_{tij}\right)+\alpha\rho g-\frac{1}{\rho}F \tag{2}$$

where α, ρ_g, u, p , g are continuous phase (gas) volume fraction, gas density, gas velocity vector, gas pressure, and gravity acceleration respectively. $\tau_{\mu ij}$ is the tensor viscosity stress, τ_{dij} is the tensor drift, and τ_{tij} is the stresses on the grid scale. Eventually, F represents the momentum transfer between continuous and discrete phase.

The tensor viscosity stress $\tau_{\mu ij}$ is defined by the following equation:

$$\tau_{\mu ij} = 2\mu\bar{S}_{ij} \tag{3}$$

where $\bar{S}_{ij}$ is the mean tension rate and μ is the gas molecular viscosity.

The two others tensor stresses τ_{dij} and τ_{tij} are solved by following equation (4) and (5) successively.

$$\tau_{dij} = \sum_{p=1}^{n} \alpha_p \rho_p u_{pmi} u_{pmj} \tag{4}$$

$$\tau_{tij} = \mu_{sgs}\left(\frac{\partial u_i}{\partial x_j} + \frac{\partial u_j}{\partial x_i}\right) \tag{5}$$

The term μ_{sgs} indicates subgrid-scale eddy viscosity which is expressed as:

$$\mu_{sgs} = \rho l_s^2 \sqrt{2\bar{S}_{ij}\bar{S}_{ij}} \tag{6}$$

where l_s is the intermediate local grid.

2.2. Mathematical model for discrete phase

The discrete phase in MP-PIC method is treated differently with the continuous phase. Here, Liouville equation (8) is used to elucidate the discrete phase kinematic behaviour by solving for $f(x, u, m, t)$ which is called the particle distribution function [17].

$$\frac{\partial f}{\partial t} + \frac{\partial f u_{si}}{\partial x_i} + \frac{\partial f A}{\partial u_i} = \left(\frac{\partial f}{\partial t}\right)_{coll} \tag{7}$$

The right-hand side of the above equation represents the collision term, where the detail discussion can be found in elsewhere (O'Rourke and Snider, 2012; O'Rourke and Snider, 2010; Snider, 2001).

$$\left(\frac{\partial f}{\partial t}\right)_{coll} = \frac{f_D - f}{\tau_D} + \frac{f_G - f}{\tau_G} \tag{8}$$

The term u_s is the time derivative of the solid particle position and the term A is the time derivative of the solid particle velocity, which in turn is given by:

$$A = D(u_g - u_s) - \frac{1}{\rho_s}\frac{\partial p}{\partial x_i} + g - \frac{1}{\alpha \rho_s}\frac{\partial \tau}{\partial x_i} \tag{9}$$

where D, ρ_s, and τ are the drag function, the solid particle density, and the isotropic solid stress sequentially. Drag function is expressed as:

$$D = C_d \frac{3}{8}\frac{\rho_g}{\rho_s}\frac{|u_g - u_s|}{R} \tag{10}$$

where C_d is drag coefficient which in this research is modelled by using sphere drag model and R is particle radius. Moreover, isotropic solid stress in equation (9) is obtained by adopting the expression from (Harris and Crighton, 1994) and is given in equation (10) below.

$$\tau = P_s \frac{\alpha^{\beta}}{max[\alpha_{cp} - \alpha, \omega(1 - \alpha)]} \tag{11}$$

In the equation (10), P_s is a constant with units similar to pressure and α_{cp} is the particle-phase volume fraction when particle is the closed pack condition. β is a constant which value is ranging from 2 to 5 and ω is a small magnitude number in the order of 10^{-7}. The relationship between the particle volume fraction and the distribution function is given by:

$$\alpha_{cp} = \iint f \frac{m}{\rho_s} dm du_p \tag{12}$$

The total volume fraction for both phases must be equal to 1, hence

$$\alpha_{cp} + \alpha = 1 \tag{13}$$

Finally, in order to close the governing equations for gas-solid flow, the interphase momentum transfer function F can be defined as:

$$F = \iint fm\left[D(u_g - u_s) - \frac{1}{\rho_s}\nabla p\right] dm du_p \tag{14}$$

2.3. Mathematical model for analytical solutions
The derivation of analytical solutions for cyclone separator had been conducted by several scientists to calculate pressure drop, efficiency, and cut-off particle diameters.

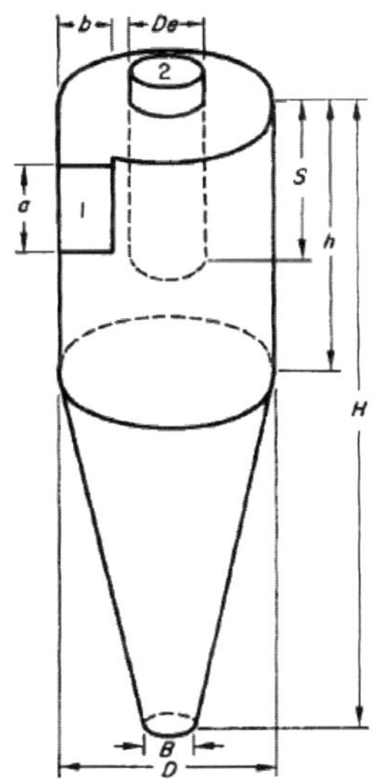

Figure 1. Cyclone with dimensions [2].

The pressure drop in a cyclone will be considered as that between points 1 and points 2 in Figure 1. Based on the calculated correlation coefficients [2], the Shepherd and Lapple method appear to give results as good as the others despite its simplicity. Shepherd and Lapple calculate the pressure drop results with the following equation:

$$\Delta H = K\frac{ab}{De^2} \tag{15}$$

where $K = 16$ for a cyclone with standard tangential inlet, and $K = 7.5$ for a cyclone with an inlet vane.

The cyclone pressure loss is expressed as inlet velocity head (ΔH). The inlet velocity head can be converted to pressure drop in terms of static pressure head, ΔP, by:

$$\Delta P = \frac{{v_g}^2 \rho_g \Delta H}{2g\rho_l} \tag{16}$$

where v_g is the gas inlet velocity, ρ_g is the gas density, g is the gravity, and ρ_l is the fluid density that is used to calculate the head pressure.

Furthermore, the capacity of the cyclone to collect particles is measured by its efficiency η, defined as the fraction of the inlet flow of solids separated in the cyclone. Leith and Licht [2] formulated the collection efficiency as:

$$\eta = 1 - \exp\left[-2(C\psi)^{1/(2n+2)}\right] \tag{17}$$

where η is the efficiency, C is the cyclone dimension factor, ψ is the impaction parameter, and n is the vortex exponent. Further, the cyclone dimension factor is the dimensionless ratio that follows:

$$C = \frac{8K_c}{K_a K_b} \tag{18}$$

K_a and K_b are the cyclone inlet height and weight divided by cyclone diameter, while K_c is the cyclone volume constant that is defined the following equation:

$$K_c = \frac{V_s + {V_{nl}}/{2}}{D^3} \tag{19}$$

With V_s is the annular shaped volume above exit duct to midlevel of entrance duct and V_{nl} is the volume of cyclone below gas outlet to the natural length (l). Both V_s and V_{nl} are defined as the function of geometry only as follows:

$$l = 2.3De\left(\frac{D^2}{ab}\right)^{0.333} \tag{20}$$

$$V_s = \frac{\pi\left(S - \frac{a}{2}\right)(D^2 - De^2)}{4} \tag{21}$$

$$V_{nl} = \begin{cases} \frac{\pi l}{4}(D^2 - De^2), & l < (h-S) \\ \frac{\pi D^2}{4}(h-S) + \left(\frac{\pi D^2}{4}\right)\left(\frac{S+l-h}{3}\right)\left(1 + \frac{d}{D} + \left(\frac{d}{D}\right)^2\right) - \frac{\pi De^2}{4}(l), & (h-S) < l < (H-S) \\ \frac{\pi D^2}{4}(h-S) + \left(\frac{\pi D^2}{4}\right)\left(\frac{H-h}{3}\right)\left(1 + \frac{B}{D} + \left(\frac{B}{D}\right)^2\right) - \frac{\pi De^2}{4}(H-S), & l > (H-S) \end{cases} \quad (22)$$

with

$$d = D - (D - B)\frac{(S + l - h)}{H - h} \quad (23)$$

Furthermore, ψ is defined as:

$$\psi = \frac{\rho_p d_p^2 v_g}{18\mu_g D}(n + 1) \quad (24)$$

with

$$n = \frac{(12D)^{0.14}}{2.5} \quad (25)$$

2.4. Simulation methodology

Simulation of gas-solid flow inside a small-scale cyclone separator was conducted by first creating the geometry and transform it into a number of discrete non-overlapping cells which is dominated by hexahedral shape. In order to ensure that the generated mesh is sufficient to provide the accurate solution, mesh independency study was performed by varying number of cells of 18628, 29829, and 43819. The physical properties for both gas and solid particles were then defined as shown in table 1 and followed by imposing boundary conditions. The gas-solid flow was solved by implemented MPPIC model in the CFDSOF® software. The calculation of turbulent velocity flow employed large eddy simulation (LES). The PIMPLE algorithm (combination of PISO and SIMPLE) was used to solve the pressure-velocity coupling problem. The simulations were carried out for the 90 μm, 110 μm, and 130 μm solid particles diameter throughout 10 s with time-step of 0.001 s.

Table 1. Properties for physical modelling

Properties	Value	Units
Particle Properties		
Particle density	2400	kg/m^3
Void fraction	0.6	
Particle size	90-130	μm
Gas properties		
Gas density	0.363	kg/m^3
Gas temperature	800	C^o
Gas viscosity	7.47 x 10^{-5}	kg/m s
Operating conditions		
Cyclone diameters, D	0.2030	m
Throughput, Q	0.1	m^3/s
Temperature, T	800	C^o

2.5. Geometry and meshing
The geometry of a small-scale cyclone separator is created by considering the non-dimensional parameters as shown in table 2 below. The definitions of the parameters can be referenced to figure 1.

Table 2. Non-dimensional parameters for defining cyclone geometry

Non-dimensional parameters of cyclone separator		
Parameters	Value	Units
a/D	0.500	-
b/D	0.200	-
S/D	0.500	-
De/D	0.500	-
h/D	1.500	-
H/D	4.000	-
B/D	0.375	-

The mesh is generated using snappyHexMesh, which is an automatic mesh generator that takes and chisels a base hexahedral mesh to fit the mesh with the geometry shape. The resulting mesh is illustrated in figure 2.

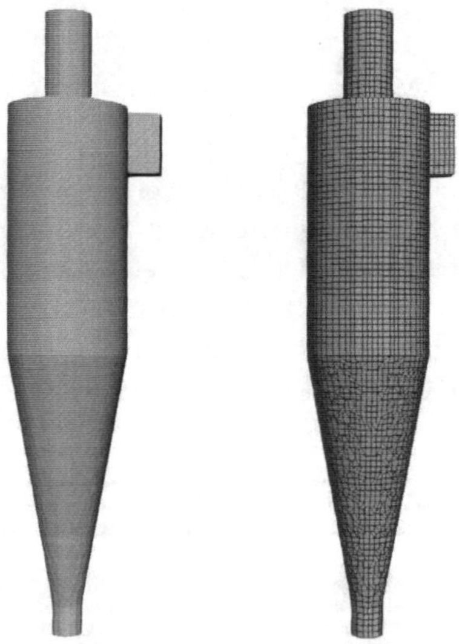

Figure 2. Cyclone geometry and mesh

2.6. Boundary conditions
Figure 2 shows that the computational domain consists of three boundaries as the inlet, outlet gas, and outlet particle. Velocity of 7.82 m/s for gas and particle was defined at the inlet. Outlet gas was considered to be an atmospheric condition which pressure was defined as zero-gauge pressure. Outlet particle was defined as a patch where the particles leave the domain. Particle interaction with wall was modeled by rebound boundary condition with restitution coefficient value of 0.0.

3. Results

3.1. Mesh independency study
Mesh independency study is a necessity in CFD simulation to convince that the solutions are assumed to be free from numerical error due to the mesh configurations. It is done by first taking the small number of cells and monitor the value of the variable interest. Then, increase the number of mesh and recheck

the interest value. This work must be undertaken until the interest value does not change with respect to the number of cells. In this study, mesh independency study was performed by simulating gas flow only to reduce the computational cost and consider the pressure drop between inlet and gas outlet as the variable to be monitored. The result is given in figure X

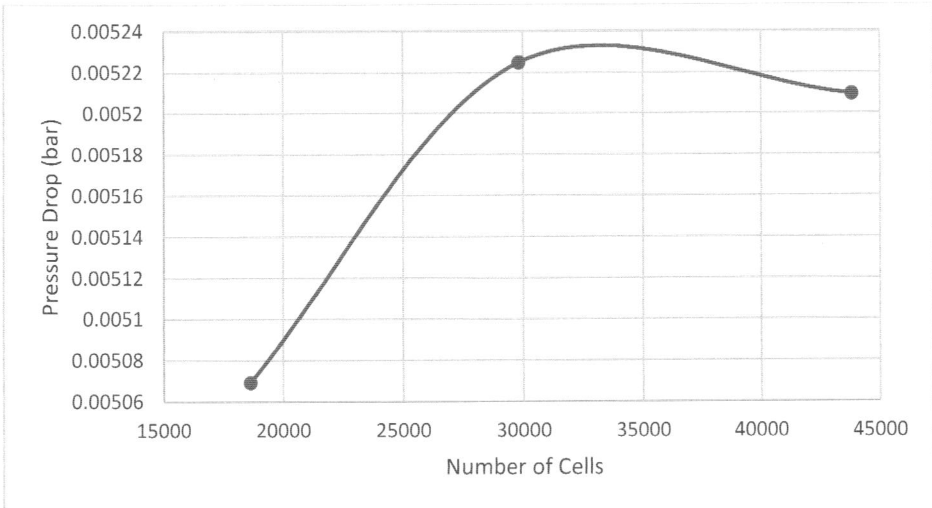

Figure 3. Pressure drop curve for mesh independency study

According to this figure, the difference between the mesh of 18628 cells and 29829 cells is noticeable, while increasing number of cells from 29829 to 43819 would only produces 0.3% difference. Hence, the mesh with 29829 cells is selected in this simulation.

3.2. Analytical solutions

Pressure drop of the cyclone can be calculated through equation (1) and (2). Moreover, the efficiency and cut-off size can be obtained by following equation (3) through (11). Below are the analytical solutions for this cyclone.

Table 3. Analytical solutions of cyclone performance

Variable	**Value**	**Units**
Pressure drop	0.008	bar
Cut-size particles	1.62	μm
Efficiency at 90 $\boldsymbol{\mu m}$	99.99	%
Efficiency at 110 $\boldsymbol{\mu m}$	100.00	%
Efficiency at 130 $\boldsymbol{\mu m}$	100.00	%

IOP Conf. Series: Materials Science and Engineering **694** (2019) 012015 doi:10.1088/1757-899X/694/1/012015

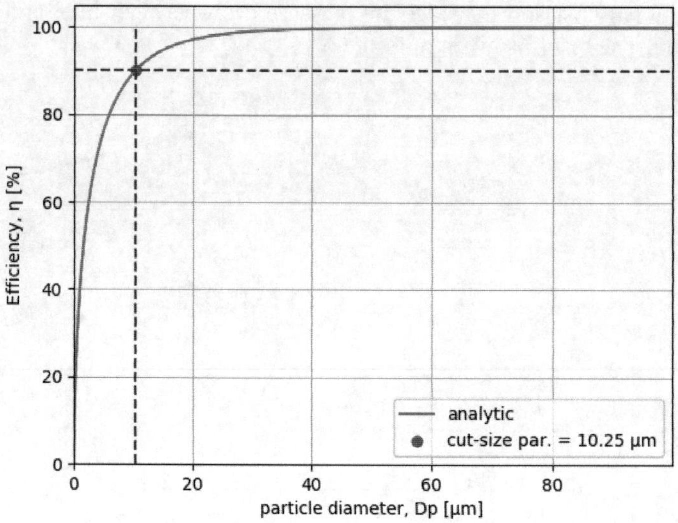

Figure 4. Cyclone efficiency curve from analytical solutions

3.3. Numerical Solutions
The cyclone efficiency was obtained from number of particles fraction between outlet particles and cyclone inlet. The visualization of particle trajectories for 3 diameters value is illustrated in figure 5.

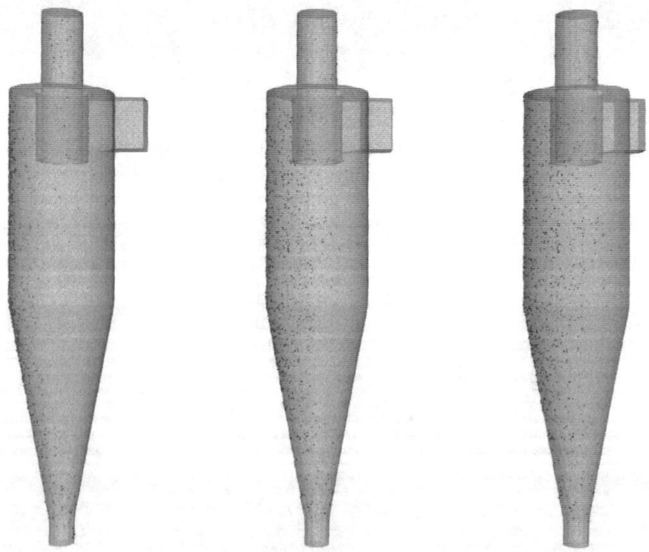

Figure 5. Particle trajectory inside cyclone separator. Left to right: 90 μm, 110 μm, 130 μm

Table 4. Efficiency of 3 different particle diameters

Diameter (μm)	Efficiency (%)
90	100
110	100
130	100

Moreover, pressure drop from CFD was calculated by subtracting average value of pressure on the cyclone inlet patch and outlet gas patch.

Table 5. Pressure drop from CFD

Diameter (μm)	Pressure Drop (bar)
90	0.0071
110	0.0074
130	0.0074

4. Conclusions

The numerical simulations of gas-solid flow in a small-scale cyclone separator have been done by using multiphase particle-in-cell (MP-PIC) method. The resulting efficiency and pressure drop from numerical simulations were compared with analytical solutions. It was found that the pressure drop from both methods showed a good agreement while the efficiency value from numerical simulations exhibited lower values than the analytical solutions. However, this deviation was not profound with order of 3 to 4 %. This research concludes that the multiphase particle-in-cell (MP-PIC) method could be adopted for simulating gas-solid flow in some circumstances.

5. Acknowledgement

The authors wold like to show the gratitude to CCIT Group Indonesia for providing the CFDSOF® license during this research.

References

[1] Sadeghbeigi, Reza (2000). *Fluid catalytic cracking handbook second editions design, operation, and troubleshooting of FCC facilities*. Houston: Gulf Publishing Company.

[2] Leith, D., & Mehta, D. (1973). Cyclone Performance and Design. *Atmoshperic Environment Pergamon Press*, 527-549.

[3] Shepherd, C. B., & Lapple, C. E. (1939). Flow Pattern and Pressure Drop in Cyclone Dust Collectors. *Ind. Eng. Chem.* **31,** 972-984.

[4] Shepherd, C. B., & Lapple, C. E. (1940). Flow Pattern and Pressure Drop in Cyclone Dust Collectors. *Ind. Eng. Chem.* **32,** 1246-1248.

[5] First, M. W., (1950). *Fundametal Factors in the Design of Cyclone Dust Collectors*. Doctoral thesis, Harvard University.

[6] Alexander, R. McK. (1949). Fundamentals of cyclone design and operation. *Proc. Australas Inst. Min. Met.* (New Series) **152-3**, 203-228.

[7] Stairmand, C. J., (1949). Pressure Drop in Cyclone Separators. *Engineering*, **16B,** 409-411.

[8] Barth, W., (1965). Design and layout of the cyclone separator on the basis of new investigations. *Brenn. Warme Kraft* **8,** 1-9.

[9] Stern, A. C., Capplan, K. J., & Bush, P. D., (1955). Cyclone Dust Collectors, *Am. Petrol. Inst. Report.*

[10] Marunic, M., & Rus, F., (2011). The Effect of Particle Size and Input Velocity on Cyclone Separation Process. *Bulleting of the Transilvania Univeristy of Brasov-Series II vol. 4(53) no. 2*, **117-122.**

[11] Lapple, C. E., (1951). Processes Use Many Collector Types. *Chem. Engng.* **58**, 144-151.

[12] Sproull, W. T., (1970). *Air Pollution and Its Control*. Exposition Press, New York.

[13] Leith, D., & Mehta, D. (1972).Collection Efficiency of Cyclone Type Particle Collectors, a New Theoretical Approach. *A.I.Ch.E. Symposium Series:* Air-1971.

[14] Stairmand, C. J., (1951). The Design and Performance of Cyclone Separators. *Trans. Instn. Chem. Engrs*, **29,** 356-383.

IOP Conf. Series: Materials Science and Engineering **694** (2019) 012015 doi:10.1088/1757-899X/694/1/012015

[15] Peterson, C. M., & Whitby, K. T., (1965). Fractional Efficiency Characteristics of Unit Type Collectors. *ASHRAE J.* **7(5),** 42-49.

[16] Ariyaratne, W.K.H., Manjula, E.V.P.J., Ratnayake, C., & Melaaen, M.C. (2016). CFD Approaches for Modelling Gas-Solids Multiphase Flows – A Review. *Proceedings of the 9th EUROSIM & the 57th SIMS (pp. 680-686).*

[17] Chandra, K.A., Ariyaratne, W.K.H., & Melaaen, C.M. (2016). Prediction of Dilute Phase Pneumatic Conveying Characteristics using MP-PIC Method. *Proceedings of the 9th EUROSIM & the 57th SIMS (pp. 639-645).*

[18] Snider, D., Williams, K., & Johnson, R.A., (2004). Multiphase Particle-In-Cell Simulations of Dense-Phase Flows In Cyclone Separators. *Proceedings of HT-FED04 (pp. 1-9).*

[19] Razmi, H., Goharrizi, A.S., Mohebbi, A. (2019). CFD simulation of an industrial hydrocyclone based on multiphase particle in cell (MPPIC) method. *Journal of Separation and Purification Technology, 209.* 851-862.

Heat and Mass Transfer in 2-Propanol Droplet Evaporation: Analysis Modified Stagnant Film Model

D Prasetyo[1] and E A Kosasih[1]

[1]Department of Mechanical Engineering, Universitas Indonesia, 16424, Depok, Indonesia

E-mail: kosri@eng.ui.ac.id

Abstract. Heat and mass transfer is an evaporation process from liquid droplet that occur in spray drying technology. So, it becomes important to understand the basic elements of spray drying, namely droplet. The formation technology involves drying is used in many industries. Various models of analogies of heat and mass transfer analogies have been formulated, with the aim of obtaining the characteristics that occur in the system. The analogy model most often used to obtain heat and mass transfer coefficients from droplet is the analogy model Ranz W. E. and Marshall W. R. The analogy has requirements, including the Lewis number (Le) which ranges in the number one. The purpose of this experiment is to compare the results obtained from the analogy of the Ranz-Marshall equation model applied to the droplet by making a modification of the C1 and C2 values that exist in the stagnant film model. In this experiment, the analysis results obtained from the modification of the stagnant film model is that the heat transfer phenomenon that occurs in droplet does not yet fully have good results. Where the value of C1 still cannot represent the value that can be used to obtain C2 properly.

1. Introduction

In recent years, the rapid growth of attention to theoretical and experimental studies of the behavior a single droplet evaporation is increasing [1]. One process that has been used from the past until now we still use to support our daily needs is the evaporation process. The process of evaporation in liquid fuels is still the world's main energy source because the density of its energy is fairly high, and also its handling is fairly easy. When liquid fuel is used, spray combustion is also commonly used. This spray combustion is used in many industries such as turbine engines and diesel engines [2]. Thus, it is important to understand the characteristics that occur in spray combustion [3]. Testing on droplets is one of the steps used for these needs. The dynamic nature possessed by droplets, as well as the heat transfer from droplets to hot surfaces, is one of the important factors that will affect the mass transfer from droplets [4].

In order to increase the effectiveness and efficiency of the spray combustion, a study is needed to determine the characteristics that exist [5]. Various models of the analogy to heat and mass transfer analogy have been formulated to show the characteristics that occur in the system. The analogy model of heat and mass transfer which is generally used is the analogy model of Ranz W. E. and Marshall W. R. [6, 7], through which we can obtain the value of the heat transfer and mass droplet. The analogy will have a valid value if it has a value of the Lewis number equal to one. This value will guarantee the similarity of the temperature profile and the concentration profile of the tested liquid [8]. In addition to the Ranz-Marshall analogy, there are 3 other models used, namely the Stagnant film model, the penetration model, and the laminar boundary model [9].

Based on the results of research conducted by E.A. Kosasih and M.I. Alhamid by making a modification to the stagnant film model, good compatibility is obtained. The results are obtained from water fluids, where the Lewis number owned by water is less than one [8]. But the results obtained are inversely proportional to E.A. Kosasih and D. Prasetyo [7] and also E.A. Kosasih and I. Agung [6], where the value of their Lewis number has a value of more than one and obtains a less good correlation. Experimental investigation has been conducted to study the effect [10].The purpose of this experiment is to obtain a modification of the stagnant film model that will be used to obtain a good correlation with more than one Lewis value. In this experiment, the modification is located on the values of C1 and C2 used. The value of C1 represents the integral of mass transfer and the value of C2 represents the value of heat transfer.

2. Literature Review

In conducting experiments, various kinds of literature are used that will underlie these experiments. The literature will also be used to obtain the origin of the modified formula.

2.1. *Volume of Sphere*

To get the volume sphere from what was previously in the form of prolate spheroid, the following formulas are used:

2.1.1. Prolate Spheroid. A prolate spheroid is a spheroid that has a polar radius greater than its equatorial radius. The formula used in prolate spheroid is written as follows [11]:

$$V_{prolate} = \frac{4\pi}{3} a^2 c \tag{2.1}$$

$$A_{prolate} = 2\pi\, a^2 + 2\pi \frac{ac}{e} sin^{-1} e \tag{2.2}$$

$$e_{prolate} = \sqrt{1 - \frac{a^2}{c^2}} \tag{2.3}$$

2.1.2. Spheroid. A spheroid is an ellipsoid that has the same length of radius on both axes. The formula of the spheroid is written as follows [11]:

$$d_{sphere} = \frac{6V}{A} \tag{2.4}$$

$$V_{sphere} = \frac{\pi d^3}{6} \tag{2.5}$$

$$A_{sphere} = 4\pi r^3 \tag{2.6}$$

2.2. *Ranz-Marshall Analogy*

Based on the formula obtained from Ranz, W. E., & Marshall, W. R. [12], two equations are used as follows, where:

- Equation for heat transfer:

$$Nu = 2 + 0{,}6 Re^{1/2} Pr^{1/3} \tag{2.7}$$

- Equation for mass transfer:

$$Sh = 2 + 0{,}6 Re^{1/2} Sc^{1/3} \tag{2.8}$$

2.3. *Stagnant Film Model*

The stagnant film model can be illustrated with figure 1. That picture shows an evaporation of a cold liquid surrounded by a hot environment that carries the heat temperature of the heater to the liquid.

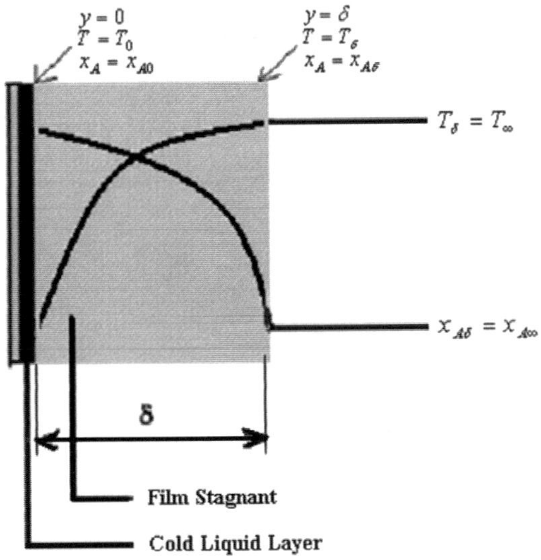

Figure 1. Illustration of stagnant film model

The total mass transfer rate of material A defined as N_{A0}, where δ_c is the thickness of concentration film, C is total concentration of A and B, D_{AB} is mass diffusivity, X is mole fraction, and where k_c is the mass transfer coefficient [8].

$$exp\left(\frac{N_{A0}\delta_c}{C.D_{AB}}\right) = 1 + \frac{X_{A0} - X_{A\infty}}{1 - X_{A0}} \tag{2.9}$$

$$N_{A0} = k_c(X_{A0} - X_{A\infty})/(1 - X_{A0}) \tag{2.10}$$

Because of $X_\delta = X_\infty$, eq. (2.9) and eq. (2.10) become,

$$exp\left(\frac{N_{A0}\delta_c}{C.D_{AB}}\right) = 1 + \frac{N_{A0}}{k_c} \tag{2.11}$$

When N_{A0} is approaching 0 on k_c, so k_c for low mass transfer approach to k_{cL}. we will obtain,

$$k_{cL} = \lim_{N_{A0}\to 0} k = \lim_{N_{A0}\to 0} \frac{N_{A0}}{exp\left(\frac{N_{A0}\delta_c}{C.D_{AB}}\right) - 1} \tag{2.12}$$

From eq. (2.9), can be simplified into,

$$N_{A0} = k_{cL}\left[ln\left(\frac{1 - X_{A\infty}}{1 - X_{AB}}\right) - C1\right] \tag{2.13}$$

Total heat transfer is defined as below where T is the temperature, k thermal conductivity, $\overline{C}_{pA}$ the molal specific heat.

$$q_0 = \frac{N_{A0}\overline{C}_{pA}(T_\infty - T_0)}{exp(-N_{A0}\overline{C}_{pA}\delta_T/k) - 1} \tag{2.14}$$

Using heat transfer coefficient, *h*, total heat transfer can be also defined as,

$$q_0 = h(T_0 - T_\infty) \tag{2.15}$$

From equation (2.14) and eq. (2.15) can be simplified into,

$$exp\left(-\frac{N_{A0}\overline{C}_{pA}\delta_T}{k}\right) = 1 - \frac{N_{A0}\overline{C}_{pA}}{h} \tag{2.16}$$

When N_{A0} is approaching 0 on *h*, so h for low mass transfer approach to h_L. we will obtain,

$$h_L = \lim_{N_{A0}\to 0} h = \lim_{N_{A0}\to 0} \frac{N_{A0}\overline{C}_{pA}}{-exp\left(-\frac{N_{A0}\overline{C}_{pA}\delta_T}{k}\right)+1} \tag{2.17}$$

From eq. (2.14), can be simplified into

$$q_0 + q_{cond} = \frac{N_0 C_P(T_\infty - T_0)}{\exp\left(-\frac{N_0 C_P}{h_L} - \frac{C_2 C_P}{k_L}\right)-1} \tag{2.18}$$

Since we have the value of N_0 and q_0, variation of droplet radius and temperature can be expressed as,

$$N_0 = -\left(\frac{\rho}{M}\right)\frac{dr}{dt} \tag{2.19}$$

$$\frac{dT}{dt} = 3\left(\frac{q_0 - N_0 h_{fg}}{N_0 r C p_A}\right) \tag{2.20}$$

By using all the formulas above, C1 and C2 then will be varied to overlook the best value of the constant for the modified stagnant film model

3. Methodology

In this experiment, the equipment that will be used in the experiment is arranged as shown in figure 2. The test was conducted in heat transfer laboratory, 6th floor MRC building, faculty of engineering, Universitas Indonesia. The main equipment used in this test consists of:

1. Inverter, used to adjust the speed that will be blown by the blower by changing the input frequency of electricity.
2. Shimaden SR 94, used to control the heat that will flow into the heater so that the heat that will be given as input can be controlled.
3. National Instrument (NI), used to calculate the actual data from the temperature at which the droplets are collected.

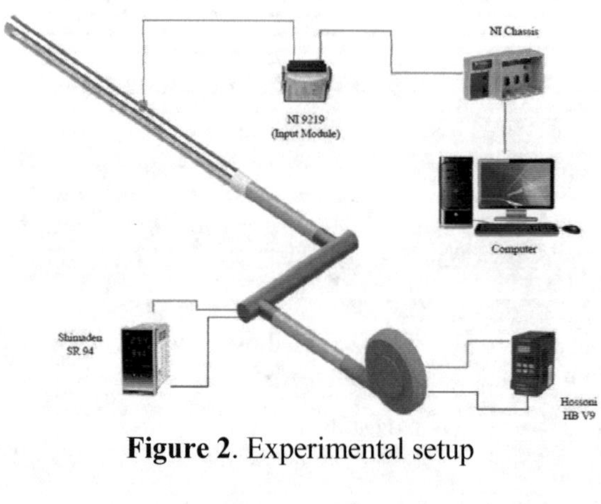

Figure 2. Experimental setup

In conducting data retrieval, several variations of input are needed that will be needed to obtain information from the tests to be performed. The following is input information used in this experiment is shown in table 1 below,

Table 1. Input parameter

Liquid	Temperature	Speed
2-Propanol	35°C	1 m/s
	35°C	7 m/s
	55°C	1 m/s
	55°C	7 m/s

After taking data using the input parameters as in table 1, then taking pictures per frame is obtained to get diameter information from the tested droplet. The pictures taken in this experiment were taken using a digital camera as shown in figure 3 below. Data from the diameter will then be processed to obtain correlations from this experiment

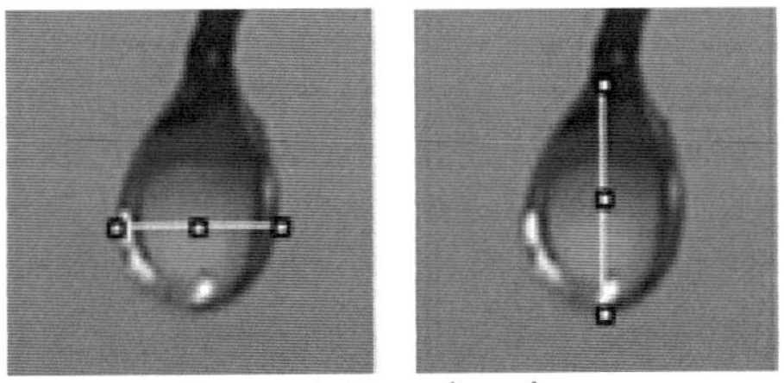

Figure 3. Droplet picture

After obtaining data from these diameters, data processing is then performed by calculating the values of C1 and C2 determined as follows in table 2:

Table 2. The value of C1 and C2

	35°C 1 m/s	35°C 7 m/s	55°C 1 m/s	55°C 7 m/s
C1 Value on:				
C1	-0.025	-0.025	-0.025	-0.025
C1*	-0.035	-0.013	-0.032	-0.02
SF	0	0	0	0
C2 Value on:				
C1	-5E-08	-5E-08	-8E-08	-7E-08
C1*	-5E-08	-5E-08	-8E-08	-7E-08
SF	0	0	0	0

4. Results and Discussions

After completing the research, the results obtained are the diameter and temperature of the droplet. With these results, it was calculated using the Ranz-Marshall formula as the basis for calculating the Nusselt and Sherwood number in this experiment. The value of the calculation results will certainly be used to obtain the coefficient values of h and k latent evaporation. With this value, then we will get the value of N_{A0} and also the value of q_0. Which will then be used as a basis for determining variations with the modification of the values of C1 and C2. The value of C1 and also C1* has the result of a predetermined modification by having a value not equal to zero. Then for the value of C1 on the Stagnant Film Model has a value of 0 because of the stagnant film. Then for the C2 value determined at -5E-08; -8E-08; and -7E-08. And the C2 value on the Stagnant Film Model has a value of 0.

And based on these data, the results will be obtained as in Figure 4 and Figure 5 below. The graph shown below is a graph of the time to temperature and also radius^2 of the 2-propanol liquid which is being tested.

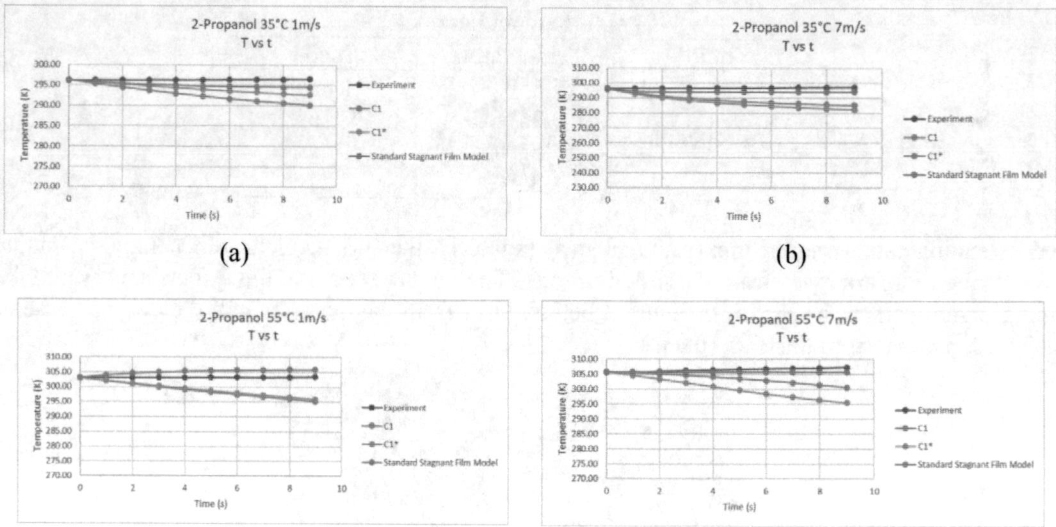

(a) (b)

(c) (d)

Figure 4. Graph temperature vs time of 2-Propanol for (a) 35°C 1 m/s, (b) 35°C 7 m/s, (c) 55°C 1 m/s, and (d) 55°C 7 m/s.

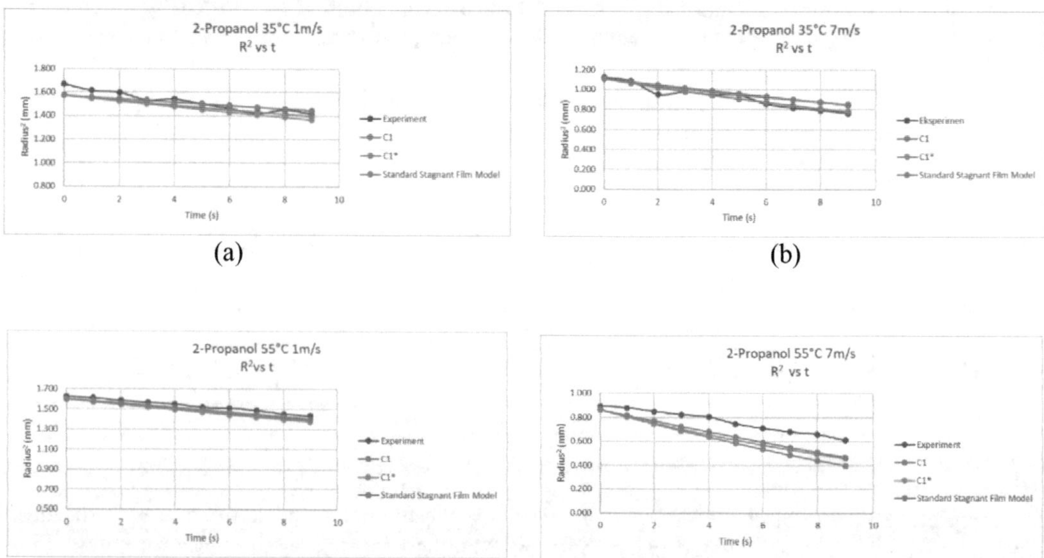

(a) (b)

(c) (d)

Figure 5. Graph radius^2 vs time of 2-Propanol for (a) 35°C 1 m/s, (b) 35°C 7 m/s, (c) 55°C 1 m/s, and (d) 55°C 7 m/s.

Based on the graph shown in figure 4, the graph represents the value of temperature over time that tends to increase. The characteristic findings obtained from all the variations tested have a feature that increases with time. But on the graph shown in figure 5, the results obtained are against the opposite.

Can be seen from that graphs in figure 5, where in this study the results obtained from the radius^2 has decreased over time.

This phenomenon can occur because the heat exhaled by the blower is increasing. So, with this increase, the value of Reynolds number will also increase. This will make the heat transferred from the heater to the droplet surface also increases. With the increase in the speed of hot air that is exhaled to the droplets, there will be an increase in latent heat and sensible heat. Sensible heat will increase due to an increase in temperature variations. And with the increasing temperature, latent heat will also increase which affects the phase change that will be faster.

Based on the results obtained in figure 4 and figure 5, can be seen that the deviation is possessed on the whole graph. In figure 4, the modified C1 value in the calculation results in an unsatisfactory temperature graph against time. These results are indicated by the results of the deviation on the graph which deviates greatly. However, the results in figure 5 show different results, where the resulting graph shows a satisfactory correlation with the results of the deviation on the radius^2 graph against time which is not far apart.

Based on this research, it was found that the value of the coefficient of the stagnant film has a value that is not equal to zero. That is because the formula of the following film stagnant:

$$exp\left(\frac{N_{A0}\delta_c}{C.D_{AB}}\right) = 1 + \frac{X_{A0} - X_{A\infty}}{1 - X_{A0}}$$

Based on the formula used from the stagnant film model, the values that exist in the stagnant film model will not have a value equal to zero. That is because the value of C1 will represent the droplet mass transfer integral, and the value of C2 will represent the droplet heat transfer integral. The value at C1 will intersect with the value of r^2 so that the stagnant film model will not have a value equal to zero.

5. Conclusion

From this experiment, concluded that modifying the value of C1 in the stagnant film model would give an unfavorable result in temperature changes graph against time. But the results obtained will be opposite with the results of the radius^2 graph against time. Modifications to the stagnant film model do not yet have a good C1 value, so the value at C2 also does not get optimal results. Then we need more in-depth research to obtain equations that will produce good correlations.

References

[1] Mezhericher, M., A. Levy, and I.J.D.T. Borde, *Theoretical models of single droplet drying kinetics: a review*. 2010. **28**(2): p. 278-293.

[2] Kitano, T., et al., *Evaporation and combustion of multicomponent fuel droplets*. 2014. **136**: p. 219-225.

[3] Harada, T., et al., *A numerical investigation of evaporation characteristics of a fuel droplet suspended from a thermocouple*. International Journal of Heat and Mass Transfer, 2011. **54**(1): p. 649-655.

[4] Nikolopoulos, N., A. Theodorakakos, and G. Bergeles, *A numerical investigation of the evaporation process of a liquid droplet impinging onto a hot substrate*. International Journal of Heat and Mass Transfer, 2007. **50**(1): p. 303-319.

[5] Liu, H., et al., *Investigation on the potential of high efficiency for internal combustion engines*. 2018. **11**(3): p. 513.

[6] Kosasih, E.A., et al. *Heat and mass transfer of ethanol and methanol droplets with temperature and air flow rates variations*. in *AIP Conference Proceedings*. 2019. AIP Publishing.

[7] Kosasih, E.A., et al. *Heat and mass transfer droplets of n-Hexane and n-Heptane towards temperature and rate of air flow rate variations*. in *AIP Conference Proceedings*. 2019. AIP Publishing.

IOP Conf. Series: Materials Science and Engineering **694** (2019) 012016 doi:10.1088/1757-899X/694/1/012016

[8] Kosasih, E. and M. Alhamid. *Heat and Mass Transfer in Water Droplet Evaporation: A New Approach on Film Stagnant Model*. in *The 9th Quality in Research (QIR) International Conference, Indonesia*. 2006.
[9] Bird, R.B., W.E. Stewart, and E.N. Lightfoot, *Transport phenomena*. 2007: John Wiley & Sons.
[10] Qian, Y., et al., *Experimental study on evaporation characteristics of lubricating oil/gasoline blended droplet.* 2019. **103**: p. 99-107.
[11] Beyer, W.H.J.B.R., Fl.: Chemical Rubber Co., and t.e. c, edited by Beyer, William H., *CRC standard mathematical tables and formulae.* 1991.
[12] Ranz, W. and W.R.J.C.e.p. Marshall, *Evaporation from drops.* 1952. **48**(3): p. 141-146.

Optimization of Nozzle Number, Nozzle Diameter and Number of Bucket of Pelton Turbine using Computational Fluid Dynamics and Taguchi Methods

O Y Leman, R Wulandari, and R D Bintara

Mechanical Engineering Department, Engineering Faculty, Universitas Negeri Malang, Indonesia

E-mail: retno.wulandari.ft@um.ac.id

Abstract. Along with the depletion of fossil fuel reserves in Indonesia, water resources can be used as an alternative source of energy. Pelton turbines are one type of water turbine that can be used. However, Pelton turbines still have weaknesses in terms of determining the optimal number of buckets and nozzles. This study aims to optimize the experimental variations in the number of buckets (N_b), the diameter of the nozzle (D_j), and the number of nozzles (N_j) on the efficiency of the Pelton Turbine. The number of buckets was varied namely 24 and 27 pieces, nozzle diameter 5.5 mm and 7 mm, and number of nozzles using a single nozzle and double nozzle with horizontal direction. The research design began with *Design of Experiment* (DoE) using Taguchi method in Minitab 18.0. The next step was to calculate and then designed turbine using Autodesk Inventor 2016. The turbine design was then simulated using ANSYS Fluent 17.0 Workbench. The calculation of the largest efficiency using the number of 27 buckets double nozzle 5.5 mm producing an efficiency of 64,69%. Based on Taguchi optimization *delta* result, the most influential variable on the efficiency was the nozzle diameter (M=0.1943).

1. Introduction

The increasing population in Indonesia is directly proportional to the electrical energy needs. The main energy source of power generation is still dependent on energy resources such as petroleum, natural gas and coal. The fossil energy resources are limited in number since fossil fuel is not a renewable energy and dwindling numbers because it continues to be exploited [1].

Water is one of the abundant renewable energy sources in number, thus it could potentially be a source of electrical energy generation alternative. According to the total water potential in Indonesia, which can be exploited by 691.3 million m^3/year, only 175.1 million m^3/year, or about 25.3% that had been utilized. While it is still 516.2 million m^3/year, or about 74.7% of untapped [2]. One of the tools that use water as an energy source for the plant is water turbines.

Pelton turbine is one type of water turbine that right to be used, because it does not require a large flow rate, relatively high efficiency, and a simple construction suitable for use as a power plant on picohydro scale (100-5000 watts) [3]. However, on the other hand Pelton turbines still have weaknesses in terms of determining the optimum number of bucket and nozzle on the efficiency of the turbine. According to this term required further research on Pelton turbine.

Research using computational approaches with methods *Computational Fluid Dynamics* (CFD) and optimization using the Taguchi method chosen in this study because it is an appropriate method when viewed from the side of efficiency in terms of time and cost. CFD methods used to find the value of the torque and velocity of the tangential water. While optimization using the Taguchi method is used to determine the most optimal experimental variation and the influence of each variable on the efficiency of turbine [4,5].

2. Methodology of Research

This study used quasi-experimental research that begins with a Design of Experiment (DoE) on Minitab 18.0 to obtain a variation experiments on variables that have been determined. These variables include the number of buckets 24 and 27, the number of single and double nozzle, and the nozzle diameter of 5.5 mm and 7 mm. The next step is conducting computational design of Pelton turbine using Autodesk Inventor Professional 2016 with a bucket and runner dimensions are calculated by using theoretical calculations. Each turbine section is formed that can represent the condition of the actual model.

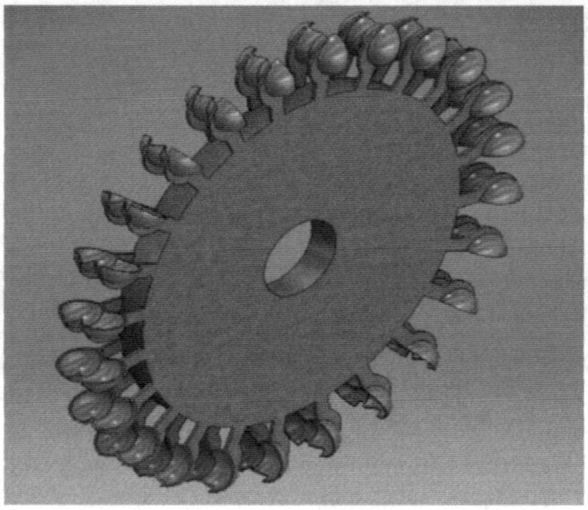

Figure 1. Computational Design of Turbine Pelton

The following step is conducting CFD analysis using software FLUENT in ANSYS Workbench 17.0 in order to obtain the value of torque and tangential velocity. This value is used as a parameter equation of Pelton turbine performance. In conducting CFD analysis using ANSYS Workbench 17.0 is divided into three steps: Pre-processing, Solving (Processing), and Post Processing.

The first step is pre-processing. In this step, the geometry of Pelton turbine model that designed is imported by using CAD software (Computer Aided Design). The following step conducting meshing or divide the volume into small parts that can be analyzed in ANSYS Workbench 17.0. At this step also set the boundary conditions are applied according to the conditions in the field. The boundary conditions include general, models, materials, cell zone conditions, boundary conditions, mesh, interfaces, dynamic mesh, references values.

The second step is Solving (Processing). Solving step is the core step of CFD analysis to calculate a solution based on the boundary conditions have been applied to the pre-processing step. At this step it is determined solution methods, solution controls, solution initialization, calculation activities, run calculation, and the number of iterations used.

The third step is post-processing. Post-Processing step is the last step in conducting CFD with interpretation of simulation data. The data generated can be a number, contours, vectors, curves, histograms, etc. [5].

CFD analysis in this study is used to find the results of torque and tangential velocity values that are used to calculate the performance of Pelton turbine includes a turbine wheel, power turbines, and turbine efficiency. The value of tangential velocity (v) is used in equation (1) to calculate the rotation of turbine wheel (n).

$$n = \frac{v \text{ x } 30}{\pi D} \quad (1)$$

While the torque value (T) is used in equation (2) to calculate the Turbine Power (Po).

$$Po = T \text{ x } 2\pi \text{ x } \frac{n}{60} \quad (2)$$

The result of the calculation of turbines power (power output) is used in the calculation of turbine efficiency. After obtaining all the efficiency values of each variation on experiment, the optimization is conducted by using the Taguchi method in each of experimental variation on the efficiencies generated.

DoE is an experimental design that contains of several experimental variables that can examine the changes because of the efficiency of Pelton turbine. By using Taguchi optimization can shorten experiment variation.

Table 1. Design of Experiment (DoE)

	Single-level designs		
Designs	**2 level**	**3 level**	**4 level**
L4	2-3		
L8	2-7		
L9		2-4	
L12	2-11		
L16	2-15		
L16			2-5
L25			
L27		2-13	
L32	2-31		

In Taguchi's design on Table 1, this study optimized using Minitab 18.0 produced four variations of the experiment are composed of three factors design and two level designs. Three factors include the number of bucket design, the number of nozzles and nozzle diameter, whereas the two-level design is 24 and 27 (the number of buckets), 1 and 2 (the number of nozzles), 5.5 mm and 7 mm (diameter nozzle). The experimental design based optimizations presented in Table 2.

Table 2. Variation of experiment

	Variables			Efficiency (%)
	Bucket Number	Total Nozzle	Nozzle diameter (mm)	
1	24	1	7	
2		2	5.5	
3	27	1	5.5	
4		2	7	

IOP Conf. Series: Materials Science and Engineering **694** (2019) 012017 doi:10.1088/1757-899X/694/1/012017

To validate and compare the results of the research design has been optimized using the Taguchi method, the study also made calculations on the eight variations of the experiment. The calculations include the performance of a turbine with full factorial design data retrieval as shown in Table 3.

Table 3. Full factorial design of data retrieval

	Variables			T	*Vt*	Rpm	P_i	*Po*	μ
	N_b	N_j	D_j						
1	24	1	5.5						
2			7						
3		2	5.5						
4			7						
5	27	1	5.5						
6			7						
7		2	5.5						
8			7						

3. Results and Discussion

Torque is a measure of rotational force that causes the turbine runner rotates around the axis point. The value of the torque is calculated using the calculator function in FLUENT software. When opening the solution menus, calculator function option can be found on the menu. Torque calculation results shown in Figure 2 shows the results of torque calculation by calculator function and shows the wall solid section where the torque is calculated, which is on the body turbine that is blue.

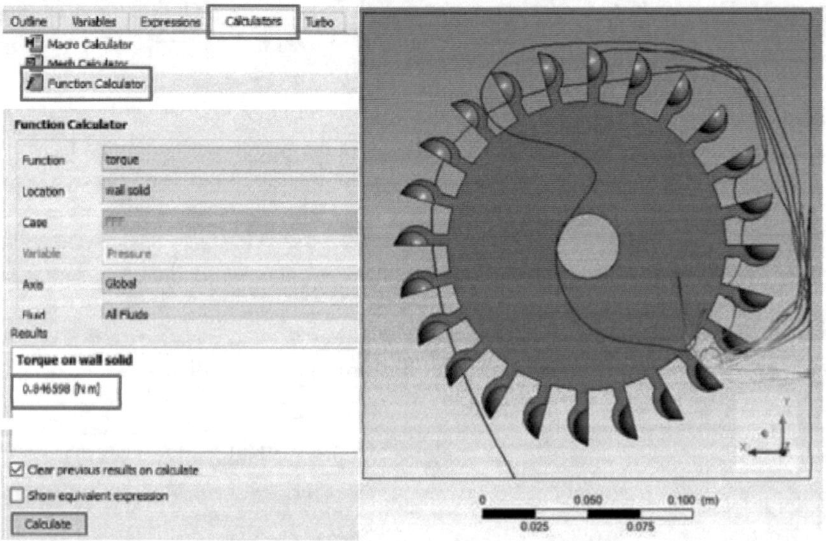

Figure 2. Torque on Bucket Number Variation of 24 with Single Nozzle of 5.5 mm

Variations experiment with bucket number of 24 with single nozzle of 5.5 mm generate torque values of 0.846598 Nm. The torque value is calculated on a wall solid that is all part of the Pelton turbine blue with global axis point is in the center of the design of the runner with the direction of the Z axis (blue axis) as the axis of rotation torque calculation.

The tangential velocity is the velocity of water coming out of the nozzle and mashing the inside of the bucket, causing the Pelton turbine spinning divided cos (90°-θ). The flow velocity is obtained by using the CFD method by using contour on result to bring its flow. Meanwhile, in order to determine the value of the tangential velocity is known to use a plane on the ground inside the bucket of water flow from the nozzle is divided cos angle in the bucket (Figure 3). Pounded bucket angle is 45° nozzle flow turbine with a number of nozzles 24 and 42° in the turbine nozzle number 27.

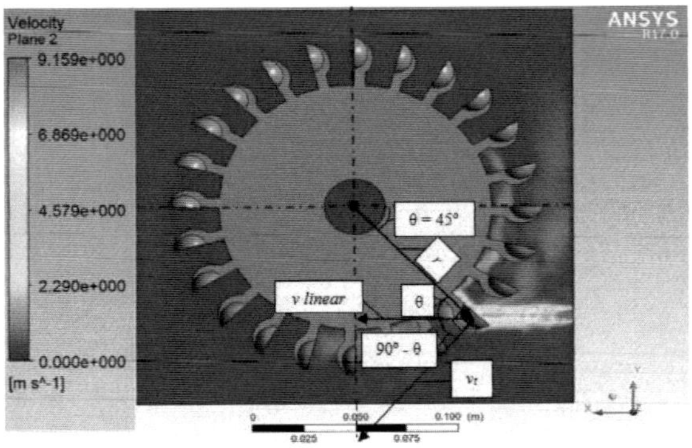

Figure 3. Tangential angle on Bucket 24

Tangential velocity (v) is calculated using trigonometry equation of $\cos\theta = \frac{v\ linear}{v}$, then the value of v can be calculated using Equation (3).

$$v = \frac{v\ linear}{\cos\theta} \tag{3}$$

Furthermore, the value of the tangential velocity is used as a variable to calculate the magnitude of the Pelton turbine rotational velocity expressed in rotations per minute (rpm).
Turbine power (Po) is the power generated by the turbine and the applicability used to turn a generator. Calculation of efficiency at each variation of the experiment is calculated using Equation (4).

$$P = T \text{ x } 2\pi \text{ x } \frac{n}{60} \tag{4}$$

To facilitate the identification of the calculation results power Pelton turbine, then presented a graph of the eight trials against power turbine in Figure 4.

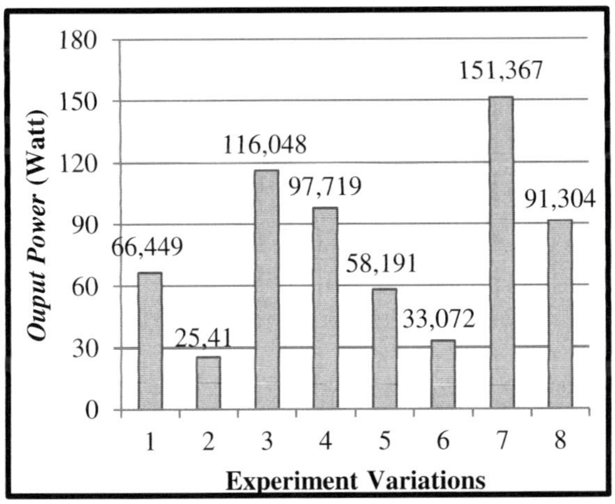

Figure 4. Graph of Turbine Power Calculation Results

Based on the graph in Figure 4 it can be seen that the largest turbine power values obtained in the experiment-7 with the number of buckets 27, double nozzle with diameter of 5.5 mm which results in a value of 151.367 watts of power turbines. The smallest turbine power value obtained in experiment 2 with the number of bucket 24, single nozzle with diameter of 7 mm turbine that generates power value of 25.41 watts. Referring to Equation (2), a large power turbine power output equation is affected by the value of the torque and rotation of the turbine. Great torque and a turbine wheel that will quickly produce output power (power turbine) that is greater to turn a generator.

Pelton turbine efficiency (η) is the ratio between the power generated by the turbine (Po) with the incoming water power (Pi) and expressed in units of percent (%). Water power is calculated using Equation (5).

$$P_i = \rho \cdot g \cdot C_n^2 \cdot H \cdot Q \cdot N_j \tag{5}$$

While the calculation of the efficiency is calculated using Equation (6).

$$\eta = \frac{P_o}{P_i} \text{ x } 100\% \tag{6}$$

To facilitate the identification of the calculation results efficiency Pelton turbine, then presented a graph of the eighth trial against efficiency turbine in Figure 5.

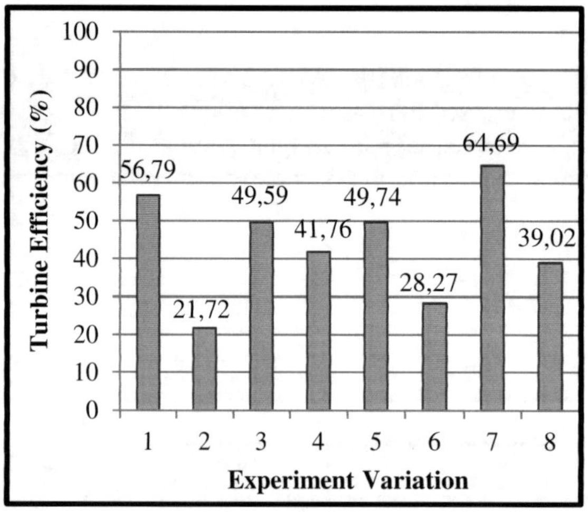

Figure 5. Graph of Turbine Efficiency Calculation Results

Based on the graph in Figure 5 can be seen that the largest turbine efficiency values obtained in the experiment-7 with the number of buckets 27, double nozzle with diameter of 5.5 mm which produces turbine efficiency value of 64.69%. The smallest turbine efficiency value obtained in experiment 2 with the number of bucket 24, single nozzle with diameter of 7 mm which produces turbine efficiency value of 21.72%. In a study [6] the efficiency of 64.69% at a reaction turbine is a high efficiency when the value obtained in experimental research. However, this study is a research using computational approach that many ignore factors outside interference or losses which are prevalent in the generator, the nozzle penstock and transmission system that caused the fall of the value of efficiency in experimental research.

Optimization testing using the Taguchi method is used to determine the factors most giving effect to the efficiency as well as determine the most optimal experimental variation. Optimization begins with the Design of Experiment (DoE) that produce a response in the form of Means.

Response main effect plot for means is average analysis of each variable of the study. Value Response Means test variation which has been optimized using the Taguchi method at each level presented in Table 4, while the value Response Means full factorial experiment variations presented in Table 5.

Table 4. *Response Table for Means* (Taguchi)

Level	Bucket Number	Total Nozzle	Diameter Nozzle
1	.4833	.4312	.6073
2	.5370	.5891	.4130
Delta	.0537	.1578	.1943
Rank	3	2	1

Table 5. *Response Table for Means* (Full Factorial)

Level	Bucket number	Total Nozzle	Diameter Nozzle
1	.5813	.5534	.7565
2	.6205	.6484	.4453
Delta	.0393	.0949	.3113
Rank	3	2	1

Table 4 and Table 5 shows the average ratio for each variation of the research that the number of buckets, the number of nozzles and nozzle diameter on the efficiency of the Pelton turbine. The results produced by the response means taking the average at every level variation resulting in an average difference. The magnitude of the effect of three parameters described in the Delta.

Table 4 shows the response sequence Means (M) is the highest to the lowest, namely Diameter Nozzle (M = 3.612), Total Nozzle (M = 3.022), and the amount Bucket (M = 1.530). Based on the results Means response to any variation of the research note that the parameters of the nozzle diameter has the highest influence on the efficiency of the Pelton turbine for generating value Delta is the greatest of the number of nozzles, and the number of buckets. Election results level setting of optimal factors in Table 4, namely Total Bucket level 2 (27 pieces) with a value of -5.405 Means response, Total Nozzle level 2 (double nozzle) by Means response value of -4.659, and the diameter of the nozzle level 1 (5.5 mm) with a value of -4.364 Means response.

In Table 5 the order of highest to lowest Means response showed the same order as Table 4 but with a different value Delta. The sequence is Diameter Nozzle (M=4.785), Total Nozzle (M=1.859), and the amount Bucket (M=0.714). Election results level setting of optimal factors in Table 5 also shows the selection of the same level as Table 4 with values different response. Selection of these levels are Total Bucket level 2 (27 pieces) with a value of -4.507 Means, Total Nozzle level 2 (double nozzle) by Means response value of -3.934, andLevel 1 Nozzle diameter (5.5 mm) with a value of -2.471 Means response.

Based on the results of the response means in Tables 4 and 5, the design of the experiment variations using optimization Taguchi or full factorial showed the same results in the order of the variables most affect the efficiency of the Pelton turbine, the diameter of the nozzle in the first place, the number of nozzles in the second, and the number of bucket in third. Taguchi or full factorial optimization also showed the same results in the selection of the most optimal level, ie the number of bucket level 2 (27 pieces), the number of nozzles level 2 (double nozzle), and level 1 nozzle diameter (5.5 mm).

IOP Conf. Series: Materials Science and Engineering **694** (2019) 012017 doi:10.1088/1757-899X/694/1/012017

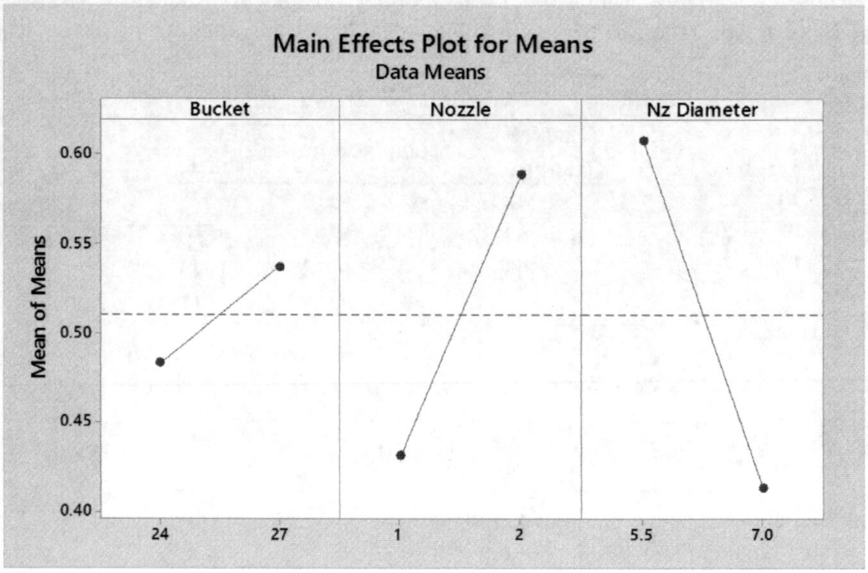

Figure 6. Graph of Main Effect Plots Means on the Research Variation (Taguchi)

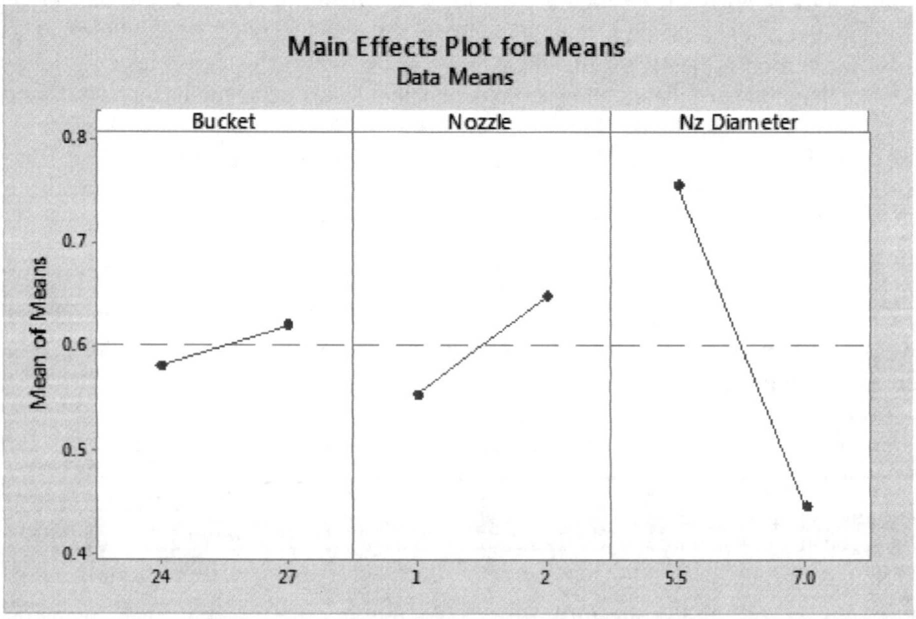

Figure 7. Graph of Main Effect Plots Means on the Research Variation (Full Factorial)

ANOVA (Analysis of Variance) is one of the statistical methods used to investigate the relationship between the response variable/dependent with one or several independent variables. In this study, ANOVA was used only to determine the influence of the variables on the efficiency of the Pelton turbine. Table 6 shows a large percentage of research variables influence on the efficiency of the variations of the experiment based on the Taguchi method, while Table 6 shows a large percentage of variable influence on the efficiency study on the variation of Full Factorial experiments.

Table 6. ANOVA on Experiment Taguchi

Analysis of Variance Taguchi							
Source	DF	Seq SS	Contribution	Adj SS	Adj MS	F-Value	P-Value
Bucket number	1	0,002889	4,41%	0,002889	0,002889	*	*
Total nozzle	1	0,024914	38,00%	0,024914	0,024914	*	*
Nozzle diameter	1	0,037762	57,59%	0,037762	0,037762	*	*
Error	0	*	*	*	*		
Total	3	0,065564	100,00%				

Table 6 shows the percentage contribution of the study variables (factors) on the efficiency in Orthogonal Array (OA) L4 or four variations of the experiment based Optimization of Taguchi. In ANOVA with OA L4 shows that the percentage of contribution is the largest diameter nozzle that has contributed to the response percentage of 57.59%, followed by Total nozzle that produces the second largest percentage of contribution is equal to 38%, and Total Bucket produce the lowest percentage of contribution that is equal to 4.41%.

Table 7. ANOVA of Full Factorial Experiment

Source	DF	Seq SS	Contribution	Adj SS	Adj MS	F-Value	P-Value
Bucket Number	1	0,003082	1,20%	0,003082	0,003082	0,29	0,620
Total Nozzle	1	0,018023	7,00%	0,018023	0,018023	1,69	0,264
Nozzle Diameter	1	0,193792	75,23%	0,193792	0,193792	18,16	0,013
Error	4	0,042695	16,57%	0,042695	0,010674		
Total	7	0,257592	100,00%				

Table 7 shows the percentage of contribution to the efficiency of research variables on Orthogonal Array (OA) L^8 or 8 variations of the experiment. In Full Factorial ANOVA with OA showed that the largest percentage of contribution resulting diameter nozzle that has contributed to the response percentage of 75.23%, followed by Total nozzle that produces the second largest percentage of contribution that is equal to 7%, and Total Bucket produce the lowest percentage of contribution that is equal to 1.2%.

4. Conclusion

The efficiency value is generated at full factorial experiment variations to-7 using a double nozzle with a nozzle diameter of 5.5 mm, and the number of bucket 27 pieces which produce an efficiency of 64.69%. Taguchi methods of optimization results based on the responses of Means (M) shows the most optimal experimental variation is the variation of the number of bucket 27 (M=0.5370), a double nozzle (M=0.5891) with a diameter of 5.5 mm (M=0.6073). Based on the Delta, the variables that most influence on the efficiency of the largest to the smallest diameter of the nozzle (M=0.1943), the number of nozzles (M=0.1578), and the number of buckets (M=0.0537). In further research can be done using other optimization methods such as RSM, etc. using research variable else to validate these results.

Acknowledgments
Thanks to the Department of Mechanical Engineering, Universitas Negeri Malang, especially the Simulation and Design Laboratory, which helped carrying out this research and facilitated hardware and Ansys software as a research tool.

References

[1] Indonesia Energy Outlook. Indonesia Energy Outlook 2018, 2018, Jakarta: BPPT (Agency for the Assessment and Application of Technology).

[2] Central Bureau of Statistics (BPS). 2017. Environmental Statistics Indonesia, 2017 Jakarta: The Central Bureau of Statistics.

[3] Sularso & Kiyokatsu, S. 1987. Basic Element Selection Planning and Engineering. Jakarta: Pradya Paramita.

[4] Wulandari, R., Maharani, D., Permanasari, AA, 2017. Design and Analysis of goose-leg of the waterwheel next-G using computational fluid dynamics (CFD). AIP Conference Proceedings 1887, 020 050.

[5] Židonis, A., & Aggidis, G. 2016. Pelton turbine: Identifying the optimum number of buckets using CFD. Journal of Hydrodynamics Vol.28 (1). (75-83). Information From http://www.sciencedirect.com/science/journal/10016058.

[6] Sentanuhady, J., Sanchorehan, D., 2016. Analysis and Numerical Simulation for Low Head Pelton Turbine Operated in Grojogan Sewu Waterfall. Proceedings of the National Symposium on Thermofluids VIII. Yogyakarta.

IOP Conf. Series: Materials Science and Engineering **694** (2019) 012018 doi:10.1088/1757-899X/694/1/012018

Development and application of a biomass burner using nyamplung seed cake as feedstock for pyrolysis process

M Syamsiro[1,2], N A Putra[1], J Winarno[1] and S R Ika[2,3]

[1]Department of Mechanical Engineering, Janabadra University, Jalan Tentara Rakyat Mataram 55-57 Yogyakarta 55231 Indonesia
[2]Center for Waste Management and Bioenergy, Janabadra University, Jalan Tentara Rakyat Mataram 55-57 Yogyakarta 55231 Indonesia
[3]Faculty of Economics, Janabadra University, Jalan Tentara Rakyat Mataram 55-57 Yogyakarta 55231 Indonesia

E-mail: syamsiro@janabadra.ac.id

Abstract. The main issue in running the pyrolysis plant for waste plastics and waste tire is how to operate the plant efficiently with the minimum energy required for heating the reactor. Therefore, the alternative heating source for the pyrolysis reactor should be applied to reduce the energy cost. Development of a novel biomass burner has been successfully fabricated, tested and applied for pyrolysis system. The experiments have been conducted to study the effect of air supply and fuel on the combustion performance of the burner. The results show that the thermal efficiency of the burner was higher by utilizing nyamplung seed cake as a feedstock compared to wood pellet. However, the maximum flame temperature of nyamplung seed cake was lower than that of wood pellet. This is due to the lower density of nyamplung seed cake compared to wood pellet which was lowering the combustion rate in the burner. The increase of the air supply generates the increase of the maximum flame temperature and the decrease of thermal efficiency. Moreover, the higher air supply the higher the flame length would be in which it will be one of the important parameters when designing larger scale of biomass burner and the pyrolysis reactor.

1. Introduction

Plastic waste has become a major stream in municipal solid waste increasing every year. This is due to the superior characteristics of plastics such as lightweight, corrosion resistant and good insulation properties. The characteristic of plastic waste that is different from organic waste is the difficulty of decomposing in the soil, it takes tens or hundreds of years to be completely degraded. To minimize the environmental impact of plastic waste, this material must be recycled to recover the material or to produce other products such as fuels and chemicals. There are several methods to recycle the plastic waste, namely mechanical recycling, feedstock recycling and energy recovery [1].

Feedstock or chemical recycling is a promising technology in which plastic waste is converted into smaller molecules in the form of liquid or gas with pyrolysis technology to produce fuels and chemicals. Pyrolysis is the process of thermal decomposition of polymeric materials by heating without involving oxygen in the process. It generally takes place at temperatures between 400-800 °C depending on the type of plastic and the product target. The product of this pyrolysis consists of gas, liquid and solid residue fractions. During the process, the long chain of hydrocarbons will be cracked into short chains. This liquid product will later become fuel, both in the form of gasoline and diesel fuel. Some plastic waste has been pyrolyzed such as CD case [2], polypropylene (PP), polystyrene (PS) [3] and other materials. The liquid product has also been tested in a diesel engine as a partial substitute for diesel fuel [4].

Therefore, converting plastic waste into fuels with pyrolysis technology is a very prospective choice for plastic recycling which cannot be recycled mechanically due to economic considerations. The main issue in running the pyrolysis plant for waste plastics and waste tire is how to operate the plant efficiently with the minimum energy required for heating the reactor. Therefore, the alternative heating source for the pyrolysis reactor should be applied to reduce the energy cost.

Biomass is one of the alternative source for heating the pyrolysis reactor which is abundant in Indonesia. Biomass has the potential to become one of the main energy sources in the future, and the modernization of the bioenergy system is suggested as an important contributor to the development of sustainable energy in the future, especially for sustainable development in industrialized and developing countries [5]. As a result, there will be a massive mobilization of biomass supply in an effort to meet energy needs in each region [6].

One of the biomass source which has the potential as an alternative energy is nyamplung (*Calophyllum inophyllum*) mostly found in the southern Central Java province such as Cilacap and Kebumen. Utilization of nyamplung as alternative fuels has been carried out, especially nyamplung seed oil as a substitute for kerosene and diesel fuel [7]. The pyrolysis process has been used by some researchers to produced bio-oil from nyamplung [8][9][10]. The oil yield depends on some parameters such as moisture content and particle size [11]. Fuel oil production from nyamplung will generate the solid residual waste produced after pressing nyamplung in the machine. The nyamplung seed cake still has some energy content which can be used as fuel for burner.

The challenge in developing pyrolysis of plastic waste in developing countries such as Indonesia is how to design cheap and easily fabricated technologies, and use independent or abundant energy sources. The implementation of this technology is more focused on small and medium scale by looking at the current characteristics of plastic waste management that involve waste banks at the village.

The integrated plastic waste pyrolysis system consists of several main parts, namely the chopper machine, pyrolysis machine, charcoal pellet machine and burner. The complete scheme of this system can be seen in Figure 1. The chopper machine reduces the particle size of the plastic waste before it is put into the pyrolysis machine in order to maximize the reactor capacity. The energy source used to heat the pyrolysis machine is taken from the gas, residual solids from the pyrolysis and biomass.

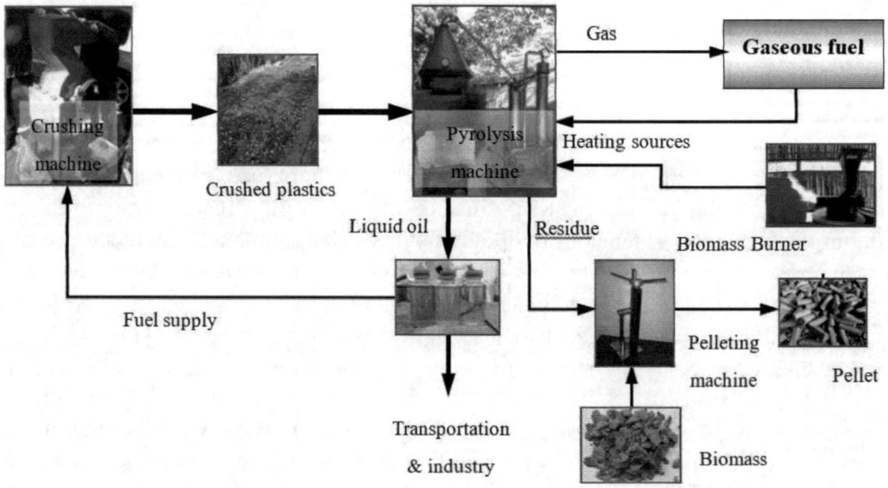

Figure 1. Integrated waste plastic pyrolysis system for small and medium scale application.

2. Materials and Methods

2.1. Materials

The biomass used as feedstocks in these experimental works was nyamplung seed cake and wood pellet. Nyamplung seed cake was collected from nyamplung processing plant in Badan Usaha Milik Desa (BUMDes) Panggung Lestari, Bantul producing oil and raw materials for cosmetics and pharmacy industries. Wood pellet has been used as comparative feedstock supllied from PT. Mahya Bioenergi, Boyolali. The physical appearances of these feedstocks can be seen in Figure 2.

Figure 2. Nyamplung seed cake and wood pellet samples used in the experiments.

2.2. Biomass burner design

The burner has been designed for the application using solid fuel such as biomass and waste. The burner consists of hopper, feeding system, air intake and blower, ash removal and flame output. Burner lid was designed with a water seal to prevent the smoke and air coming out from this part. The feeding system can be used to prevent backflow of the air. The air supply was coming from the blower. A schematic design of the biomass burner used in these works is shown in Figure 3.

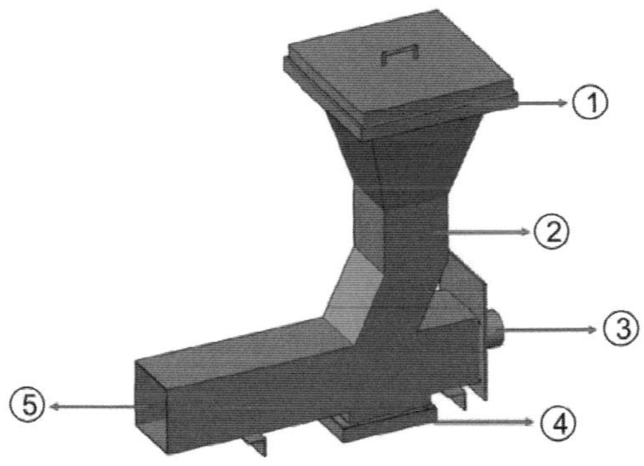

1. Fuel hopper
2. Fuel feeding system
3. Air intake
4. Ash removal
5. Flame output

Figure 3. Burner Design

One of the important part of the burner is ash removal system. This part was used to remove the ash produced in combustion of biomass. The ash would be droped and collected to the bottom part of burner during the process. Water had been filled at the bottom to inhibit the ash from spreding into the air. This system was very important in continuous combustion system since the ash can block the flame during combustion if still remain in combustion chamber. Figure 4 shows the ash removal system used in the burner.

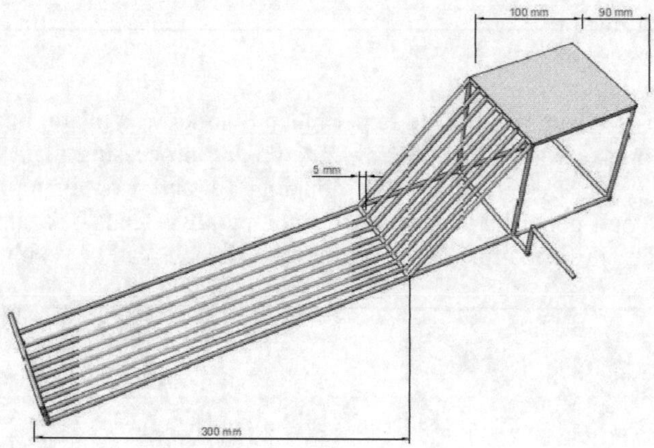

Figure 4. Ash removal system.

2.3. Experimental design

The experimental work had been carried out using nyamplung seed cake and wood pellet as burner fuels. This study aims to investigate the effect of air flow rate on the fuel consumption, thermal efficiency, and flame temperature of the burner for each feedstock. The air flow rate was measured using digital anemometer. The thermal efficiency was calculated based on water boiling test (WBT) method which commonly used for testing the cookstove. A temperature probe (K-type thermocouple) was put on the flame output to measure the flame temperature of each fuel and air flow rate.

The thermal efficiency of the burner was calculated based on WBT test method using the following formula :

$$\text{Thermal Eff.} = \frac{4{,}186\ x\ (Pcl - Pc)x(Tcf - Tci) + 2260\ x\ (Wcv)}{Fci\ x\ LHV} \text{x}100\% \quad (1)$$

where:

LHV = Lower Heating Value of biomass (kJ/kg)
Fci = mass of solid fuel (kg)
Pcl = initial mass of pot and water (kg)
Pc = mass of pot (kg)
Tci = initial mass of water (kg)
Tcf = mass of water after testing (kg)
Wcv = mass of water vapour (kg)

3. Results and Discussion

3.1. Effect of air flow rate on fuel consumption

The experimental works had been conducted at the air flow rate of 7, 10.5, and 15 m/s for each feedstock to study the effect of oxygen supply on the performance of the burner. The results show that the increase of the air flow rate will increase the fuel consumption of the burner for both feedstocks as shown in Figure 5. The reason is when the air flow rate increase, it means that the oxygen supply will also be increased. This condition will effect to more combustion reaction occured in the combustion chamber. As a result, the fuel combusted in the chamber will increase.

The figure also shows that nyamplung seed cake has higher fuel consumption rate compared to wood pellet. The density of the fuel might be the key factor for this trend. Nyamplung seed cake has lower density than wood pellet since it has not been mehanically pressed yet. The density of the fuel would affect the supplying of the feedstock into the chamber. This condition would reduce the feedstock rate to the burner.

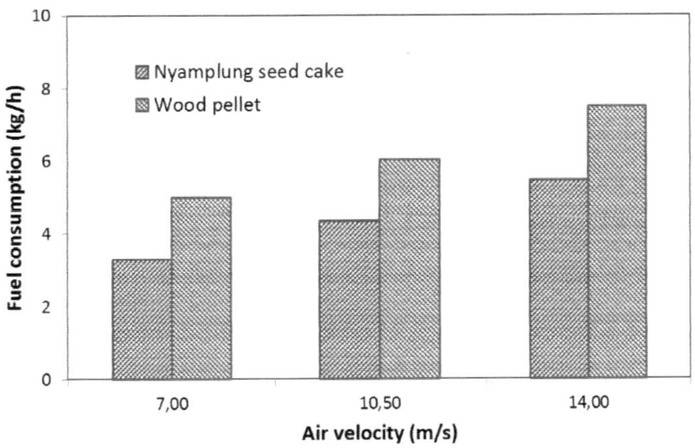

Figure 5. Effect of air flow rateon the fuel consumption of biomass burner.

3.2. Effect of air flow rate on thermal efficiency

Figure 6 showed the thermal eeficiency of the burner as the effect of air flow rate. From the figure, we can see that the maximum thermal efficiency of the burner can be achieved at the air flow rate of 10.5 m/s for both fuels. The lower and higher air flow rate will reduce the thermal efficiency of the burner. This is due to the lack of oxygen supply in the lower air flow rate which affected to the combustion process in the chamber. Some fuel may be not combusted completely. When the air flow rate increase, more oxygen will be supplied which lead to reduce combustion temperature since the oversupply of the air.

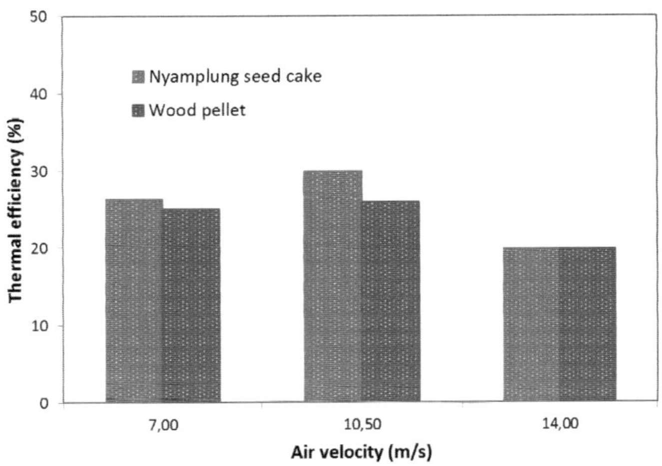

Figure 6. Effect of air flow rate on the thermal efficiency of biomass burner.

3.3. Effect of air flow rate on flame temperature

The most important thing when using burner for pyrolysis process is flame temperature. During the pyrolysis process, the minimum temperature is required to keep the pyrolysis reaction occured. Figure 7 shows the flame temperature of burner using nyamplung seed cake at variuous air flow rate. It shows that the increase of air flow rate increase the flame temperature. More oxygen supply affect the reaction to become more intensive.

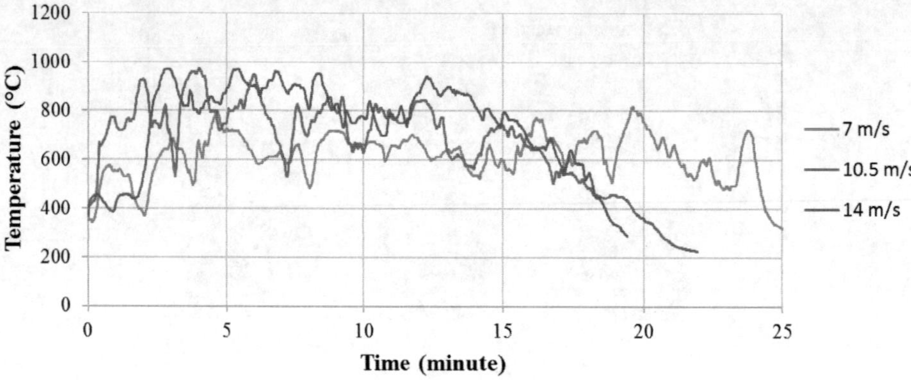

Figure 7. Effect of air flow rate on the flame temperature of burner using nyamplung seed cake.

The effect of air flow rate on the flame temperature of burner using wood pellet as a fuel is shown in Figure 8. From the figure, we can also see that the highest flame temperature can be achieved at the air flow rate of 10.5 m/s. When the air flow rate was increased to 14 m/s, the flame temperature was slightly reduced. The reason is that wood pellet has higher density compared to nyamplung seed cake. The high density of fuel will make the air more difficult to penetrate inside the pellet. Therefore, the reaction rate was lower than that of nyamplung seed cake.

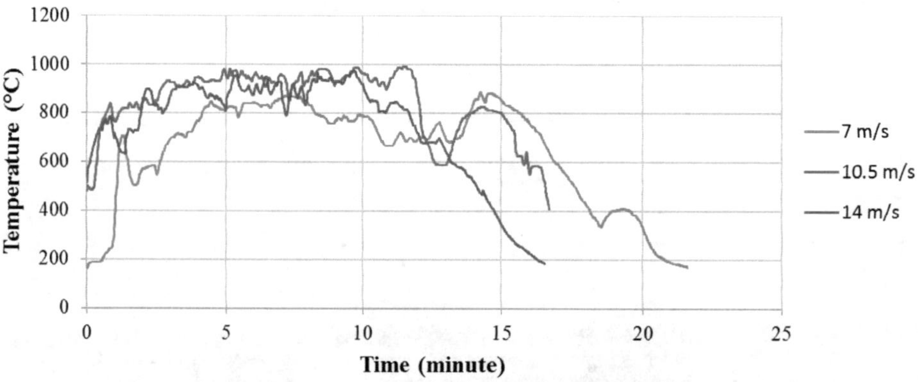

Figure 8. Effect of air flow rate on the flame temperature of burner using wood pellet.

Figure 9 shows the comparison of the flame temperature between nyamplung seed cake and wood pellet. The result shows that wood pellet generate higher flame temperature compare to nyamplung seed cake. Higher density and higher heating value of wood pellet resulted in higher flame temperature. The figure also shows that the flame temperature of wood pellet was more stable than that of nyamplung seed cake. The uniform characteristics of wood pellet will make steady combustion process.

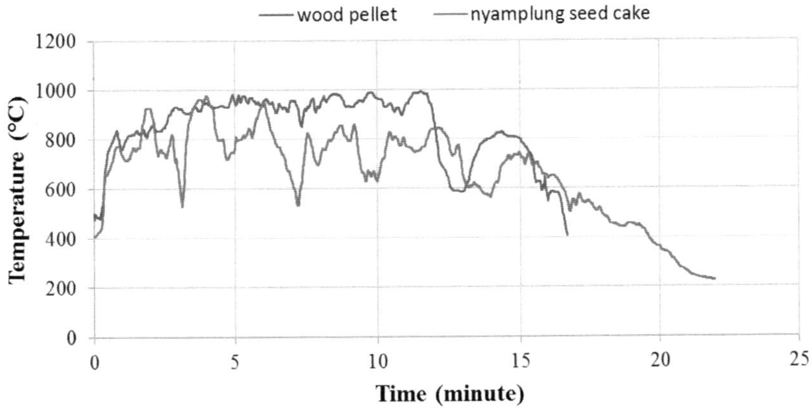

Figure 9. Comparison of flame temperature of burner using wood pellet and nyamplung cake seed at the air flow velocity of 10.5 m/s.

The maximum and average flame temperature of burner using nyamplung seed cake and wood pellet was summarized in Table 1. Wood pellet produced higher maximum and average flame temperature compared to nyamplung seed cake. The average flame temperature was high enough for pyrolysis process which requires the temperature around 400-500°C. The highest average flame temperature of burner by using nyamplung seed cake can be achieved at the air flow rate of 14 m/s.

Table 1. Flame temperature of burner using wood pellet and nyamplung seed cake.

Feedstock	Air flow rate (m/s)	Maximum temperature (°C)	Average temperature (°C)
Nyamplung seed cake	7	818	629.8
	10.5	975.2	783.9
	14	972.6	792.1
Wood pellet	7	887	748.4
	10.5	991.9	859
	14	973.4	837.5

4. Conclusion

Development of a biomass burner has been successfully fabricated, tested and applied for pyrolysis system. The results show that the increase of the air flow rate would increase the fuel consumption of the burner for both feedstocks. Nyamplung seed cake had lower fuel consumption rate compared to wood pellet. The density of the fuel might be the key factor for this trend. The contrary results show that the thermal efficiency of the burner was higher by using nyamplung seed cake as a feedstock compared to that of wood pellet. However, the maximum flame temperature of nyamplung seed cake was lower than that of wood pellet. This is due to the lower density and lower heating value of nyamplung compared to wood pellet which lowering the combustion rate in the burner. The increase of air supply increase of the maximum flame temperature but decrease the thermal efficiency. The flame temperature of wood pellet was also more stable than nyamplung seed cake. The uniform characteristics of wood pellet would make steady combustion process.

Acknowledgments
This study is financially supported by the Ministry of Research, Technology and Higher Education, Republic of Indonesia through Program Kemitraan Masyarakat (PKM) 2019. The authors would also like to thank BUMDes Panggung Lestari for technically supporting the program.

References

[1] S. M. Al-Salem, P. Lettieri, and J. Baeyens, "Recycling and recovery routes of plastic solid waste (PSW): A review," *Waste Manag.*, vol. 29, no. 10, pp. 2625–2643, 2009.

[2] M. Syamsiro, D. Y. Mathias, H. Saptoadi, D. R. Sawitri, A.-S. Nizami, and M. Rehan, "Pyrolysis of Compact Disc (CD) Case Wastes to Produce Liquid Fuel as a Renewable Source of Electricity Generation," *Energy Procedia*, vol. 145, pp. 484–489, 2018.

[3] M. Syamsiro *et al.*, "Liquid and Gaseous Fuels from Waste Plastics by Sequential Pyrolysis and Catalytic Reforming Processes over Indonesian Natural Zeolite Catalysts," *Waste Technol.*, vol. 2, no. 2, pp. 44–51, 2014.

[4] M. Syamsiro, H. Saptoadi, M. Kismurtono, Z. Mufrodi, and K. Yoshikawa, "Utilization of waste polyethylene pyrolysis oil as partial substitute for diesel fuel in a DI diesel engine," *Int. J. Smart Grid Clean Energy*, vol. 8, no. 1, pp. 38–47, 2019.

[5] G. Berndes, M. Hoogwijk, and R. Van Den Broek, "The contribution of biomass in the future global energy supply: A review of 17 studies," *Biomass and Bioenergy*, vol. 25, no. 1, pp. 1–28, 2003.

[6] A. Welfle, P. Gilbert, and P. Thornley, "Increasing biomass resource availability through supply chain analysis," *Biomass and Bioenergy*, vol. 70, no. 0, pp. 249–266, 2014.

[7] S. Rajamohan and R. Kasimani, "Studies on the effects of storage stability of bio-oil obtained from pyrolysis of Calophyllum inophyllum deoiled seed cake on the performance and emission characteristics of a direct-injection diesel engine," *Environ. Sci. Pollut. Res.*, vol. 25, no. 18, pp. 17749–17767, 2018.

[8] N. Ashwath, H. Nam, and S. C. Capareda, "Optimising pyrolysis conditions for thermal conversion of beauty leaf tree (Calophyllum inophyllum L.) press cake," *Green Energy Technol.*, no. 9789811006951, pp. 267–280, 2018.

[9] S. Rajamohan and R. Kasimani, "Analytical characterization of products obtained from slow pyrolysis of Calophyllum inophyllum seed cake: study on performance and emission characteristics of direct injection diesel engine fuelled with bio-oil blends," *Environ. Sci. Pollut. Res.*, vol. 25, no. 10, pp. 9523–9538, 2018.

[10] H. Dewajani, Rochmadi, S. Purwono, and A. Budiman, "Catalytic cracking of Indonesian nyamplung oil (Calophyllum inophyllum) for bio-fuels production using ZSM-5 catalyst," *J. Eng. Sci. Technol.*, vol. 10, no. Spec.issue7, pp. 61–69, 2015.

[11] M. Fadhlullah, S. N. B. Widiyanto, and E. Restiawaty, "The potential of nyamplung (Calophyllum inophyllum L.) seed oil as biodiesel feedstock: Effect of seed moisture content and particle size on oil yield," *Energy Procedia*, vol. 68, pp. 177–185, 2015.

IOP Conf. Series: Materials Science and Engineering **694** (2019) 012019 doi:10.1088/1757-899X/694/1/012019

Comparison of Jet Diffusion Flame Characteristics and Flame Temperature of Dimethyl Ether (DME) and Liquefied Petroleum Gas (LPG)

R Anggarani[1,2], C S Wibowo[1,2], Maymuchar[2] and I M K Dhiputra[1]

[1]Mechanical Engineering Dept., Universitas Indonesia, Jl.Kampus UI, Depok 16424
[2]Research and Development Center for Oil & Gas Technology "LEMIGAS", Jl.Ciledug Raya Kav.109, Kebayoran Lama, South Jakarta 12230

E-mail: riesta.anggarani@esdm.go.id

Abstract. Dimethyl Ether (DME) is currently considered as alternative to be used as energy source. Its similar physical properties with LPG made it possible to be used in household gas stove using same apparatus. Concern comes on its combustion properties regarding its lower calorific value than LPG. This research aimed to investigate the jet diffusion flame characteristics of DME compare to LPG in terms of flame height and lift-off flame, with correlation being made with fuel flow rate and burner hole diameter. The flame temperature is also measured on the same experiments. The results show that DME produce lower flame height than LPG. The flame height produced by both fuels were independent to burner hole diameter. The transitional regime in term of Reynold number is differ significantly between DME and LPG where DME reach the transitional regime at lower R_e. The flame temperature of DME at nozzle tip and dark zone is higher than LPG.

1. Introduction

Dimethyl Ether (DME) is investigated intensively around the world as its similarity properties with Liquefied Petroleum gas (LPG) made it possible to substitute LPG [1]. DME can be made from various feedstock such as coal, natural gas, biomass, and from crude oil. All of these are attractive alternative energy sources. Many research were conducted to investigate the conversion process of renewable source to produce DME. A wind-based electrolytic hydrogen and CO_2 captured from ethanol fermentation process are among the newest renewable feedstock to produce DME [2]. The biomass that can be vary in a wide range based on local vegetation is also a very attractive source for DME. Rice straw bio-DME is studied in Thailand [3], black-liquor gasification to produce DME already established in Sweden [4], and a two-stage gasification concept is proposed to utilize low-grade residual biomass from agriculture and forestry to produce wide range of synthetic fuels including DME [5].

The use of DME as alternative fuel to substitute current fuels especially crude oil derivatives also becomes interesting topics among researcher. One of the most widely studied are the characteristics of Jet Diffusion Flame (JDF) of DME. The study of this field is very useful on the application of DME for gas fuels for domestic gas stoves, kilns and burners [6]. The JDFs of DME is investigated analytically by proposing a prediction model for Flame Length (L_f) and Width (W_f) and verify the result by sets of experiment [7]. The prediction model of flame length and flame width is very useful for the requirement of burner design and operation condition. The theory of predicting the flame dimension model can be looked back through on the paper of Hottel and Hawthorne [8]. A relation of the nozzle diameter, molecular diffusivity, time of flow from nozzle to tip, mol ratio of air and fuel are developed empirically to predict flame length. Another approach to predict flame dimension is made by Kang et al [6] and [7] by analysing the flame structure and combine it with the flow field with mass and momentum conservation. Wang et al also study the flame dimension in term of flame radiant fraction and flame volume under sub atmospheric condition and found there is relationship between flame volume and ambient pressure [9].

On the JDFs, flame lift-off (LO) also considered an important parameter. On designing burner, the lift off height is an important parameter to keep flame stable [10]. The mechanism of lift off flame is studied by An et al [11] by elucidating the transition mechanism between attached and lifted swirl flames. Other approach was made by investigate the role of chemical kinetics in turbulent lifted jet flames using multi-flamelet generated manifold approach [12].

One of the drawbacks on DME combustion is higher NO_x emission, Kim et al observed this finding in Compression Ignition (CI) engine [13]. The NO_x emission is expected to be higher than other gases because DME has higher adiabatic flame temperature. A study by Kang et al [14] investigate the mechanism of NO_x formation in DME flame under moderate or intense low-oxygen dilution (MILD). Under MILD condition, the NO_x formation of DME combustion happened by NNH and N_2O pathways, it differ from the traditional DME jet diffusion flame.

This study is aimed to investigate the JDFs characteristics in terms of flame height (F_H) of DME compare to LPG in correlation with fuel nozzle diameter (D_f) and fuel volumetric flow rate (Q_F) using the correlation build by Hottel & Hawthorne [8] and Kang et al [6] for the baseline. In relation with factors affecting the flame height, temperature of the flames also measured to give the analytical reasoning in constructing the flame height equation. All of the experiments conducted in diffusional combustion using a cylindrical burner which can be replaced of its upper part to change the fuel nozzle diameter, D_f.

2. Experimental Set Up

2.1 Materials

The DME used in this experiments is the high purity DME with purity of 99.8% produced by PT Bumi Tangerang Gas Industry in Banten. LPG used in this experiments is bought from PT PERTAMINA (persero) which has the composition of 60% m/m N-Propane and 40% m/m N-Butane. The physical properties of DME and LPG presented in Table 1.

Table 1. Physical properties of DME and LPG

Physical Properties	Unit	DME	LPG
Gas Density (ρ_G), @1atm, 25^{0}C	kg/m^3	1.92	2.08
Dynamic Viscosity (μ_G), @1atm, 25^{0}C	kg/m.s	$9.1.10^{-6}$	$7.85.10^{-6}$
Saturated Pressure (P_{sa})	MPa	0.61	0.63
Specific Heat (C_p)	kJ/kg.K	1.43	1.73
Auto Ignition Temp.	K	508	717
Adiabatic Flame Temp.	K	2293	2263
LHV	MJ/kg	28.8	46.1

2.2 Diffusion Flame Experiment Set Up

The experiments were performed in Combustion Laboratory of Research and Development Centre for Oil and Gas Technology "LEMIGAS". A working table equipped with fitted cylindrical burner, supporting bar for assembling the ruler, a line of stainless pipeline for delivering the fuel from gas cylinder into burner and the measuring instrument is used to conduct the jet diffusion flame experiments. This apparatus is designed to be used in atmospheric pressure without any closure to simulate the jet diffusion flame in an external combustion burner such as domestic gas stove.

The experimental schematic is shown in Figure 1.

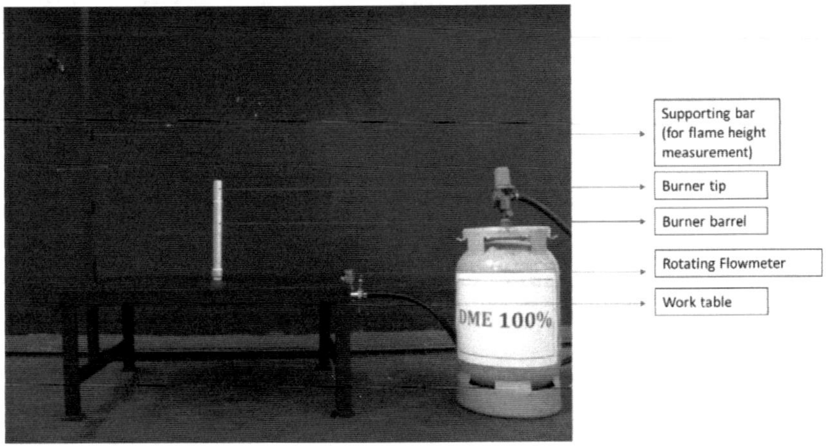

Figure 1. Experimental set up for jet diffusion flame measurement

The cylindrical burner used in this experiment is divided into two (2) parts; the burner barrel and burner head. Both parts are connected using screw connection. The burner head can be replaced to get the variation of fuel nozzle diameter, because the fuel nozzle is placed on the center top of burner head. Figure 2 shows us the design of cylindrical burner used.

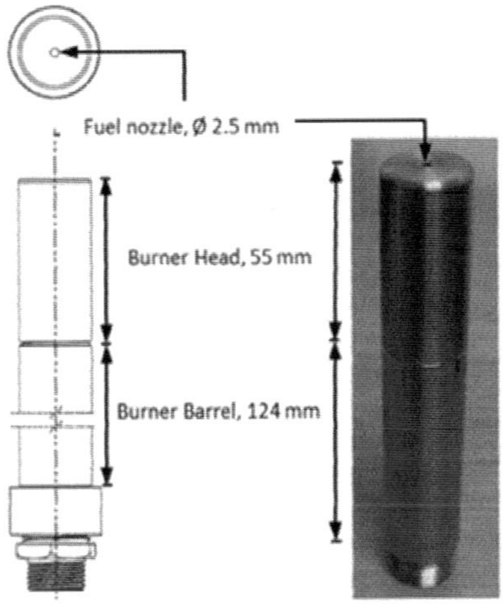

Figure 2. Schematics of cylindrical burner

Data of the F_H is measured for each fuels by using direct measurement by a stainless ruler fitted right beside the burner, with 0 scale is fitted align with fuel nozzle position. We use 6 variations of fuel nozzle diameter; 2.5 mm, 3 mm, 3.5 mm, 4 mm, 4.5 mm, and 5 mm. For each fuel nozzle diameter, we measured the F_H at variation of fuel flow rate which measured using rotameter type flowmeter. Each measurement is taken 3 times after the flames stabilized at approximately 2 minutes after changing the fuel flow rate. The data that is used for calculation is the means of the repeated value.

Temperature measurement is performed using K-type thermocouple which coupled with APPA 51 data logger to convert the physical data into digital data series. The data processing is done using APPA 50 Virtual DTM. Measurement of temperature were appointed on 2 points, T_1 which located right on the center of fuel nozzle and T_2 which located on the center of dark zone. T_2 position depends on the fuel and Q_f, so it can be moved along the Y axis by the change of both variable. Each measurement is taken for 1 minute where the data is recorded each second on data logger. The presented data is the mean of the recorded data in 1 position.

3. Result & Discussions

Comparison of the F_H as the result of change on Q_f is made between DME and LPG in Figure 3. We may observe that the change of Q_f only affect the F_H in the laminar regime (LR) area where the fuel jet velocity is quite low. Another finding is that the change of fuel nozzle diameter (D_f) do not affect the F_H, or we can say that the F_H is independent to D_f. These phenomenon happen for both DME and LPG flames, and this is inline with the finding of Hottel & Hawthorne [8].

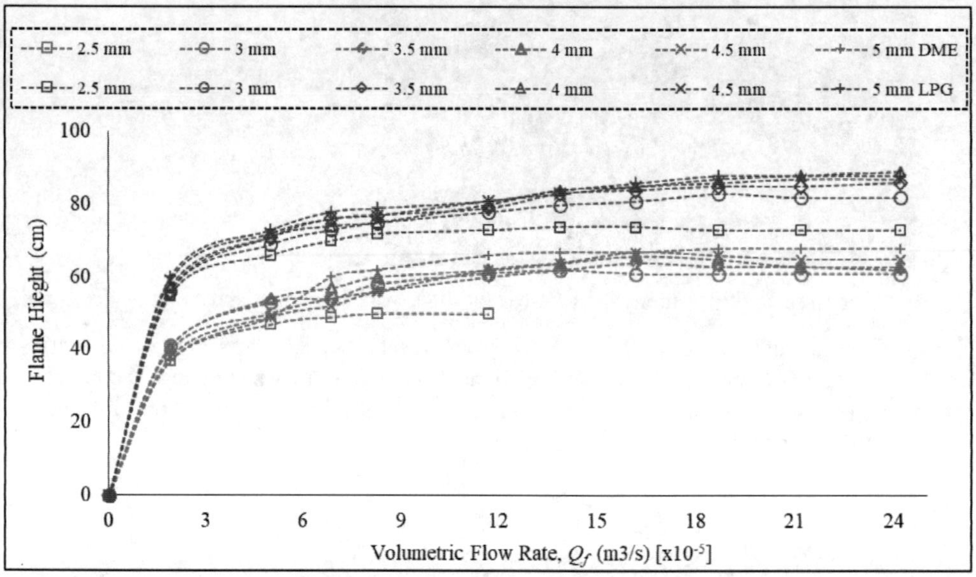

Figure 3. F_H of DME and LPG at various Q_f

The interesting finding is the significant different of F_H between DME and LPG. On a paper based on research by Kang et al [6], it is said that the Fluid Regime (FR) that is classified into Laminar regime (LR), Transitional Regime (TR) and the Fully Turbulent Regime (FTR) between DME and LPG is similar. The similarity FR of DME and LPG is inline in this research, from Figure 3 we can observe that both DME and LPG curves for various D_f developed same pattern. The difference comes when we observe the F_H of DME is significantly lower than that of LPG, this is differ with Kang's finding where the F_H of DME is similar with LPG.

One way to analyse the difference is use the R_e number that is a dimensionless number to characterize FR since the R_e is an indicator of the change in the mass transferring patterns, where in laminar is controlled by molecular collision, and in turbulent is significantly intensified by Eddy mixing. R_e is defined as a ratio of momentum to shear stress inside the fluids. Equation (1) gives the formula of R_e.

$$R_e = \frac{ud}{v} = \frac{\rho du}{\mu} \quad (1)$$

On equation (1), ρ is the fuel density, D_f is the fuel nozzle diameter, u is fuel jet velocity, v is kinematic viscosity of fuel and μ is molecular viscosity ($v = \mu/\rho$). Here we use the R_e to analyze FR in the TR, since in this regime the flow pattern showing critical change from the LR to FTR and it is good acceptance to analyze since there is difference approach to solve the LR and the FTR.

Figure 4 shows us that the TR limit in term of R_e between DME and LPG is quite different and shows that along with the increasing D_f, the difference is getting higher. In typical manner, the transitional regime from laminar to turbulent range in Reynold number of 2300-4000 [6], but in this study, the Reynold number at TR higher than the typical. This happen because DME and LPG has high adiabatic flame temperature (T_{ad}) as listed in Table 1. As we notice that molecular viscosity μ is approximately proportional to T, consequently the DME and LPG has higher transitional R_e number. The differences of behavior of R_e to D_f from this study to the finding of Kang where find that the transitional regime of

DME and LPG is similar might come to the differences in the experimental set up. Kang et al provide co-flow using air that surrounded the fuel stream in steady velocity, while in this study only fuel stream injected from fuel nozzle. By injecting air co-flow right inside the fuel stream, the flow pattern of the fuel will be changed and this also affected the temperature of the unburned fuel.

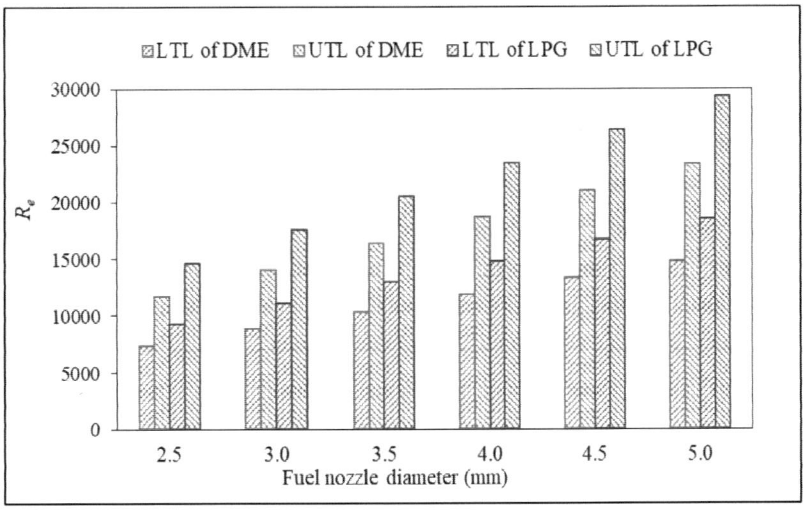

Figure 4. Comparison of the Transitional Regime of DME to LPG under different D_f (LTL ; Lower Transitional Limit, UTL; Upper Transitional Limit)

Analyzing the effect of temperature to the flow pattern of flames is also important. In this study, we measured the temperature of the DME and LPG flames to help us verify that the behaviour of Jet Diffusion Flame of DME is comparable to that of LPG because the temperature is not differ significantly. In Figure 5 the comparison of flames temperature between DME and LPG is shown.

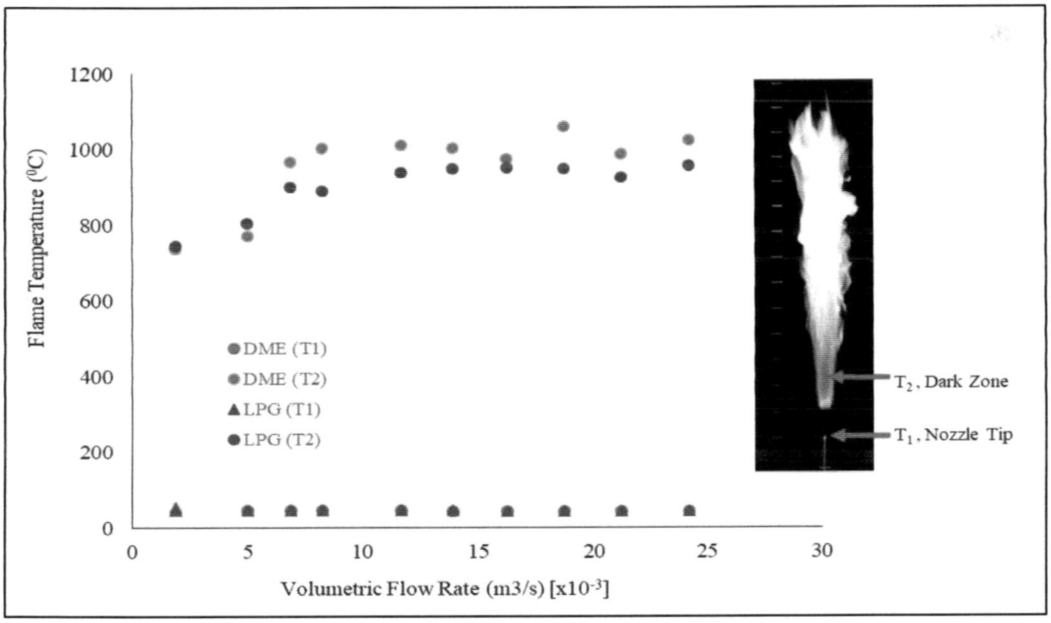

Figure 5. Temperature measurement at nozzle tip and dark zone of flame of DME and LPG

The temperature at nozzle tip position between DME and LPG is similar, it shows us that in term of the heat dissipation from the combustible gas is not affecting the temperature at the bottom part of flame. In the dark zone point, the flame of DME shows higher temperature than LPG. Dark zone area is the part of the flame where no Oxygen can enter, so only the fuel itself that being pyrolized and the temperature is depend on the properties of the fuel. Table 1 verify this result, where adiabatic flame temperature of DME is higher than the LPG.

4. Conclusion

The correlation between fuel volumetric flow rate, fuel nozzle diameter and the jet diffusion flame characteristic in term of flame height is compared for DME and LPG flame. The experiment is performed in diffusional combustion with only fuel is injected from the fuel nozzle and contacted to the surrounding air at atmospheric condition. Six (6) fuel nozzle diameter (2.5 mm, 3 mm, 3.5 mm, 4.0 mm, 4.5 mm, 5.0 mm) are used and the fuel volumetric flow rate is varied to get the FR ranging from LR, TR and FTR. From the presented result, we can draw the following conclusion:

1. The F_H of DME is lower than LPG for all Q_f and D_f variation. The analysis using R_e to construct the flame regime shows that the transitional regime of DME is reached at lower R_e number than LPG.
2. The flame temperature of DME at the point of nozzle tip and dark zone is higher than LPG. The heat dissipation effect is not seen on the nozzle tip for both DME and LPG.

Acknowledgments
This research is conducted under fully support of Thermodynamic and Combustion Laboratory of Mechanical Engineering of Universitas Indonesia and was performed using the support of facility in Combustion Laboratory of LEMIGAS.

References

1. Marchionna, M., et al., *Fundamental investigations on di-methyl ether (DME) as LPG substitute or make-up for domestic uses.* Fuel Processing Technology, 2008. **89**(12): p. 1255-1261.
2. Matzen, M. and Y. Demirel, *Methanol and dimethyl ether from renewable hydrogen and carbon dioxide: Alternative fuels production and life-cycle assessment.* Journal of Cleaner Production, 2016. **139**: p. 1068-1077.
3. Silalertruksa, T., et al., *Life cycle GHG analysis of rice straw bio-DME production and application in Thailand.* Applied Energy, 2013. **112**: p. 560-567.
4. Pettersson, K., et al., *Integration of next-generation biofuel production in the Swedish forest industry – A geographically explicit approach.* Applied Energy, 2015. **154**: p. 317-332.
5. Dahmen, N., et al., *The bioliq process for producing synthetic transportation fuels.* Wiley Interdisciplinary Reviews: Energy and Environment, 2017. **6**(3): p. e236.
6. Kang, Y.-H., et al., *Experimental and theoretical study on the flow, mixing, and combustion characteristics of dimethyl ether, methane, and LPG jet diffusion flames.* Fuel Processing Technology, 2015. **129**: p. 98-112.
7. Kang, Y., et al., *On predicting the length, width, and volume of the jet diffusion flame.* Applied Thermal Engineering, 2016. **94**: p. 799-812.
8. H.C Hottel, W.R.H., *Diffusion in Laminar Flame Jets*, in *Symposium on Combustion & Flame and Explosion Phenomena 3* 1948. p. 254-266.
9. Wang, J.-w., et al., *Flame volume and radiant fraction of jet diffusion methane flame at sub-atmospheric pressures.* Fuel, 2016. **167**: p. 82-88.
10. Gattei, L., *A study on the fluid dynamics of domestic Gas burners*, in *Il Facolta di Ingenegneria Aerospaziale*2008, Universita di Bologna: Forli, Italy.
11. An, Q., et al., *Coupled dynamics of lift-off and precessing vortex core formation in swirl flames.* Combustion and Flame, 2016. **168**: p. 228-239.

12. Saini, R., et al., *Investigation of the Role of Chemical Kinetics in Controlling Stabilization Mechanism of the Turbulent Lifted Jet Flame Using Multi-flamelet Generated Manifold Approach*, in *Energy for Propulsion : A Sustainable Technologies Approach*, A.K. Runchal, et al., Editors. 2018, Springer Singapore: Singapore. p. 293-314.
13. Kim, M.Y., et al., *Combustion and emission characteristics of DME as an alternative fuel for compression ignition engines with a high pressure injection system.* Fuel, 2008. **87**(12): p. 2779-2786.
14. Kang, Y., et al., *Detailed multi-dimensional study on NOx formation and destruction mechanisms in dimethyl ether/air diffusion flame under the moderate or intense low-oxygen dilution (MILD) condition.* Energy, 2017. **119**: p. 1195-1211.

IOP Conf. Series: Materials Science and Engineering **694** (2019) 012020 doi:10.1088/1757-899X/694/1/012020

The Potential of Hydroelectric Power of Kawiwi River in Ratahan District, North Sulawesi, Indonesia

B L Maluegha, C S C Punuhsingon and G D Soplanit

Mechanical Engineering Department, Faculty of Engineering, Sam Ratulangi University, Jalan Kampus Unsrat Bahu, Manado, 95115, Indonesia

E-mail: benny_maluegha@unsrat.ac.id

Abstract. This research was aimed at calculating the hydro energy potential for generating electricity from Kawiwi river in Ratahan District, North Sulawesi, Indonesia. The required data were the estimation of daily electrical loads of the buildings around Kawiwi river and the monthly average flow velocity of the river in a year. Data of electrical loads were obtained from questionnaires given to the people who is living by the river. Meanwhile, the data of the monthly average flow velocity for June, July and August 2019 were directly measured, and for other months were estimated based on monthly average rainfall in South East Minahasa regency, where Ratahan district is located. The monthly average flow velocity of Sungai Kawiwi are as follows: Januari 1.35 m/s, February 2.45 m/s, March 2.12 m/s, April 2.69 m/s, May 1.10, June 0.76 m/s, July 0.70 m/s, August 0,60 m/s, September 0,16 m/s, October 0,96 m/s, November 1,17 m/s, and December 2.20 m/s. The data, then, were analyzed by using HOMER 2.68 Beta with an assumption that the electricity produced from Kawiwi river was generated by Darrieus Hydro Turbine, a kinetic turbine. By HOMER's calculation, it was found that the electricity produced from Kawiwi river was 16.880 kWh/year.

1. Introduction

Indonesia's potential energy from hydro for generating electric power is quite immense that is more than 75 GW [1]. However, because of lack of reliable data, the assessment of the hydro energy potential for producing electricity becomes very limited. As a result, the use of hydro energy for power generation is much smaller than its potential. For that reason, research for identifying the potential energy of hydro resources, especially from rivers, should be increased.

Ratahan, a district in South East Minahasa regency in North Sulawesi province, Indonesia, is crossed by two rivers: Kawiwi and Palaus. The objective of this research is to calculate the hydro energy potential for generating electricity from one of the rivers, that is Kawiwi. The results of this research can be used to design an energy generating system with a kinetic turbine to increase the role of renewables to meet the electricity need of the people in Ratahan district.

There were some research on hydro energy and its technology have been made. Doda & Mohamad [2] conducted a study to calculate the potential of hydropower in Bone Bolango regency, Gorontalo, Indonesia which can produce 65 kW. The potential hydropower from a micro-hydro power station in Ngantang district, Malang regency, East Java has been investigated by Hanggara and Irvani [3]. The power station can provide electricity for 47 households. Anaza, et al. [4] overviewed the estimation of micro-hydro energy potential which depends on head and flow rate. Bhat [5] stated that hydropower can have a significant role in supplying electricity in the grid.

Studies on the technology of hydro energy were also conducted by some researchers. Kumar and Srivastava [6] analyzed the performance of Darrieus Hydro Kinetic Turbine with aerofoil blade profile and found that the aerofoil blade can be applied to a turbine for generating power in small scale. Chen and Engeda [7] investigated the use of ultra-low-head hydro energy and they found some satisfactory results. Sunoko, et al. [8] investigated the effect of runner angle positions on the efficiency of eight curved bladed kinetic water turbine. Boedi, et al. [9] conducted a study on the performance of a vertical axis hinged blade kinetic turbine by using Response Surface Methodology.

2. Research methods

This research was carried out in several main steps: data collection, data analysis and calculation. Those steps are explained in the following.

2.1. Data collection

The required data in this research were

1. estimation of daily electrical loads in buildings around Kawiwi river;
2. the monthly average flow velocites of the river in a year.

2.1.1. Daily electrical loads. These data were obtained from questionnaires given to the people living by Kawiwi river. The questionnaires included a question about the type of buildings based on their usage (such as: residential home, worship place, business place, etc.) and a series of questions about electrical equipment usage in the buildings. The questions were arranged in a form of a table which contains

- type of electrical equipment
- power of the electrical equipment (Watt)
- operational hours of the electrical equipment in a day
- total numbers of operational hours of the electrical equipment in a day (hours)
- total energy consumption of the electrical equipment in a day (Watt hours)

The data about electrical loads of the buildings around Kawiwi river was one of the required inputs to calculate the power that can be generated by the river.

2.1.2. Monthly average flow velocities. Since the equipment to measure flow velocity is not available, the monthly average flow velocities were determined by "float method". The duration of a light object (i.e. a medium-sized half-filled plastic bottle) floating through a 10 meters distance was timed, and the velocity was determined by dividing the duration (in seconds) into the distance (10 meters). The data collection was conducted 20 times for every measurement in the river. Accordingly, the velocity of a month was the average number of the 20 velocities that have been previously calculated. Actually, the measurement of the average velocity should be carried out at least once for every month. However, due to the time limitation, the measurement could not be done for all month. The direct measurements were conducted only for three months, i.e. June, July, and August 2019. For other months, the average flow velocities were estimated by comparing data of average monthly rainfall from Statistics of South East Minahasa regency [10] with the calculated average velocity of a determined month.

2.2. Data analysis and calculation

Those data, then, were inputted into HOMER 2.68 Beta. HOMER can analyze the data and calculate the electric energy that can be generated from Kawiwi river. Beside those data, some components should be selected in the software so that the analysis and calculation could be performed by HOMER 2.68 Beta. The selected components would be used to form a power generation system that produced the electricity. In this research, it was assumed that the electricity would be generated by a kinetic turbine. So, the Darrieus Hydro Turbine from HOMER's turbine database was selected to drive a generator to produce electricity. Since the objective of this research only to obtain the hydroelectric power generated by Kawiwi river, the default values provided in HOMER 2.68 were used as the inputs of other

IOP Conf. Series: Materials Science and Engineering **694** (2019) 012020 doi:10.1088/1757-899X/694/1/012020

parameters. Furthermore, other technical and economic aspects were inputted based on a rough estimation. These will not affect the calculation result of the hydroelectric power in this research.

3. Results and discussions

3.1. Daily electrical loads

From the questionnaires, only data about electrical consumption from residential houses were given. The data from other types of buildings (such as shops, offices, worship places, etc.) were not available. Therefore, the profile of daily electrical usage of the buildings around Kawiwi river was determined by multiplying the data of one residential house by the number of residential houses around the river. A village where Kawiwi river cross in Ratahan district is West Tosuraya village which has 524 residential houses [12]. Accordingly, the electrical energy consumption from a questionnaire of a residential house was selected and multiplied by 524 to obtain the electrical loads of buildings around Kawiwi river.
The daily electrical usage of a selected residential house in West Tosuraya village is shown in Table 1.

Table 1. Daily electrical usage of a selected residential house in West Tosuraya village.

No	Electrical Equipment	Power (Watt)	Operational Hours	Number of Operational Hours	Energy Consumption (Wh)
1	Lamp A	20	18.00 – 23.00	5	100
2	Lamp B	20	18.00 – 22.00	4	80
3	Lamp C	5	19.00 – 05.00	10	50
4	Lamp D	10	18.00 – 22.00	4	40
5	Iron[a]	300	15.00 – 16.00	0,86	258
6	Television	150	08.00 – 15.00	7	1050
7	Refrigerator	100	00.00 – 24.00	24	2400
8	Magic jar (cooking)	350	05.30 – 06.00	0,5	175
9	Magic jar (heating)	60	06.00 – 20.00	14	840
	Total of Daily Electrical Energy Consumption of a Household				4993

[a] From the questionnaire, it was found that the iron is used three times in a week for two hours for each usage. To simplify the analysis, it was assumed that the iron is used for 0,86 hours every day (six hour divided by seven days) from 15.00 to 16.00.

After multiplying by 524, the daily electrical usage must be presented in hours as can be seen in the following table.

Table 2. Profile of hourly electrical consumption in West Tosuraya village in a day.

Hour in a day	Electrical consumption	
	(W)	(kW)
00.00 – 01.00	55.020	55,02
01.00 – 02.00	55.020	55,02
02.00 – 03.00	55.020	55,02
03.00 – 04.00	55.020	55,02
04.00 – 05.00	55.020	55,02
05.00 – 06.00	144.100	144,10
06.00 – 07.00	83.840	83,84
07.00 – 08.00	83.840	83,84
08.00 – 09.00	162.440	162,44
09.00 – 10.00	162.440	162,44
10.00 – 11.00	162.440	162,44
11.00 – 12.00	162.440	162,44
12.00 – 13.00	162.440	162,44
13.00 – 14.00	162.440	162,44
14.00 – 15.00	162.440	162,44
15.00 – 16.00	219.032	219,03
16.00 – 17.00	83.840	83,84
17.00 – 18.00	83.840	83,84
18.00 – 19.00	110.040	110,04
19.00 – 20.00	112.660	112,66
20.00 – 21.00	81.220	81,22
21.00 – 22.00	81.220	81,22
22.00 – 23.00	65.500	65,50
23.00 – 24.00	55.020	55,02
Total	2.616.332	2.616,33

The profile of hourly electrical consumption in West Tosuraya village in a day from Table 2 can be presented in the form of a graph as shown in Figure 1.

IOP Conf. Series: Materials Science and Engineering **694** (2019) 012020 doi:10.1088/1757-899X/694/1/012020

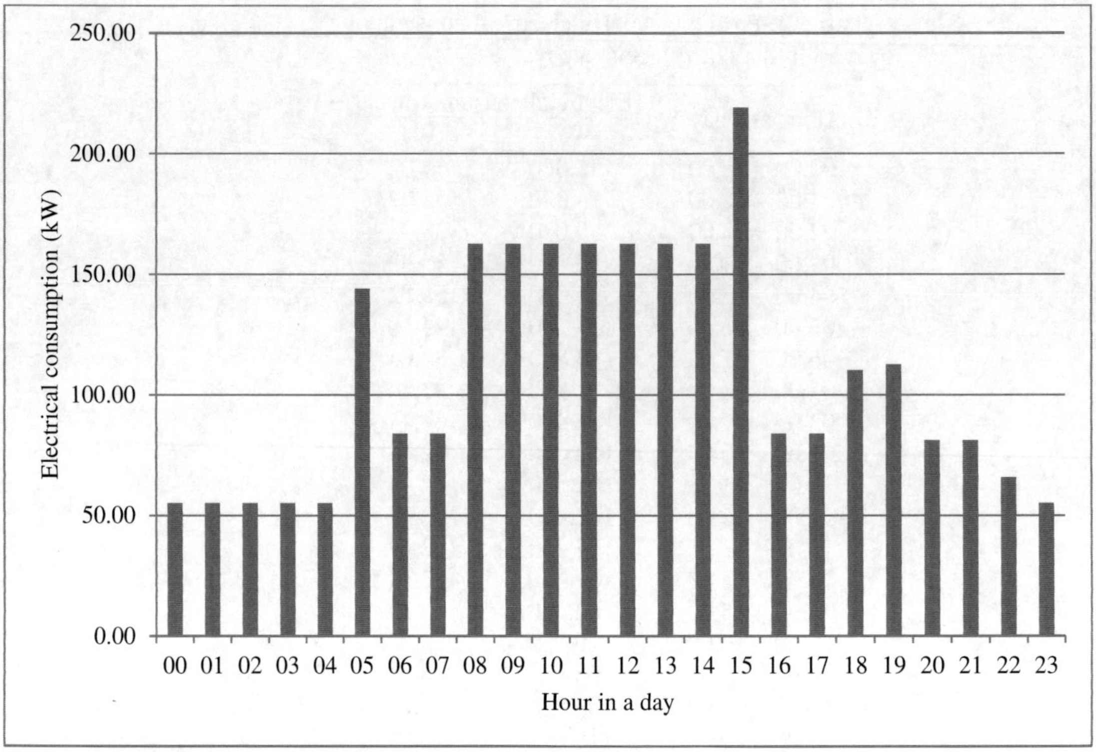

Figure 1. Profile of hourly electrical consumption in a day in West Tosuraya village.

Figure 1 depicts that the highest consumption occurs at 15.00 when iron is used. Whereas the lowest happens from 23.00 to 04.00 when refrigerators and some lamps are still on.

3.2. Monthly average flow velocities

The data of monthly average flow velocities taken from direct measurement at Kawiwi river (for June, July and August 2019) and also from the calculation based on the rainfall data in South East Minahasa regency in 2018 (other months) [10] are presented in Table 3. In the calculation, the average flow velocity of June was used as the basis of comparison.

Table 3. Monthly average flow velocities of Kawiwi river.

Month	Rainfall (mm^3)	Monthly average flow velocity of Kawiwi river (m/s)
January	156	1,35
February	284	2,45
March	246	2,12
April	312	2,69
May	127	1,10
June	88	0,76
July	98	0,70
August	46	0,60
September	19	0,16
October	111	0,96
November	135	1,17
December	255	2,20

Table 3 shows that the highest average velocity is in June, and the lowest is in the September. The variation of the average velocity is mainly caused by seasonal effects.

3.3. Data analysis and calculation by using HOMER 2.68 Beta
Data analysis and calculation using HOMER 2.68 Beta were carried out in the following steps.

3.3.1. Selecting loads and components. After opening the software, the load and components for electrical generation system must be selected. As can be seen in Figure 2, the selected load and equipment are
- Primary Load 1
- Darrieus H Turbine
- Converter

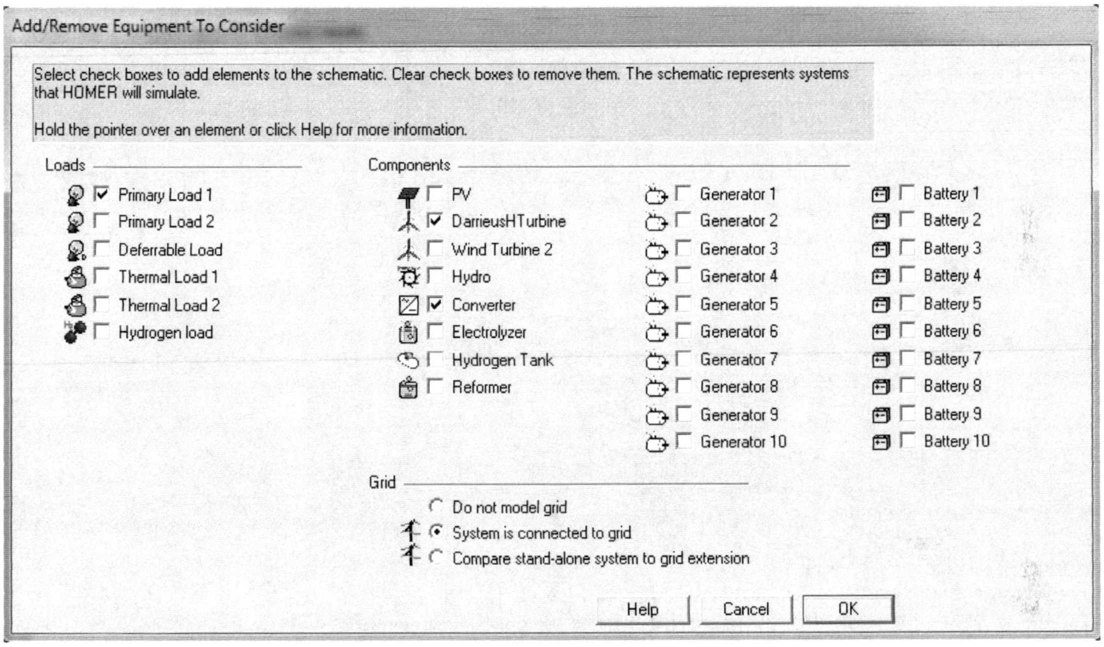

Figure 2. The selected load and components for generating electricity.

Since Tosuraya Barat village has connection with grid from State Electricity Company (PLN), the option of "System is connected to grid" must be chosen.

3.3.2. Inputting primary load. The primary load inputted in the software is the profile of hourly electrical consumption in West Tosuraya village in a day as presented Table 2 and Figure 1. The result of inputting primary load can be seen in Figure 3.

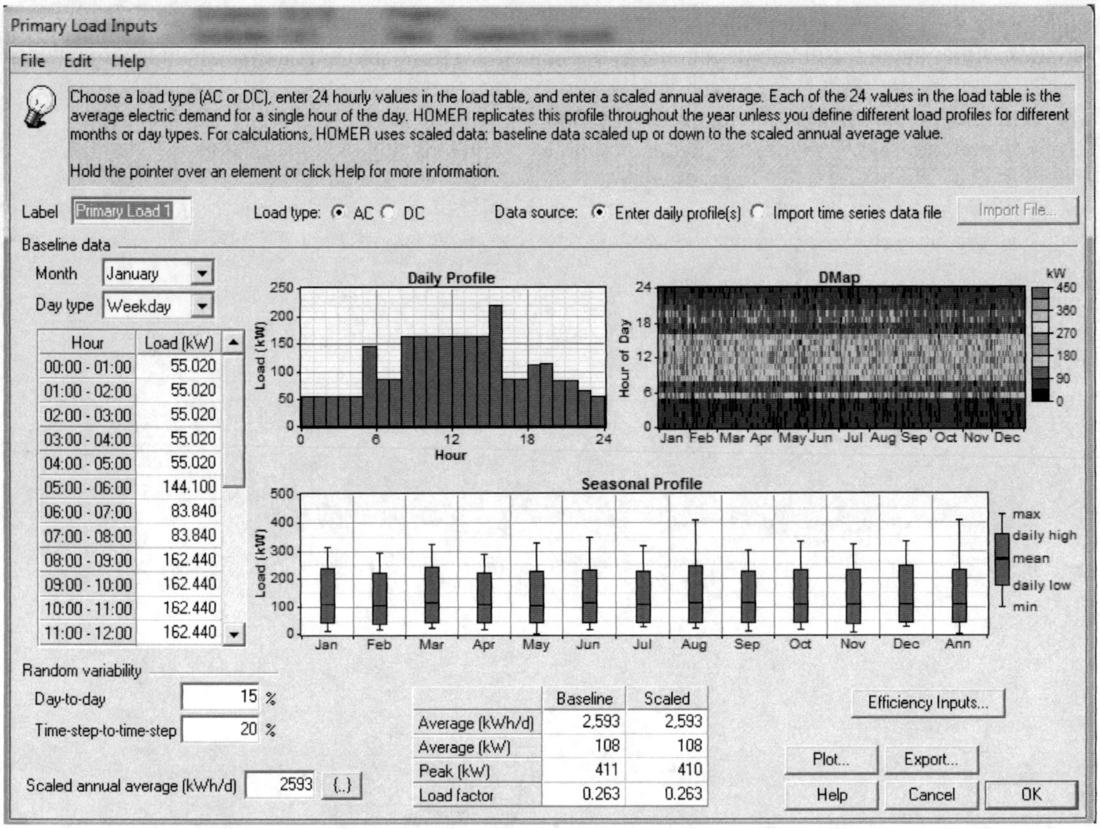

Figure 3. Primary load input

3.3.3. Inputting resource data. As explained earlier, it was assumed that the electricity would be produced by a hydro kinetic turbine. Because the feature of kinetic turbines and wind turbines is quite similar and also the resource input for kinetic turbine is not available in HOMER 2.68 Beta, "Wind Resource Inputs" was used to input the monthly average flow velocities of Kawiwi river. However, the data must be prepared in a Microsoft Excel file and, then, was inputted through "Import time series data file". This must be done so that HOMER will not consider other advanced parameters for wind turbines (such as Weibull k). The result of resource input can be seen in Figure 4.

IOP Conf. Series: Materials Science and Engineering **694** (2019) 012020 doi:10.1088/1757-899X/694/1/012020

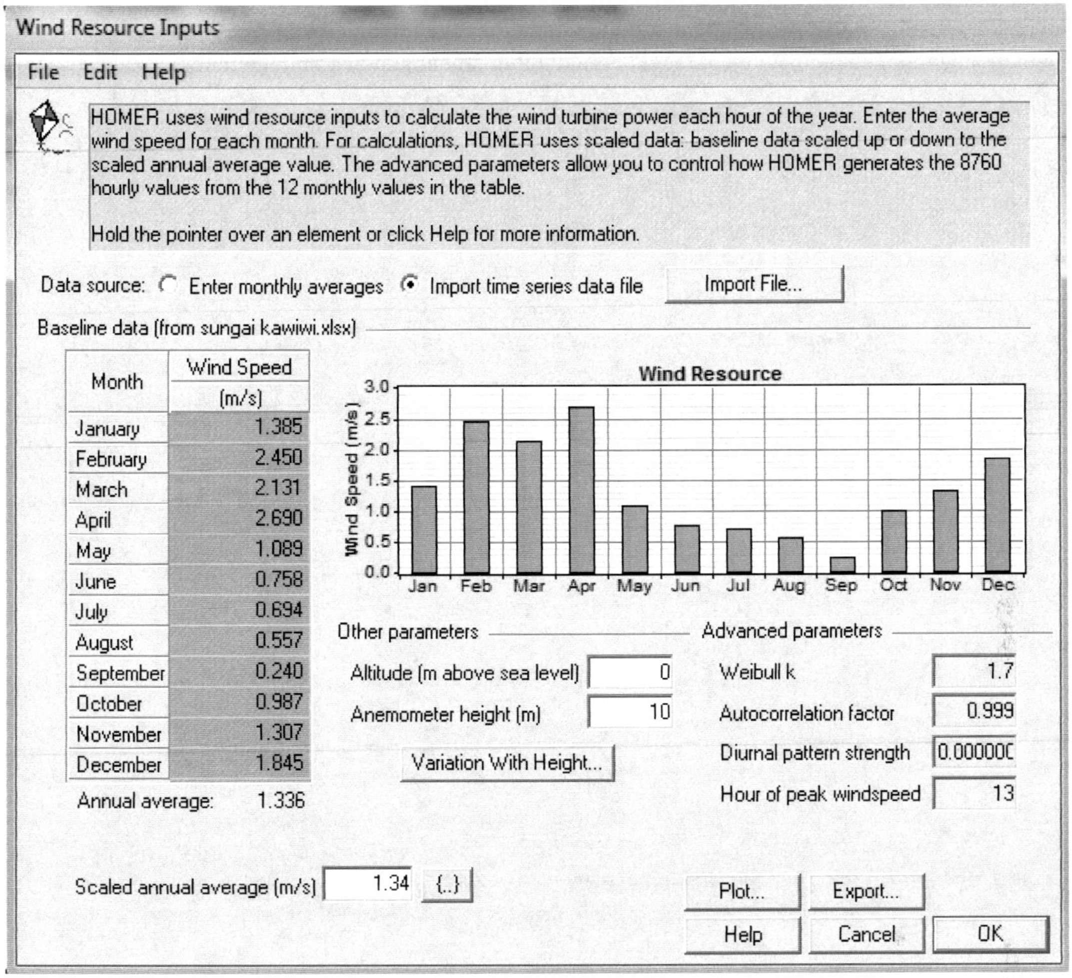

Figure 4. Monthly average flow velocities input

3.3.4. Inputting turbine data. Darrieus Hydokinetic Turbines (DHT) is the only kinetic turbine available in the database of HOMER 2.68 Beta. Figure 5 shows the result of the turbine input.

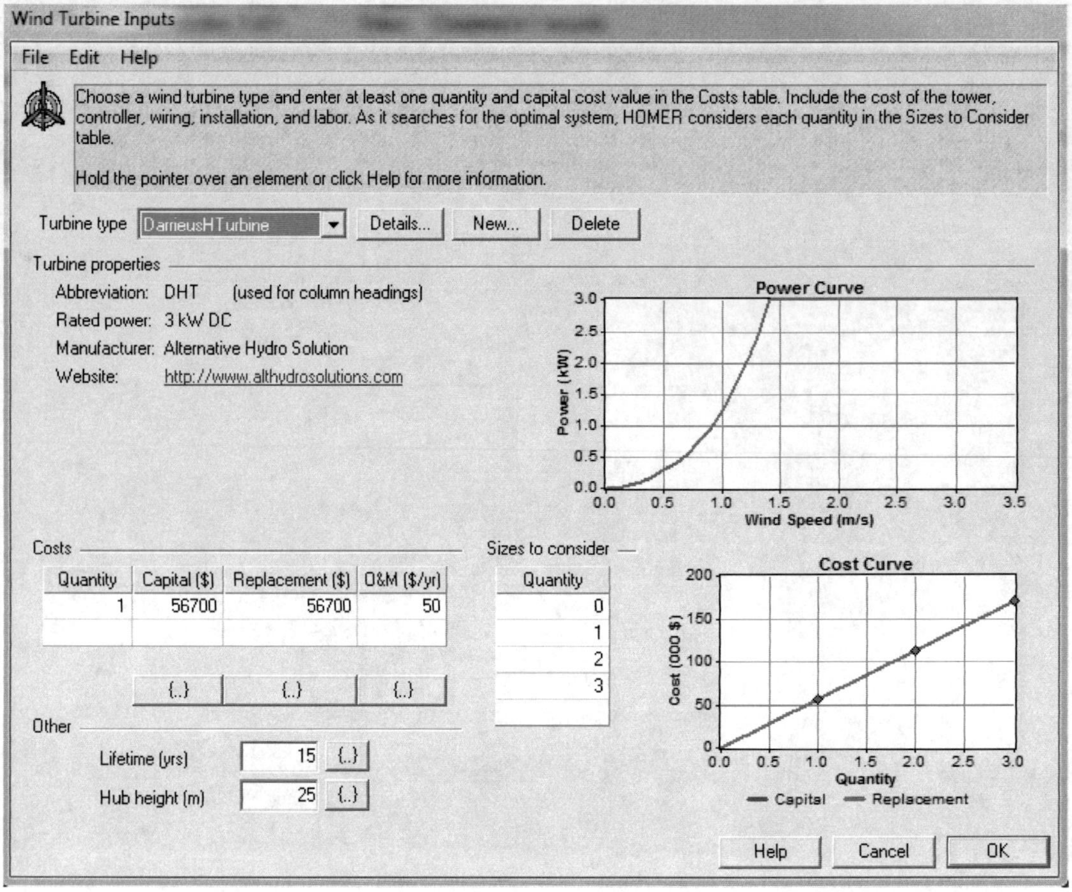

Figure 5. Turbine data input

3.3.5. Inputting converter data. The electricity produced from Darrieus Hydokinetic Turbines (DHT) is DC. Therefore, a converter is required to change the DC current to AC current. The input of converter data is shown in Figure 6.

IOP Conf. Series: Materials Science and Engineering **694** (2019) 012020 doi:10.1088/1757-899X/694/1/012020

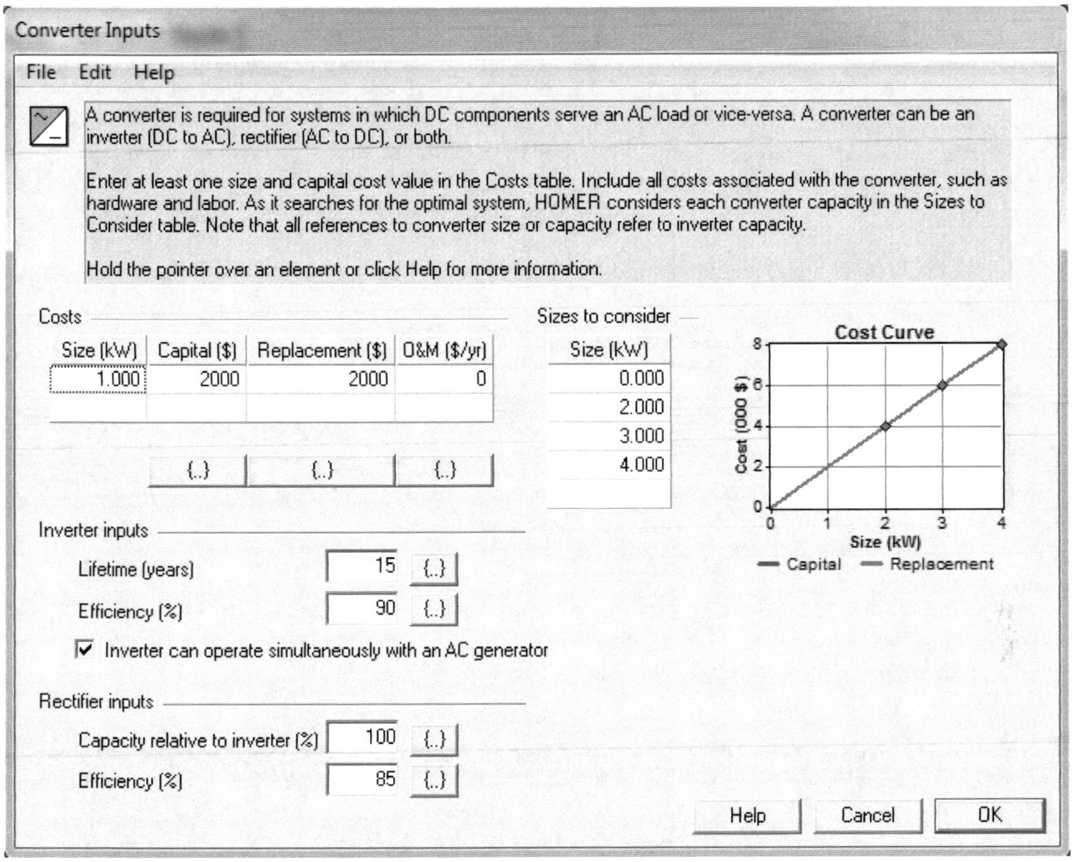

Figure 6. Converter data input

3.3.6. Calculation. After all data were inputted, the schematic of the power generation system was displayed in HOMER 2.68 Beta as can be seen in the following figure.

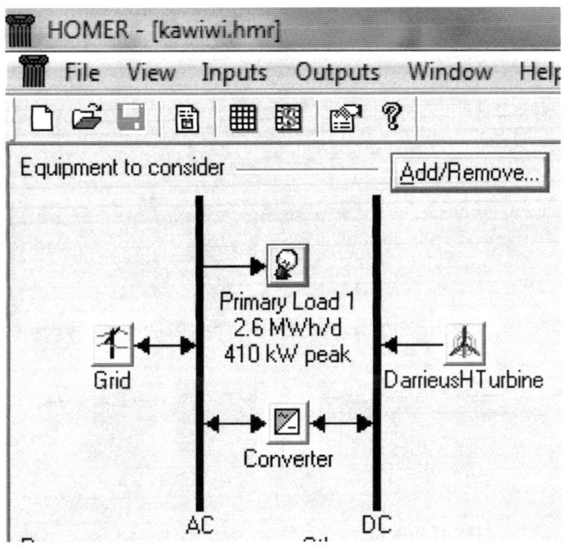

Figure 7. Schematic of the power generation system

Figure 7 depicts the power generation system from the river consisted of the Darrieus Hydro Turbine and a converter, and it is shown that the electricity was also supplied by grid. The load is 2.6 MWh/day with 410 kW peak. Afterwards, the calculation was performed, and two type of power generation systems for satisfying all the loads will be displayed (Figure 8). The first system shows that the grid is the only supplier of the electricity. This system was ignored because it is not in accordance with the objective of this research.

Double click on a system below for simulation results.

	DHT	Conv. (kW)	Grid (kW)	Initial Capital	Operating Cost ($/yr)	Total NPC	COE ($/kWh)	Ren. Frac.
			1000	$ 0	94,645	$ 1,209,875	0.100	0.00
	1	3	1000	$ 62,700	94,841	$ 1,275,088	0.105	0.02

Figure 8. Power generation systems resulted from the calculation

Finally, the number of electricity produced by the turbine was obtained in the "DHT" tab after double-clicking the second system. As shown in Figure 9, the total production of electricity from Darrieus Hydro Turbine was 16.880 kWh/year.

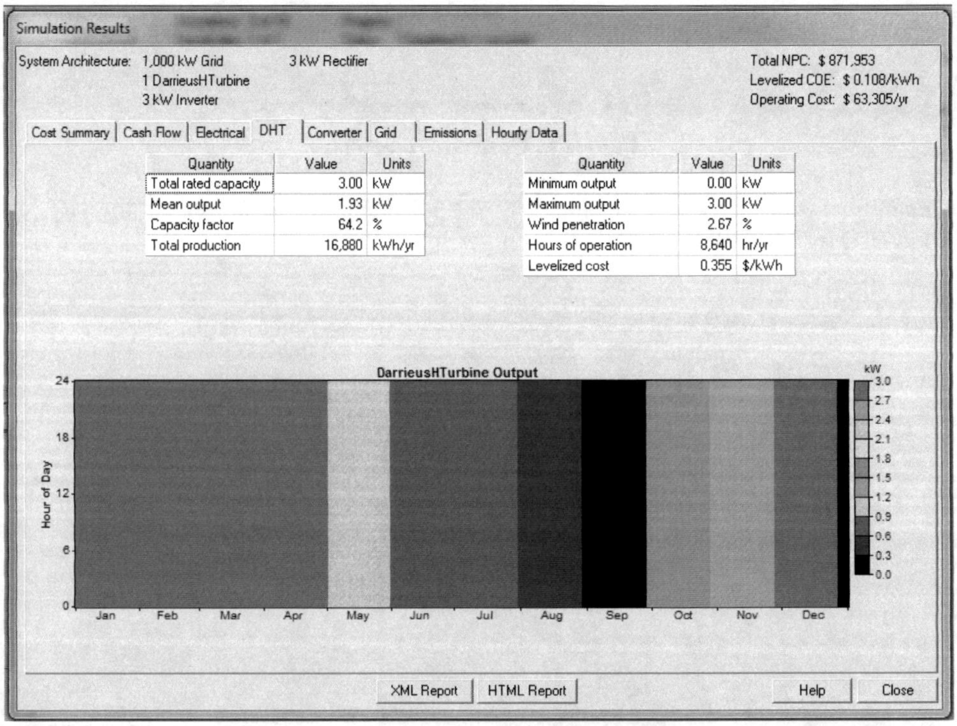

Figure 9. Electricity produced from Darrieus Hydro Turbine

4. Conclusion

In conclusion, the power generation system for Kawiwi river which consisted of the Darrieus Hydro Turbine and a converter can only supply electrical energy as much as 16.880 kWh/year.

IOP Conf. Series: Materials Science and Engineering **694** (2019) 012020 doi:10.1088/1757-899X/694/1/012020

References

[1] Suyono 2017 *Kajian Penyediaan dan Pemanfaatan Migas, Batubara, EBT dan Listrik* (Jakarta: Pusat Data dan Teknologi Informasi Energi dan Sumber Daya Mineral Kementerian Energi dan Sumber Daya Mineral) p 16

[2] Doda N and Mohammad H April 2018 Analisis potensi pengembangan pembangkit listrik tenaga mikrohidro di Kabupaten Bone Bolango *Gorontalo J. Inf. & Sci.* **1** 1-10

[3] Hanggara I and Irvani H March 2017 Potensi pltmh (pembangkit listrik tenaga mikro hidro) di kecamatan ngantang kabupaten malang jawa timur *J. Reka Buana* **2** 149-55

[4] Anaza S O, Abdulazeez M S, Yisah Y A, Yusuf Y O, Salawu B U and Momoh S U 2017 Micro hydro-electric energy generation - an overview *American J. of Eng. Res. (AJER)* **6** 5-12

[5] Bhat S S. 14 January 2018 Hydropower in new roles in changed context *Hydro Nepal J. Wat., En. and Env.* **22** 48-50

[6] Kumar A and Srivastava P July 2016 Performance analysis of darrieus hydro kinetic turbine with aerofoil blade profile *J. Bas. and Appl. Eng. Res.* **3** 935-39

[7] Chen J and Engeda A 24 October 2018 Design and development of an ultra low head axial hydro turbine for electricity supply: part I *ASME*

[8] Soenoko R, Setyarini P H and Gapsari F March 2018 Eight curved bladed kinetic water turbine performance *ARPN J. of Eng. and Appl. Sci.* **13** 2138-47

[9] Boedi, S.D., R. Soenoko, S. Wahyudi, M. A. Choiron. 2018. *A Vertical Axis Hinged Blade Kinetic Turbine Performance Using Response Surface Methodology*. Journal of Engineering Science and Technology Vol. 12, No. 8 (2017) 2187 - 2201 © School of Engineering, Taylor's University

[10] Statictic of Minahasa Selatan Regency 2018 *Ratahan in Figure 2018* p 27

IOP Conf. Series: Materials Science and Engineering **694** (2019) 012021 doi:10.1088/1757-899X/694/1/012021

Analysis of sea wave power plant design in Bangka Island Indonesia

R Priyoko Prayitnoadi[1], Firlya Rosa[1] Mu'alimah Hudatwi[2], M Ubed Nurhadi[1], Ari Febrianto[1] and Nova Roliana[1]

[1] Jurusan Teknik Mesin, Fakultas Teknik, Universitas Bangka Belitung, Bangka 33172, Indonesia

[2] Jurusan Ilmu Kelautan, Fakultas Pertanian Perikanan dan Biologi, Universitas Bangka Belitung, Bangka 33172, Indonesia

E-mail: f105a@yahoo.com

Abstract. Indonesia has a vast ocean of energy resource potential to be utilized including ocean wave energy. Sea waves can be used to drive electricity generation as an alternative to replacing electricity from fossil fuels. This research was conducted to determine the potential rotation of the electric motor drive shaft in two alternative designs of sea wave power plants (PLTGL). The analysis of shaft rotation results was done by collecting sea wave data in the coast of Bangka and making a PLTGL design in the Solid Works application. Next, the sea wave data was included as a parameter of the motion study simulation. Of the two alternative designs of PLTGL, the circular type gets the highest shaft rotation, which is 29.16 rpm. Although this research ignores the factors of material strength and load resistance, this research can be continued to find out more, either in the form of electricity generated or modified mechanism design.

1. Introduction

Indonesia as an archipelago with an area of 1,904,556 km2 consisting of; 17,508 islands, 5.8 million km2 of ocean and 81,290 million km of beach length, the potential of ocean energy, especially ocean waves, is potential to be empowered as a new and renewable alternative primary energy especially for power generation. Hasan, et al [1] has conducted studies that explain the investigation of pattern power plants and emissions in Indonesia for the past 23 years from 1987-2009. This study shows that Indonesia's electricity sector is still very dependent on fossil fuel resources, especially diesel oil and coal as primary energy sources and this is also a major contributor to greenhouse gas emissions in the country. Several power plant designs that utilize sea wave energy have been studied [2], [3], [4]. According to the National Energy Management Plan 2005-2025 targets the use of renewable energy, one of which is wave energy, amounting to 0.2% of the national energy needs [5].

Seeing the topographic of Indonesian islands, Indonesia needs to take a role in terms of research and development of energy from the sea, in an effort to realize the use of new renewable energy. In more detail, the problem can be formulated by asking question, how to create a mechanism that can produce electricity by utilizing ocean wave energy. As a first step, the PLTGL design begins by analyzing two PLTGL mechanism designs using the Solid Works application with sea wave data in Bangka island waters.

The purpose of this research was to obtain an alternative technology design that can produce electricity that is clean, inexpensive, practical and efficient and utilizes abundant ocean wave energy. The benefit of this research is to contribute to science in the form of a mechanical design of sea wave power plants. The results of the study will also provide input to the development of research for new renewable energy alternatives that can always be developed and are considered by parties related to regional and national energy policy makers.

2. Research Methodology

The study began by observing the condition of sea wave height in the coast of Bangka, namely in Tanjung Berikat, Central Bangka Regency. At the same time, two PLTGL designs were made using a CAD application, i.e. Solid Works. The design has the principle of utilizing the motion of ocean waves to rotate the driving shaft of the electric generating motor. Up and down movements of the sea waves move the lever which rests on a fixed frame. The lever is connected by a mechanism that connects the lever with the pinion gear as part of the rack and pinion system. Alternatives are made on the mechanism, including linear and circular combinations. Furthermore, the rack and pinion system will drive the shaft and be connected to an electric generating motor. Analysis of the results will be obtained in the form of the amount of rotation at the end of the shaft that will rotate the electric generating motor.

Two PLTGL designs that have been completed then proceed with making a simulation of the movement mechanism by entering sea wave data in the motion study in the Solid Works application. Regardless of the material strength and other loads in the PLTGL design, shaft rotation can be produced by first converting the angular velocity unit to the number of rotation on the shaft. Sea wave data obtained from the Geospatial Information Agency (BIG) in Tanjung Berikat, Central Bangka Regency. Data in the form of sea wave height and sea wave were for one month period, in January 2019. There were five data to be used as parameter data in the Solid Work application, i.e. the highest value, the lowest value, and three values in between.

The predetermined data values were entered into the Solid Works application in the motion study application. The design was simulated and produced the value of the shaft rotation in the form of angular speed. Angular speed values were converted into rotations per minute. To get the radians value per second the formula was used:

$$rad/s = \frac{\omega.\pi}{180} \quad (1)$$

Where ω = angular velocity (deg / s). After knowing the angular velocity, then to find the rotation, it can be searched by the following equation:

$$n = \frac{\omega}{2\pi} \quad (2)$$

Where n = rotation (rps). ω = angular velocity (rad / s). After getting the results of rotation on the shaft in units of rotation per second (rps) then the next conversion to the unit of rpm will be obtained through the following equation:

$$n = n(rps) \times 60\ second \quad (3)$$

Where n = Rotation (Rpm).

3. Result and Discussion

3.1. Sea wave data

Sea wave height data obtained from the Geospatial Information Agency in the form of average wave height data in the coast of the island of Bangka (Table 1) to enter data into Solid Works, Period data (T) must be converted to Frequency (Hz). Furthermore, data representation is needed to represent the values that will be entered into Solid Works, i.e. The maximum, minimum, and three intermediate values (Table 2).

Table 1. Data on sea level and wave period at Tanjung Berikat Beach in January 2019

Date	Sea Waves	
	Height (m)	Period (T)
1/1/2019	**0,561**	**6,730**
2/1/2019	0,675	7,060
3/1/2019	0,603	6,651
4/1/2019	0,584	6,118
5/1/2019	0,717	5,048
6/1/2019	**0,722**	**5,288**
7/1/2019	0,607	5,554
8/1/2019	0,471	6,013
9/1/2019	0,319	6,141
10/1/2019	**0,285**	**4,457**
11/1/2019	0,373	3,975
12/1/2019	0,327	5,211
13/1/2019	0,326	5,784
14/1/2019	0,322	5,017
15/1/2019	0,369	4,272
16/1/2019	0,388	5,445
17/1/2019	0,491	5,852
18/1/2019	0,502	6,314
19/1/2019	0,732	5,162
20/1/2019	0,836	5,290
21/1/2019	0,663	5,428
22/1/2019	0,516	5,512
23/1/2019	0,798	4,691
24/1/2019	1,110	5,100
25/1/2019	**1,220**	**1,226**
26/1/2019	1,174	5,182
27/1/2019	**0,956**	**4,897**
28/1/2019	0,770	4,637
29/1/2019	0,689	4,467
30/1/2019	0,378	4,895
31/1/2019	0,765	4,926

The data chosen to be data that will be entered into the motion study application are as follows:

Table 2. The data chosen for Solid Works application

No	Wave Height (m)	Period (T)	Frequency (Hz)
1	0,285	4,457	0,224
2	0,561	6,730	0,148
3	0,722	5,288	0,189
4	0,956	4,897	0,204
5	1,22	1,226	0,190

3.2. Explanation of two PLTGL design

The system in the design of a PLTGL with a linear and a circular rack gear and pinion combination system are to make use of the motion of the ocean waves that will be in contact with the buoy which will provide a linear motion that is a vertical up and down motion which is then continued by the buoy lever. The float lever is connected to the linear rack gear (Fig.1) and circular rack gear (Fig.2) so that those rack gears rotates the pinion connected to the shaft or changes the vertical motion into rotational motion. The resulting rotation on the shaft is one-way rotation thanks to the ratchet mechanism on the pinion gear. The movement of the pinion gear will give only one direction force when the buoy moves up. When the buoy is down, there will be no transfer of force between the rack and pinion gears. Next on the shaft mounted a flywheel which is useful when the pinion gear does not receive force, rotation on the shaft will not drop significantly. Next to continue the rotation on the shaft to the motor, a belt is attached. The resulting rotation is expected to be converted into electrical energy.

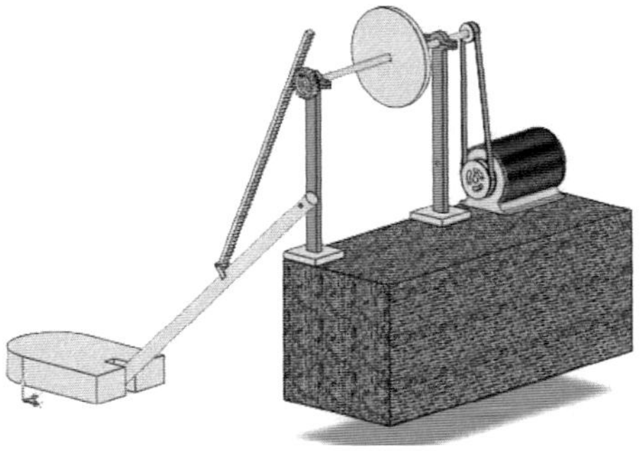

Figure 1. Linear Rack Gear Design

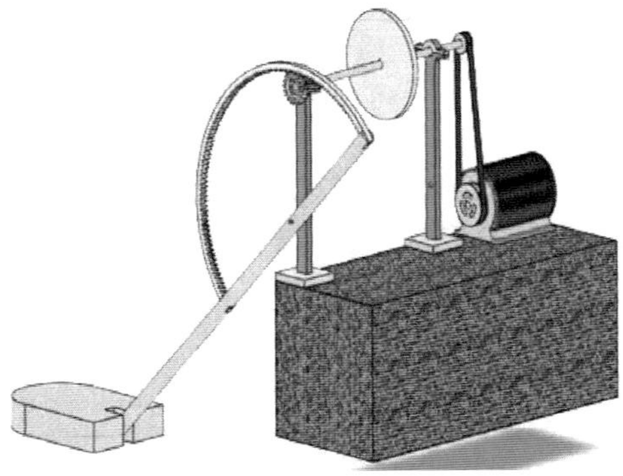

Figure 2. Circular Rack Gear Design

3.3. Simulation of two PLTGL design

To find the value of rotation per minute of the electric motor drive shaft, we need the data entered into the Solid Works application in the form of Frequency data. The output value was angular speed in degree per second. To find out rotations per minute, it was first converted into angular speed radians per second and finally converted to rotations per minute.

.

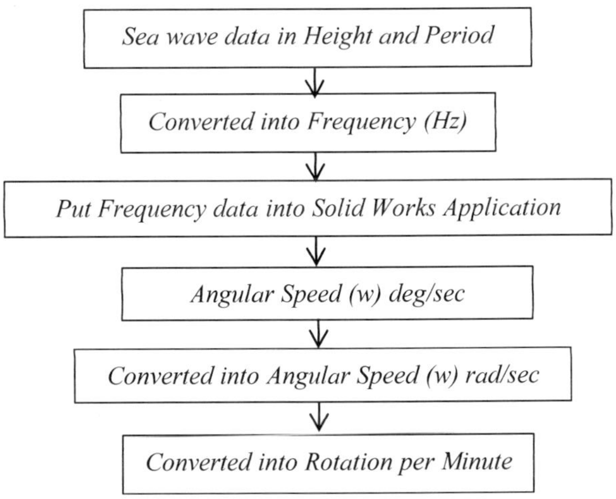

Figure 3. Simulation analysis procedure

For circular rack gear design, the wave height will be the same as the path length of the circular rack gear mechanism. Meanwhile for linear rack gear design, the length of the path traveled will depend on the tilt angle between the buoy and linear rack gear so that the rotation of the pinion is smaller than that of the circular rack gear.

Table 3. Resulted data from simulation on Solid Works

No	Sea Waves Height (m)	Frequency (Hz)	Linear Rack Gear Design: Angular Speed Deg/s (Max) (Min)	Linear Rack Gear Design: Rotation (n) (rad/s)	Linear Rack Gear Design: Rotation (n) (rpm)	Circular Rack Gear Design: Angular Speed Deg/s (Max) (Min)	Circular Rack Gear Design: Rotation (n) (rad/s)	Circular Rack Gear Design: Rotation (n) (rpm)
1	0,285	0,224	37	0,102	0,980	54	0,150	9,00
			21	0,058	0,556	33	0,091	**5,46**
2	0,561	0,148	46	0,127	1,219	66	0,183	10,98
			26	0,072	0,689	45	0,125	7,5
3	0,722	0,189	76	0,211	2,014	107	0,297	17,82
			56	0,155	1,484	89	0,247	14,82
4	0,956	0,204	109	0,302	2,890	149	0,418	25,08
			93	0,258	2,466	133	0,369	22,14
5	1,22	0,190	131	0,363	3,474	175	0,486	**29,16**
			111	0,308	2,944	138	0,383	22,98

4. Conclusion

Based on the analysis of sea wave power plants (PLTGL) with a combination of circular rack gear and pinion in the bonded coast, the highest rotation is 29.16 rpm at height and wave frequencies 1.22 m and 0.190 Hz respectively. The lowest rotation is 5.46 rpm at wave height of 0.285 m and frequency of 0.224 Hz. So it can be concluded with the wave data obtained at the bonded coast and taken 5 wave data with the ratio of the lowest and highest waves by analysing the simulation, the PLTGL design mechanism can produce rotation ranging from 5.46 rpm to 29.16 rpm.

Acknowledgment

We gratefully acknowledge the support from Universitas Bangka Belitung (UBB) through the Penelitian Dasar Tingkat Universitas (PDTU) program – Lembaga Penelitian dan Pengabdian Masyarakat (LPPM) in the year 2019.

References

[1] Hasan M H, Muzammil W K, Mahlia T M I, Jannifar A and Hasanuddin I 2012 A review on the pattern of electricity generation and emission in Indonesia from 1987 to 2009 *Renew. Sustain. Energy Rev.* **16** 3206–19

[2] Al-Habaibeh A, Su D, McCague J and Knight A 2010 An innovative approach for energy generation from waves *Energy Convers. Manag.* **51** 1664–8

[3] Uihlein A and Magagna D 2016 Wave and tidal current energy - A review of the current state of research beyond technology *Renew. Sustain. Energy Rev.* **58** 1070–81

[4] Wahyudie, A, Susilo, T.B, Aryani, F.A, Nandar, C, Jama, M., Daher A 2017 Ocean wave power potential assessment along the South Coast of Central Java Island Indonesia *Ocean. 2017 - Anchorage* 1–6

[5] Government of the Republic of Indonesia 2007 Law of the Republic of Indonesia Number 17 of 2017 on Long Termn National Development Plan of 2005-2025

IOP Conf. Series: Materials Science and Engineering **694** (2019) 012022 doi:10.1088/1757-899X/694/1/012022

The Effect of Biogas Pressure in the Performance and Emission of Spark Ignition Engine

Suyitno[1], Ridwan[1], Z. Arifin[1] and S H Pranolo[2]

[1]Mechanical Engineering Department, Universitas Sebelas Maret, Jl. Ir. Sutami 36 A, Surakarta, Indonesia, 57126
[2]Chemical Engineering Department, Universitas Sebelas Maret, Jl. Ir. Sutami 36 A, Surakarta, Indonesia, 57126

E-mail: suyitno@uns.ac.id

Abstract. Biogas is a renewable energy which can be used for fuel in the internal combustion engine and produce electricity by using a generator. On a small scale, electricity generation from biogas is important and attractive for special purposes especially for the areas which are not connected to the main electricity network or which are plenty of sources of biomass waste. Therefore, the purpose of this study is to investigate the performance and emissions of the internal combustion engine using the biogas fuel. The study was conducted by a simulation using ASPEN HYSYS and by experiment. The biogas pressure was varied from 0 to 4 barg. The results show that the highest power and efficiency were obtained at 700 W and 14.2%, respectively. The use of pressurized biogas can increase efficiency by 1%. Low CO and HC emissions and high CO_2 emissions indicate that combustion occurs completely.

1. Introduction

Alternative fuels such as biogas as a substitute for fossil fuels have been developed for many years. In areas with problems of some animal manures and biomass wastes, we can try to overcome them by processing them into biogas. The well-known processing process is carried out by anaerobic digestion (AD) [1]. The processing of natural dyes from indigofera plants and cow dung by AD technique has also been carried out by previous studies [2, 3].

The use of biogas to generate electricity has also been carried out [4-6]. On a fairly large scale, the power plant from biogas with a capacity of 685 kW was able to supply 10% of the city's electricity needs [6]. On a large enough scale, the choice of power plants can be done between the gas turbine or gas engine. Meanwhile, for small scale less than 10 kW, the main selection is done on the gas engine using an internal combustion engine (ICE).

Various attempts have been made to improve the performance of ICE using biogas fuels. Some important variables that affect to the performance of biogas engines are the ratio of methane and CO_2 in biogas [7], combustion mode, and air fuel ratio [7] where the engine used is diesel type. While the use of Otto engines has also been investigated where the variables investigated were the compression ratio [8] and swirl [9]. Unfortunately, the influence of biogas pressure itself is rarely examined even though it is predicted that the biogas pressure can increase the density and therefore the energy entering to the combustion chamber also increases. The use of pressurized biogas in an ICE may certainly affect the performance and emissions quality of ICE. So that, the purpose of this research is to investigate the performance and emissions of ICE when using biogas fuel at various biogas pressures.

2. Methods

Biogas is produced from the AD reactor where Indigofera plant waste was mixed with cow dung. Biogas characteristics have been reported previously [3]. Biogas was then flowed into the internal combustion engine which was connected to the electric generator.

The ICE has 4 stroke, 1 cylinder, and 160 cc. The compression ratio is 8.5: 1. The engine is equipped with an automatic voltage regulator (AVR) so that the AC voltage is 220 V with a frequency of 50 Hz. The loads from the generator were obtained from electric lamps. The amount of biogas that enters the ICE was measured using the gas meter. The electric energy produced was then measured with a DL69-2047 digital multi meter.

The ICE performance is expressed in terms of power and efficiency. The efficiency of the ICE generator system can be calculated as the ratio between the output power and the fuel flow rate multiplied by the lower heating value of the fuel used [10] as written in Equation 1.

$$\eta = P/(\dot{m} \times LHV) \quad (1)$$

Specific fuel consumption (sfc) is the ratio between the amount of fuel used (kg) and the energy produced (kWh), or equal to the ratio between the instantaneous fuel flow rate and the power produced [10] as follows.

$$sfc = m/(P \times t) \quad (2)$$

HYSYS ASPEN simulation was used to investigate in more detail the combustion process of biogas in the ICE. The software has various modules than can be combined to provide a full representation of various reactors. However, ICE modules are not available in ASPEN HYSYS. Therefore, the ICE model was constructed and composed of a compressor module, a RGibbs reactor, and two turbines. The compressor module was used to simulate the compression stroke with 80% isentropic efficiency. The RGibbs reactor module was used to simulate combustion stroke in the combustion chamber. The amount of energy produced in the combustion process was then flowed into the turbine module to represent the expansion stroke. Air and biogas were mixed by the mixer before being put into the compressor as can be seen in Figure 1. The power and efficiency can then be calculated by ASPEN HYSYS and compared with experimental results.

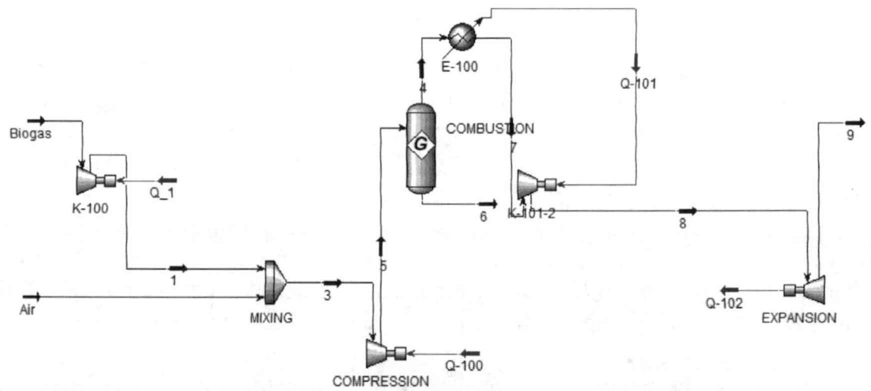

Figure 1. The process scheme of ICE in ASPEN HYSYS

To provide the pressurized biogas, the compressor was vacuum first and the biogas was streamed to the compressor. The pressurized biogas was then flowed into the mixer and flowed into the ICE. Biogas pressure was varied from 0 to 3 barg. To see whether biogas pressure affected to the power and efficiency of ICE, an analysis of variance of one variable was carried out [11].

3. Results and Discussion

The results of the output power and efficiency of the ICE using biogas fuel can be seen in Figure 2. It is noted that the highest power that can be obtained with the pressurized biogas fuel was 700 W at 3 bar pressure and 2400 rpm. Figure 2a shows that the ups and downs of the output power generated at different biogas pressures were influenced by the biogas consumption. In the greater biogas consumption, the power produced was also high. In general, it can be seen that the performance of ICE by simulation resulted a higher output power compared to the power generated from ICE by experiments. It is caused that the simulation was done with a perfect combustion model so that the temperature and pressure of the combustion chamber were higher than the actual temperature and pressure.

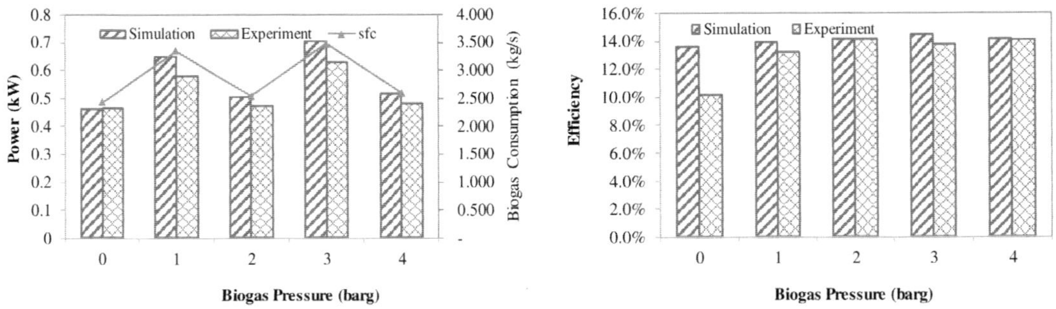

(a) (b)

Figure 2. Power, fuel consumption, and efficiency of ICE

Table 1. Analysis of variance for pressure factor effect to the output power of ICE

Source of variation	Sum of Squares	Degrees of Freedom	Mean Square	F0
Pressure	0.096	4	0.024	41.603
Error	0.006	11	0.001	
Total	0.103	15		

In Figure 2b it can be seen that the highest efficiency of the ICE was 14.2% achieved at a biogas pressure of 3 barg. Improving the efficiency of ICE by increasing biogas pressure is about 1%. The enhancing efficiency of ICE is quite significant from the analysis of variance test as can be seen in Table 1 where the $F_{0.05,\ 4,\ 11}$ is higher than 3.36.

With higher biogas pressures, the quality of combustion is better [12] due to the phenomenon that the pressurized biogas has an easier characteristic to ignite. Meanwhile, if the efficiency of ICE found from the study is lower than that of the efficiency of ICE from previous studies, i.e. 26.2% [13]. When compared with the highest efficiency of 23.3% [13] on spark ignition engines made from raw biogas with 65% methane content, the maximum efficiency in this study remains lower.

The sfc of the ICE with the biogas fuel in this study was ranging from 5.37 to 6.73 kg/kWh. Meanwhile, when compared with previous studies on the spark ignition engine with methane-enriched biogas which has a minimum sfc of 0.471 kg/kWh [13], the lowest sfc found in this study remained far higher. When compared with a minimum sfc of 0.651 g/kWh [13] on the ICE with a raw biogas with 65% methane content, the minimum sfc of the ICE in this study was still higher. The reason is because the biogas in this study had 8% methane content and 26% carbon monoxide content. As a result, the energy produced from the burning of the biogas is low so that to produce a high power, a bigger amount of biogas is needed with not too good quality, so that the sfc increases and efficiency decreased.

IOP Conf. Series: Materials Science and Engineering **694** (2019) 012022 doi:10.1088/1757-899X/694/1/012022

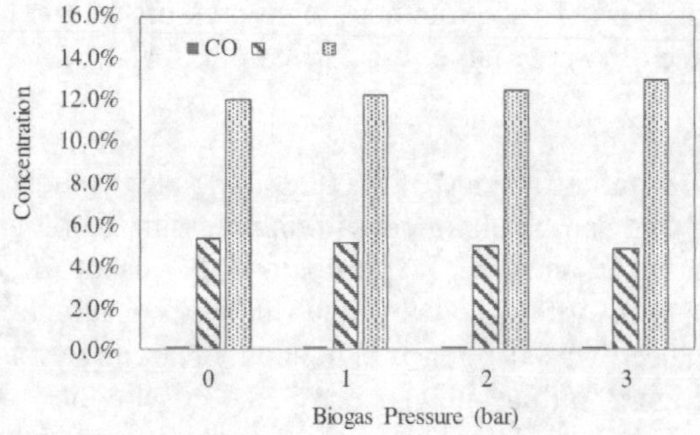

Figure 3. Flue gas emission from ICE with biogas fuel

Figure 3 shows the exhaust emissions from ICE with the biogas fuel. The CO, CO_2 and O_2 concentrations found in the flue gas were ranging from 0.09 to 0.11%, from 4.8 to 5.3%, and from 11.96 to 12.97%, respectively. Hydrocarbon (HC) content in flue gas was ranging from 866-875 ppm. The higher the biogas pressure, the content of CO and CO_2 was slightly decreasing. The presence of a very low content of CO and HC and a high content of CO_2 indicates that the combustion in the ICE occurs completely. While the high content of oxygen in the flue gas showed that the air to fuel ratio (AFR) mixture of air-biogas that enters the ICE was higher than the stoichiometric AFR. The high AFR can be carefully identified by the analysis with ASPEN HYSYS and it can be seen that the work of compression increased sharply while the input energy tended to be the same so that the net output of work generated from ICE tended to be low. Therefore, besides being influenced by the quality of biogas, the ICE efficiency of 14.2% in this study was also influenced by the high use of AFR.

4. Results and Discussion

The study on the effect of pressure on biogas-based ICE performance and emissions has been conducted. The output power and efficiency have been compared between the results of the experiments and the results of the ASPEN HYSYS simulation. The highest power and efficiency were obtained at 700 W and 14.2%. The use of pressurized biogas can increase the efficiency of ICE by 1%. A low content of CO and HC emissions and a high content of CO_2 emissions indicated that combustion occurred completely. Therefore, it can also be known that the cause of the low efficiency of the engine aside from the biogas quality factor is also influenced by the use of excess air when burning the biogas in the ICE. Therefore, efforts to improve the performance of ICE with biogas fuel can be done by improving the quality of biogas and controlling the proper use of AFR are interesting for future works.

References

[1] Suyitno M and Dharmanto N 2010 Teknologi Biogas. Graha Ilmu, Yogyakarta.

[2] Ridwan, Prasetyo S D, Kusuma A C, Rahman R A and Suyitno 2019 Recent progress of biogas produced from the waste of natural indigo dyes for electricity generation. In: *American Institute of Physics Conference Series*.

[3] Sihta F, Suyitno, Heru Wibowo A and Tanding R 2018 Enhancing biogas quality of indigofera plant waste through co-digestion with cow dung. In: *MATEC Web of Conferences*

[4] Murphy J D and McKeogh E 2004 Technical, economic and environmental analysis of energy production from municipal solid waste *Renewable Energy* 29 1043-57

[5] Pöschl M, Ward S and Owende P 2010 Evaluation of energy efficiency of various biogas production and utilization pathways *Applied Energy* 87 3305-21

[6] Santos I F S d, Gonçalves A T T, Borges P B, Barros R M and da Silva Lima R 2018 Combined use of biogas from sanitary landfill and wastewater treatment plants for distributed energy generation in Brazil *Resources, Conservation and Recycling* **136** 376-88

[7] Henham A, Makkar M J E C and Management 1998 Combustion of simulated biogas in a dual-

IOP Conf. Series: Materials Science and Engineering **694** (2019) 012022 doi:10.1088/1757-899X/694/1/012022

fuel diesel engine **39** 2001-9

[8] Porpatham E, Ramesh A and Nagalingam B J F 2012 Effect of compression ratio on the performance and combustion of a biogas fuelled spark ignition engine **95** 247-56

[9] Porpatham E, Ramesh A, Nagalingam B J E c and management 2013 Effect of swirl on the performance and combustion of a biogas fuelled spark ignition engine **76** 463-71

[10] Silva F P, de Souza S N M, Kitamura D S, Nogueira C E C, Otto R B J R and Reviews S E 2018 Energy efficiency of a micro-generation unit of electricity from biogas of swine manure **82** 3900-6

[11] Montgomery D C 2017 *Design and analysis of experiments*: John wiley & sons.

[12] Bora B J, Saha U K, Chatterjee S and Veer V 2014 Effect of compression ratio on performance, combustion and emission characteristics of a dual fuel diesel engine run on raw biogas *Energy Conversion and Management* **87** 1000-9

[13] Chandra R, Vijay V K, Subbarao P M V and Khura T K 2011 Performance evaluation of a constant speed IC engine on CNG, methane enriched biogas and biogas *Applied Energy* **88** 3969-77

IOP Conf. Series: Materials Science and Engineering **694** (2019) 012023 doi:10.1088/1757-899X/694/1/012023

Ergonomics in Mining: Participatory Approach in Indonesian Tin Mining

R Priyoko Prayitnoadi[1], Glyn Lawson[2], Brendan Ryan[2] and Setia Hermawati[2]

[1]Jurusan Teknik Mesin, Fakultas Teknik, Universitas Bangka Belitung, Bangka 33172, Indonesia

[2]Human Factors Research Group, University of Nottingham Nottingham NG7 1RD, United Kingdom

E-mail: priyokoprayitnoadi@gmail.com

Abstract. This paper presents the investigation of the acceptability and suitability of Participatory Ergonomics (PE) interventions within tin mining companies. The study was conducted by scenario-based-design (SBD) method and questionnaire with participants from small medium enterprises (SMEs) tin companies. Five companies and fifteen participants were involved in this study. The acceptance of participatory ergonomics through SBD study revealed positive in Indonesian tin mining companies The study shows that support from occupational safety and health (OSH) authorities are needed in order to strengthen workers' knowledge in ergonomics and good safety practices and to allow them to implement ergonomics through training and the participatory approach. The results of this study may contribute towards the implementation of PE. As a consequence of improved implementation, tin mining accidents may be reduced.

1. Introduction

Tin mining in Indonesia has been operating for more than 300 years. Located in Bangka Belitung islands, tin mining is continue to produce around 90% of Indonesia's tin and fulfills 30% of the world's tin requirements [1,2]. Indonesia is the world's second-largest tin exporter after China [3]. Since 2000, Bangka and Belitung Island became new province and through the regional autonomy, illegal miners have emerged developed as local government has the ability to issue permission where central government has not [4]. Tin mining activity can be done by small scale miners and even by individuals who do not have experience and knowledge of mining best practices. Consequences to illegal mining practice with unsafe devices, lack of knowledge about the tin mining exploitation process and limited funds result in dangerous incidents, accidents and even fatalities. This could be the reason why the number of occupational accidents in the mining industry sector is relatively high when compared to other industry sectors [5].

One of the approaches to support ergonomics intervention is participatory ergonomics (PE). Wilson et.al [6] defined participatory ergonomics as 'the involvement of people in planning and controlling a significant amount of their own work activities, with sufficient knowledge and power to influence both processes and outcomes in order to achieve desirable goal'. The implementation of PE has been applying in many fields and industries such as in military, manufacturing, production and processing plants, service industries, construction and transportation [7,8].

In review of ergonomics applications in industrially developing countries (IDCs), Hermawati et al. [9] posited the view that a participatory approach is needed as a bottom-up approach in implementing ergonomics at workplace. They also suggested that by 'having an appropriate adaptation and modification of ergonomics tools will definitely be beneficial in the future'. Hermawati et al [9] found that there are few studies addressing the health and safety domain in Indonesia, and the mining sector, which was shown to have the least number of ergonomics application.

So far there has been little research on PE in Indonesia; the research work that has been done has tended to focus on the physical ergonomics rather than organizational ergonomics [9]. To the best of authors' knowledge, there are no other scientific publications that discuss the implementation of participatory ergonomics (PE) in occupational safety in Indonesian tin mining.

The aims of this research were to investigate the acceptability and suitability of participatory ergonomics interventions within tin mining companies in Indonesian tin mining. The approach for this study i.e. measuring an interaction between people during a discussion was developed during this study. To the extent of authors' knowledge, this approach had never been applied before. However, this work should be evaluated to review the novel contribution against other works. This paper's main contribution is on the identification by the actual condition of PE issues, new approach to measure participation, and future research recommendation in addressing identified problems Although this paper is focus on Indonesian tin mining companies, it is expected that this findings will also be useful for other IDCs where tin mining SMEs is one source of employment.

2. Research Methods

The study used a Scenario-Based-Design (SBD) and Questionnaire methods. Those methods are described as follow.

2.1 Scenario-Based-Design (SBD)

SBD was used to identify claims and trade-offs that may impact PE in the study of participatory ergonomics intervention. It consisted of two problem scenarios, i.e. workers problems in onshore tin mining, and in a tin smelter.

2.1.1 Story in onshore tin mining; the story said that there are six workers at the 'front' (the mention of a place where a huge mixed water and sand are collected to be sucked up with suction machine unit). Some workers hose down the land with high-pressure water, whilst some of them are manage the sand. The workers work in a hot, wet, and noisy environment, with no personal protective equipment (PPE). The workers are working 8 hours a day and are paid based on tin result. The participants were asked some questions i.e. why do not they use PPE? What are the consequences of not using PPE? How do you suggest preventing those consequences? Moreover, what kinds of PPE do you think suitable for them?

2.1.2 Story in smelter; where it said that the participants imagined work for a tin smelter small medium enterprise in town. In the plant, there are activities in the workplace as the participants can see in the picture given. It is a very dusty, hot, noisy and dangerous work environment. The participants were asked to discuss within the group to make a list at least three most important things that they could discuss it in more detail in order to identify the consequences and to find the solutions to overcome those problems. Moreover, the group would discuss about any experience in similar problems in their workplace, explain what their experienced and what they could do to solve these problems.
These two scenarios were given to a group consisting of owner, manager, and worker (n = 15). The groups (n = 5) were tasked to discuss safety issues through those problem scenarios and it recorded by video recorder.

SBD was used to measure participation and to find ergonomics issues through SBD discussions. To measure participation, the author collected data from videos and calculated the duration each participant's spoke. The data was translated into a line chart, which is shown the relationship between time and duration. This line chart allows the author to analyze how much time each participant contributes in a discussion in terms of time durations and time intervals. Another chart, a pie chart, was made to displays a total duration of each participant in percentage. Meanwhile, to find ergonomics issues, the author transcribed and translated the audio from the video and it grouped for each question in the scenarios to analyze through content analysis.

2.2 Questionnaire

Questionnaire was used to identify how participants found PE process, i.e. did you get a chance to contribute in problem solving?, and how likely success of further PE if it applicable, i.e. do you think this type of participatory approach would be suitable way to improve your work? The questionnaire, which consisted of twelve questions, was given to the participants after finished their scenario-based-design (SBD) study. Questionnaire responses were translated into English and content analysis was used to analyze collected data by roles categories. Tabulations of content analysis were made and managed according to the questions. The summary of study has been made for each participant category.

3. Results

3.1 Participatory Measurement

A Microsoft Excel database of intermittent voice duration was formed during participatory analysis. The data was translated into a line chart, which is shown the relationship between time and duration. This line chart was analyzed how much time each participant contributes in a discussion in terms of time durations and time intervals. Another chart, a pie chart, was made to displays a total duration of each participant.

The acceptance of participatory ergonomics through SBD study revealed positive in Indonesian tin mining companies. Based on participatory measurement, it shows that across all companies, owner contributed 35%, manager and worker contributed 38% and 27% respectively (See Figure 1).

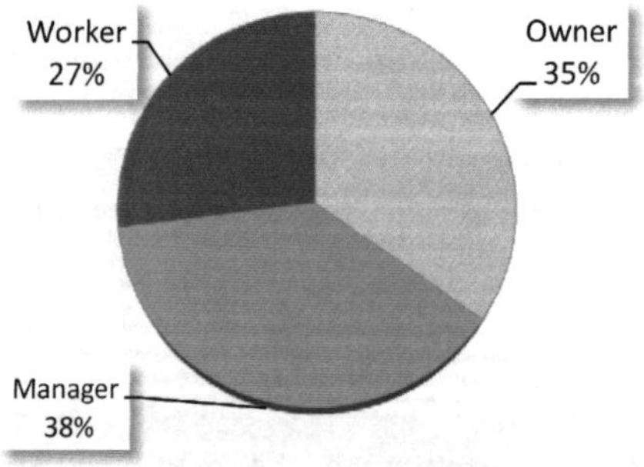

Figure 1. Participation analysis on all companies

Nevertheless, the participatory distribution for each role in every company is varying. For example, company 1 and 2 were found that managers have greater contribution in discussion. Company 3 indicates that owner has dominant in discussion with more than 50% talk time. Worker has slightly more than manager does in company 5. Meanwhile, company 4 showed a good balance of talk time, where there are small spreads among the roles, i.e. owner is 26%, manager 36% and worker is 38% (Figure 2)

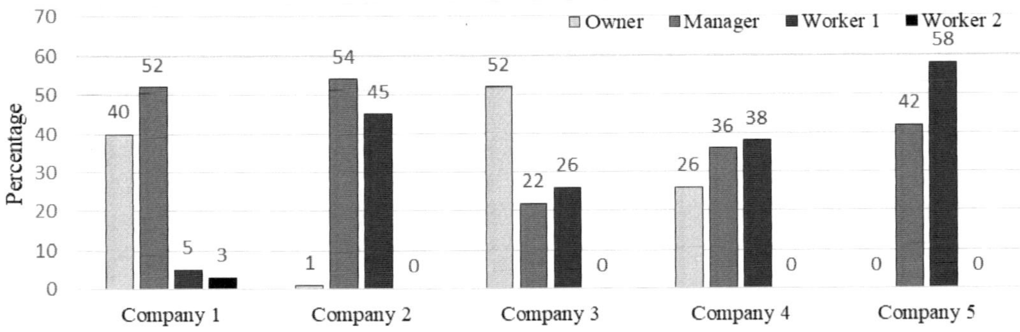

Figure 2. Participatory measurement for all companies

3.2 Ergonomics Issues

Generally, participants were expressing the same arguments from the question about using personal protective equipment (PPE) during work, i.e. lack of knowledge and attitude of workers in use of PPE. This situation happens due to external and internal factors. External factors are mostly come from company such as less information about safety regulations including no adequate training provided, inappropriate of PPE, and lack of daily safety supervision whilst internal factor is a problem of human cultural such as sense of 'unnecessary' of using PPE, and take for granted a wrongdoings.

Since ergonomics is important to fitting workplace conditions and job demands to the capabilities of the working population, ergonomist or industrial experts must work together with non-experts to improve human-machine working relationship and working safety. This means that participatory ergonomics approach for disseminating ergonomics information is needed.

3.3 Participatory Issues

Most of participants expressed that PE approach is better than the existing approach. Most of owners said this approach is new and involving negotiation among roles. Some managers argued this approach is more practical and through team working. Indeed, PE intervention approach will be an effective and efficient way to implement, but it will not work if participants are reluctant. Opportunities to say something, share knowledge and team based solution are the reasons why this approach is more effective and efficient way. Moreover, this type of participatory approach would be a suitable way because it involves all levels, increasing safety awareness and self-responsibility.

The barriers in implementing PE in Indonesian tin mining are concluded as communication among roles, which is hard to do. Specific communication problems are misunderstanding about the work place problem, and disagreeing or forcing arguments that could happen.

These barriers could bring this approach into a long-term solution become slower. In fact, long-term solution needs dissemination of PE to improve understanding, improving quality of communication among roles, and making easier to accept new information/approaches.

4. Discussion

The acceptance of PE through scenario based design (SBD) study revealed positive in Indonesian tin mining companies. Based on participatory analysis, it shows that owner contribute 35%, manager and worker are contribute 38% and 27% respectively. Nevertheless, the participatory distribution for each role in every company is varying.

Although almost all workers admitted that they contribute in solving problem, this finding confirmed the idea that there are workers who reluctant to talk and only work as boss asked them. In contrast, Owner in one company showed only 1% contribution. This is inconsistent with the questionnaire result that all owners have chances to contribute in problem solving. It is presumably that the owner reluctant or fears to say something and it is likely that 'management see participation as threat to their right to manage rather than as an aid' [10]. Moreover, they said that group or individual resistances are said to be one of the reasons why workforce hesitate to get involved and the workforce feel lack, sufficient motivations, time and energy. They also said that 'they might defer too much either to management or to technical experts'.

All participants were happy with the group's chosen solutions in SBD discussion. Owner said that solution solved through group with different roles and the consensus was taken. Manager, in this case, said that this could improve occupational safety awareness since it involves floor workers. Meanwhile, worker was happy because they had opportunities to say something that bosses should know the workplace problems directly and formally. Besides, this makes 'togetherness' in working was implemented as Haines & Wilson [10] mentioned that 'work teams participating together to identify problems and implement improvements'. Through all those findings, it can be concluded that participating in solving problem in-group, which consist of all roles in Indonesian tin mining companies, is generally acceptable with the terms that deepening study to investigate problems in implementing PE is conducted to generate a comprehensive suggestions in supporting this acceptance.

5. Conclusion

It is found that despite the floor workers are accept the PE approach to improve occupational safety, the study distinctly prove that support from OSH authorities are needed in order to strengthen workers' knowledge in ergonomics, good safety practices and allow them implemented ergonomics through training and participatory approach. The findings of this study offered new information on specific barriers to implementation of participatory ergonomics intervention.

Acknowledgment

This research was funded by the Directorate General of Higher Education, Ministry of Research and Higher Education, the Republic of Indonesia through the Overseas Postgraduate Scholarship Program.

References

[1] Hodal K 2012 Death metal: tin mining in Indonesia *Guard.*

[2] PT.Timah (Persero) Tbk 2013 PT.Timah (Persero) Tbk. Annual Report 2013

[3] Brown T J, Bide T, Hannis S D, Idoine N E, Hetherington L E, Shaw R A, Walters A S, Lusty P A J and Kendall R 2012 *World mineral production 2006-2010* (Nottingham: British Geological Survey)

[4] Aspinall C 2002 *Small-Scale Mining in Indonesia* (Jakarta)

[5] Markkanen P . 2004 *Occupational Safety and Health in lndonesia* (Manila)

[6] Wilson J R, Haines H and Morris W 2005 Participatory Ergonomics *Evaluation of Human Work* ed J R Wilson and C Nigel (Taylor & Francis) pp 933–49

[7] Hignett S, Wilson J R and Morris W 2005 Finding ergonomic solutions - Participatory approaches *Occup. Med. (Chic. Ill).* **55** 200–7

[8] Kuorinka I 1997 Tools and means of implementing participatory ergonomics *Int. J. Ind. Ergon.* **19** 267–70

[9] Hermawati S, Lawson G and Sutarto A P 2014 Mapping ergonomics application to improve SMEs working condition in industrially developing countries: a critical review

[10] Haines H and Wilson J R 1998 *Development of a framework for participatory ergonomics* (Health and Safety Executive)

IOP Conf. Series: Materials Science and Engineering **694** (2019) 012024 doi:10.1088/1757-899X/694/1/012024

Modification of coolant fluid controller for minimum quantity lubrication (mql) methods with use of arduino uno-based control system

Zulkifli Amin[1], Meiki Eru Putra[2], Ismet Hari Mulyadi[1] and Ardi Desfiandra[1]

[1]Mechanical Engineering, Engineering Faculty, Andalas University, Kampus Limau Manis, Padang-25163, Indonesia

[2]Mechanical Engineering, Dharma Andalas University, Simpang Haru, Padang-25123, Indonesia

E-mail: zul270171amin@gmail.com

Abstract. Coolant is important in the machining process. Excessive and uncontrolled use of coolant can cause high production costs, affect the human health and the waste of coolant polluting the environment. One method that has been used to reduce the amount of coolant is the limiting coolant distribution method or commonly called MQL (Minimum Quantity Lubrication). The MQL equipment which is available in the market has a relatively expensive, so a prototype of a coolant control system of MQL method has been designed and made. However, the prototype made not meet the desired specifications then a modified control system is added to the prototype. Prototype performance is compared before and after the control system being added. The duration of fluid drop and the output capacity of the fluid sprayed are measured to test the function of the device. It is found that with the addition of modified control systems, the performance of MQL cooling regulator is improved, but it is applied mostly for fluids which have viscosity ranging about viscosity of oil and cooking oil. For cooking oil fluid, the MQL tool performance increase as the capacity of the fluid decreases due to the control system being applied at all case of air valve openings for the diameter orifice of 2 mm. It is also found that at the aperture of the air valve of ¼, with a 10 second delay time and an orifice diameter of 2 mm, the capacity coolant sprayed is reduced from 107.348 ml/hr to 89.609 ml/hr when the modified control system is added. For the Bromus fluid, the capacity of the fluid decreases as the control system being applied. The MQL tool performance increase only occur in the smallest valve opening of ¼ that is about 9 to 26%. However, the capacity of Bromus types of fluid increase as the air valve openings increase.

1. Introduction

The use of coolant in the machining process by using a machine tool is an important part. The coolant has a role in reducing friction so that the tool will last longer and be able to maintain the quality of the machining product. During this time many people argue that the more liquid given, the better. Excessive use of coolant in the machining process has a negative impact on the environment as revealed by the coolant supplier company, Blaser. Economically, the use of coolant had already exceeded the costs for procurement of tools [1]. Excessive use of coolant will cause high production costs and affect the health and waste of polluting the environment [2,3].

One method that has been used to reduce the effects of coolant is the method of limiting the coolant distribution or commonly called (Minimum Quantity Lubrication). Various studies have proven that this method is able to exceed the ability of excessive use of coolant in extending tool life [4,5]. Some large companies have successfully used it, such as the Ford Motor Co., America and General Motor Co., USA production lines [6,7]. This tool is commercially available in the market, but in terms of investment cost, this tool is still expensive, especially for small manufacturing companies.

In a previous work, it has been designed and made a prototype of a coolant regulator using a simple MQL method [8]. The prototype was functionally worked, but the prototype has not been able to produce the desired coolant flow rate range of 1 to 500 ml/hour at a pressure of 30 MPa. Then a control system is created and applied to the MQL tool. Modification of the tool by adding the control system is expected to be able to eliminate deficiencies that exist in the previous MQL tool.

MQL tools are usually used in the machining process to reduce the working temperature in the cutting process. The cutting process is carried out using a cutting tool that is mounted on a machine tool. Cutting the workpiece by utilizing the relative motion between the cutting tool with the workpiece so as to produce a product in accordance with the desired product geometry. This process has the waste part from machining on a product commonly referred to as chip [9].

Reducing the temperature in the machining process can use several methods, one of which is by flowing the cutting fluid. In general, coolant is a cooling medium used to cool workpieces and cutting tools during the machining process. The coolant is also used to lubricate the cutting tool so that it has a longer service life. Coolant is a liquid resulting from a mixture of ethylene or propylene glycol and water. Usually the ratio of mineral substances is about 50/50 [10].

Minimum Quantiy Lubrication (MQL) is a cooling system in the machining process that allows the use of a limited amount of coolant. In the MQL method the distribution of coolant is by combining liquid fluid and air so as to form aerosols or fogging in other words the use of coolant can be minimized. MQL system consists of two types, namely external and inernal systems.

In this external system has 2 air channels, the first functions to create an air casing as a mixing chamber, while the second provides atomization (fogging) of the air supply, and the misted oil is supplied through the middle section. This external system with ejector nozzle is the cheapest and simplest method [11]. Whereas in the internal system, aerosols are prepared in sprayers and then supplied through shafts or internal channels on the device. When this system is used in machining centers or manufacturing cells, the aerosol supply unit must react to the frequency of changes in the tool which only takes 1 or 2 seconds. The proper aerosol parameter settings for each tool are given.

This principle is applied to MQL tools and can be controlled as required. The control system itself is the process of regulating or controlling one or several quantities (variables or parameters) so that they are at a certain value or range.

Open loop control system (Open Loop) is a control system in which the amount of output does not have an effect on the amount of input, so the controlled variable output cannot be compared to the desired value of the initial setting variable as figure 1.

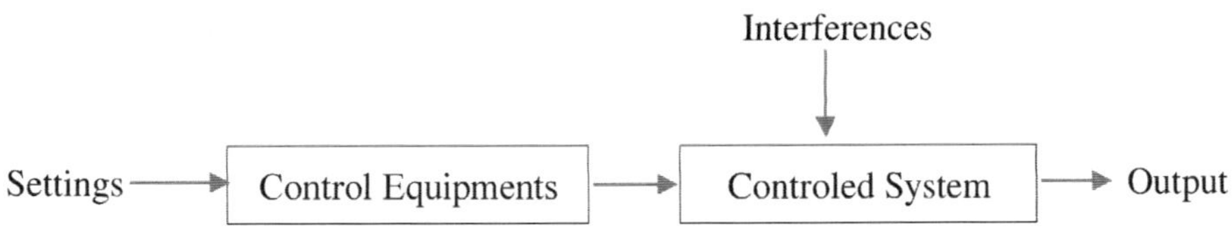

Figure 1. Open loop control system [12]

Closed loop control system (closed loop) is a control system where the amount of output has an effect on the amount of input, so that the amount of output can be compared to the desired input [12]. Furthermore, the value difference between the value output and the initial setting input is used as a correction which is the control target as seen in figure 2.

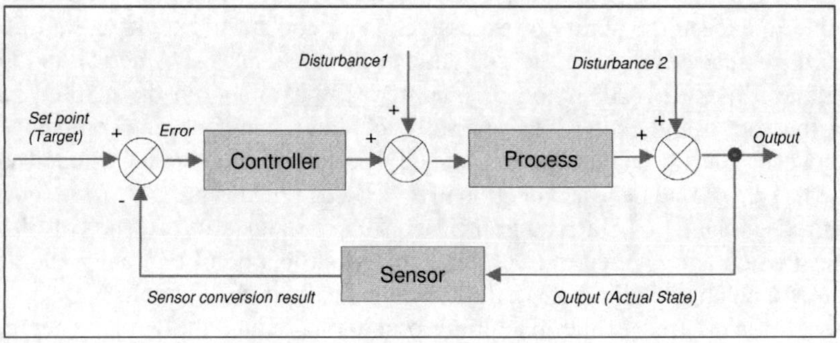

Figure 2. Close loop control system [12]

2. Methodology

2.1. Tool Design Specifications

In this study, it is planned that a cooling fluid regulating system will be created using the Minimum Quantity Lubrication (MQL) method. Based on the analysis results obtained by looking at products that are already on the market, it can be concluded that the existing equipment is still expensive and they have a complicated system and limitations in fluid distribution. In order to fulfill these requirements, it is planned that the tools can meet the needs as listed in table 1.

Table 1. Specifications of the MQL prototype tool [8].

Specification	
Flowrate	1 - 500 ml / hour
Air pressure	30 Pa
Capacity	1 L.
Fluid Down System	Gravity
Mixing Type	Internal mixing
Valve Open Range	Aperture, ¼, ½, ¾, full
Viscosity Range	SAE 40 or SAE 20

The desired flowrate is from 1 to 500 ml/hour. The range of flowrate chosen is very wide because each type of material from the workpiece has a different level of hardness, so the greater the range of flowrate the more types of material can be tested. The MQL tool uses air to drive the desired fluid and air pressure which ranges from 1 to10 Pa, depending also on the flowrate to be used. If the flowrate is high, then the air pressure is set to be high accordingly.

2.2. System Design

In this study an internal system is used, the form of system design that will be made as shown in figure 3.

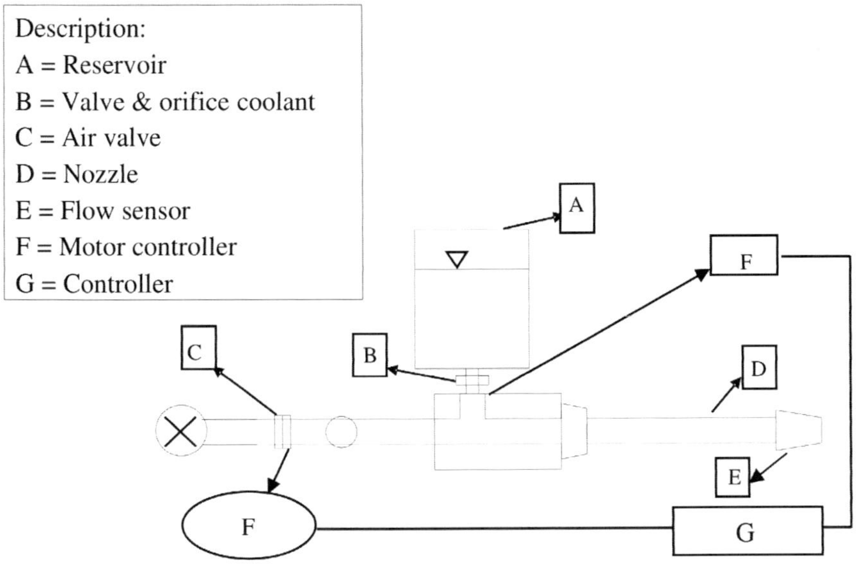

Figure 3. System Design

The working principle of the MQL tool is that the fluid in reservoir *A* passes through valve *B* which is open as specified. Fluid valves can be open ¼, ½, ¾, or fully open. Fluid valves will be able to open and close based on a set amount of time. Then after the fluid passes through the valve, then the air will blow to push the fluid out the nozzle. The air that comes out of the compressor will pass through valve *C*, just like valve *B*, the air valve can also be adjusted to aperture of ¼, ½, ¾, or even a full aperture.

The mixture of air and fluid will pass through the nozzle *D*. On the nozzle *D* there is a flow sensor (flowensor) to read the flow rate of the discharge fluid. This measurement signal is sent to Arduino. If the measured flow rate is not the same as what has been set in the program, then the Arduino will make an adjustment.

Arduino UNO is an Arduino product that is used as a controller of an electronic or mechatronic system. Aduino UNO consists of an electronic board containing an ATmega 328 microcontroller [13]. Arduino consists of two main parts namely a physical circuit board (often also called a microcontroller) and a software or IDE (Integrated Development Environment) that runs on a computer. This software is often called Arduino IDE which is used to write and upload code from a computer to the Arduino hardware board as shown in figure 4.

Flowsensors consist of plastic valve bodies, water rotors, and hall effect sensors. When water flows through, the rotors wind up [14]. The speed is converted from the difference in the rotation of the timer on the sensor rotor, changes will be seen at different rates of flow. The advantage of this sensor is that it only requires 1 signal (GIS) in addition to the 5V dc and ground lines as shown in figure 5.
DC servo motors are widely used in the design of mechatronics because the torque characteristics and speed of DC motors in accordance with the configuration of the electronic circuit in this case is a servo motor [15]. The physical form of a DC servo motor can be seen as figure 6.

IOP Conf. Series: Materials Science and Engineering **694** (2019) 012024 doi:10.1088/1757-899X/694/1/012024

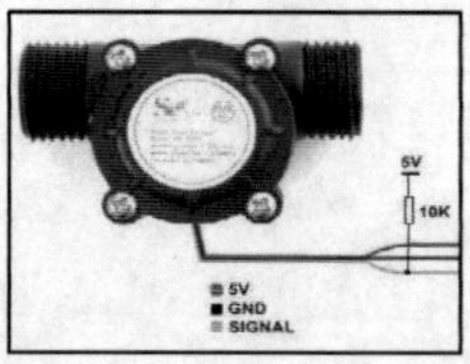

Figure 4. Arduino uno [13] **Figure 5.** Flow sensor [14] **Figure 6.** Servo motor [15]

2.3. Tools and Materials
To test the capabilities of the MQL tool, a measuring media and device are required. To find out the ability of the tool (in this case how much volume of fluid is coming out) then a sponge is used as shown in figure 7. Sponge is used to capture the fluid that comes out so that it can be weighed. This media is used to ease measuring the amount of fluid released. The fluid mass measured is the fluid mass used for 10 minutes of operating time of the tool. 10 minutes will be converted into units of hours so that the measured flowrate becomes ml/hour. To measure how much fluid is trapped inside the sponge, a digital scale that has a precision of up to 1/100 gram will be used. The scales used are BL 4100s type Setra with the specifications of being able to read weights to the lowest of 0.01 grams as shown in figure 8. This scales will be used to measure dry and wet sponge (weight after trapped fluid).

Figure 7. Sponge

Figure 8. Digital scales

To get the mass of fluid, use Equation (1)

$$\text{Fluid mass} = \text{Weight of sponge after} - \text{weight of initial sponge} \quad (1)$$

After the fluid mass is obtained, it will be converted to ml using Equation (2):

$$\text{Density } (\rho) = \text{mass} / \text{volume} = \text{gr} / \text{cm3} \quad (2)$$

2.4. Measurement Method
After the tool is finished, several tests will be conducted to determine the lowest and highest capabilities of the tool. For this reason, the parameters set are the aperture of the valves for fluid flow and for air flow. The apertures are ¼, ½, ¾ and full aperture for both fluid flow and air flow valves. While the desired output is the amount of fluid trapped in the sponge.

2.5. Testing the MQL Tools
To ensure that the equipment planned and manufactured meets the initial criteria, the MQL tool is tested. The procedure for testing the MQL tool is as follows. First, the MQL tool is prepared; determined the initial sponge's weight, then place the sponge at the nozzle's mouth and fill fluid into the reservoir. Next, the control system is turned on. After that, activate the MQL tool, then adjust the fluid valve opening as desired. Operate the machine for 10 minutes. Weigh the weight of the sponge

after the process. Repeat the procedure for next valve opening variation. After getting the data, then it is processed and analyzed.

3. Result and Discussion

3.1. Modification of the MQL Tool

In this research, the cooling fluid regulator has been successfully modified in the machining process using the MQL (Minimum Quantity Lubrication) method by utilizing an Arduino Uno-based control system. The modified tool can be seen in figure 9.

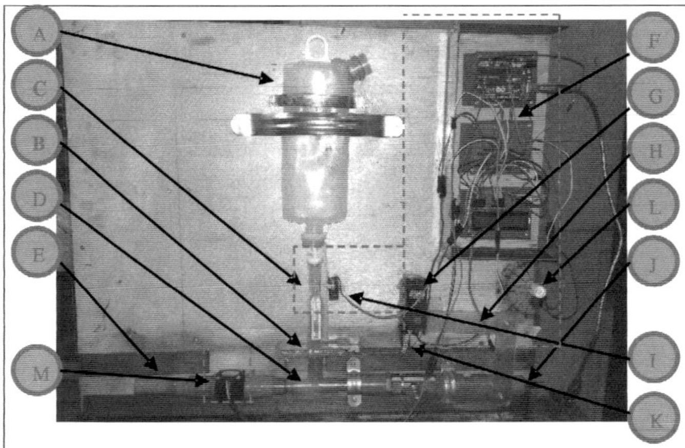

Description:
A. *Reservoir*
B. Fluid control (Orifice)
C. Droplet control
D. T Joint
E. Nozzle
F. Arduino circuit
G. Servo motor 1
H. Air valve drive arm
I. Servo motor 2
J. Air valve
K. Motor servo 3
L. Air regulator
M. Flow sensor

Figure 9. Modified MQL equipment

In figure 9, the modified part is marked by a dashed red line. The emphasis on modifying the MQL tool is done on the part to adjust the droplet when the fluid enter the orifice to the fitting T and on the air valve control.

3.2. Functional Test of the Modified MQL Tool

3.2.1. Test of the duration of the fluid droplet. This test aims to find out how long the fluid drops from the bottle to the fitting T. The test is done by putting oil paper into the fitting T, then calculated with a stopwatch the time required by the fluid dropping from the reservoir into the oil paper. The test is done by varying the number of different volumes in the bottle which are 100 ml and 300 ml with 2 repetitions each. The tests are carried out before and after applying of the control system on the tool. The time span obtained is useful to know whether the amount of fluid volume in the bottle affects the flow rate of fluid down to the fitting T. The data in table 2 are the data obtained after testing. Applying the control system resulted in a longer fall of the fluid from the reservoir.

IOP Conf. Series: Materials Science and Engineering **694** (2019) 012024 doi:10.1088/1757-899X/694/1/012024

Table 2. Duration of the fluid droplet

Fluid Types	Fluid Amount	Data 1 (without a control system) Time (s)	Data 2 (with a control system) Time (s)
Oil	100 ml	12.87	13.82
		12.35	13.35
	300 ml	11.19	10.35
		10.52	9.25
Cooking oil	100 ml	5.17	4.95
		5.55	4.75
	300 ml	4.29	3.35
		4.57	4.17
Bromus	100 ml	2.26	2.46
		2.28	2.56
	300 ml	1.29	2.03
		1.93	2.18

In table 2 it can be seen that the amount of fluid volume in the bottle affects the drop time of the fluid. The more volume in the bottle, the faster the fluid dripping down. Apart from the amount of volume, the thing that also influences the fluid drop time is the vicosity of the fluid. In theory, the fluid sequence from the thick to the thin one is oil, cooking oil, and the most watery bromus. Tests are carried out before and after the existence of a control system, the difference can be seen in figure 10.

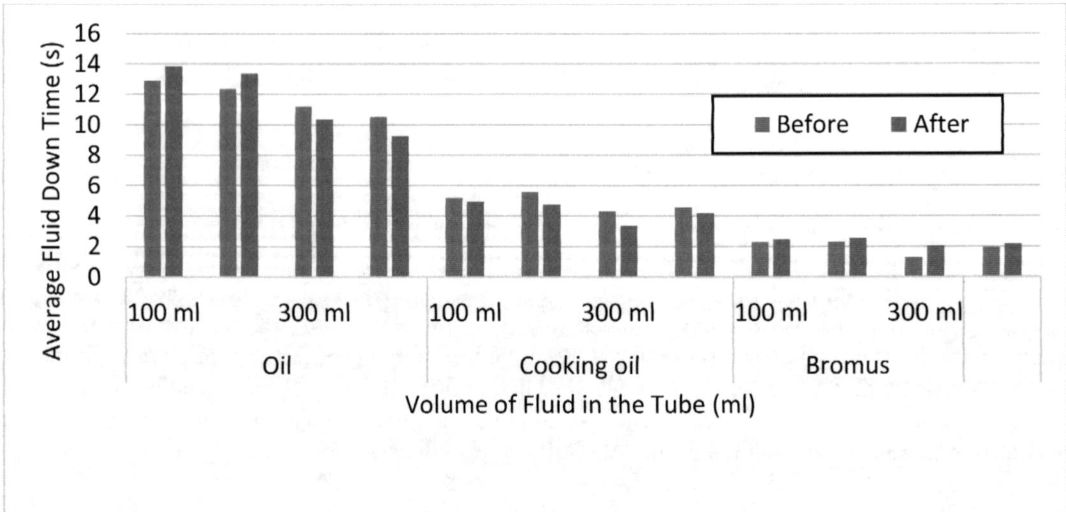

Figure 10. Graph of data of the droplet lag time of the fluid without using and using a control system

There is a difference, that is, the droplet lag time of the fluid will be longer if using a control system, it is influenced by the addition of a droplet regulator before the fluid enters the orifice and the fitting T. The graph shows that not all the droplet lag time of the fluid increase but there are decrease of the droplet lag time. The droplet lag time of the fluid increase namely the droplet lag time of the oil fluid for 100 ml volume and the the droplet lag time for bromus fluid. From these data, the discrepancy occurs because it should be that with the addition of a control system causes a slowdown in the time the fluid falls. The difference in data is suspected to occur because when data collection with variations in the type of fluid, the tool is not first cleaned from the remaining of the previous fluid, then the mixing of the fluid occurs so that the oil fluid will be thin and runny and the time of the fluid fall becomes faster. However, in general the MQL tool with the addition of a control system slowing down the fluid drop. This is because the falling fluid will be regulated by a droplet scheme before enter the Orifice and fitting T.

3.3. Tool Performance Test

This test is carried out to get how much volume (capacity) of each fluid within 1 hour. It is expected that the addition of the control system will cause an increase in the performance of the MQL tool that has been made in achieving the desired specifications.

The graph of the results of testing the tool performance before and after adding the control system for cooking oil fluid can be seen in figure 11. The graph show that at all cases for the diameter orifice of 2 mm, the capacity of the fluid decreases as the control system being applied, so the tool performance increase. Meanwhile for the diameter orifice of 5 mm, the capacity of the fluid decreases only for ¾ and full air valve openings.

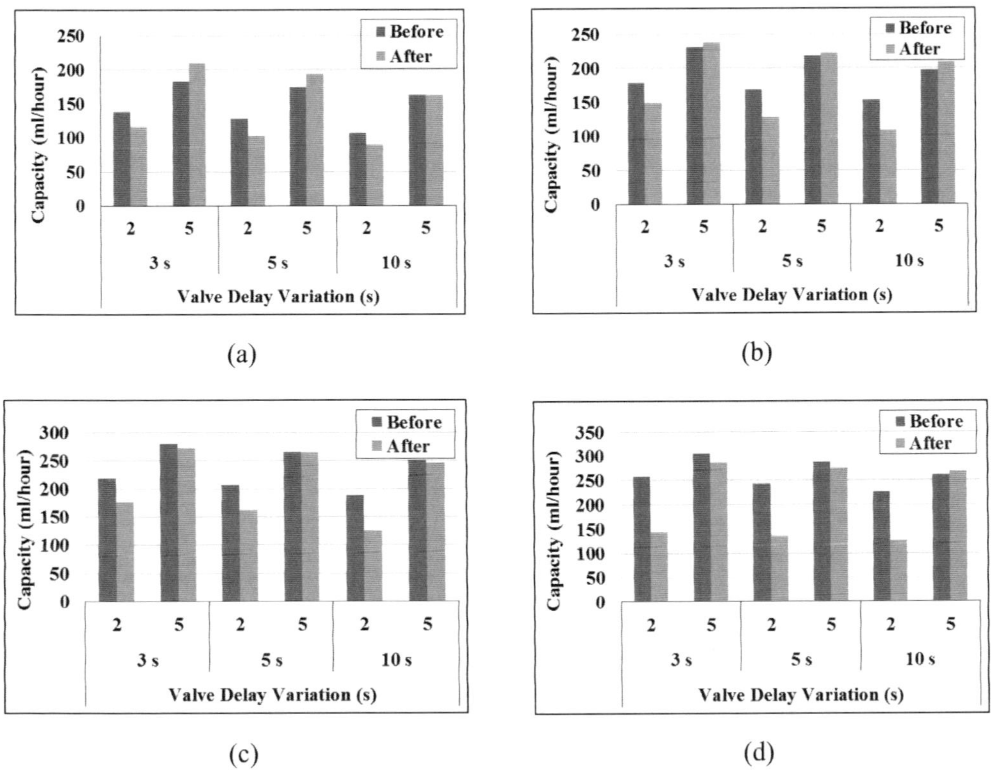

Figure 11. Capacity of cooking oil fluid without using and using a control system, a) For air valve openings ¼, b) For air valve openings ½, c) For air valve openings ¾, dan d) For full air valve openings

The capacity of cooking oil types of fluid has increased only in the cases of air valve openings in a range of about 0 to 45% and for the diameter orifice of 5 mm. The decrease in performance for cooking oil fluids after using a control system tends to occur in small air valve openings and long delay times, this is due to the long delay making droplets and orifice too long so that the fluid will flow in large amounts. This large amount of fluid flowing causes an increase in the fluid capacity of the MQL tool.

The graph of the results of testing the performance of the MQL tool before and after adding the control system for oil fluid can be seen in figure 12. In this Figure, all the data that has been obtained as expected. There is an increase in the performance of the MQL tool after the implementation of the control system. Improved performance on this oil fluid is the best, with a range of about 1 to 67% in all air valve openings. This is suspected because of the role of the control system applied to the MQL tool. Performance of oil type fluids is improved because oil has a high viscosity compared to bromus and cooking oil fluids. The fluid flow down to the fitting T drop by drop. The fall of the oil droplet is set to coincide with an air push. This arrangement makes the capacity of the fluid decreases and the MQL tool performance will increase.

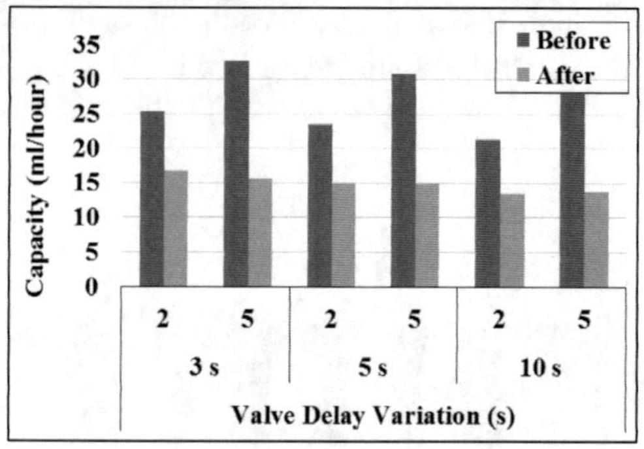

(a)

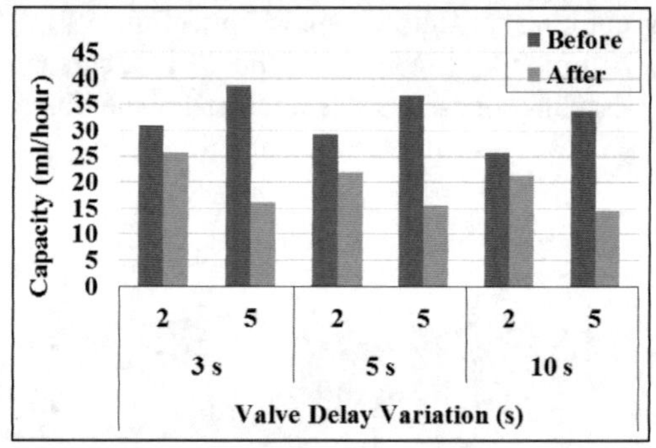

(b)

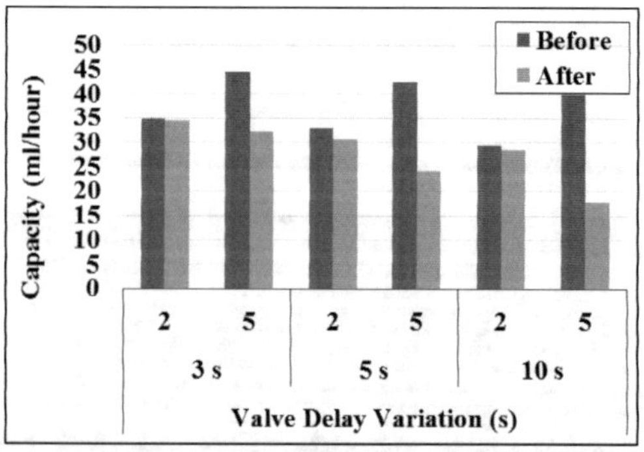

(c)

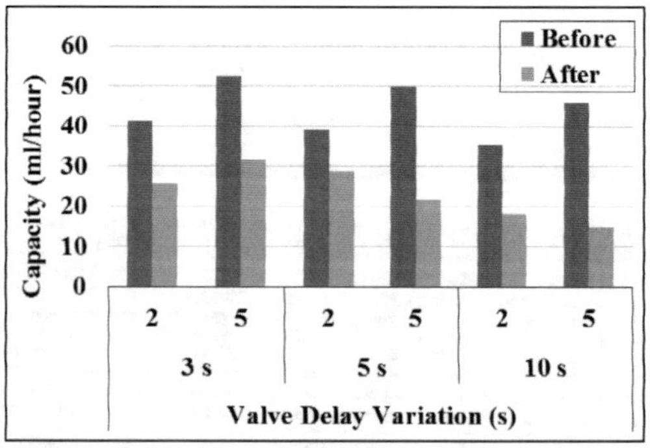

(d)

Figure 12. Oil fluid capacity before and after using a control system, a) For air valve openings ¼, b) For air valve openings ½, c) For air valve openings ¾, and d) For air full valve openings

In figure 13 the following will show the difference in the capacity of the bromus type fluid. The graph of the results of testing the tool performance before and after adding the control system for bromus type of cutting fluid can be seen in figure 13. The graph a show that it is only for air valve openings ¼, the capacity of the fluid decreases as the control system being applied, so the tool performance increase. Meanwhile, the capacity of bromus types of fluid has increased as the air valve openings increase.

Broadly speaking, the bromus type fluid capacity has decreased performance, but in the smallest valve opening ¼ there was an increase of about 9 to 26% and in the air valve opening ¾ and full air valve openings also increased, but only at 5mm orifice with delay 3s and 5s. From the data it can be seen that the tool performance increases only at small air valve openings because at small air valve openings air pressure will be large and bursts occur in the form of aerosols (a mixture of liquid and gas fluid), while for large valve openings air pressure drops so that the fluid will continue flows in the nozzle interval and cannot be controlled by droplet and orifice schematics. That is caused by the viscosity of the bromus which is smaller than the oil fluid and cooking oil, so that the flowing fluid cannot be controlled and the capacity becomes high.

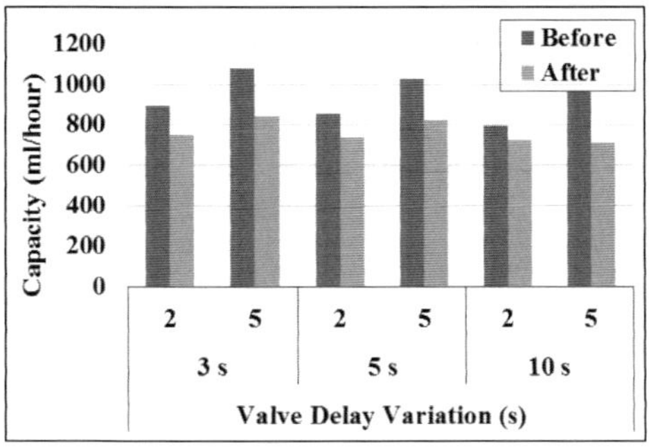

(a)

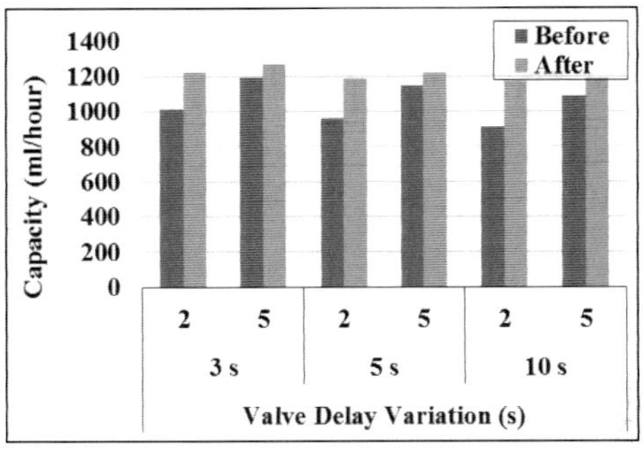

(b)

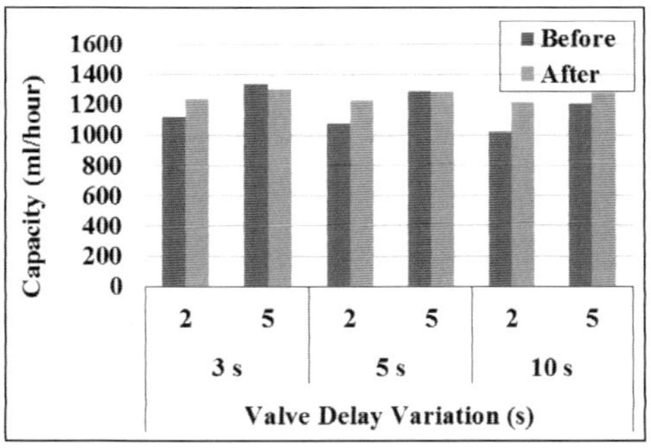

(c)

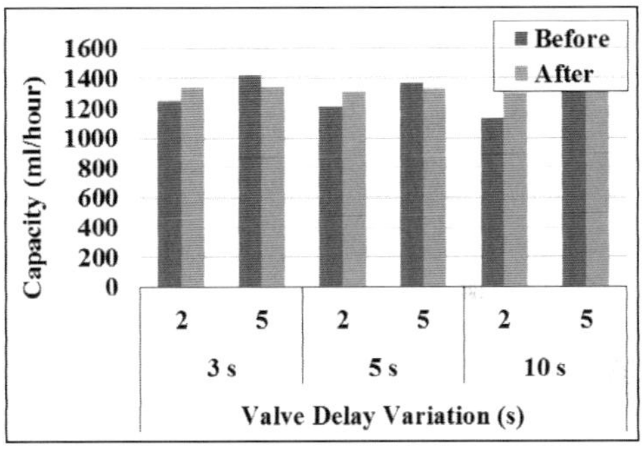

(d)

Figure 13. Bromus fluid capacity before and after using the control system, a) For air valve openings ¼, b) For air valve openings ½, c) For air valve openings ¾, and d) For air full valve openings.

In the following table 3 we can see the difference between the performance of MQL tools before and after using a control system for one type of air and orifice valve openings.

Table 3. Performance of MQL tool

Fluid Types	**Time (s)**	***Orifis Diameter* (mm)**	**Capacity (ml/hour)**	
			Before	**After**
Oli	10	2	21.273	13.500
Bromus	10	2	796.576	723.061
Cooking Oil	10	2	107.348	89.609

From table 3, it can be seen a decrease in fluid capacity, in other words an increase in performance from the previous condition. The increase is quite significant, ranging between 65% to 67%, but from this data, the performance of this MQL tool is still not in accordance with the design plan in table 1.

The improved performance does not meet the desired specifications. This is due to the inappropriate selection of material and mechanical systems of the MQL tool components. For example, the fluid valve material is made of acrylic. This material cannot withstand the falling fluid and causes the fluid still leaking down to the fitting T. The channel used is made of PVC pipe with a fairly large size, with a diameter of ¼ inches. The PVC material is larger than the channel on the MQL tool that is available on the market, so the air sprayed is not optimal to push the fluid that falls from the reservoir.

4. Conclusions and suggestions

4.1. Conclusions

After obtaining data from several tests that have been carried out, it can be concluded that:

1. It is found that with the addition of modified control systems, the performance of MQL cooling regulator is improved, but it is applied mostly for fluids which have viscosity ranging about viscosity of oil and cooking oil.
2. The tool performance increase as the capacity of the fluid decreases due to the control system being applied for oil fluid. Improved performance on this oil fluid is the best, with a range of about 1 to 67% in all air valve openings. Improved performance in oil type fluids because oil has a high viscosity compared to bromus and cooking oil fluids.
3. For cooking oil fluid, the MQL tool performance increase as the capacity of the fluid decreases due to the control system being applied at all case of air valve openings for the diameter orifice of 2 mm. It is also found that at the aperture of the air valve of ¼, with a 10 second delay time and an orifice diameter of 2 mm, the capacity coolant sprayed is reduced from 107.348 ml/hr to 89.609 ml/hr when the modified control system is added.
4. For the Bromus fluid, the capacity of the fluid decreases as the control system being applied. The MQL tool performance increase only occur in the smallest valve opening of ¼ that is about 9 to 26%. However, the capacity of Bromus types of fluid increase as the air valve openings increase. That is caused by the viscosity of the bromus which is smaller than the oil fluid and cooking oil, so that the flowing fluid cannot be controlled and the capacity becomes high.

4.2. Suggestions

The results of this research still has some shortcomings. For further research in order to minimize these deficiencies, the authors provide the following suggestions:

1. It is recommended that the mechanical system of MQL distribution can be improved so that the performance of the MQL tool can be maximized.

2. It is recommended that the size of the fluid channel pipe is smaller so that the channel is more maximum and the pressure used is also maximum.

3. It is recommended that the air pressure used in this MQL tool be adjusted so that it is constant according to the desired specifications.

References

[1] Blaser, https://www.blaser.com/en_ID
[2] L, Lillienberg et al. 2008. Exposure to Metalworking Fluid Aerosols and Determinants of Exposure. Ann Occup Hyg, Vol. 52, No. 7, pp. 597–605.
[3] L, Godderis et al. 2008. Exposure to Metalworking Xuids and Respiratory and Dermatological Complaints in a Secondary Aluminum Plant. Int Arch Occup Environ Health 81: 845–853.
[4] WF, Sales et al. 2002. Cooling ability of cutting fluids and Measurement of the Chip-tool

Interface Temperatures. Industrial Lubrication and Tibology Volume 54, Number 2, pp, 57–68.

[5] YS, Liaoa et al. 2006. Feasibility Study of The Minimum Quantity Lubrication in High-Speed End Milling of Nak80 Hardened Steel By Coated Carbide Tool. International Journal of Machine Tools & Manufacture 47, (1667–1676).

[6] NR, Dhar et al. 2006. Effect of Minimum Quantity Lubrication (Mql) on Tool Wear and Surface Roughness in Turning Aisi-4340 Steel. Journal of Materials Processing Technology 172, 299-304.

[7] L, Bruce Tai et al. 2014. Minimum Quantity Lubrication (MQL) in Automotive Powertrain Machining. Procedia CIRP Volume 14, Pages 523-528.

[8] Rezfozano, Frissandy. 2015. Design and Manufacture of MQL (Minimum Quantity Lubrication) Tools. Department of Mechanical Engineering Andalas University, Padang.

[9] Rochim, T. 1993. Machining Process, Theory and Technology, Production Engineering Laboratory, Department of Mechanical Engineering, FTI, ITB.

[10] NN 2. 2015. Effect of Cooling Liquid on Machining Process. Available at: http://www.teknikmesin.org/p affect- liquid- cooling- on- machining process /. Accessed August 13, 2017.

[11] B.Arifianto.2010. microcontroller in the world of automation. Available at http://students.ukdw.ac.id/~robotic/apApps. Accessed July 12, 2017.

[12] Adriansyah, Andi. 2010. Basic Control Systems. "UMB" Teaching Material Development Center. Jakarta.

[13] Kadir, Abdul. 2012. A Practical Guide to Learning Microcontroller Application and Programming Using Arduino. Yogyakarta: Andi Publisher.

[14] Electroschematic.2015.Characteristic flowsensor and Half Effect. Available at: www.electroschematic.com/12145/working-with-water-flow-sensors-arduino/. Accessed September 12, 2017

[15] Andy. 2011. Servo Motors. Available at: http://andydharmalau.com/?p=85. Accessed September 22, 2017.

The Effect of Speed and Distance on the Achievement of Austenitization Temperature in Low carbon Steel with Thickness Variations on the Flame Hardening Process

N Iskandar[1*], Sulardjaka[1], Munadi[1], R Y Hanasta[1], D F Fitriyana[2] and M Dzulfikar[3]

[1] Department of Mechanical Engineering, Faculty of Engineering, University of Diponegoro, Jl. Prof. Sudarto. SH, Tembalang, Semarang, 50275, Indonesia
[2] Department of Mechanical Engineering, Faculty of Engineering, Semarang State University, Jl. Taman Siswa Sekaran, Gunungpati, Semarang, 50229, Indonesia
[3] Department of Mechanical Engineering, Faculty of Engineering, University of Wahid Hasyim, Jl. Menoreh Tengah X/22, Sampangan, Semarang, 50236, Indonesia

E-mail*: norman.undip@gmail.com

Abstract. Austenite temperature must be achieved in the surface hardening process. In the process of surface hardening by the method of flame hardening, failure is often encountered, for example, such as increasing the value of hardness that is not optimal, and the value of hardness that is less uniform The type of fire used, the distance of fire to the specimen, the speed of fire and the level of thickness of the material are factors that are very influential in the process of flame hardening. The purpose of this study is to measure and observe the profile or rate of the trend of increase and decrease in temperature in the flame hardening process in low carbon steel material. The experiment method that is carried out is by giving the variation of the rate. The measurement process uses several thermocouples mounted on the specimen and connected to a computer for the process of data recording and process monitoring. The conclusion of this study is the greater the value of the thickness of low carbon steel material that is processed by flame hardening, so to reach the temperature of austenitization by the method of flame hardening the speed of fire must be slowed. An improper combination between the distance of fire and specimens and the speed of fire can cause austenite temperature not to be reached. The heat reduction rate takes place more slowly than the rate of increase in heat the speed of fire and distance of fire to low carbon steel which has varying levels of thickness.

1. Introduction

The value of surface hardness in metals, for several reasons, is often sought to be increased. One area that utilizes the surface hardening process a lot is the metal industry for agricultural equipment. In Indonesia, there are quite a number of small and medium-sized metal industries (IKM) engaged in the manufacture of agricultural equipment. One of the problems faced by metal IKM is quality problems related to the surface hardness value of their products. One of the quality references for agricultural equipment products is the Indonesian National Standard (SNI).

SNI requires that the surface hardness value meets at least three criteria, namely the hardness value meets the minimum limit, the level of hardness that is evenly owned in the specified area and the third is for some products such as knives, the hardness value and the level of hardness must be met in one third of the product surface area .

IKM that use the surface hardening method with the flame hardening method with oxy-acetylene welding machines have problems to meet SNI standards. The value of violence can be achieved but the level of uniformity in the value of violence in the area required cannot be met. Covering one-third of the area must have a violence value according to SNI standards is also a problem in itself. This is further complicated when materials have different thickness levels.

Flame hardening process is a useful technique for a surface modification that can be readily applied to extend the service life of many machine steel parts. Such various applications are brought about by the ability to harden regions locally, the simple and easy operation, the high-speed and low-cost processing, and the improved mechanical properties [1]. There is so many studies have been mostly focused on the improvement of hardness on the surfaces and the automation of the flame hardening process [2–5] and details on the flame hardening apparatus [6] .

Several other studies related to flame hardening and specifically flame hardening in low carbon steel have been carried out by S. Sharkar et al. In particular they examined the application of lasers to this process where the analysis of temperature changes to the austenite temperature was studied in depth [7].

Studies on the process of how to achieve austenite temperatures and their characteristics for various types of materials have also been carried out such as for stainless steel materials by Y. Zhang et al. And on ultra-high strength steels by R. Veerababua et al [8,9].

This research tries to discover and explore phenomena related to the use of flame hardening in similar materials but has different thicknesses. Variations in the speed of the nozzle and the distance of the nozzle to the workpiece will also be examined for their effect on the surface hardening process. The results of this study are expected to be used as a reference for fire hardening process operators and can also be used as materials for designing automation in flame hardening by creating machines instead of human operators.

2. Materials and Methods

2.1. Materials

The test material used is the type of AISI 1026 Carbon steel (AISI, n.d.) which is included in the low carbon steel category. The following table 1 shows the chemical composition data of the test material.

Table 1. Composition of materials element

Element	Test Materials	
	6 mm (wt%)	10 mm (wt%)
C	0,24	0,24
Si	0,46	0,32
Mn	0,88	0,62
Cr	0,46	0,31
P	<0,01	<0,01
S	<0,05	0,31

Based on the results of the ingredients composition test in Table 1, it can be determined the Ac_3 temperature for each material. The temperature of Ac_3 or the temperature at which the material undergoes the austenite phase as a whole can form a martensitic phase if it is cooled quickly. The Ac_3 temperature can be predicted by entering material composition data into the Material Algorithm Project (MAP) application. This prediction is used to compare whether the nozzle speed of each speed variation will reach the temperature of Ac_3. In Table 2 are the results of the estimated Ac3 temperature for each material. This result is consistent with Amendee's study in 1985 which stated that the Ac_3 temperature of mild steel was 790-910 with three variations in heat levels [10].

Table 2. Results of prediction of Ac_3 temperature with MAP

Thickness 6 mm		Thickness 10 mm	
Ac_3 Point (°C)	*Heat Rate* (°C/s)	Ac_3 Point (°C)	*Heat rate* (°C/s)
900	5	876	5
907	10	882	10
913	15	887	15
918	20	892	20

2.2. Methods

The process of testing the flame heating is carried out accompanied by quenching treatment with water cooling media. This test utilizes a flame from oxy-acetylene welding with a neutral flame type to heat the surface of the test specimen. The oxy-acetylene welding flame pattern can be seen in Figure 1b. i.e. go left one way. Fire is fired in the middle area between the distance of the thermocouple and the edge of the test material, or 10 mm each from the edge of the test object and from the position of the thermocouple. Thermocouple installation as in Figure 1a.

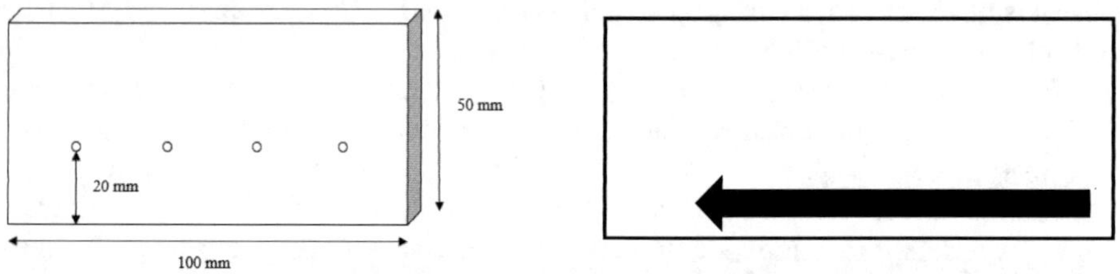

(a) (b)

Figure 1. (a) Dimensions of the test specimen, (b) Direction of heating of the flame hardening

The testing procedure carried out is to prepare low carbon steel material with the composition as listed in Table 1. Each material is given four holes in single line position to be filled the thermocouple, where the distance of each hole is 20mm. Thermocouple coded T1, T2, T3, and T4. The line of fire from the nozzle is depicted on the material as a guide in the form of a straight line. The distance of the fire path to the thermocouple hole perpendicular is 10 mm.

The distance of the nozzle to the workpiece specimen is arranged in two variations, namely 30mm and 50mm. The speed of the moving nozzle rate is also varied ie 10mm/min, 15mm/min and 20mm/min. The thermocouple is connected to the computer directly so data can be recorded directly.

3. Result and Discussion

3.1. Temperature profile on material with a thickness of 6mm and nozzle distance to the workpiece 30mm.

Temperature profile measured by a thermocouple mounted on a material with a thickness of 6mm, a distance of 10mm from the fire line fired from the nozzle, with 30mm distances of the test specimen and with variations in the speed of the nozzle movement of 10mm/min, 15mm/min and 20 mm/min are shown in Figure 2 to Figure 4.

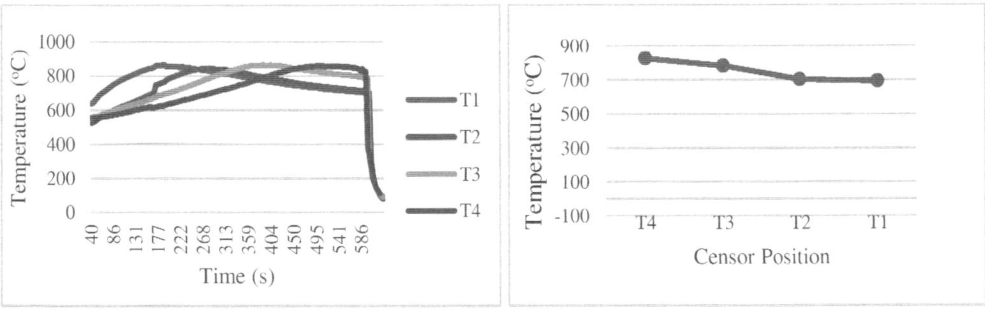

(a) (b)

Figure 2. (a) Profile of temperature - time for materials with a thickness of 6mm, nozzle distance of 30mm and nozzle velocity of 10mm / min. (b) the temperature when quenching.

Figure 2 is a profile of temperature versus time for a material with a thickness of 6mm, nozzle distance of 30mm and nozzle velocity of 10mm / min. The highest and lowest temperatures at T1 are 40oC at 1st second and 863,2°C at 190th seconds. The position of the T4 sensor is located at the end of the heating nozzle so it has the highest temperature of 826,5°C. Then the temperature drops further on the T3 and T2 sensors while the sensor position T1 is the initial position of the nozzle so that it has the smallest temperature which is 693,6°C because in the T1 section it has cooled with the longest time.

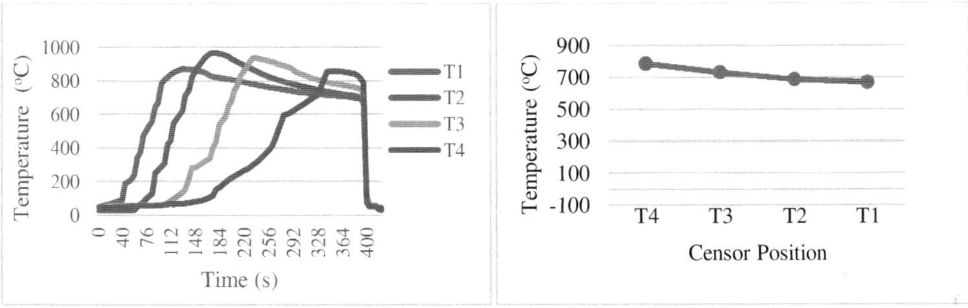

(a) (b)

Figure 3. (a) Profile of temperature - time for materials with a thickness of 6mm, nozzle distance of 30mm and nozzle velocity of 15mm / min. (b) the temperature when quenching.

Figure 3 is the temperature profile of a material with a thickness of 6mm, the distance of the nozzle to the test object by 30mm and the velocity of the nozzle to 15mm / min. The highest and lowest temperature at T1 is 30°C at the 1st second and 872° C at the 130th second. The temperature at which the material is quenched at the 400th second is shown in Figure 3b. The T4 sensor position is the last closest position to the flame which has the highest temperature of 785,6°C. The temperature decreases further on the T3 and T2 sensors and the lowest is at T1 where the measured temperature is 667,8°C.

Figure 4 is a plot of the temperature profile on the material with a thickness of 6mm with a spacing of 30mm with a nozzle speed of 20mm / min

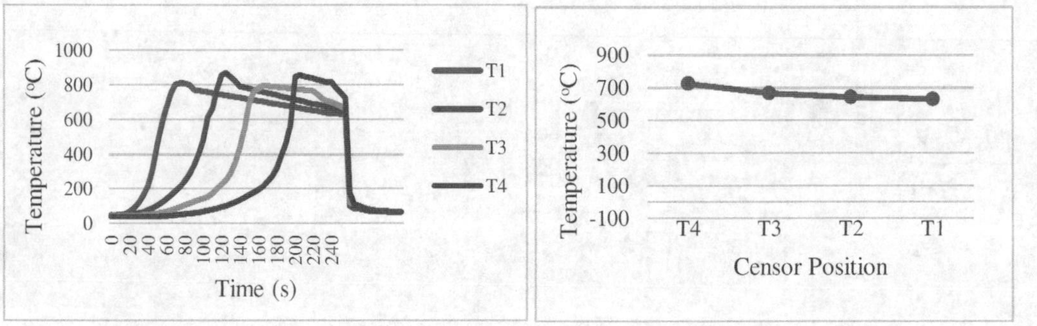

(a) (b)

Figure 4. (a) Profile of temperature - time for materials with a thickness of 6mm, nozzle distance of 30mm and nozzle velocity of 20mm / min. (b) the temperature when quenching.

The highest and lowest temperature at T1 is 47,4°C at the 1st second and 810°C at the 75th second. The temperature when the material is quenched at 300th seconds as shown in Figure 4b. At position T4 which is located at the end of the heating process has the highest temperature of 724.2°C. T1 is the sensor at the initial position of the nozzle so that it has the lowest temperature of 630.7°C.

3.2. Temperature profile on material with a thickness of 6mm and nozzle distance to the workpiece 50mm.

Temperature profile measured by a thermocouple mounted on a material with a thickness of 6mm, a distance of 10mm from the fire line fired from the nozzle, with 50mm distances of the test specimen and with variations in the speed of the nozzle movement of 10mm/min, 15mm/min and 20 mm/min are shown in Figure 5 to Figure 7.

Figures 5a and 5b below are the results of the plot of the material temperature profile which has a thickness of 6mm and the distance of the nozzle to the test specimen is 50mm with a nozzle speed of 10 mm/min

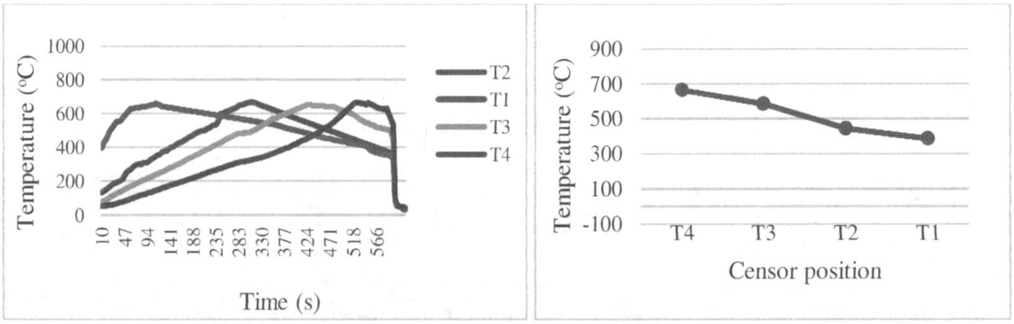

(a) (b)

Figure 5. (a) Profile of temperature - time for materials with a thickness of 6mm, nozzle distance of 50mm and nozzle velocity of 10mm / min. (b) the temperature when quenching.

The highest and lowest temperatures at T1 are 32°C at the 1st second and 659.4°C at the 110th second. The temperature at which the material is quenched at the 600th second will be shown in Figure 5b. The temperature reads at T4 which is 663.5°C. Then the temperature at the sensor position T1 is the initial position of the nozzle so that it has the smallest temperature which is 385.2°C because in the T1 section it has cooled with the longest time.

Figures 6a and 6b below are the results of the plot of the material temperature profile which has a thickness of 6mm and the distance of the nozzle to the test specimen is 50mm with a nozzle speed of 15 mm/min

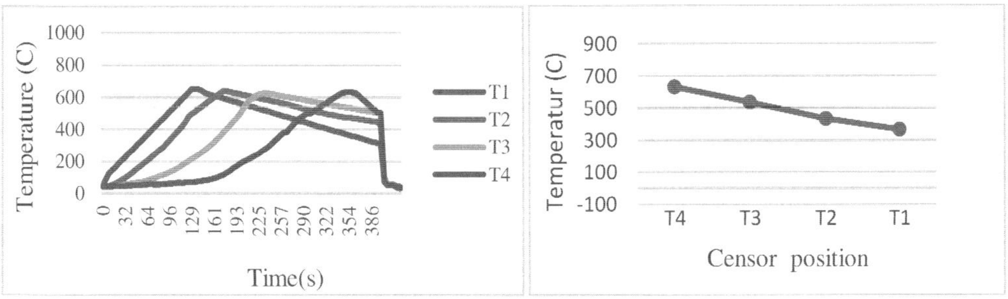

(a) (b)

Figure 6. (a) Profile of temperature - time for materials with a thickness of 6mm, nozzle distance of 30mm and nozzle velocity of 15mm / min. (b) the temperature when quenching.

The highest and lowest temperature at T1 is 41.7°C at the 1st second and 650°C at the 110th second. The temperature at which the material is quenched at the 400th second will be shown in Figure 6b where the temperature measured by the T4 sensor is 632°C and T1 which has the lowest temperature of 365.9°C.

Figures 7a and 7b below are the results of the plot of the material temperature profile which has a thickness of 6mm and the distance of the nozzle to the test specimen is 50mm with a nozzle speed of 20 mm/min

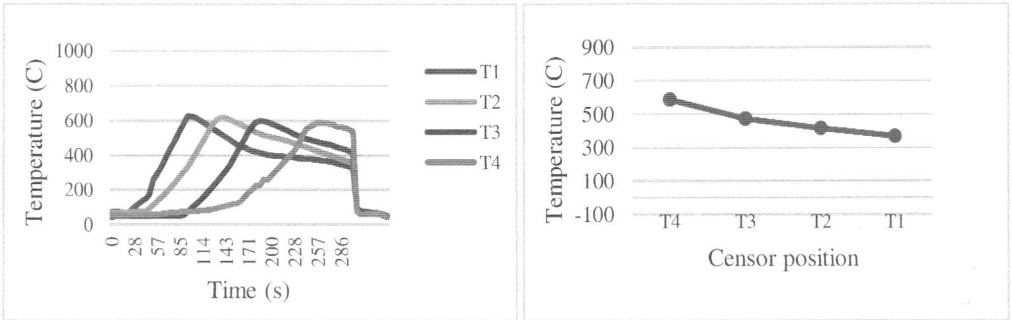

(a) (b)

Figure 7. (a) Profile of temperature - time for materials with a thickness of 6mm, nozzle distance of 50mm and nozzle velocity of 20mm / min. (b) the temperature when quenching.

The highest and lowest temperature at T1 is 72.9°C at the 1st second and 626.5°C at the 94th second. The temperature when the material is quenched is at the 300th second, at the T4 sensor position, the measured temperature is the highest temperature at 587°C. the lowest temperature measured by the T1 sensor is 368.9°C.

3.3. Temperature profile on material with a thickness of 10mm and nozzle distance to the workpiece 30mm.

Temperature profile measured by a thermocouple mounted on a material with a thickness of 10mm, a distance of 10mm from the fire line fired from the nozzle, with 30mm distances of the test specimen and with variations in the speed of the nozzle movement of 10mm/min, 15mm/min and 20 mm/min are shown in Figure 8 to Figure 10.

Figures 8a and 8b below are the results of the plot of the material temperature profile which has a thickness of 10mm and the distance of the nozzle to the test specimen is 30mm with a nozzle speed of 10 mm/min

IOP Conf. Series: Materials Science and Engineering **694** (2019) 012025 doi:10.1088/1757-899X/694/1/012025

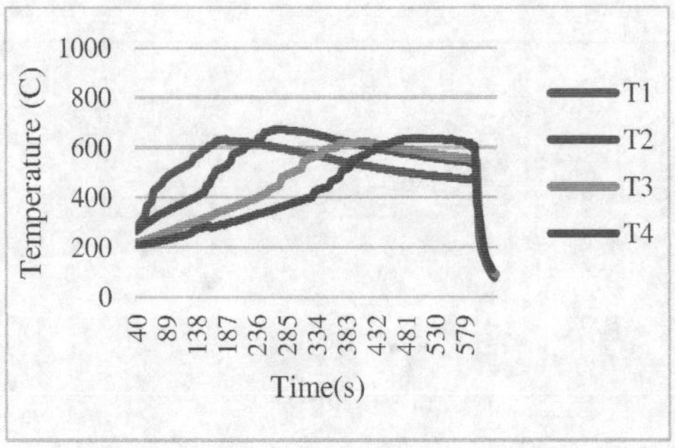

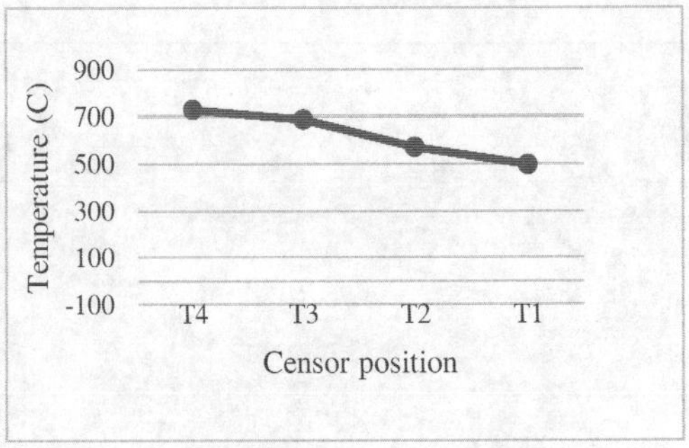

(b) (b)

Figure 8. (a) Profile of temperature - time for materials with a thickness of 10mm, nozzle distance of 30mm and nozzle velocity of 10mm / min. (B) the temperature when quenching.

The highest and lowest temperatures at T1 are 32°C at the 1st second and 632.5°C at the 94th second. The temperature when the material is quenched is at the 600th second, at the T4 sensor position, the highest temperature measured is 726.8°C. At the T1 sensor position, the measured temperature is the smallest temperature, which is 494.6°C.

Figures 9a and 9b below are the results of the plot of the material temperature profile which has a thickness of 10mm and the distance of the nozzle to the test specimen is 30mm with a nozzle speed of 15 mm/min

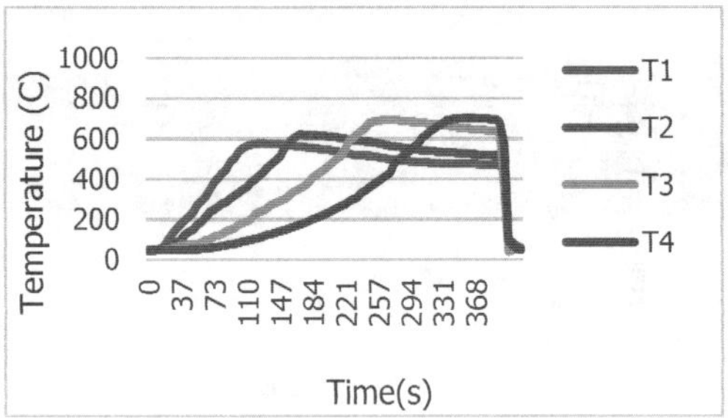

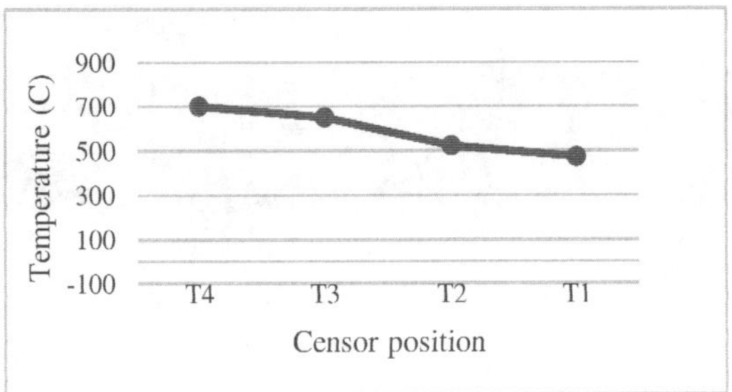

(c) (b)

Figure 9. (a) Profile of temperature - time for materials with a thickness of 10mm, nozzle distance of 30mm and nozzle velocity of 15mm / min. (b) the temperature when quenching.

The highest and lowest temperature at T1 is 44°C at the 1st second and 575°C at the 124th second. The temperature when the material is quenched is at the 400th second on the T4 sensor, the measured temperature is the highest temperature that is 701.1°C. At T1 Sensor the measured temperature is the smallest temperature that is 472.9°C.

Figures 10a and 10b below are the results of the plot of the material temperature profile which has a thickness of 10mm and the distance of the nozzle to the test specimen is 30mm with a nozzle speed of 20 mm/min. Suhu tertinggi dan terendah pada T1 adalah 44°C pada detik ke-1 dan 501°C pada detik ke-116. Suhu saat material di *quenching* yaitu pada detik ke-300, pada posisi sensor T4 terbaca memiliki suhu paling tinggi yaitu 621,3°C. Pada posisi sensor T1 memiliki suhu yang paling kecil yaitu 424°C.

IOP Conf. Series: Materials Science and Engineering **694** (2019) 012025 doi:10.1088/1757-899X/694/1/012025

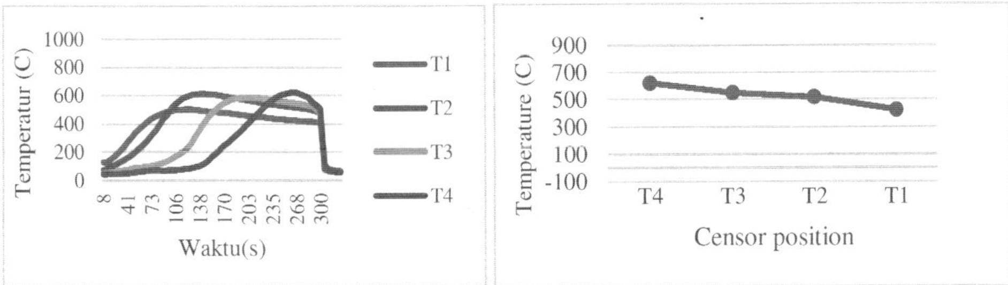

(a) (b)

Figure 10. (a) Profile of temperature - time for materials with a thickness of 10mm, nozzle distance of 30mm and nozzle velocity of 20mm / min. (B) the temperature when quenching.

3.4. Temperature profile on material with a thickness of 10mm and nozzle distance to the workpiece 50mm.

Temperature profile measured by a thermocouple mounted on a material with a thickness of 10mm, a distance of 10mm from the fire line fired from the nozzle, with 50mm distances of the test specimen and with variations in the speed of the nozzle movement of 10mm/min, 15mm/min and 20 mm/min are shown in Figure 11 to Figure 13.

Figures 11a and 11b below are the results of the plot of the material temperature profile which has a thickness of 10mm and the distance of the nozzle to the test specimen is 50mm with a nozzle speed of 10 mm/min

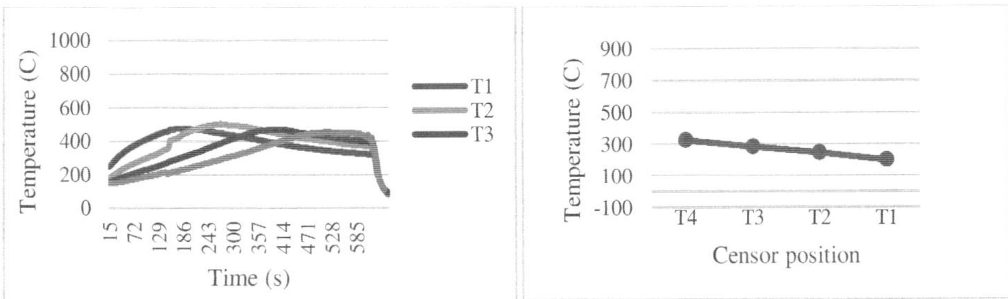

(a) (b)

Figure 11. (a) Profile of temperature - time for materials with a thickness of 10mm, nozzle distance of 50mm and nozzle velocity of 10mm / min. (b) the temperature when quenching.

The highest and lowest temperature at T1 is 44°C at the 1st second and 474°C at the 178th second. The temperature when the material is quenched is at the 600th second, on the T4 sensor the measured temperature is the highest temperature of 323.5°C. On the T1 sensor the smallest temperature value measured is 198.7°C.

Figures 12a and 12b below are the results of the plot of the material temperature profile which has a thickness of 6mm and the distance of the nozzle to the test specimen is 50mm with a nozzle speed of 15 mm/min

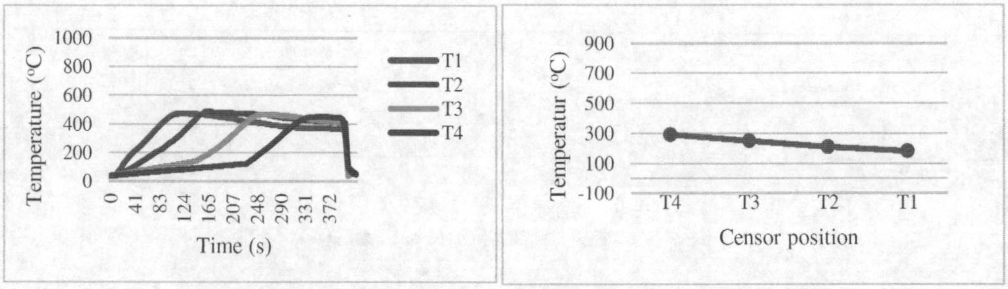

(a) (b)

Figure 12. (a) Profile of temperature - time for materials with a thickness of 10mm, nozzle distance of 50mm and nozzle velocity of 15mm / min. (b) the temperature when quenching.

The highest and lowest temperature at T1 is 39°C at the 1st second and 472°C at the 124th second. The temperature when the material is quenched is at the 400th second, at the T4 sensor the measured temperature is 292°C. At the T1 sensor position the smallest measured temperature is 182.2°C.

Figures 13a and 13b below are the results of the plot of the material temperature profile which has a thickness of 10mm and the distance of the nozzle to the test specimen is 50mm with a nozzle speed of 20 mm/min

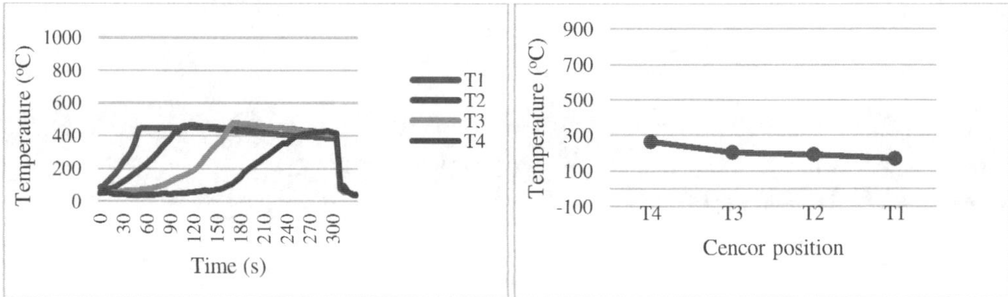

(a) (b)

Figure 13. (a) Profile of temperature - time for materials with a thickness of 10mm, nozzle distance of 50mm and nozzle velocity of 20mm / min. (B) the temperature when quenching.

The highest and lowest temperature at T1 is 51.9°C at the 1st second and 445°C at the 115th second. The temperature when the material is quenched is at the 300th second, on the T4 sensor, the highest measured temperature is 265°C. At the T1 sensor position the smallest measured temperature is 170.2°C

4. Conclusions

At the condition of the distance of the nozzle to the same test object, the speed of the nozzle's speed of movement is the same, and by using the same type of fire, then in the same material with different thickness, a different temperature increase rate will be achieved at the same position at a certain distance from the path of the fire. So that the austenite temperature is more difficult to achieve for materials that have more thickness

References

[1] M.K. Lee, G.H. Kim, K.H. Kim, W.W. Kim, "Effects of the surface temperature and cooling rate on the residual stresses in a flame hardening of 12Cr steel," Journal of Materials Processing Technology 176 (2006) 140-145.
[2] J. Grum, S. Bozic, M. Zupancic, J. Mater. Proc. Technol. 114 (2001) 57–70.
[3] I. Kyoji, Japanese Patent 61-113712 (1986).

[4] S. Toyoaki, Japanese Patent 64-39327 (1989).

[5] R.J. Ackert, R.W. Witty, P.A. Crozier, US Patent 4,486,248 (1982).

[6] M.K. Lee, G.H. Kim, K.H. Kim, W.W. Kim, Surf. Coat. Technol. 184 (2004) 239–246.

[7] S. Sarkar, M. Gopinath, S.S. Chakraborty, B. Syed, and A.K. Nath, "Analysis of temperature and surface hardening of low carbon thin steel sheets using Yb-fiber laser," Surface & Coatings Technology 302 (2016) 344-358.

[8] Y. Zhang, D. Zhan, X. Qi, Z. Jiang, "Austenite and precipitation in secondary-hardening ultra-high-strength stainless steel," Materials Characterization 144 (2018) 393-399.

[9] R. Veerababua, K.S. Prasada, P.S, Karamcheda, R. Balamuralikrishnana, S. Karthikeyan, "Austenite stability and M_2C carbide decomposition in experimental secondary hardening ultra-high strength steels during high temperature austenitizing treatments," Materials Characterization 144 (2018) 191-204.

[10] Amende, W. (1985). Harten _on Werkstoffen und Bauteilen des Maschinenbaus mit dem Hochleistungslaser, Technologie Aktuell, (VDI Verlag, Du¨ sseldorf).

Effect of Welding Current and Welding Speed on Weld Geometry and Distortion in TIG Welding of A36 Mild Steel Pipe with V-Groove Joint

M Panji, A S Baskoro and A Widyianto

Mechanical Engineering Department, Faculty of Engineering, Universitas Indonesia, Kampus Baru UI Depok, 16424, Indonesia

E-mail: ario@eng.ui.ac.id

Abstract. In this study, the effect of welding current and welding speed on the weld geometry and distortion in TIG welding process was investigated. The material used in this experiment is A36 mild steel pipe. The dimensions of specimen were 114.3 mm of outer diameter and 6 mm of thickness. The wire feeder used was ER70-S 6 with a V-groove joint as a type welding joint. The welding parameters varied were the welding current and welding speed. Welding results will be tested for weld geometry (outer bead width) and distortion after the welding process. The weld geometry was measured using Digital Microscope and the distortion was measured using outside micrometers before and after the welding process. The results show that increasing the welding current can widen the weld bead and increase distortion. While increasing the speed has little effect on the weld bead and distortion. Welding parameters with a welding current of 170 A and a welding speed of 0.9 mm / s produces the widest outer bead and the biggest tapers and distortion. In addition, the results of tapers and distortion in the axial direction were greater than the transverse direction of the pipe.

1. Introduction

Welding has an important role in an industry, one of which is pipe welding. The gas, nuclear, oil, water and chemical industries are examples of an industry that applies pipe welding. In the industry, pipes are an effective means of transportation to flow from one place to another place. One of the problems in pipe welding is the depth of penetration and distortion that occurs after the welding process. If the welding penetration is too deep then the flow in the pipe will be interrupted. Distortion in welding pipes is interpreted as a change in the shape of the diameter of the pipe such as being oval, tapers, smaller or irregular, as shown in Figure 1. This will certainly interfere if the pipe will be installed in pipeline construction.

Many studies have been done by previous researchers about distortion. Sattari-Far [1] reported the results of his research that the welding sequence has an influence on the distortion that occurs. This type of joint uses a V-groove joint and the results of the experiment are compared with the FE model. In addition to verification the results are compared with FE modelling. Many studies use models such as finite element, ANSYS and ABAQUS to predict distortion and residual stress results from welding [2-5]. Kalyankar [6] made a review of distortion and methods to reduce the distortion that occurs. Distortion can be reduced before the welding process, during the welding process and after the welding process by several methods.

The objective of this study to determine the effect of welding current and welding speed on weld geometry and distortion on A36 mild steel pipe. The base metal has a diameter of 114.3 mm with a thickness of 6 mm and joined using a V-groove joint. ER 70S-6 is used as an added material to joint pipes [7]. Weld geometry measurements include the width of weld bead. While distortion is measured by changes in pipe diameter before and after the welding process.

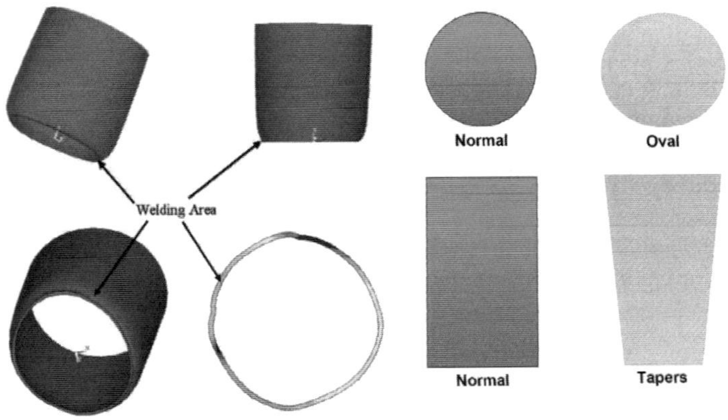

Figure 1. Distortion in pipe [1]

2. Experiment Method

In this experiment use A36 mild steel pipe with outer diameter 114.3 mm and thickness of 6 mm. The specimen was joint using wire feeder ER 70S-6 of 1.6 mm diameter. V-groove joint method was chosen with no gap between pipes. Figure 2 shows the schematic of V-groove joint of pipes. The chemical composition of base metal and wire feeder were listed in Table 1 and Table 2, respectively.

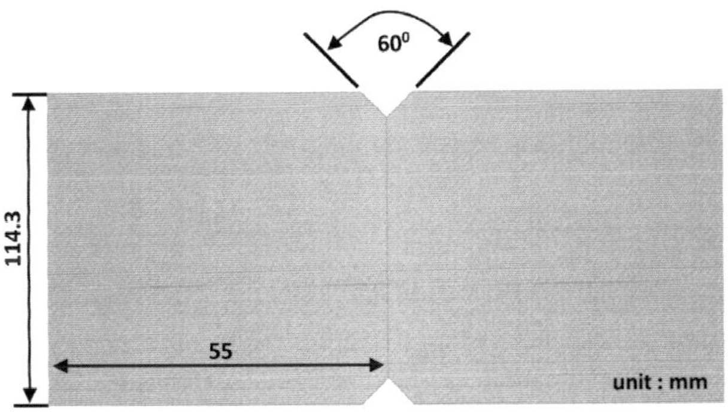

Figure 2. Dimension of V-groove joint of pipe

Table 1. Chemical composition (wt %) of A36 pipe

A36	C	Si	Mn	P	S	Cr	Mo
	0.183	0.233	0.392	0.013	0.007	0.017	<0.005
	Ni	Al	Cu	Nb	Ti	V	Fe
	<0.005	0.003	0.007	<0.002	<0.002	<0.002	bal.

IOP Conf. Series: Materials Science and Engineering **694** (2019) 012026 doi:10.1088/1757-899X/694/1/012026

Table 2. Chemical composition (wt %) of ER-70S 6

	C	Si	P	S	Cu
ER-70S 6	0.06	0.8	0.025	0.03	0.5

All of the specimens were cut with a size of 55 mm and cleaned with sandpaper from grits 200-1000. Acetone solution was used to clean the remaining impurities. In this study, the type of welding used was Tungsten Inert Gas (TIG) welding using Dynasty 210 DX welding machine. Argon gas was used to protect the weld surface from oxidation with a flow rate of 11 L/min. A tungsten electrode was used EWTh-2 with a red code AWS with a diameter of 2.4 mm and a sharpening angle of 25^0. TIG welding parameter variations used were welding current and welding speed. Welding position was 5G where the pipe was idle while the welding torch rotates automatically and welding starts at an angle of 330 degrees. The welding parameters were performed in Table 3.

Table 3. Welding parameters GTAW process

No	Code	Welding current (A)	Welding speed (mm/s)	Arc efficiency (%)	Voltage (V)	Heat input (kJ/mm)
1	P-1511	150	1.1	60	11.3	924.55
2	P-1510	150	1	60	11.3	1017.00
3	P-1509	150	0.9	60	11.3	1130.00
4	P-1611	160	1.1	60	12.45	1086.55
5	P-1610	160	1	60	12.45	1195.20
6	P-1609	160	0.9	60	12.45	1328.00
7	P-1711	170	1.1	60	13.4	1242.55
8	P-1710	170	1	60	13.4	1366.80
9	P-1709	170	0.9	60	13.4	1518.67

Weld geometry can be determined from width of weld bead (WoWB). The WoWB being measured was the outside of the pipe. The WoWB measurements start from 0 to 285 degrees with an increase of 15 degrees. This measurement uses a digital microscope (Dino-Lite).

Pipe distortion measurement can be measured by calculating the change in diameter before welding and after welding. Measurements were made with a variation of 4 sections from the welding point of 10, 15, 30 and 50 mm and measured from the axial and transverse direction of the pipe position, as shown in Figure 3. Each measurement location was measured using an outside micrometer with an accuracy of 0.01 mm. This measurement was carried out before and after welding at the same pipe position on both pipes to be welded. The difference from this measurement was used as the value of the distortion that occurs after the welding process. There are several types of deformation in the pipe after welded such as the oval, irregular and tapers in the direction of the axial and transverse pipe.

The value of transverse distortion (TD) and axial distortion (AD) can be found by using Equation 1:

$$\begin{aligned} TD &= D_{a1} - D_{b1} \\ AD &= D_{a2} - D_{b2} \end{aligned} \tag{1}$$

where D_{a1}, D_{a2} were pipe diameter after welded while D_{b1}, D_{b2} were pipe diameter before welded in the transverse and axial direction. In addition to seeing the change in diameter of the pipe into an oval can be calculated using Equation 2 in each section of the pipe. If the calculation results for each section of the pipe were not the same then the diameter of the pipe has become oval. Tapers were the diameter of the pipe at one end with the other end of the pipe not the same in the direction of the axial or transverse pipe. It can be calculated using Equation 3:

$$\text{Oval} = \text{AD} - \text{TD} \tag{2}$$

$$\text{Tapers} = \text{AD}_{max} - \text{AD}_{\min}/\text{TD}_{max} - \text{TD}_{\min} \tag{3}$$

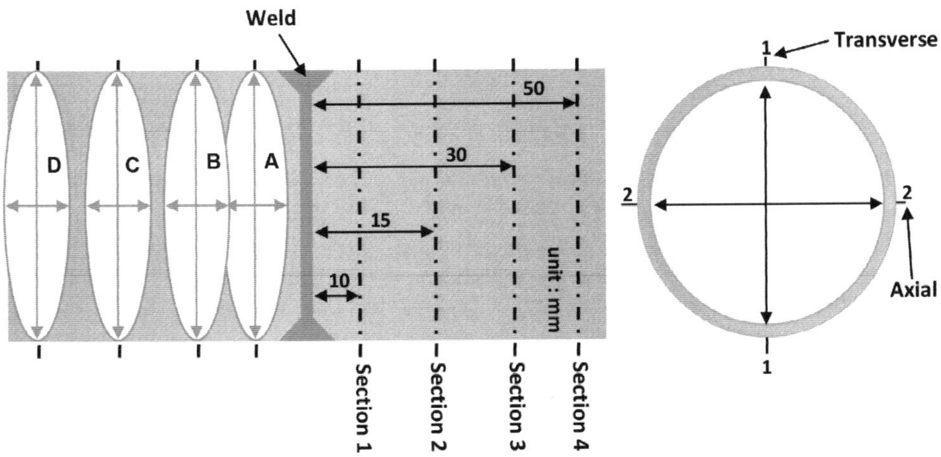

Figure 3. Schematic measurement of distortion in the pipe

3. Results and Discussion

3.1. Weld Geometry

One of the welding geometries can be determined from the width of weld bead (WoWB). Figures 4a and 4b show the weld surface profile of pipe welding in parameter code P-1611 and the average of outer bead width. Increasing the welding current can raise the average of outer bead width while increasing the welding sped can decrease the average of outer bead width. The widest of outer bead width at welding current of 170 A and welding speed of 0.9 mm/s and the smallest of outer bead width at welding current of 150 A and welding speed of 1.1 mm/s.

Increasing the welding current will indirectly increase the temperature, while the welding speed will affect the rate of temperature received by the material. If the welding speed slows down the temperature will also slow down which will widen the weld bead [8, 9].

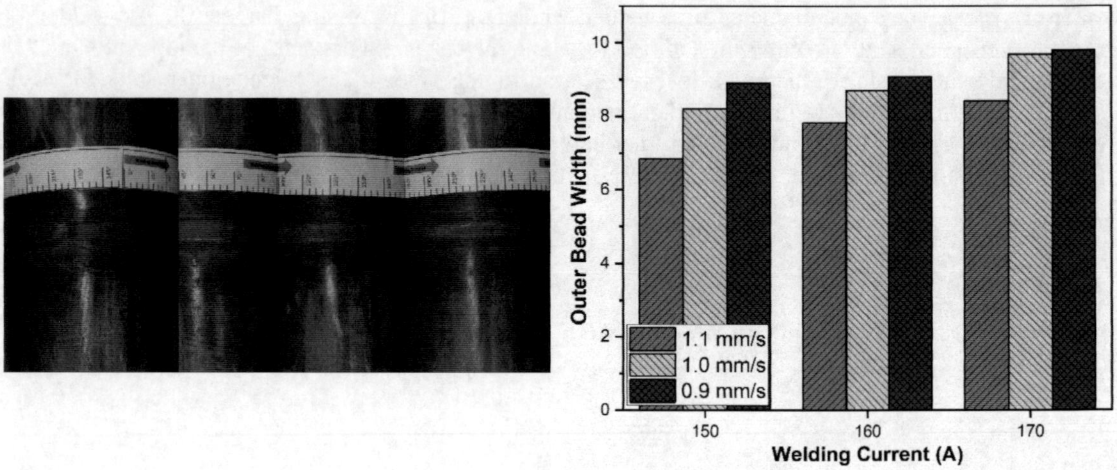

(a) (b)

Figure 4. (a) Photograph of the welding surface profile and (b) The average of outer bead width

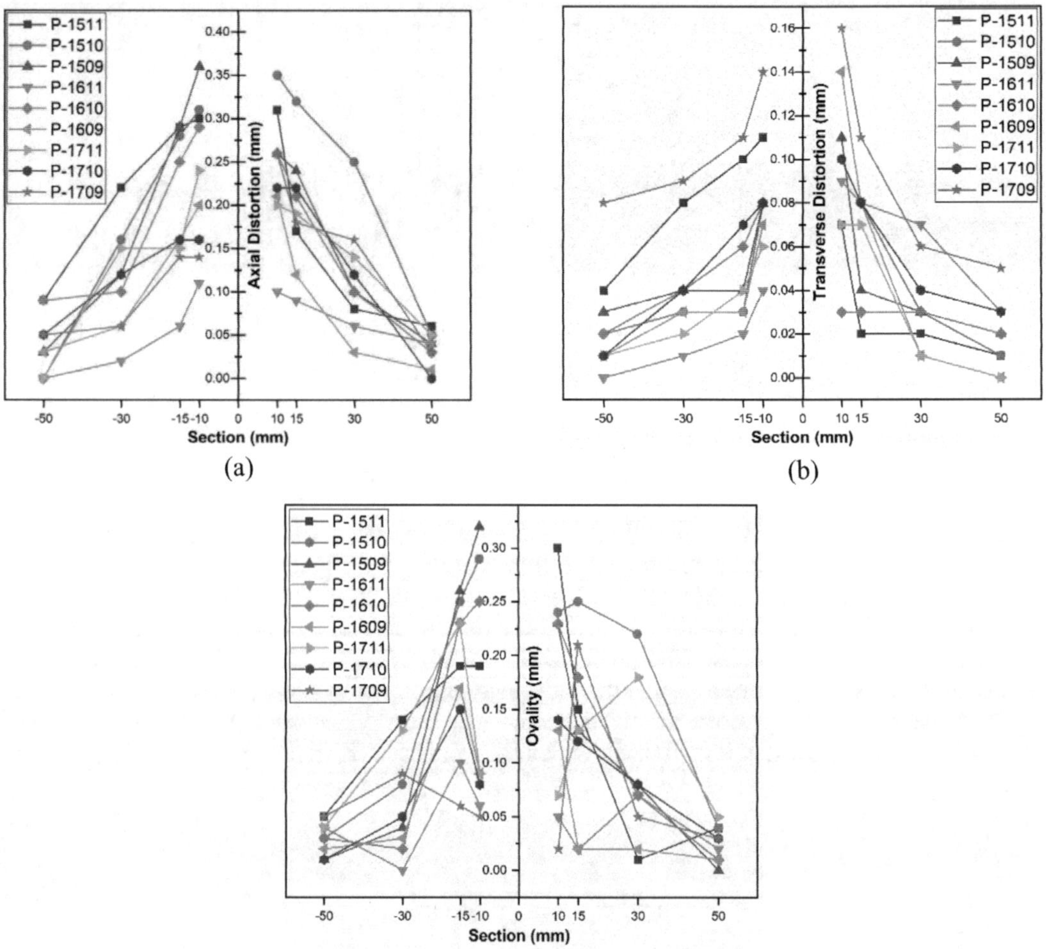

(a) (b)

(c)

Figure 5. Result of distortion test (a) axial distortion, (b) transverse distortion and (c) ovality in each section of pipes

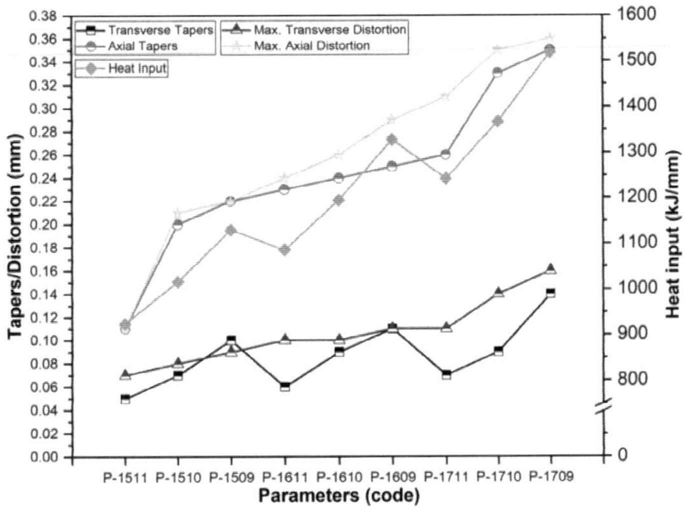

Figure 6. Heat input vs. tapers/distortion on all parameters

3.2. Distortion

The calculation result of axial distortion, transverse distortion and ovality in millimeter (Equation 1-3) as shown in Figures 5a, 5b and 5c, respectively. The left pipe and the right pipe were measured in diameter in each section (see Figure 3) in the axial and transverse direction. The results show that axial distortion and transverse distortion decrease in each section. Ovality in the pipe can be determined from the difference between AD and AT in each section. The results show that after welding the pipe was ovality (see Figure 5c).

The closer from the welding point, the heat input and temperature will increase so that the distortion that occurs is greater while the farther from the welding point, the heat input and temperature will decrease so that the distortion that occurs is smaller [6, 10].

Figure 6 shows the comparison between heat input, tapers and maximum distortion in pipe welding. Heat input has a significant effect on tapers and the distortion that occurs. Raising the heat input can increase tapers and distortion. Axial tapers and axial distortion were greater than transverse tapers and transverse distortion. This was due to the welding point starts at 330 degrees or approaches transverse direction so that the heat received is not large enough but in the axial direction the heat received will be even greater. As a result, the tapers and distortion in the axial direction were getting bigger.

4. Conclusions

Generally increasing the welding current (hence, increasing heat input) can widen the weld bead and increase distortion whereas increasing the welding speed (hence, decreasing heat input) can reduce the weld bead and reduce distortion. The results show that welding current of 170 A and welding speed of 0.9 mm/s produces the widest outer weld bead. Based on the calculation of distortion, axial distortion and transverse distortion will decrease if getting further from the welding point. After the welding process, the pipe will also ovality. In addition, the results of tapers and distortion in the axial direction were greater than the transverse direction of the pipe. The highest tapers and distortion occur at 170 A welding current, welding speed 0.9 mm/s and the axial direction of the pipe.

Acknowledgments

This research is supported by the Master Program to Doctorate for Scholar Excellent (PMDSU) program of the Ministry of Research & Technology and High Education (RISTEK DIKTI) 2019 with contract number NKB-1855/UN2.R3.1/HKP.05.00/2019.

IOP Conf. Series: Materials Science and Engineering **694** (2019) 012026 doi:10.1088/1757-899X/694/1/012026

References

[1] I. Sattari-Far, Y. J. I. J. o. P. V. Javadi, and Piping, "Influence of welding sequence on welding distortions in pipes," vol. 85, no. 4, pp. 265-274, 2008.

[2] G. Fu, M. I. Lourenço, M. Duan, and S. F. J. M. S. Estefen, "Influence of the welding sequence on residual stress and distortion of fillet welded structures," vol. 46, pp. 30-55, 2016.

[3] K. Ganesh *et al.*, "Modeling, prediction and validation of thermal cycles, residual stresses and distortion in type 316 LN stainless steel weld joint made by TIG welding process," vol. 86, pp. 767-774, 2014.

[4] M. Hemmatzadeh, H. Moshayedi, I. J. I. J. o. P. V. Sattari-Far, and Piping, "Influence of heat input and radius to pipe thickness ratio on the residual stresses in circumferential arc welded pipes of API X46 steels," vol. 150, pp. 62-71, 2017.

[5] W. Jiang, K. J. I. J. o. P. V. Yahiaoui, and Piping, "Effect of welding sequence on residual stress distribution in a multipass welded piping branch junction," vol. 95, pp. 39-47, 2012.

[6] V. Kalyankar and P. J. M. T. P. Shah, "A review on methodologies to reduce welding distortion," vol. 5, no. 11, pp. 24741-24749, 2018.

[7] A. Alves de Resende, F. Keocheguerians, and L. O. J. W. I. Vilarinho, "The influence of CO2 and O2 content on globular spray transition current when using argon-based blends in GMAW of ER 70S 6 wire," vol. 24, no. 8, pp. 593-601, 2010.

[8] F. Madadi, F. Ashrafizadeh, M. J. J. o. A. Shamanian, and Compounds, "Optimization of pulsed TIG cladding process of stellite alloy on carbon steel using RSM," vol. 510, no. 1, pp. 71-77, 2012.

[9] B. Qi, M. Yang, B. Cong, and F. J. T. I. J. o. A. M. T. Liu, "The effect of arc behavior on weld geometry by high-frequency pulse GTAW process with 0Cr18Ni9Ti stainless steel," vol. 66, no. 9-12, pp. 1545-1553, 2013.

[10] T. S. Kumar, V. Balasubramanian, M. J. M. Sanavullah, and design, "Influences of pulsed current tungsten inert gas welding parameters on the tensile properties of AA 6061 aluminium alloy," vol. 28, no. 7, pp. 2080-2092, 2007.

IOP Conf. Series: Materials Science and Engineering **694** (2019) 012027 doi:10.1088/1757-899X/694/1/012027

The effect of groove welding on mechanical properties and microstructure of wear resistance steel plate by Submerged Arc Welding

Zuldesmi Mansjur[1], **Hendro M Sumual**[2] **and I P Tamba**[3]

[1] Mechanical Engineering Study Program, Engineering Faculty, Manado State University

[2,3] Mechanical Engineering Education Department, Engineering Faculty, Manado State University

E-mail: zuldesmi@unima.ac.id

Abstract. Aim of this research is to analyze the mechanical properties and microstructure of wear resistance steel plates joined by Submerged Arc Welding at different grooves welding. The welding current of 300 Ampere, voltage of 28 Volt and travel speed of 38 cm/min was applied. The microscopical examination methods were used for weld zones. Impact toughness and rockwell hardness test were obtained from the weld metal, heat affected zone and the base metal samples. The fracture surfaces were examined by using optical microscope. As results, there was no difference of hardness and microstructure among three grooves welding. The significant differences were detected at the weld zones. Base metal had the highest hardness value. Ferrite and pearlite phases were obtained at three weld zones, but the heat affected zone had coarser grain structure than the base metal. Weld metal volume and electrode length significantly affected the impact toughness of joint, where the larger of the weld metal volume and length of electrode fill in the groove welding, the higher of impact toughness. The lowest impact energy and toughness value were attained at V groove and the highest was at I groove. The fracture surfaces at all groove welds showed ductile fracture dominantly.

1. Background

The wear resistant steels are typically used in various challenging high-stress wear conditions, such as mineral haulage and crushing, slurry transportation, dredging, demolition of concrete structures, and forest industry. In many applications, the wear environment is very complex and contains different wear mechanisms. Moreover, the wear resistant steels are not standardized and the generally used hardness based grading does not completely describe nor predict their wear behavior [1]. Thus, well-planned and multifaceted application oriented wear testing of the steels as a part of the material selection process is highly recommended for achieving a proper correlation to practical applications.

Sometimes, repairs of the wear resistant steels are required due to possible damage during during application and service. These parts repairs which involve welding are particularly critical in applications where high quality and/or precision are required. The submerged arc welding process is commonly used for repairing surface due to its easy applicability, high current density and its ability to deposit multiple filler wire at the same time, especially in restoration of worn parts, which is of great importance to manufacturers [2]. Submerged arc welding is suitable for welding of all weldable steels.The successful application of a welding the process mainly depends on the effective understanding of the temperature distribution in the weldment that can be helpful to estimate the evolution of the metallurgical microstructures and its subsequent influence on the mechanical properties [3]. However, many premature failures of a component, such as micro-fissures and residual stress corrosion cracking, are related to the microstructure and strength of the weld [4,5]. Thus, it is important to evaluate the microstructure and mechanical properties of joints made from wear resitant plate.

In the wear resistant carbon steels, the decomposition of austenite leads to textured microstructure composed of ferrite and pearlite bands. This is connected to the higher concentration of manganese in interdendritic regions that stabilize the austenite at lower temperatures. Consequently, ferrite grains are preferably nucleated in regions with a low concentration of manganese. Additionally, the carbon is segregated from the ferrite as it grows, concentrating on the regions of high manganese, where the perlite is formed [6]. This microstructure usually contains elongated inclusions that are responsible for the anisotropy of ductility and impact energy. Some properties may also be affected, such as: i-machinability, ii- hydrogen-induced embrittlement, in particular,in the heat affected zone of weld beads. The occurrence of heterogeneous microstructures (bands of ferrite and pearlite) may lead to the breaking of teeth during the fabrication of gears in the forging process [7]. It is crucial to recognize the microstructures of the weld zone in order to predict the hardness and impact toughness of the welded joint.

In this experimental investigation and relevant analyses, the influence of groove welding on mechanical properties of wear resistant plate was systematically investigated. In order to determine these parameters; the current, voltage and wire speed between the applicable ranges were tried in the welding laboratory, and then according to the penetration and visual tests the acceptable welding parameters were selected. Thus, the microstructural and mechanical properties of submerged arc welding (SAW) welds of 10 mm thick wear resistant steel plate according to the selected parameters were found out, instead of comparing the different weld zones.

2. Materials and Methods

Wear resistance plates were welded by using submerged arc welding process. For the microstructural investigation, microscopical examination methods were used in weld zones (weld metal, heat affected zone and base metal). Different samples obtained from the weld metal and the base metal, were subjected to hardness and impact toughness tests. Fracture surfaces were examined by optical microscope. Charpy impact toughness tests were made for the samples with three types of welding grooves.

2.1. Materials

The material used in this study was wear resistance plates of 10 mm thick with three groove welding types (I, V and U). Table 1 lists the chemical composition (wt %) of the wear resistant plate.

Table 1. Chemical composition (wt %) of the wear resistant plate

C	S_{max}	P_{max}	Cr	Mn	Mo
0.14	0.01	0.01	0.5	1.2	1.25

2.2. Welding conditions

The welding process for each groove was otomatic submerged arc welding with parameters as shown in Table 2.

Table 2. Welding parameters of the SAW.

Voltage (V)	Current (A)	Welding speed (cm/min)	Heat input energy (kJ/cm)
28	300	38	13.263

Before welding, the samples were preheated at temperature of 100⁰C in order to prevent an initial hardening on the heat affected zone and residual stress on the thick plates. Then, they were polished with a brush and thoroughly washed with acetone to remove any greasy dirt and impurities.

The SAW process was realized automatically in the laboratory setting for hardness, impack toughness and microstructural investigations by using reversed polarity and multilayer welding at V and U grooves. For I groove, single layer of filling process was applied on two sides (upper and lower) of surface with root gap of 2 mm.

2.3. Microstructural investigation

Submerged arc welded joints of wear resistant plate were cross-sectioned perpendicular to the welding direction. In the first step, specimens were prepared with standard metallographic preparations like grinding, polishing and etching with Picral (4gr picric acid + 100 ml ethanol) for 10–15 seconds at room temperature. In the second step, the weld zones were investigated as micrographs by Optical Microscope with magnifications of 20X.

2.4. Hardness

As an additional destructive test method, Rockwell C hardness measurement (HRc) was carried out under a load of 150 Kg over cross sections on vertical (the welding centerline) and horizontal lines 1 mm distance between each indentation on metallographic specimen taken from each welded plate with Tokyo testing machine type RH-3N MFG. The Figure 1 shows the point of hardness testing.

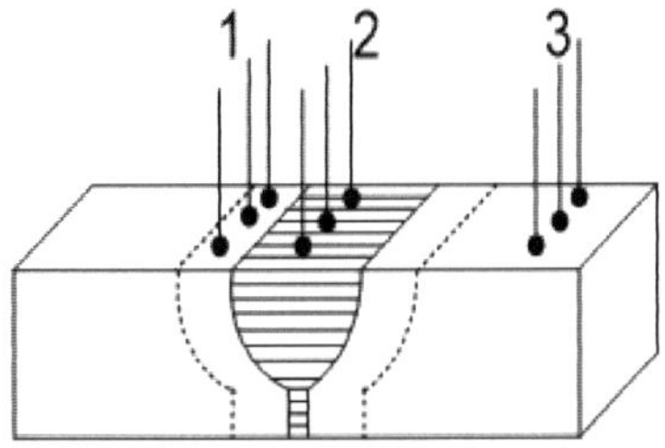

Figure 1. Test point of the hardness test samples ; (1). Heat Affected Zone (HAZ), (2). Weld Metal (WM) and (3). Base Metal (BM)

2.5. Notch impact toughness test

The direct notch impact testing was implemented to all welded specimens with three groove types as shown in Figure 2.

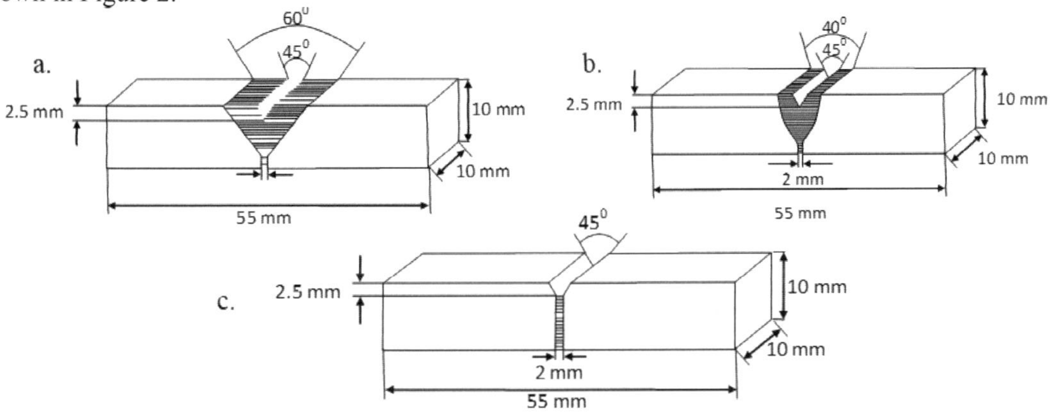

Figure 2. Welded specimen with various groove types; a. V groove, b. U groove and c. I groove

IOP Conf. Series: Materials Science and Engineering **694** (2019) 012027 doi:10.1088/1757-899X/694/1/012027

3. Results and Discussions

3.1. Rockwell Hardness test

The hardness evaluation was made on metallography specimens for single layer (I groove) and multiple layer (V and U grooves), in order to present results from the point of heat input effect on hardness as shown in Figure 3.

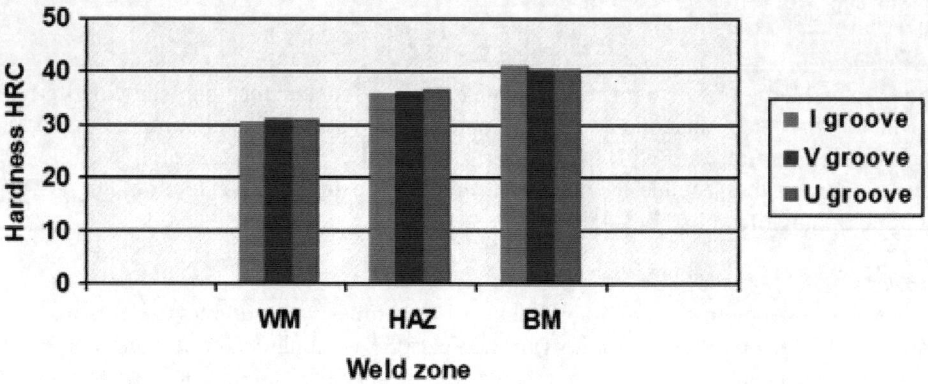

Figure 3. Hardness comparison among three welded zones: WM, HAZ and BM

In Figure 3, the variation of grooves welding are not contribute differences of hardness significantly for each weld zones. However, the differences are clearly distinguished in comparing to the three weld zones. Base metal hardness of the welds varied between 40.16 and 41.38 HRc and this value was the highest among the weld metal zone and HAZ. This conditions is owing to occurring of recrystallization condition and cooling rate. Besides, at the base metal zone, there was no heat affect occured at the samples and material compositions also influence the changes of hardness. The hardness of HAZ (36-36.66HRc) are close to the base metal. The grain growth was observed in HAZ due to during the welding process, the high heat transfer from the base metal through to the heat affected zone until the austenite phase was formed and slowly cooling was taken place. Because of the grain coarsening occurred at the HAZ interface, the hardness values were decreased. On the other hand, minimum hardness values 30.38-31.05 HRc were measured at the weld metal.

3.2. Microstructural investigation

The microstructural examination on the samples of welded joint was carried out using Optical Microscope and with 20x magnification. As displayed in Figure 4, ferrite (light color) and pearlite (dark color) phases were noticed in all weld zones, but the base metal had finer grain structure than the others. This fine grain structure lead to higher hardness of the base metal. The pearlite phase was appeared dominantly in this zone. Meanwhile, the grain became coarse and ferrite phase become visible primarily in HAZ. The change of grain size was probably because of hot operation during welding process with lower cooling rate. Meanwhile, in the weld metal zone, the cooling rate become lower till to room temperature and the phase turn into an initial phase. The ferrite phase be revealed mainly with a clear grain boundary in the weld metal zone.

The grain size is a parameter which directly affects the mechanical properties of metals and alloys. In fine grained materials, mechanical properties are high due to increase of grain boundary. During the welding process, high temperature values in HAZ are observed because of the high heat input. The grain coarsening is seen in the microstructure and also the mechanical properties decrease. At the same time, the alloying elements in the structure of these steels form finely dispersed carbides. The dissolution of alloy carbides takes place on account of the high heat input effect during welding. It causes a reduction in mechanical properties. Additionally, because of the carbide dissolution, grain boundaries are released and grain growth during welding becomes easier. It is seen that the micro hardness results are compatible with the literature [8,9].

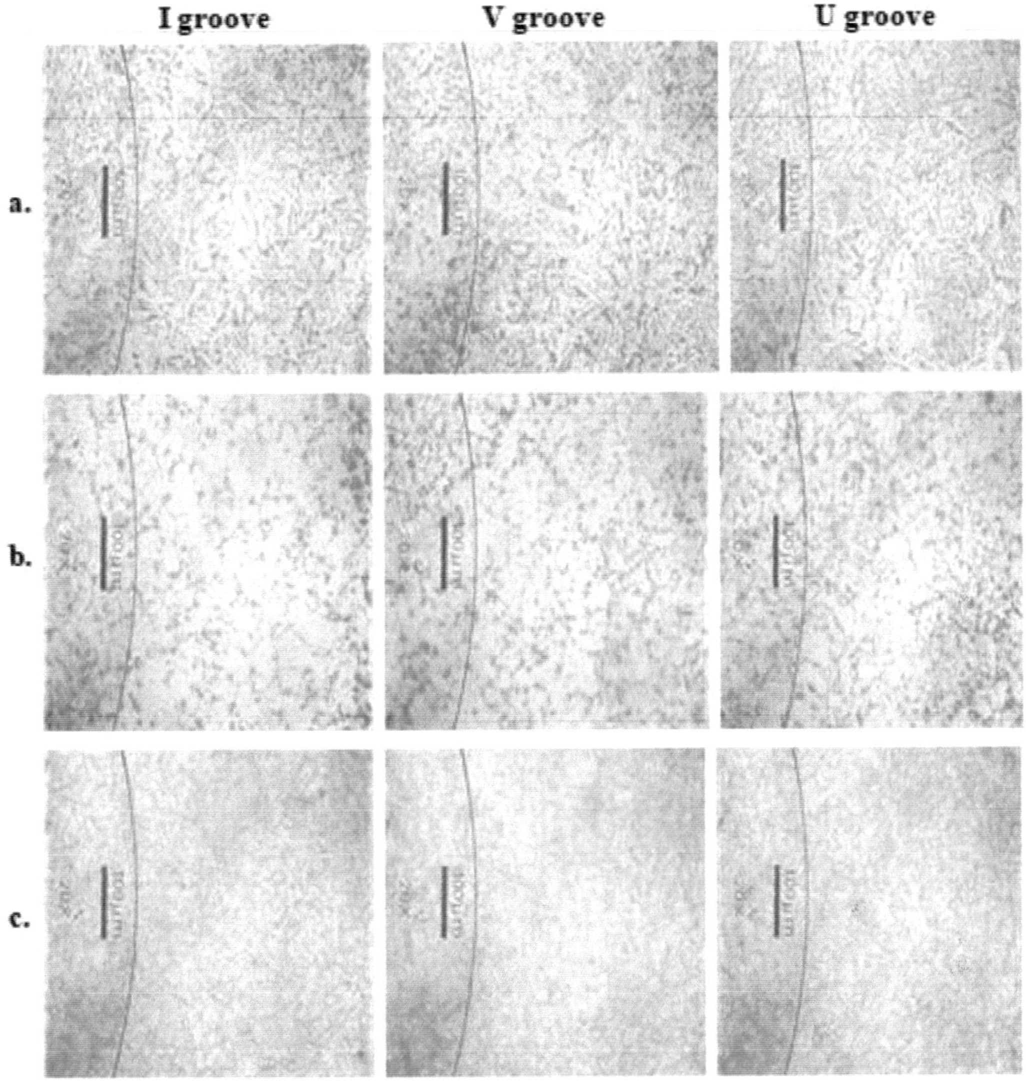

Figure 4. Microstructure photograph of specimen with three types of groove welding: I, V and U grooves at three welded zones: a. BM, b. HAZ and c. WM

3.3. Notch impact toughness test

In order to compare three grooves welding, 55mm × 10mm × 10mm test pieces were used. The tests were carried out at room temperature (~28°C). The notch impact values pointed out in joule/mm^2 are presented in Table 3.

Table 3. The impact test results of three groove welding samples

No	Specimen	α (⁰)	β (⁰)	T (⁰C)	IE (Joule)	Average IE (Joule)	IT (J/mm^2)	Average IT (J/mm^2)
1	U. 1		62		205.96		2.74	
2	U. 2		63	28	204.10	205.34	2.72	2.73
3	U. 3		62		205.96		2.74	
4	V. 1		64		201.77		2.69	
5	V. 2	156	63	28	204.10	203.32	2.72	2.71
6	V. 3		63		204.10		2.72	
7	I. 1		30		265.60		3.54	
8	I. 2		28	28	268.25	266.48	3.57	3.55
9	I. 3		30		265.60		3.54	

The specimen with I groove contributed the highest impack toughness values of 3.55 J/mm^2 and it achieved the best absorbed energy about 266.48 J, following U groove (impack toughness of 2.73 J/mm^2 and absorbed energy of 205.34 J) and V groove (impack toughness of 2.71 J/mm^2 and absorbed energy of 203.32 J). Weld metal volume and length of electrode influence the strength and toughness of joint crucially, especially for V and U groove samples. By increasing the welding metal volume and length of electrode, the strength of welding become increase. However, these variables were not affect to I groove sample. The highest impack toughness of I groove sample was mainly caused by the welding process. During the welding process, filler metal was applied onto both sides (upper and lower) of surface with root gap of 2 mm. It was carried out because at I groove, a penetration was not trough the surface actually.

The fractured surfaces are presented in Figure 5. The I groove samples with high impact energy showed almost fully ductile fracture (~89%) and , as shown in Figure 5c. The fractured specimens of U groove (Figure 5b) had two different fracture surfaces in a ductile (~78.6%), and a brittle manner over part (~ 21.4%) and its impact value was relatively lower. The V groove specimens gave the lowest values, and it had smaller ductile zone about ~75% and broader brittle zone about ~ 25% , as presented in Figure 5a.

Figure 5. Fracture surface of impack thougness of test samples with (a) V, (b) U and (c) I grooves

During the welding process of wear resistant steel plate, preheating, temperature between passes, heat input should be controlled strictly in keeping with the declared standards, in order to keep away from cold cracks and possible failures, and to achieve the required properties of welded structures. Inadequate heat input mostly influences hardness and the strength of these structures. Moreover, the hydrogen dissolution and residual stresses in the weld may result in cold cracks, while minimizing deformability and enhancing sensitiveness to embrittlement [10].

4. Conclusions

The effects of groove welding on the microstructure and mechanical properties of wear resistant plate were studied. The following are the main conclusions of this study:

1. By varying groove welding, hardnesses and microstructures of samples were not change significantly. However, their changes were appeared at three weld zones (base metal, heat affected zone, and weld metal)
2. The highest hardness was obtained at base metal zone (40.16-41.38 HRc) followed by HAZ (36-36.66 HRc) and weld metal zone (30.38-31.05 HRc).
3. As microstructure investigation, the base metal had finer grain size than HAZ and weld metal. The grain coarsening was occurred during the weld process because of the high heat input in HAZ.
4. The highest impact energy and toughness value were attained at I groove and the lowest was at I groove. The fracture surfaces at all groove welds showed ductile fracture dominantly.

References

[1] N. Ojala, K. Valtonen, V. Heino, M. Kallio, J. Aaltonen, P. Siitonen, V.-T. Kuokkala, Effects of composition and microstructure on the abrasive wear performance of quenched wear resistant steels, Wear 317 (2014) 225–232, https://doi.org/10. 1016/j.wear.2014.06.003

[2] B Gulenc, N Kahraman. Wear Behaviour of Bulldozer Rollers Welded using a Submerged Arc Welding Process. Materials and Design. 24 (2003) 537–542.

[3] D. V. Kiran, J. Cheon, N. Arif, H. Chung, S. J. Na, Three-dimensional Finite Element Modelling of Pulsed AC Gas Metal Arc Welding Process. Int. J. Adv. Manuf. Technol. 86 (2016) 1453–1474.

[4] A. Chamanfar, M. Jahazi, A. Bonakdar, E. Morin, A. Firoozrai, Cracking in fusion zone and heat affected zone of electron beam welded Inconel-713LC gas turbine blades, Mater. Sci. Eng. A 642 (2015) 230–240.

[5] A.H.S. Iyer, K. Stiller, G. Leijon, H.C.M. Andersson-Östling, M. Hörnqvist Colliander, Influence of dwell time on fatigue crack propagation in Alloy 718 laser welds, Mater. Sci. Eng. A 704 (2017) 440–447

[6] R.A. Grange, Effect of microstructural banding in steel, Metall. Trans. 2 (1971) 417–426, https://doi.org/10.1007/BF02663328.

[7] J. Krawczyk, B. Pawłowski, P. Bała, Banded microstructure in forged 18CrNiMo7-6 steel, Metall. Foundry Eng. 35 (2009) 45, https://doi.org/10.7494/mafe.2009.35. 1.45.

[8] Porter, D. A. Weldable High-Strength Steels: Challenges and Engineering Applications. IIW International Conference, 2-3 July, Helsinki, (2015) 1-14.

[9] Ada, H. Weldability of Pipes Produced for Crude Oil and Natural Gas Pipelines with The Method of Submerged and Spiral Arc Welding and Researching Their Mechanical Properties, M.Sc. Thesis, p. 145, Gazi University, Ankara. (2006).

[10] Dunđer, M., Vuherer, T., Samardžić I., 2014. Weldability of microalloyed high strength steels TStE 420 and S960QL. Metalurgija 53, 3, 335-338.

IOP Conf. Series: Materials Science and Engineering **694** (2019) 012028 doi:10.1088/1757-899X/694/1/012028

In-Service Welding Simulation of 28” Pipeline Using Finite Element Method

Sahrudin Tambunan[1], Ardiyansyah Yatim[2], Putu Wira Sanjaya[1] and Michael Simon P H[1]

[1]PT 3S Engineering, Bukit Rivaria, Depok, Indonesia.
[2]Mechanical Engineering, Universitas Indonesia, Depok, Indonesia.

E-mail: Sahrudin.tambunan@3S-eng.co.id, ardiyansyah@eng.ui.ac.id

Abstract. To maintain the integrity of the gas pipeline, the repairing of pipeline system with common method such as cut and replaces method by conducting hot tap and bypass or clamp method is conducted. Both methods require in-service welding in the process. During in-service welding, the process affects the structure strength. The structure then needs to be analysed during the welding process. There are two common problems associated with in-service welding. Firstly, the high gas flow in pipe causes the weld rapidly to cool due to the convective transfer of heat. The other problem is burn-through during in-service welding. Pressurized natural gas imposes a significant stress on the pipe wall, and since the pipe strength is decreased during welding, this cause failure in pipe wall. Burn-through occurs when the region around the weld pool has insufficient strength to withstand the internal gas pressure. This paper shows the finite element analysis procedure of In-Service Welding Analysis to avoid burn-through failure. A thermal-mechanical based Finite Element model had been conducted to assess the risk. The model is simulated using three-dimension (3D) mechanical, thermos-elastic-plastic-metallurgical finite element computational procedure. The temperature, stress and strain parameters of inner surface were used to assess the model. The temperature and strain check are below the allowable value while the stress check with some applicable method which are based on either stress or strain based.

1. Introduction

In-service welding is commonly used for the fabrication of structures ranging from, large to small and simple to complex especially when the system cannot be shut down during the repair process. Fusion welding is a significant engineering process because of the unparalleled advantages it has over other joining methods. However, welding process causes the material properties to change that result in weaker material. To capture this changes, finite element analysis need to be conducted.

Finite element method has been used widely to simulate the in-service welding process especially when the analytical models hit its limitations. Heat source distribution applied in the finite element should represent the actual process. A model of heat distribution was introduced to accommodate the real welding process.

Therefore, the analysis of in-service welding process needs to be presented to show that the parameters obtained from finite element are still within allowable value. The pipeline is shown in table 1 [1-4].

Table 1. Pipeline Data

Parameter	Unit	Value
Outside Diameter	mm	711.2
Wall Thickness	mm	8.7
Material	-	API 5L X65
Operating Pressure	Psi	950
Design Pressure	Psi	1151
Operating Temperature	^{0}F	80
Design Temperature	^{0}F	150
Flow Velocity	m/s	5.5
Content	-	Natural Gas

2. Modelling and Assessment Method

2.1. Thermo-Mechanical Analysis

The in-service welding analysis was conducted with two type analysis, non-linear mechanical analysis and transient thermal analysis. Non-linier mechanical analysis was performed by applying pressure load. While the transient thermal analysis, was performed by applying thermal load from the welding process. Non-linier mechanical and transient analyses are performed simultaneously.

2.2. Thermal Boundary Condition

Radiation losses will be dominant at high temperature area and in the weld zone while convection losses has a major role in low temperature area which away from the weld pool. The combined convective and radiative heat transfer coefficient was taken into account with the following equation[5]:

$$h = \begin{Bmatrix} 0.0668T \left({}^{W}\!/_{m^2}\,°C\right) when\ 0 \leq T \leq 500°C \\ (0.231T - 82.1) \left({}^{W}\!/_{m^2}\,°C\right) when\ T \geq 500°C \end{Bmatrix} \quad (1)$$

The equation was applied to the outer surface of the 3D model. While convection heat transfer coefficient at the inner side of the pipe-wall was determined by the equation below [5]:

$$\frac{h_g D}{k_g} = 0.023\left(\frac{\rho_g v_g D}{\mu_g}\right)^{0.8}\left(\frac{C_{pg}\mu_g}{k_g}\right)^{0.4} \quad (2)$$

Where, k_g, ρ_g and C_{pg} are the thermal conductivity, gas density and specific heat respectively. μ_g and v_g are the viscosity and gas velocity respectively. D is the inside pipe diameter.

2.3. Weld Heat Source Modelling

The double ellipsoidal heat source model was used in this analysis. It based on Gaussian distribution of power density (W/m^3) that was introduced in [6] as shown in figure 1. This model is able to analyse the thermal history of shallow and deep penetration weld. It simulates welding action and capable of transporting heat well below the surface of the weld pool. This model combines two ellipsoidal sources that represent the actual experimental experience i.e. the temperature gradient in front of the source is much higher, compared to lower gradient in the rear of the heat source [7].

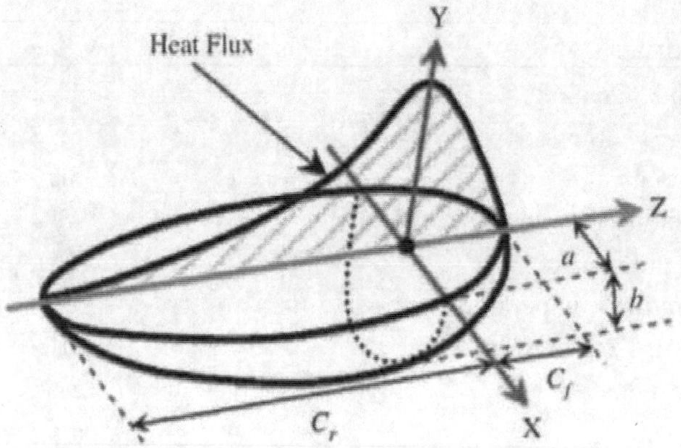

Figure 1. Double ellipsoidal heat source model

The power density distribution expressed in moving Cartesian coordinate system (x, y and z), is explained in the following equation.
Distribution inside the front quadrant:

$$q_f\,(x,y,z,t)=\frac{6\sqrt{3}f_fQ}{abc_f\pi\sqrt{\pi}}exp\left\{-3\left(\frac{x^2}{a^2}+\frac{y^2}{b^2}+\frac{(z+vt)^2}{c_f^2}\right)\right\} \quad (3)$$

For the rear quadrant:

$$q_r\,(x,y,z,t)=\frac{6\sqrt{3}f_fQ}{abc_r\pi\sqrt{\pi}}exp\left\{-3\left(\frac{x^2}{a^2}+\frac{y^2}{b^2}+\frac{(z+vt)^2}{c_r^2}\right)\right\} \quad (4)$$

DFLUX, an Abaqus user subroutine [8], was used to apply the non-uniform distributed flux as function of time, position, temperature to the model during heat transfer analysis. Parameters below as shown in table 2 [5] to be used in double ellipsoidal power density distribution equation

Table 2. Double ellipsoidal heat source parameter

Parameter	Symbol	Unit	Value	Ref
Length of front ellipsoidal	c_f	mm	12.9	[10]
Length of rear ellipsoidal	c_r	mm	10.3	[10]
Depth of the heat source	b	mm	4	[10]
Width of the heat source	a	mm	5	[10]
Heat fraction (front)	f_f	-	6	[10]
Heat fraction (rear)	f_r	-	8	[10]

2.4. Analysis Setup

The pipe was modelled as a half circle to reduce the number of elements that will reduce running time. The most critical case of in-service welding analysis is when the pipe is directly exposed to a heat source (welding), hence this analysis was focused in pipe section only. The analysis was done in the first pass sequence only because in this case all heat sources from welding will be absorbed by the pipe only.

While in other cases when the pipe and split tee are exposed to heat at the same time, it is not as critical as the first case because in that case, the pipe has shared the heat with a split tee.

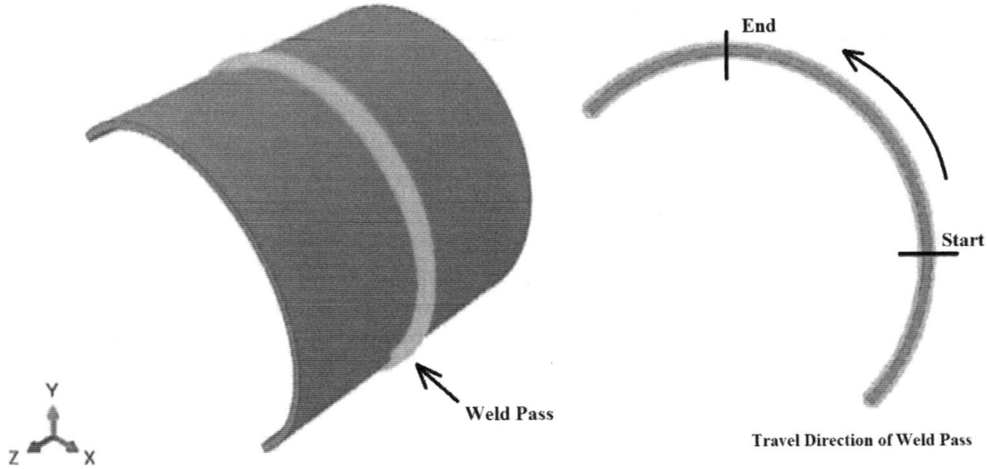

Figure 2. Pipeline model, weld pass location and weld travel direction

2.5. Assessment Criteria

The analysis was carried out by using Finite Element software. The split tee model was assessed using the following step.

Temperature criteria were checked according to the API RP2201 – 1995 [9] recommendation. The critical temperature for welding process is 980^{0}C (1250 K). If the temperature higher than recommendation then high risk of burn through will be happen.

Stress criteria were checked according to ASME B31.8 [10] and Vakili et.al. [7]. Based on the ASME B31.8, the combined stress (von Mises) shall not exceed the given value of 0.9xSMYS and the temperature of pipe shall not exceed 232^{0}C. (Table 841.1.8-1). While Vakili et.al [7], in preventing burn-through effective stress along the main pipe wall should be checked against the yield stress at the same temperature.

Strain criteria were checked according to ASME B31.8 or ASME BPVC VIII Division 2 [10]. Based on the ASME B31.8, the maximum permitted strain is limited to 2%. While ASME BPVC VIII Division 2, the calculated equivalent plastic strain (PEEQ) from FE results should be lower than the minimum of plastic strain at tensile strength and limiting triaxial strain (ε_L).

Displacement criteria were checked by measure the radial deformation. The maximum allowable radial deformation is 1 mm.

First assessment step is temperature check. If temperature check is pass then it will be continued with stress or strain or displacement check. Until one of the parameters is ok then the structure will within the allowable value.

3. Finite Element Results

The analysis was evaluated based on the design criteria in section 2.5. Those parameters are obtained from element that located on inner surface pipe and centre line of weld pass as shown in the following figure 3. History of temperature, stress, and strain at inner surface of pipe during welding are given in figure 4, figure 5 and figure 6 respectively.

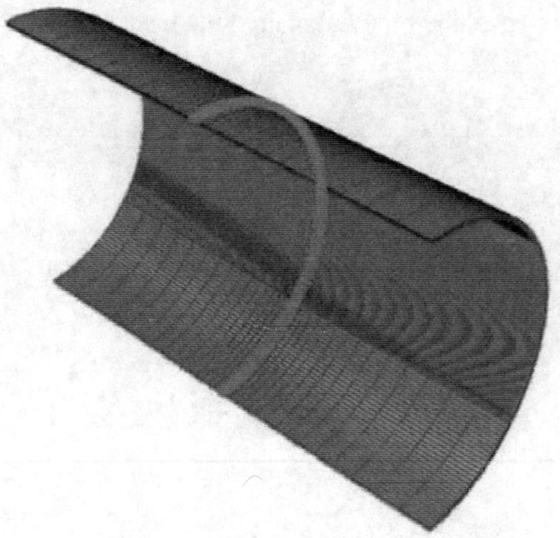

Figure 3. Location of evaluated line – Inner surface

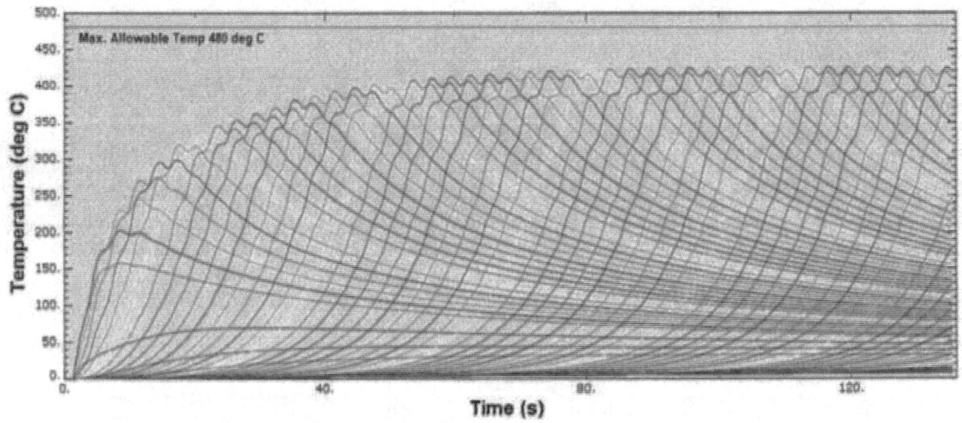

Figure 4. Temperature history at the evaluated line – Inner surface

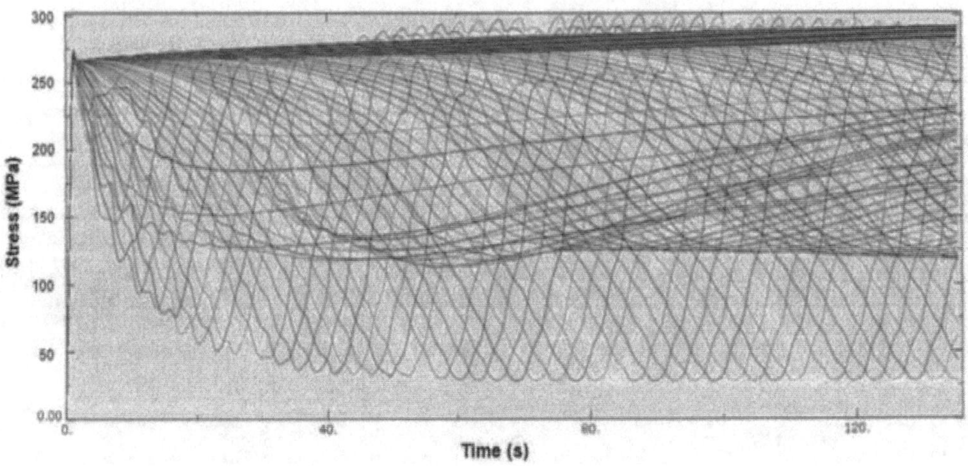

Figure 5. Combined stress history at the evaluated line – Inner surface

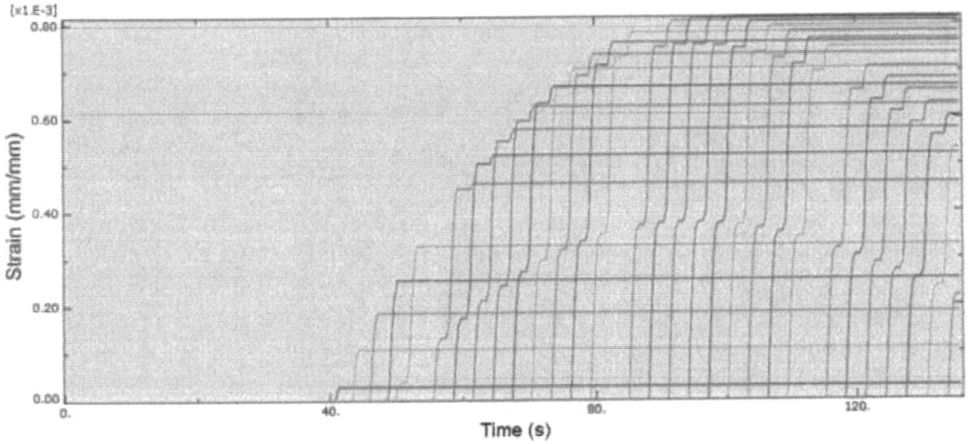

Figure 6. Strain history at the evaluated line – Inner surface

The colour variation of the curve in the figure above shows the state of the inner surface node of the pipe that is passed by the heat source right on the top surface. Based on the figure above, maximum temperature, maximum combined stress and strain during welding in the evaluated area are 425^{0}C, 300 MPa and 0.08% respectively as shown in the following table 3.

Table 3. Parameters Assessment

Design Criteria	Unit	Value	Allowable	Status
Temperature Check	^{0}C	425	980	PASS
Stress Check (Vakili et.al [7])	MPa	300	268	FAIL
Stress Check (ASME B31.8 [1]])	MPa	295	240	FAIL
Strain Check (ASME B31.8 para 833.5 [10-12])	%	0.08	2	PASS

The value of the parameters temperature check and strain check are below the allowable value. While the stress parameters are above the allowable value. Although stress parameters indicate that the value are outside of the allowable value but since temperature and strain parameter pass the assessment then the structure are within the allowable value[13-16].

4. Discussion and conclusions

Thermal-mechanical based Finite Element (FE) model had been utilized to assess the risk of burn-through during the welding of pressurized gas pipelines. In this analysis, the pipe sleeve will be assessed based on temperature, stress and/or strain parameter. The welding of Split Tee was simulated using three-dimension (3D) mechanical, thermos-mechanical and metallurgical finite element model.

The FEM pre-processing, calculation have been carried out by finite element software. The thermos-elastic-plastic-metallurgical finite element computational procedure was performed to analyse the welding temperature and the welding residual stress in pipeline which has been welded when the pipeline still in operating condition (in-service welding).

The analysis has shown that assessment parameters obtained from in-service welding simulation with internal pressure of 950 psig are within allowable value.

IOP Conf. Series: Materials Science and Engineering **694** (2019) 012028 doi:10.1088/1757-899X/694/1/012028

References

[1] 28-in Pipeline Repair Application Data Sheet.
[2] Stopple Plus Tee 28x28 in 600 lb RF.
[3] ASME BPVC: Materials, 2015.
[4] API 5L 2015 Specification of Line Pipe.
[5] ABID et.al; 2009 3D Thermal Finite Element Analysis of Single Pass Girth Welded Low Carbon Steel Pipe-Flange Joints, Turkish J. Eng. Env. Sci.
[6] Brickstad B,B.I, "Josefson 1998 A Parametric Study of Residual Stresses in Multiphase but Welded Stainless Steel Pipes Journal of Pressure Vessel 75, pp 11-25.
[7] Farid Vakili-Tahami Finite Element Analysis of the In-Service Welding of T Joint Pipe Connections.
[8] Simulia Abaqus version 6.13.
[9] API RP 2201. 1995 Safe Hot Tapping Practices in the Petroleum & Petrochemical Industries.
[10] ASME Section VIII Div.2. 2017ASME BPVC: Rules for Construction Pressure Bessel – Alternative Rules.
[11] AWSW A5.1 2004 Specification for Carbon Steel Electrodes for Shielded Metal Arc Welding.
[12] Welding Procedure Specification Pipe Sleeve API 5L X65.
[13] In-Service Welding Analysis Report of Pipe Sleeve Design – Pipeline Repair Option A.
[14] Parmar,R.S 2002 *Welding Engineering and Technology* (Delhi: Khanna Publisher).
[15] Sabapathy, P.N. 2002 Predicting Weld Cooling Rates and The Onset of Failure during In-Service Welding (Adelaide: The University of Adelaide).
[16] Karayan, A.I., Hersuni, A., Adisty D., Yatim, A. Failure analysis of seawater inlet pipe. Journal of Failure Analysis and Prevention Volume 11, Issue 5, October 2011, Pages 481-486

IOP Conf. Series: Materials Science and Engineering **694** (2019) 012029 doi:10.1088/1757-899X/694/1/012029

Corrosion Resistance of Sn-Zn Coated on Low Carbon Steel Material in Wet Gas Pipeline

Roni Hadi Wijaya and Bambang Soegijono

Department of Physics, Faculty of Mathematics and Natural Sciences (FMIPA), Universitas Indonesia, Depok 16424, Indonesia

E-mail: naufal@ui.ac.id

Abstract. Low carbon steel pipeline material is usually to transport wet gas. The presence of water in gas may lead corrosion and leakage. To enhance corrosion resistance of this material, we investigated electrodeposited of Sn-Zn on various composition Zn. Microstructure, structure, thermal properties and corrosion behaviour were analysed with Scanning Electron Microscopy (SEM), X-ray diffraction (XRD), Differential Scanning Calorimetry (DSC) and Potentiodynamic Polarization. Electrodeposited Sn-Zn with smallest Zn content shows highest corrosion rate. With various composition, XRD pattern shows tetragonal crystal structure. The various Zn content have changed the melting point and the enthalpy. Electrodeposited Sn-Zn on low carbon steel pipeline material enhance the corrosion resistance.

1. Introduction

The pipeline has transport wet gas from well head to the gas station is serious problem when using low carbon material. The presence of CO_2, H_2O, H_2S and Chloride may lead localize corrosion on material API 5L grade B that used in pipeline onshore.

The Indonesian natural gas with high CO_2 causing the failure of material especially in the presence of water or CO_2 dissolved in water or aqueous media can cause corrosive environment (carbonic acid / H_2CO_3) and accelerate the corrosion process on carbon steel material. In the pipeline, formation water that containing chloride ions is medium to distributed and can caused a uniform corrosion. The failure or success of corrosion resistance material is based on selection proper materials and their application such as identify chemicals will be present in environment (acid, neutral, or alkaline), the minimum and maximum temperature used, and fabrication and installation process.

Many researchers already published failure analysis corrosion on carbon steel pipeline and flowline. The aim of this paper was to enhance the corrosion resistance of low carbon steel pipeline with coated Sn-Zn on the surface material. Half the world's used zinc to protect steel from the corrosion for a wide application with the reason; low prices, supply zinc available and good cathodic protection with makes the coating layer when the substrate exposed to the corrosive environments. Zinc is protected the substrate by galvanic action as a sacrificial metal. Combined zinc and thin is interested to observe how both Sn-Zn protected the carbon steel from the corrosion. Most commonly used for coating is; hot dipping, electroplating, mechanical coating, sherardizing, thermally sprayed coating and dust painting. Electrodeposition was chosen in this proses due to good adhesion, the coating is uniform with limitation the power of the bath. Sn-Zn coating with high content of tin is good protection against oxidation and uniform corrosion of metals. [22]

2. Material and Methods

2.1. History of the Pipeline

This gas pipeline located in the onshore through rice field from production well to the gas station. The flow line operated since 2009, protected by external coating and ICCP (Impressed Current Cathodic Protection). This pipeline material is low carbon steel (API 5L grade B) and leak in 2016 and 2018. The operational data record show average water cut 96%, pH 7.5, gas temperature 83-110°F, CO_2 content 1.91% and partial pressure 3.57 Psi or operation below the design pressure.

2.2. Visual Inspection

The pipeline failure has been visual inspection, the external was in good condition with coated by 3 LPE shown in figure 1(b) and internal found general and localized corrosion in the bottom of the pipe as shown in figure 1(a).

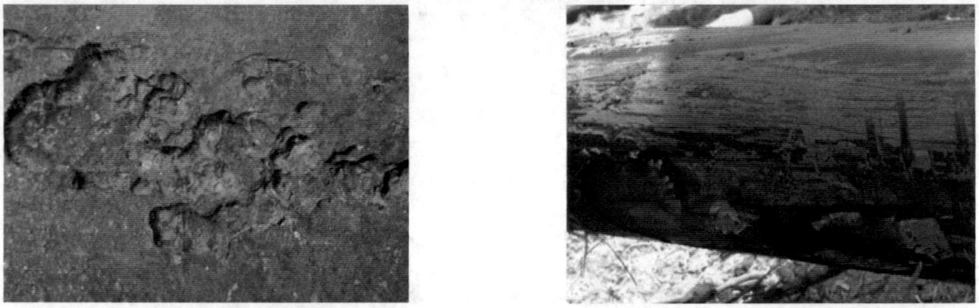

(a) Internal of the Pipe (b) External of the Pipeline

Figure 1. Visual Inspection of the Pipeline

2.3. Wall Thickness Measurement

Wall thickness measurement was conducted on the gas pipeline using ultrasonic testing and calculated for a corrosion rate described in table 1

Table 1. Pipeline Measurement, Corrosion Rate & Remaining Life

Onshore Pipeline	Position	Thickness Year 2006	Thickness Year 2018	Corrosion Rate (mm per year)
Well X	12 o'clock	7.1 mm	6.8 mm	0.023
To	03 o'clock	7.1 mm	5.3 mm	0.138
Gas Station	06 o'clock	7.1 mm	3.2 mm	0.325
	09 o'clock	7.1 mm	5.5 mm	0.123

Bottom of pipe has significant corrosion rate 0.325 mm per year shown in table 1. and need inspection strategies to mitigate the cause of failure or enhance the corrosion rate resistance of low carbon steel.

2.4. Analysis of Surface Corrosion Pipeline

From the figure 2 result, it can be seen that the intergranular corrosion of low carbon steel began on transition grain boundary when the steel polarized to the active passive transition zone. [11]. Figure

2(a)(b) general corrosion of surface low carbon steel shown IGC observed. Figure 2(c)(d) pitting corrosion of surface low carbon steel shown similar integranular corrosion observed.

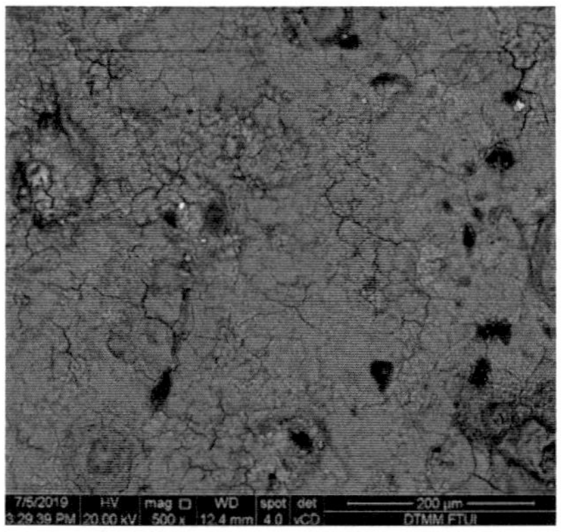

(a) General Corrosion

(b) General Corrosion

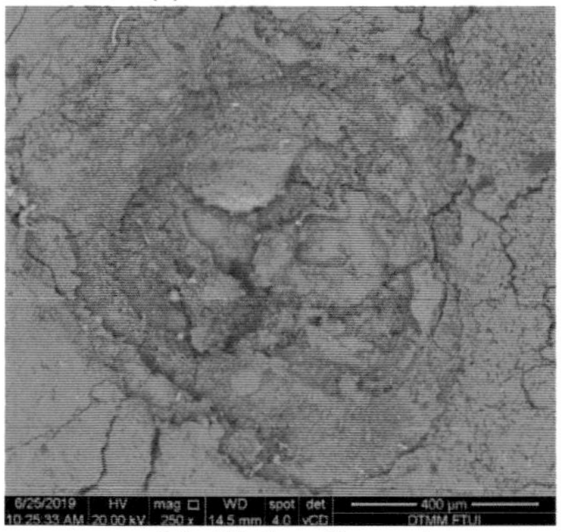

(c) Pitting Corrosion

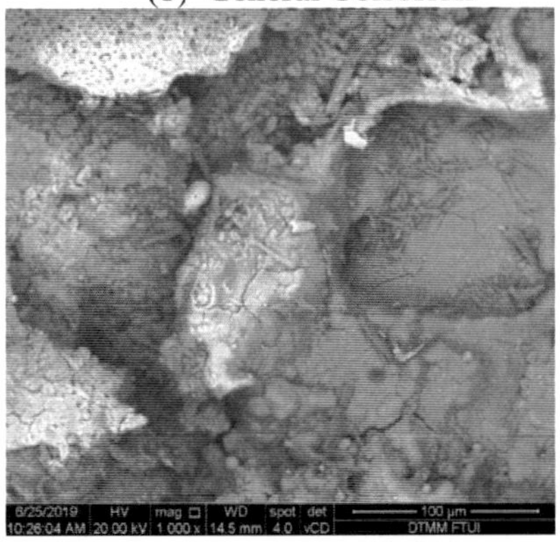

(d) Pitting Corrosion

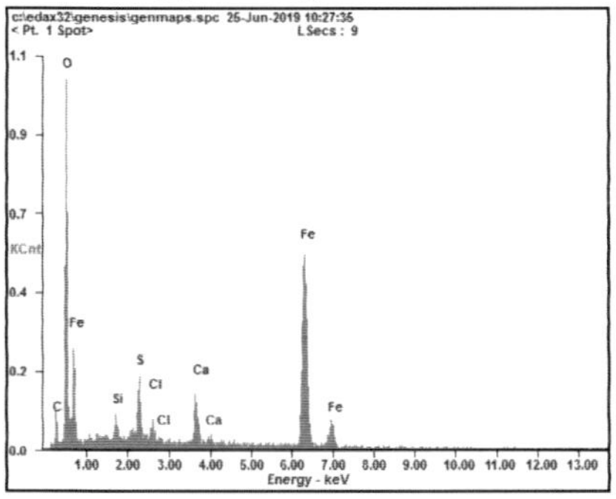

(e) EDX on Pitting Corrosion

Element	*Wt%*	*At%*
CK	06.15	13.70
OK	31.70	53.06
SiK	01.62	01.55
SK	04.19	03.50
ClK	01.44	01.09
CaK	04.11	02.74
FeK	50.79	24.36
Matrix	Correction	ZAF

(f) Pitting Corrosion EDX Result

Figure 2. SEM-EDX Morphologies on surface corrosion low carbon steel material

3. Result and Discussion

3.1. Experimental

Low carbon steel material with dimensions 10mm x 10mm mechanically polished on silicon carbide abrasive paper (500 grit) and the substrate were deposited by tin-zinc coating with different composition of Zn shown in table 2. After electrodeposition, samples were characterized by SEM (Scanning Electron Microscopy), XRD (X-ray Diffraction) and DSC (Differential Scanning Calorimetry). Corrosion

IOP Conf. Series: Materials Science and Engineering **694** (2019) 012029 doi:10.1088/1757-899X/694/1/012029

behaviour was conducted by potentiodynamic polarization at solution 3.5% NaCl to determine the corrosion currents (Icorr) by tafel plot.

Table 2. Sample Sn-xZn weight composition.

Composition (wt%)	Sn	Zn	Total
Sn-0%Zn	25 gram	0 gram	25 gram
Sn-4%Zn	25 gram	1.01 gram	26.01 gram
Sn-9%Zn	25 gram	2.24 gram	27.24 gram
Sn-10.5%Zn	25 gram	2.624 gram	27.62 gram

3.2. XRD Analysis

Sample coating of Sn-xZn on low carbon steel substrate was identify by X-ray diffraction (XRD) with weight composition shown in table 2. Diffraction pattern show peak the hexagonal Zn phase and the tetragonal β-Sn phase on 9% Zn and up. [22] The analysis used to identify the phase and structure crystal surface coating of Sn-Zn shown that no significant Zn peak observed as shown in figure 3.

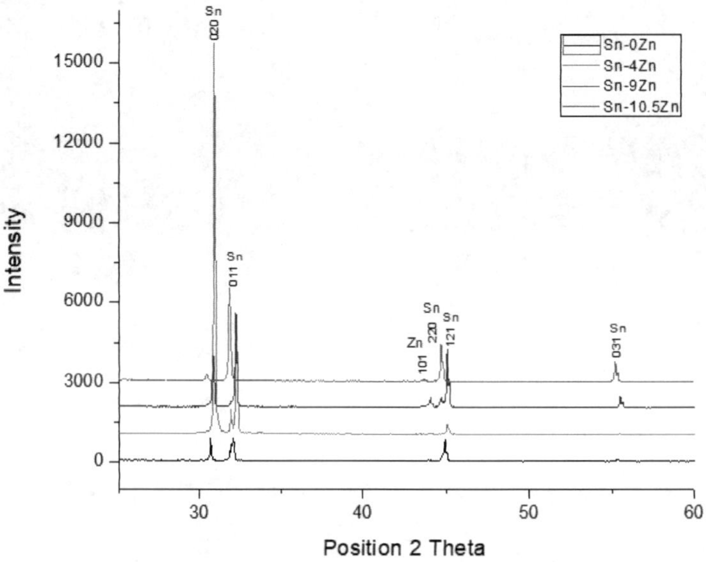

Figure 3. XRD patterns of the Electrodeposition Sn-Zn metallic coating for various Sn-xZn: Sn-0%Zn, Sn-4%Zn, Sn-9%Zn and Sn-10.5%Zn

3.3. SEM – EDS Analysis

Metallographic of Sn-Zn was analysis by Scanning Electron Microscope (SEM) to get the morphology the metal surface coating deposits and corrosion products after polarization test in 3.5 % NaCl are shown in figure 4. respectively various composition of Zn. The microstructure hypoeutectic present of 2 phases is observed; β-Sn phase in figure 4(a) shown like a crack grain in pure Sn. Sn-4%Zn majority of the microstructure shown like a pure Sn and β-Sn plus eutectic was observed in figure 4(b), the Sn-9%Zn alloy shown in figure 4(c) has elliptical morphology with light gray colour and only eutectic phase was observed which describes as fine nedles Zn (black colour). Increasing of Zn in alloy to the eutectic composition shown needles black colour disappear, Zn rich island occurred is observed in figure 4(d) because the phase is at hypereutectic zone. [22]

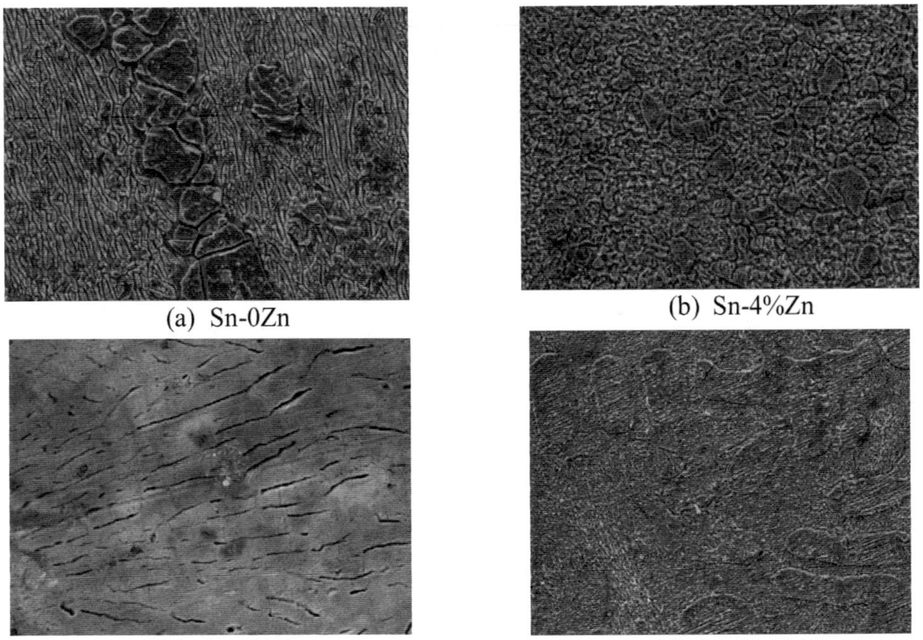

(a) Sn-0Zn (b) Sn-4%Zn

(c) Sn-9%Zn (d) Sn-10.5%Zn

Figure 4. Micrographics (a) Pure Sn-0Zn, (b) Sn-4%Zn, (c) Sn-9%Zn, and (d) Sn-10.5%Zn

3.4. DSC (Different Scanning Calorimetry)

DSC analysis to observed thermal behaviour of Sn-Zn alloy and the result shown in figure 5 where increase of % Zn weight affected decrease melting point or in pure Sn shown the highest melting point up to 230°C. The Sn-4%Zn has a highest enthalphy shown in table 3.

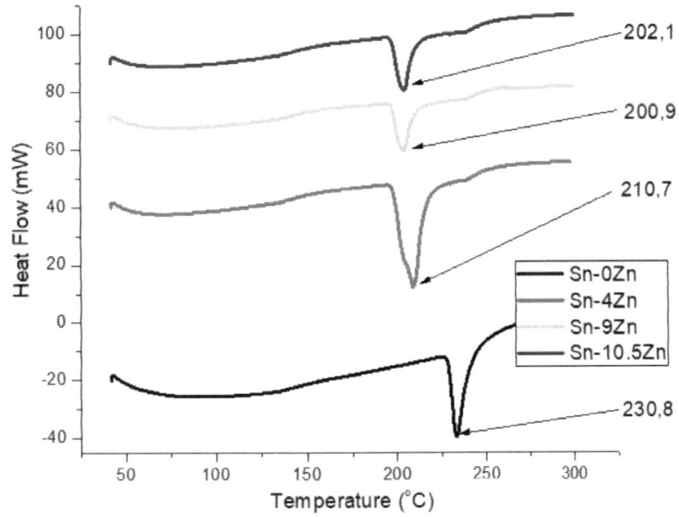

Figure 5. DSC curve for sample Sn and alloy; pure Sn, Sn-4%Zn, Sn-9%Zn and Sn-10.5%Zn

Table 3. Melting Point and enthalpy from the sample.

No	Sample	Sn (wt %)	Zn (wt %)	Different Scanning Calorimetry				
				Melting Point (°C)	Delta H (J/g)	Onset Point	Endset Point	Different
1	Sn-0Zn	100	0	230,8	54,1	229,1	238,8	9,7
2	Sn-4Zn	96	4	210,7	60,7	196,9	211,1	14,2
3	Sn-9Zn	91	9	200,9	59,6	197,0	207,9	10,9
4	Sn-10,5Zn	89,5	10,5	202,1	57,9	195,7	208,2	12,5

3.5. Potentiodynamic Polarization

Linear polarization resistance (LPR) measurement curve are generally used to characterize corrosion behaviour of sample alloy. This potentiodynamic polarization can be used to analysis of the electrochemical, electrical and mass-transport processes in the corrosion mechanism of electrochemical reactions [4]. This electrochemical measurements performed at room temperature, 3.5 wt% NaCl solutions and SCE (Saturated Calomel Electrode) was used as reference.

Table 4 shows, the addition of Zn resulting increase the corrosion rate due to Zn electrochemically more active than Sn [3].

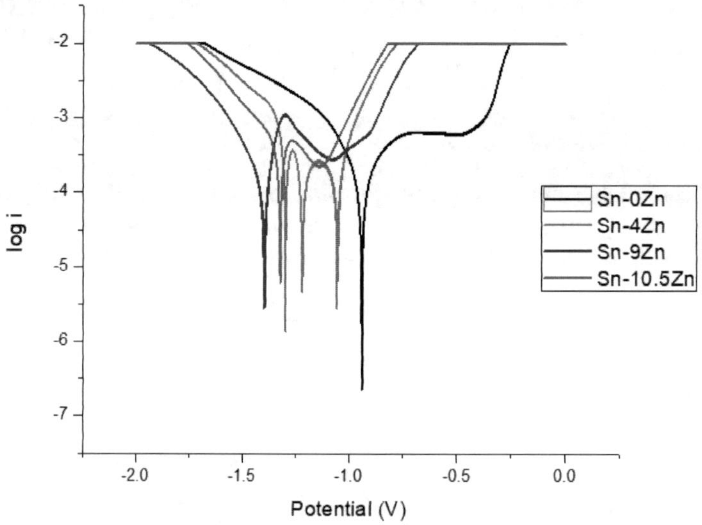

Figure 6. Linear Polarization tafel plot

Table 4. Effect of Zn addition to corrosion rate in 3.5% NaCl Solution at ambient temperature

Sample	E_{Corr} (V)	I_{Corr} (A)	Corrosion Rate (mm/year)
Sn-0Zn	-0,939	1,117E-4	0,0019
Sn-4%Zn	-1,296	5,130E-4	0,0195
Sn-9%Zn	-1,397	1,442E-4	0,0024
Sn-10.5%Zn	-1,320	3,769E-4	0,0038

The cyclic voltammetry observed at Sn-Zn coating various weight (%) composition tested with solution 3.5% NaCl as shown in figure 7 (a) (b) (c) and (d).

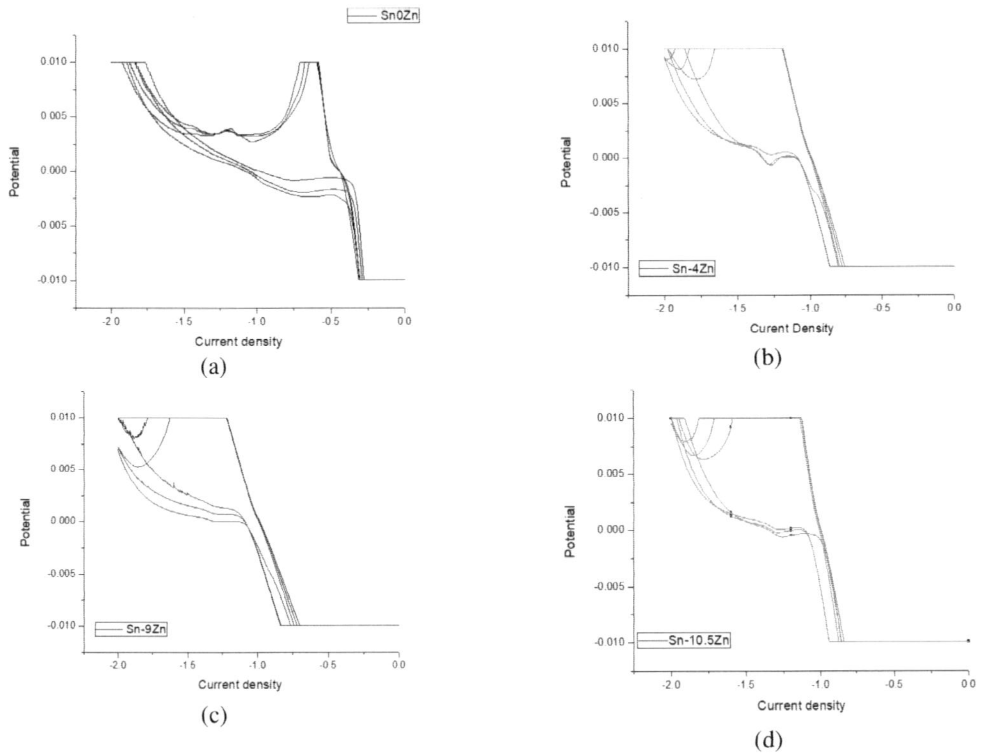

Figure 7. Cyclic Voltammetry Polarization on Sn-Zn in 3.5% NaCl Solution at ambient temperature, scan rate 0.05 V/sec

4. Conclusion

Sn-Zn coated on surface carbon steel substrate by method electroplating or electrodeposition was successfully increase resistance corrosion rate. Potensiodynamic measurement on 3.5% NaCl solution measurement shown the best corrosion rate is Sn0Zn, followed by Sn-9%Zn, Sn-10.5%Zn and Sn-4%Zn.

Coated substrate with various composition of Sn-Zn also increased resistance corrosion rate from 0.325 mm per year based on actual measurement to maximum 0.0195 mm per year.

Each metal has protect steel with a different mechanism and Tin is nobler than steel and under ordinary atmospheric exposure, protects steel by forming corrosion resistant envelop around it [4].

Acknowledgement

The authors would like to show appreciation to the Research of Center of Material Science (RCMS) and CMPFA (Center for Material and Processing Failure Analysis) universitas Indonesia for supporting the research activity for providing data XRD and SEM-EDX. This study also funded by universitas Indonesia through Hibah PITTA B No. NKB-0617/UN2.R3.1/HKP.05.00/2019.

References

[1] O.P. Oladijo, M.H. Mathabatha, A.P.I. Popoola, T.P.Ntsoanec, Characterization and corrosion behaviour of plasma sprayed Zn-Sn alloy coating on mild steel, Surfcae & Coating

Technology. 352 (2018), 654-661.

[2] Z. Ahmad, "CHAPTER 2 - BASIC CONCEPTS IN CORROSION," in Principles of Corrosion Engineering and Corrosion Control, Washington, Butterworth-Heinemann, 2006, pp. 9-56.

[3] S. Dubent, M. Mertens, M. Saurat, Electrodeposition, characterization and corrosion behaviour of tin-20wt% zinc coatings electroplated from a non-cyanide alkaline bath, Mater. Chem. Phys. 120 (2010) 371–380.

[4] Y. Salhi, S. Cherrouf, M. Cherkaoui, K. Abdelouahdi,"Electrodeposition of nanostructured Sn–Zn coatings", Applied Surface Science. 367 (2016) 64-69

[5] Edition, T. (2002) 'CORROSION ENGINEER ' S REFERENCE Third Edition'

[6] Arici, M., Nazir, H. and Aksu, M. L. (2011) 'Investigation of Sn – Zn electrodeposition from acidic bath on EQCM', 509, pp. 1534–1537. doi: 10.1016/j.jallcom.2010.10.161.

[7] Salhi, Y. *et al.* (2016) 'Applied Surface Science Electrodeposition of nanostructured Sn – Zn coatings', 367, pp. 64–69.

[8] Dini, J. W. (1993) *The Materials Science of Coatings and Substrates*..

[9] Ashiru, O. A. and Shirokoff, J. (1996) 'Electrodeposition and characterization of tin-zinc alloy coatings'.

[10] Javidi, M., Chamanfar, R. and Bekhrad, S. (2019) 'Investigation on the efficiency of corrosion inhibitor in CO2 corrosion of carbon steel in the presence of iron carbonate scale', *Journal of Natural gas Sscience and Engineering*, 61(June 2018), pp. 197–205.

[11] Zhou, Y. and Zuo, Y. (2015) 'The Intergranular Corrosion of Mild Steel in CO2 + NaNO2 Solution', *Electrochimica Acta*, 154, pp. 157–165.

[12] Edition, T. (2002) 'CORROSION ENGINEER ' S REFERENCE Third Edition'.

[13] Xu, S. *et al.* (2017) 'Failure analysis of a carbon steel pipeline exposed to wet hydrogen sul fi de environment', *Engineering Failure Analysis*, 71, pp. 1–10.

[14] Abdel Salam, M. A. (2015) *Handbook of Material Failure Analysis With Case Studies from the Oil and Gas Industry*.

[15] Mansoori, H. *et al.* (2017) 'Pitting corrosion failure analysis of a wet gas pipeline', 82(August), pp. 16–25.

[16] Oladijo, O. P. *et al.* (2018) 'Surface & Coatings Technology Characterization and corrosion behaviour of plasma sprayed Zn-Sn alloy coating on mild steel', *Surface & Coating Technology*, 352, pp. 654–661.

[17] Javidi, M. and Bekhrad, S. (2018) 'Failure analysis of a wet gas pipeline due to localised CO 2 corrosion', *Engineering Failure Analysis*. Elsevier, 89(December 2017), pp. 46–56. doi: 10.1016/j.engfailanal.2018.03.006.

[18] Yavas, D. *et al.* (2018) 'Electrochimica Acta Morphology and stress evolution during the initial stages of intergranular corrosion of X70 steel', *Electrochica Acta*, 285, pp. 336–343.

[19] Benamor, A. *et al.* (2018) 'E ff ect of temperature and fl uid speed on the corrosion behavior of carbon steel pipeline in Qatari oil fi eld produced water', 808(December 2017), pp. 218–227. doi: 10.1016/j.jelechem.2017.12.009.

[20] Peng, S. *et al.* (2010) 'Applied Surface Science Morphology and antimony segregation of spangles on batch hot-dip galvanized coatings', *Applied Surface Science*. Elsevier B.V., 256(16), pp. 5015–5020. doi: 10.1016/j.apsusc.2010.03.046.

[21] Wu, K. *et al.* (2019) 'International Journal of Greenhouse Gas Control Understanding the corrosion behavior of carbon steel in amino- functionalized ionic liquids for CO 2 capture assisted by weight loss and electrochemical techniques', *International Journal of Greenhouse Gas Control*, 83(August 2018), pp. 216–227.

[22] Me, C. M., Kociubczyk, A. I. and Ares, A. E. (2018) 'Electrochemical behavior of Sn-Zn alloys with different grain structures in chloride-containing solutions', pp. 1084–1096.

The Effect of Quench Delay on Structure and Corrosion Properties of Aluminum Lithium Alloy 2091

Claudya Ruth Fransisca and Bambang Soegijono

Department of Physics, University of Indonesia, Depok, Indonesia

E-mail: naufal@ui.ac.id

Abstract. The aerospace industry is always trying to improve aircraft performance and reduce aircraft weight. Therefore aluminum lithium alloys are targeted as advanced materials for aerospace technology because of their low density, good corrosion resistance, high toughness and lightweight. This 2091 alloy is an aluminum-lithium alloy which has a composition 2.1 wt% Cu, 2.0 wt% Li dan 0.1 wt% Zr. This alloy has been subjected to quench delay heat treatment to enhance corrosion resistance. The experiments were carried out to determine the relationship between delay quenching and its corrosion properties. Aluminum-lithium 2091 had solutionized at 525°C for 6 hours and then has been quenched in water with a variation of delay quenching (30 second, 60 second, 90 second). Characterization using x-ray diffraction were used to investigate the phase of each sample. The results show different diffraction patterns from one sample to another due to delay quenching. While electrochemical testing using a 3.5% NaCl solution show different corrosion resistance due different quench delay.

1. Introduction

Aluminum-lithium alloy is targeted as advanced materials for aerospace technology because it has a low density, high toughness, good corrosion resistance and lightweight. The aerospace industry is always trying to reduce aircraft weight to save on fuel costs and improve aircraft performance. Al-Li 2091 alloy is one way to save aircraft weight, because every 1% by weight of lithium added to aluminum can reduce density by 3% and increase elastic modulus [1]. The 2091 lithium aluminum alloy has a 7% elastic modulus and the microstructure is determined by the thickness of the material. If the material thickness is > 3.5mm has non recrystallized microstructure, while if < 3.5mm or thin sheet material has an elongated recrystallized grain structure [2].

The mechanical properties of 2091 alloy are affected by the microstructure and the rate of material cooling. If the greater the degree of non recrystallization, the more uniform the exfoliation corrosion. The more fibrous the structure, the higher the stress corrosion cracking threshold value. The melting range for Al-Li alloys is between 560 - 670°C [1]. Aluminum-lithium 2091 had solutionized at 525°C for 6 hours. The purpose of the solutionize procedure is to simply put into solution the maximum amount of alloying elements that are present in that particular alloy. This solutionized process aims to help the process of precipitation hardening where after the heat treatment is continued with quenching.

Quench delay is a heat treatment process where the process is carried out by cooling relatively quickly from austenization temperature but given a cooling pause. Quenching media used is water. The purpose of the quenching process is for the hardening process, which is expected to obtain a large price of violence, strength, and toughness but with minimal residual stress, distortion, and cracking. This experiment uses quench delay to study the effect of cooling rate. If the cooling rate is too fast it will cause distortion, but if the cooling rate is too slow, undesirable grain boundary precipitation will result[3,4]. Therefore, this study wants to investigate the most optimal cooling conditions in order to obtain minimal distortion and no corrosion.

2. Experimental Procedure

2.1. Sample Preparation

The material used in this study is aluminum lithium alloy 2091. The composition of this alloy is shown in table 1. There were 5 different samples in this experiment which is one sample of Al-Li 2091 alloy without quenching and heat treatment, and 4 samples of Al-Li 2091 alloy with heat treatment with different delay quench intervals (0 seconds, 30 seconds, 60 seconds, 90 seconds). Aluminum-lithium 2091 had solutionized at 525°C for 6 hours in a furnace and then have been quenched in water at room temperature with a variation delay quench. Before the heat treatment begins, the surface samples are cleaned using silicon carbide (SiC) paper grit-200. Heat treatment aims to make form a homogeneous solid solution that will eventually strengthen the aluminum-lithium alloy 2091.

All samples were analyzed at the UPP IPD Research Laboratory, Faculty of Mathematics and Natural Sciences, University of Indonesia. This x-ray diffraction aims to see the characterization of each Al-Li 2091 alloy treatment. The result of the XRD show different diffraction pattern from one sample to another due to delay quenching. For XRD analysis will be explained further in the part result and discussion.

Table 1. Chemical Composition of Aluminum Lithium Alloy 2091 (wt%)

Element	Li	Cu	Zr	Mg	Fe	Si	Ti	Al
Weight (%)	1.9	1.85	0.07	1.2	0.07	0.02	0.02	94.72

2.2. Corrosion Test

In the corrosion test data were collected out using potentiodynamic polarization tests to estimate local corrosion of aluminum alloys. Five types of samples were used to evaluate corrosion resistance and the dimensions of samples are 1.5 cm X 0.7 cm. Electrolytes used in corrosion test were 3.5% NaCl solution and 7% NaCl solution. The electrodes used to quantify the current and voltage are called the Working Electrode and the Reference Electrode. Three-electrode system was used to analyze the electrochemical response to the specimens. The three-electrode system was consists of an Ag/AgCl electrode as a reference electrode, the platinum sheet as an anauxiliary electrode and the Al-Li alloy as a working electrode[5].

Electrochemical test aims to investigate the corrosion rate of each sample and determine the effect of quench delay on the corrosion properties of aluminum lithium alloy. After data from Potentiostat is collected, the corrosion rate of each sample can be calculated using Faraday's Law in Determination Corrosion Rates of Metals and Alloys[6].

$$\text{Corrosion rate of alloy, } r = C\frac{Mi}{n\rho} = \Sigma C\frac{MI}{n\rho A} \quad (1)$$

Where C is corrosion rate constant (3.27 x 10^{-3} mm^3/year), M is atomic weight of each element in the alloy (gr/mol), i is current density in (A/cm^2) where the cm^2 defines the cross-sectional area exposed to NaCl, n is number of electrons involved in corrosion process, ρ is mass density of each element in the alloy (gr/cm^3)[7].

3. Results and Discussion

3.1. XRD Analysis

In Figure 1, it can be seen the XRD plot results from five samples in this experiment and the data collected from thc sample characterization results are presented in Table 2. In Figure 1, the diffraction pattern shows five peaks where the first peak has the highest intensity. For untreated Al-Li sample, five peaks were identified with angle positions of 38.024°, 44.137°, 64.880°, 77.763°, and 82.130°. For delay 0 second of quenching, five peaks were identified with angle positions of 37.805°, 44.574°, 64.662°, 77.763°, and 82.130°. For delay 30 second of quenching, five peaks were identified with angle positions of 38.024°, 44.356°, 64.630°, 77.981°, and 82.348°. For delay 60 second of quenching, five peaks were identified with angle positions of 38.242°, 44.356°, 64.662°, 77.981°, and 82.566°. For delay 90 second of quenching, five peaks were identified with angle positions of 37.805°, 44.137°, 64.880°, 78.199 °, and 82.130°.

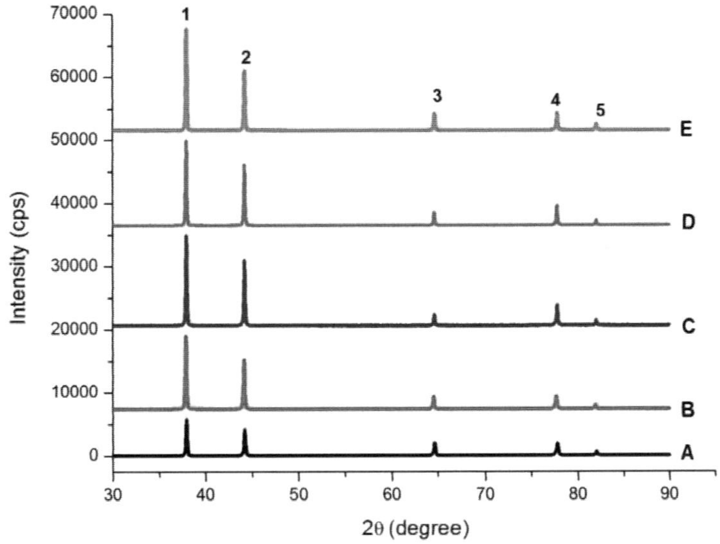

Figure 1. XRD Plot of Aluminum Lithium Alloy 2091

Figure 1 show the XRD plot results from five samples in this experiment. The graph shows that aluminum-lithium alloy 2091 with delay 90 second has the highest peak with the highest intensity. This means that at this peak aluminum content is in maximum condition. Although the diffraction pattern shows the difference of each peak is very small, but there is still a shift in the peak, where the shift in the peak affects the different microstructure of the crystal system that is seen in the difference in microstrain. From the result of XRD, shows that only one phase was detected from all samples, i.e aluminum. This is because the XRD is unable to detect phases below 5 wt%. Table 2, shows all samples have a cubic crystal structure, so the value of lattice parameters are the same for a, b, and c. Where each sample has a different cubic crystal structure (values a, b and c differ between samples). The sample with delay 0 seconds of delay quenching has the largest lattice parameters. The value of lattice parameters decreases if the delay quench is getting longer, this is because thc cooling time affects grain boundary precipitation[4,8]. Different lattice parameter values between samples indicate different crystal structure sizes so that the volume also changes. The volume of unit cell decreases if the delay time of quenching is getting longer.

Table 2. Crystallografic parameters of the aluminum lithium alloy 2091

Sample	Al-Li without heat treatment	Al-Li Delay 0	Al-Li Delay 30	Al-Li Delay 60	Al-Li Delay 90
First peak's height (cps)	486,63	1057,72	1129,24	1163,24	1420,84
First peak's d (Å)	2,3691	2,3722	2,3690	2,3688	2,3668
Lattice Parameter (Å)	a = 4,0454 b = 4,0454 c = 4,0454	a = 4,0479 b = 4,0479 c = 4,0479	a = 4,0431 b = 4,0431 c = 4,0431	a = 4,0430 b = 4,0430 c = 4,0430	a = 4,0429 b = 4,0429 c = 4,0429
Volume (Å^3)	66,208	66,327	66,09509	66,088	66,08422
Density (g/cm^3)	2,71	2,7	2,71	2,71	2,71
Concentration (%)	100	100	100	100	100

3.2. Potentiodynamic Polarization Curves Analysis

Potentiodynamic polarization plots of untreated Al-Li and quenched Al-Li at 3.5% NaCl and 7% NaCl are shown in Figure 2 and Figure 3.

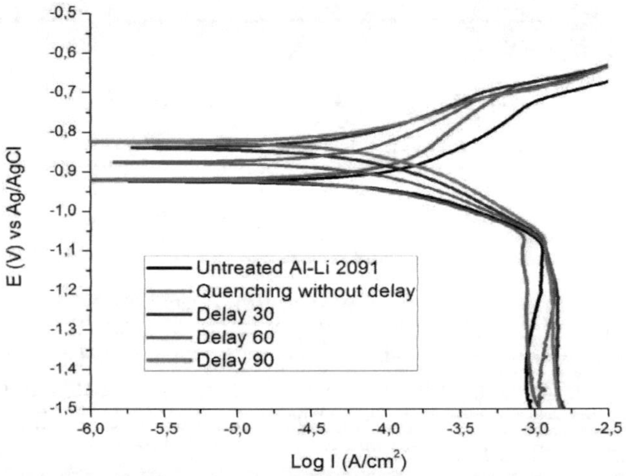

Figure 2. Potentiodynamic polarization plots of the aluminum lithium 2091 in 3.5 wt% NaCl

The results of the electrochemical test are shown in Table 3. Table 3 shows that the corrosion resistance of five Al-Li samples decreases when the chloride concentration results in anodic alloy surface dissolution. On the surface of the sample anodic dissolution occurs where aluminum-lithium is ionized and produces cations that come out into solution and then hydrolyze. Subsequently, the reaction of aluminum alloy with H2O as follows

The anodic reaction is :

$$Al \rightarrow Al^{3+} + 3e^- \quad (2)$$

The cathodic reaction occurs on the alloys surface according to the reaction below :

$$O_2 + 2H_2O + 4e^- \rightarrow 4OH^- \quad (3)$$

Main reaction :

$$2Al^{3+} + 6Cl^- \rightarrow Al_2Cl_6 \quad (4)$$
$$Al_2Cl_6 + 6H_2O \rightarrow Al_2(OH)_6 + 6HCl \quad (5)$$

Table 3. Potentiodynamic polarization result for untreated and quenched Al-Li 2091 in the 3,5 wt% and 7 wt% NaCl solution

Wt% NaCl	Sample	R (ohm)	$E_{corrosion}$ (V)	$I_{corrosion}$ (A)	Corrosion Rate (mm/year)
3.5	Without heat treatment	2.226×10^2	-0.924	1.154×10^{-4}	1.279
	Delay 0	3.144×10^2	-0.920	8.172×10^{-5}	0.9055
	Delay 30	4.019×10^2	-0.876	6.393×10^{-5}	0.7084
	Delay 60	5.136×10^2	-0.840	5.003×10^{-5}	0.5543
	Delay 90	4.131×10^2	-0.825	6.220×10^{-5}	0.6892
7.0	Without heat treatment	5.082×10^1	-1.586	5.055×10^{-4}	5.601
	Delay 0	1.866×10^2	-1.495	1.377×10^{-4}	1.526
	Delay 30	4.025×10^1	-1.577	6.383×10^{-4}	7.072
	Delay 60	6.808×10^1	-1.582	3.725×10^{-4}	4.127
	Delay 90	4.468×10^2	-1.404	5.750×10^{-5}	0.6371

Corrosion rate is influenced by potential corrosion E (V) and corrosion current I (A). Table 3 shows the greatest corrosion rate in aluminum lithium samples without heat treatment, where the sample has the most negative corrosion potential value than the other samples. This means that the more negative corrosion potential will cause low corrosion resistance. This is caused by the release of electrons to metals in anodic reactions which causes excess electrons in the metal to shift the potential of the metal in a more negative direction which results in slowing the anodic reaction and the acceleration of the cathodic reaction.

Table 3 also states the concentration of the solution can affect the rate of corrosion. It is seen that the highest corrosion rate is aluminum-lithium samples with 30 seconds of delay quench in the 7 wt% NaCl. Electrochemical test in 3.5% NaCl solution, samples with 60 seconds of quench delay have the best corrosion resistance, with a corrosion rate of 0.5543 mm/year. While in the electrochemical test in a solution of 7% NaCl, the sample that had the best corrosion resistance was a sample with a 90 second quench delay, with a corrosion rate of 0.6371 mm/ycar. This means that the right time to make aluminum-lithium has good corrosion resistance is around 60-90 seconds.

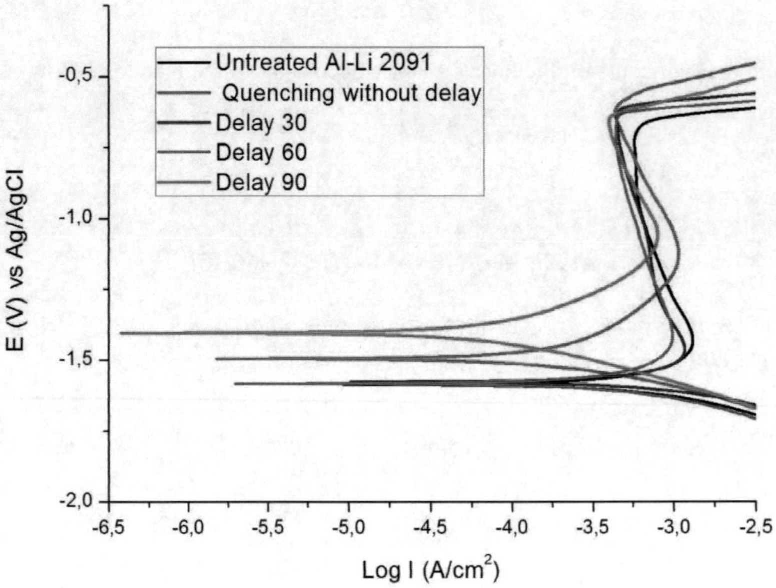

Figure 3. Potentiodynamic polarization plots of the aluminum lithium 2091 in 7 wt% NaCl

4. Conclusion

The corrosion rate is influenced by the time of quench delay, corrosion potential, and concentration of the electrolyte solution. Samples can get better corrosion resistance when subjected to cooling in air for 60-90 seconds and then water quenching process. Different lattice parameter values between samples indicate different crystal structure sizes so that the volume also changes. The volume of unit cell decreases if the delay time of quenching is getting longer.

Acknowledgement

The authors acknowledge with gratitude the financial supports from the Universitas Indonesia under grant Publikasi Internasional Terindeks untuk Tugas Akhir Mahasiswa (PITTA), No. 2239/UN2.R3.1/HKPO5.00/2018.

References

[1] Joshi A, Engineer M and Bombay I I T 1993 LITHIUM ALUMINIUM ALLOYS – The New Generation Aerospace Alloys Junior Research Fellow Commercial Al-Li Alloys Physical Properties of Al-Li alloys Property

[2] Anon 2091 2099 aluminum lithium alloy sheet plate 2091 2099

[3] Chen Z Z, Song C H and Luo W B 2018 Effect of heat treatment on mechanical properties and microstructure of rheo-diecasting aluminum alloy *Zhongguo Youse Jinshu Xuebao/Chinese J. Nonferrous Met.* **28** 518–27

[4] Totten G E 2009 Quenching of Aluminum Alloys : 25–30

[5] Li S, Dong H, Wang X and Liu Z 2019 Quenching sensitivity of Al-Zn-Mg alloy after non-isothermal heat treatment *Materials (Basel).* **12** 1–17

[6] Loto R T, Loto C A and Ohijeagbon I 2018 Effect of heat treatment processes on the localized corrosion resistance of austenitic stainless steel type 301 in chloride/sulphate solution *Results Phys.* **11** 570–6

IOP Conf. Series: Materials Science and Engineering **694** (2019) 012030 doi:10.1088/1757-899X/694/1/012030

[7] Ahmad Z 2006 Faraday ' S Laws of Electrolysis and Its Application in Determining The Corrosion Rate 58–61
[8] Mattos W S, Totten G E and Canale L C F 2017 Uphill Quenching of Aluminum Alloys *Mater. Perform. Charact.* **6** 20160125

IOP Conf. Series: Materials Science and Engineering **694** (2019) 012031 doi:10.1088/1757-899X/694/1/012031

Influence of Mangan on Structure and Corrosion Properties of Cobalt Chrome Molybdenum (Co-Cr-Mo)

Daniel Eka Perkasa, Aliya Qonita and Bambang Soegijono
Department of Physics, Universitas Indonesia, Depok, Indonesia

E-mail: naufal@ui.ac.id

Abstract. Co-Cr-Mo alloy are usually used as implant materials in biomedical application. Corrosion resistance is one of the properties needed to full fill the requirement. The nature of material resistance in the body (biocompatibility) is determined by the corrosion resistance of a biomaterial. Corrosion which emits metal ions in the tissue directed at the implant and activates the system in the body. Mangan (Mn) were added to this alloy to enhance corrosion resistance. Addition of Mangan (Mn) might also affect to the structure of this alloy. X ray diffraction and Electrochemical experiment were conducted to characterized the structure and corrosion resistance of this alloy. The addition of Manganase shows significant influence on the structure and corrosion resistance on the Co-Cr-Mo alloy.

1. Introduction

The Cobalt-Chromium-Molybdenum (Co-Cr-Mo) alloy is one of the metal alloys used in medical implant materials since the industrial revolution at the 19th century. This alloy is used in medical material implants because of its superiority in biocompatibility and mechanical properties[1][2]. Due to biocompatibility of this alloy, Co-Cr-Mo alloy can be used in orthodontic and dental applications. Co-Cr-Mo have two possible crystalline structures: face centered cubic (FCC) and close packed hexagonal (CPH) [3]. Co-Cr-Mo alloy material consisting of Cobalt (Co) alloy as the main constituent element, where this element gives elastic modulus, strength, and hardness [4]. Chromium (Cr) is the second constituent element, which gives corrosion resistance. The element chromium (Cr) in this alloy will form chromium oxide (Cr2O3) as a passive layer [5]. Molybdenum is the third constituent element, this element makes the surface of the grain finer so it increases strength [6] [7]. The problem of using Cobalt-Chromium-Molybdenum alloy (Co-Cr-Mo) in the implant is affected by corrosion and wear resistance. This alloy is considered to have a slow oxidation rate at room temperature, it implies had good corrosion resistance. However, interactions between metal materials used as implants that interact with body fluids can cause tribocorrosion (a form of metal degradation) [3]. The release of metal ions adversely affects body tissues around the implant and disrupts the system in the body. In the study a study was conducted on the effects of variations of manganese (0.85wt%, 3wt%, 5wt%, 7.5wt%) on Co-Cr-Mo on microstructure and corrosion behavior. X-ray diffraction is carried out to determine the material structure of the alloy. Electrochemical behavior using 3.5% NaCl solution, used to determine the corrosion behavior of the material. The addition of manganese (Mn) in the Co-Cr-Mo alloy can affect structural changes in the alloy[8]. Changes in the structure of the alloy can affect corrosion resistance.

2. Experiment Methods and Material

2.1. Sample Preparation
The raw material of Co, Cr, Mo and Mn were provided to fabricate samples with different weight percentages (wt%) of Mn. In Table 1, their composition of alloy can be seen. 4 types of sample were studied, which are called Mn0.85, Mn3, Mn5 and Mn7.5, containing 0.85%, 3%, 5% and 7.5% manganese, respectively. Co-Cr-Mo alloy was cut into 4 cubic shaped pieces with 0.8 cm in length, width, and their thickness.

Table 1. List of compositions in Co-Cr-Mo alloy

Sample	Co	Cr	Mo	Si	Mn
Mn0.85	Base	20.3	6.1	1.1	0.85
Mn3	Base	20.3	6.1	1.1	3
Mn5	Base	20.3	6.1	1.1	5
Mn7.5	Base	20.3	6.1	1.1	7.5

2.2. Corrosion Testing
Corrosion test were performed using Digi-Ivy © DY2300 series potentiostat. The method used as a corrosion fitting phenomenon is linear sweep voltammetry (LSV). The experiments were carried out in an electrochemical cell containing Ag/AgCl as reference electrode and Co-Cr-Mo alloy samples as working electrode (0.5 cm diameter) and platinum counter electrode. Corrosion studies were carried out in 3.5% concentration of NaCl solutions. The corrosion rate of each samples had been calculated after all data were collected from potentiostat. To determine the corrosion rate of each sample, first Faraday's law could be applied in this case [9]. Mathematically, first Faraday's law defined as:

$$\text{Corrosion rate of alloy, } r = \Sigma C \frac{Mi}{n\rho} = \Sigma C \frac{MI}{n\rho A} \tag{1}$$

where C is corrosion rate constant (3.27×10^{-3} mm^3/year was used in this research), M is atomic weight of each element in the alloy (g/mol), i is current density in Ampere/cm^2, I is current passed the samples (Ampere), A is area exposed to NaCl solutions and ethanol solutions (cm^2), n is number of electrons involved in corrosion process, ρ is mass density of each element in the alloy (g/cm^3).

3. Results and Discussion

3.1. XRD Analysis
From XRD data, cobalt Ht and cobalt Eta phases are identified. In Figure 1 (a) there are 7 main peaks that can be identified. In Figure 1 (b) it can be identified, where there are 2 peaks (B and F) which have 2 phases, namely the cobalt Ht phase and the cobalt Eta phase in the same peak and it can be observed that peaks B and F have a dominant cobalt phase with a cubic structure.

There are 9 peaks identified, for sample Mn0.85 identified 9 peaks with angular positions of 40.8629 °, 43.7273 °, 44.5850 °, 46.7972 °, 50.9094 °, 61.9868 °, 74.2559 °, 74.7773 °, 83.5773 °. For sample Mn3, it has 9 peaks with angular positions of 39.9843 °, 43.2681 °, 43.9808 °, 45.8859 °, 50.3942 °, 60.9432 °, 72.6375 °, 74.0465 °, 82.2442 °. For sample Mn5, it has 9 peaks with angular positions 41.0898 °, 43.7098 °, 44.3933 °, 46.9779 °, 50.9195 °, 62.0918 °, 74.8908 °, 74.8943 °, 83.6293 °. for Mn7.5, it has 9 peaks with angular positions 41.0146 °, 43.7650 °, 43.9028 °, 46.7979 °, 50.9914 °, 61.6665 °, 75.0307 °, 74.7726 °, 82.8472 °. In Figure 1, there were 2 identified phases of the first maximum intensity, namely the cobalt Ht (peak B) and cobalt Eta (peak C) phases. In Table 2, the maximum cobalt Ht phase intensity at Mn0.85 is 187,502 cps, at Mn3 the maximum intensity is 303,925 cps, for Mn5 the maximum intensity is 239,778, and 139,629 cps for Mn7.5. Cobalt Eta phase maximum intensity at Mn0.85 is 177.6851 cps, at Mn3 the maximum intensity is 615.8161 cps, at Mn5 the maximum intensity is 41.75745, and 90.53899 for Mn7.5.

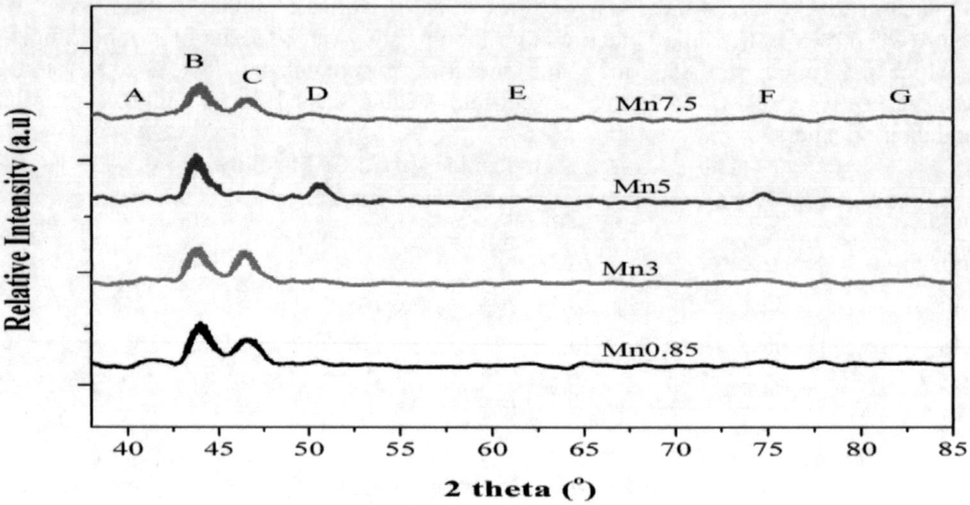

(a)

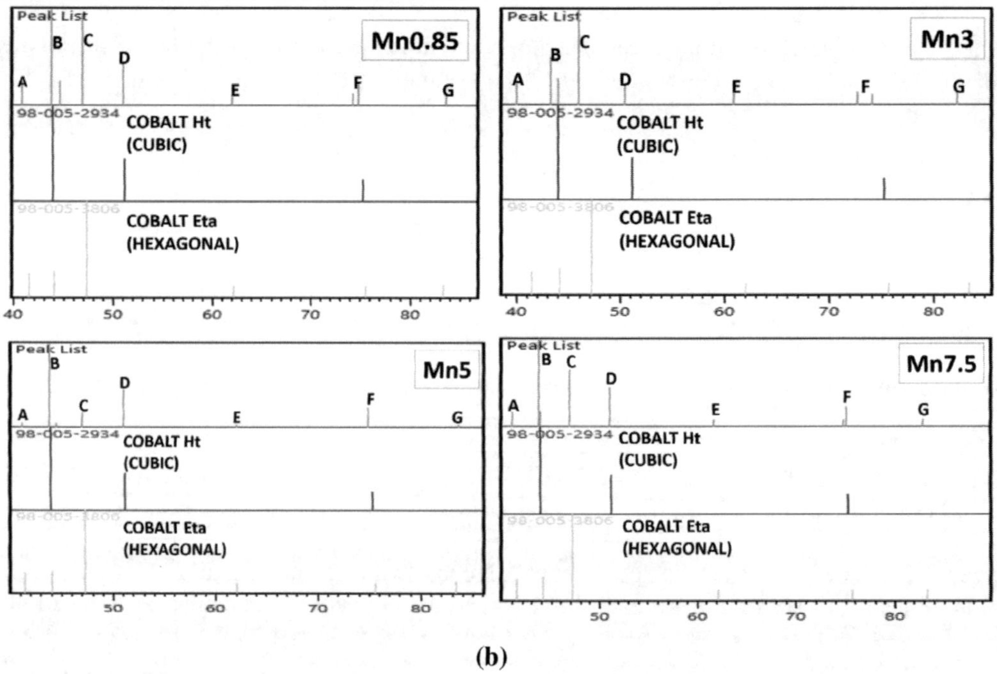

(b)

A = Cobalt Eta [010] B = Cobalt Ht [111] C = Cobalt Eta [011] D = Cobalt Ht [002]

E = Cobalt Eta [012] F = Cobalt Ht [022] G = Cobalt Eta [013]

Figure 1. (a) XRD plot of Co-Cr-Mo in Mn0.85, Mn3, Mn5, Mn7.5. **(b)** Peak list of Co-Cr-Mo on cobalt cubic and cobalt hexagon structure in Mn0.85, Mn3, Mn5, Mn7.5

Table 2. Crystallografic parameters of the Mn0.85, Mn3, Mn5, Mn7.5

Parameter		Sample			
		Mn0.85	**Mn3**	**Mn5**	**Mn7.5**
First peak's Height (cps)	Cobalt HT (Cubic)	187.5020	303.9251	239.7788	139.6297
	Cobalt ETA (Hexagonal)	177.6851	615.8161	41.75745	90.53899
Comparison Peak (cps)		1.0552	0.4935	5.7421	1.5422
First peak's d (Å)	Cobalt HT (Cubic)	2.0684	2.0893	2.0693	2.0668
	Cobalt ETA (Hexagonal)	1.9397	1.9761	1.9327	1.9397
Lattice Parameter (Å)	Cobalt HT (Cubic) a=b=c	3.5940	3.6172	3.5824	3.5756
	Cobalt ETA (Hexagonal) a=b≠c	a, b = 2.5567 c = 4.0739	a, b = 2.6007 c = 4.1130	a, b = 2.5332 c = 4.0761	a, b = 2.5356 c= 4.1163
Volume (Å^3)	Cobalt HT (Cubic)	46.4222	47.3482	45.9745	45.7080
	Cobalt ETA (Hexagonal)	23.0602	24.0917	22.6528	22.9202
Density (g/cm^3)	Cobalt HT (Cubic)	8.43	8.27	86.9	8.56
	Cobalt ETA (Hexagonal)	8.94	8.12	13.1	8.54
Concentration (%)	Cobalt HT (Cubic)	45.4	18.2	79.8	59.8
	Cobalt ETA (Hexagonal)	54.6	81.9	20.2	40.2

Based on the crystal structure, the cobalt Ht phase has a cubic shaped crystal structure with the same lattice parameters in a, b, and c. The size of the crystalline volume of the cobalt Ht phase changes for each of the largest samples is Mn3, Mn0.85, Mn5 and Mn7.5. Whereas in the cobalt Eta phase has a hexagonal-shaped crystal structure with a lattice parameter a equal to b, and not equal to c. the size of the crystalline volume of the cobalt Eta phase changes from the largest for each sample from the largest Mn3, Mn0.85, Mn7.5 and Mn5.

In Figure 1. it can be observed that the cobalt Ht and cobalt Eta phases the maximum intensity shown at peak B and C shows the change in peak marked by an increase in peak intensity. Because the intensity of the X-ray diffraction pattern is directly proportional to the concentration of the component that produces it [10], it can be seen that there is a change in the concentration of cobalt Ht and cobalt Eta in the sample. In Table 2, it can be observed at the first maximum intensity the highest concentration of the cobalt Ht phase of the Cubic structure is found in Mn3, then Mn5, Mn0.85, and

Mn7.5. For the cobalt Eta phase, the highest concentration of Hexagonal structure was found in Mn3, Mn0.85, Mn7.5, and Mn5. The comparison of concentration cobalt Ht (cubic) to cobalt Eta (hexagonal), the greatest concentration of Cobalt Ht is found in Mn5, Mn7.5, Mn0.85, Mn3. This explains that the addition of manganese to co-cr-mo alloy affects the cobalt concentration of cubic and hexagonal structures [8], where the largest cobalt concentration of cubic structure is found in Mn5 with manganese composition of 5wt%.

3.2. Potentiodynamic Polarization Curve Analysis

Corrosion mechanism in Co-Cr-Mo was analyzed using electrochemical behavior in a 3.5% NaCl solution. In Figure 2. Tafel graph is presented stating the relationship between anodic reaction and cathodic reaction. The potential at which the rate of anodic reaction is proportional to the rate of the cathodic reaction is called E corrosion [3]. In presenting the data presented in table 3 which contains the values of E (V) corrosion, I (A) corrosion, and corrosion rate (mm / year). Some of the reactions responsible for corrosion in Co-Cr-Mo alloy at NaCl electrolytes are shown below[11]

Anodic reaction : $Co \rightarrow Co^{2+} + 2e$ (2)

$$Co^{2+} + 2H_2O \rightarrow CoOH^+ + H_3O^+ \quad (3)$$

Formation of $CoOH^+$ is mainly responsible for the sudden increase in current due to the dissolution of Co metal. The electrons produced in the anodic reaction will be used in the cathodic reaction in the reduction of oxygen in solution.

Cathodic reaction : $O_2 + 2H_2O + 4e^- \rightarrow 4OH^-$ (4)

Main reaction : $Co^{2+} + 2Cl^- \rightarrow CoCl_2$ (5)

$$CoCl_2 + 2H_2O \rightarrow Co(OH)_2 + 2HCl \quad (6)$$

Table 3. Data from the corrosion test of the Mn0.85, Mn3, Mn5, Mn7.5 in the 3,5 wt% NaCl solution

Wt % NaCl	Sample	R (ohm)	$E_{corrosion}$ (V)	$I_{corrosion}$ (A)	Corrosion Rate (mm/year)
3,5 wt%	Mn0.85	9.040 x 10^1	-0.058	2.842 x 10^{-4}	0.333
	Mn3	8.216 x 10^1	-0.067	3.127 x 10^{-4}	0.366
	Mn5	2,242 x 10^3	-0.282	1.146 x 10^{-5}	0.013
	Mn7.5	2.726 x 10^1	-0.142	9.427 x 10^{-4}	1.105

Formation of $Co(OH)_2$ increases the pH of the electrolyte, which induces further corrosion process. The corrosion rate is influenced by Potential E (V) corrosion and I (A) corrosion. The table states the greatest corrosion rate if the potential I (A) corrosion is getting negative, the fastest corrosion rate is the Mn7.5, Mn3, Mn0.85, Mn5 samples. The late corrosion rate is at a potential that has a more negative potential value of E (V) and I (A), namely in sample Mn5 with a corrosion rate of 0.013 mm/ year.

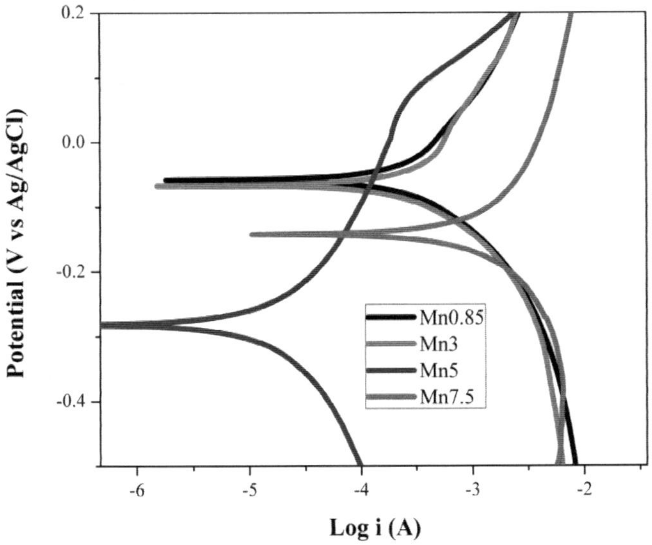

Figure 2. Potentiodynamic polarization curves of the samples in 3,5 wt% NaCl.

This is because the anodic reaction releases electrons to the metal so that excess electrons in the metal shift the metal potential towards more negatively which results in the anodic reaction being slow and accelerating the cathodic reaction [3].

4. Conclusion

Variation of manganese in the Co-Cr-Mo alloy influenced the phase composition, cobalt Ht (cubic) and cobalt Eta (hexagon), and the corrosion rate. The most corrosion resistance sample is Mn5 with 5% manganese composition, with the highest phase of cobalt cubic structure in the alloy.

Acknowledgement

This research fully funded by Universitas Indonesia through Hibah PITTA B No. NKB-0616/UN2.R3.1/HKP.05.00/2019

References

[1] Manam N S, Harun W S W, Shri D N A, Ghani S A C, Kurniawan T, Ismail M H and Ibrahim M H I 2017 Study of corrosion in biocompatible metals for implants: A review *J. Alloys Compd.* **701** 698–715

[2] Asri R I M, Harun W S W, Samykano M, Lah N A C, Ghani S A C, Tarlochan F and Raza M R 2017 Corrosion and surface modification on biocompatible metals: A review *Mater. Sci. Eng. C* **77** 1261–74

[3] Bellfontaine G 2010 The Corrosion of CoCrMo Alloys for Biomedical Applications *Thesis* 1–88

[4] Aherwar A, Singh A and Patnaik A 2018 Study on mechanical and wear characterization of novel Co30Cr4Mo biomedical alloy with added nickel under dry and wet sliding conditions using Taguchi approach *Proc. Inst. Mech. Eng. Part L J. Mater. Des. Appl.* **232** 535–54

[5] Pound B G 2014 Passive films on metallic biomaterials under simulated physiological conditions *J. Biomed. Mater. Res. Part A* **102** 1595–604

[6] Aherwar A, Singh A and Patnaik A 2018 A study on mechanical behavior and wear

performance of a metal – metal Co – 30Cr biomedical alloy with different molybdenum addition and optimized using Taguchi experimental design A study on mechanical behavior and wear performance of a metal – metal Co – 30Cr biomedical alloy with different molybdenum addition and optimized using Taguchi experimental design *J. Brazilian Soc. Mech. Sci. Eng.*

[7] Köse N 2016 Biological response to orthopedic implants and biomaterials, in: Musculoskeletal Research and Basic Science *Musculoskelet. Res. Basic Sci.* 3–14

[8] Wu P, Peng Z, Liu N, Niu M, Zhu Z and Wang X 2016 The effect of Mn content on the microstructure and properties of CoCrCu0.1Fe0.15Mo1.5MnxNi near equiatomic alloys *Mater. Trans.* **57** 5–8

[9] Ahmad Z 2016 Application of Faraday's Laws In Determination of Corrosion Rates of Metals and Alloys," in Principle of Corrosion Engineering and Corrosion Control *Butterworth-Heinemann, United Kingdom* **3** 58–61

[10] Norrish K 1962 Quantitative Analysis by X-ray Diffraction *Clay Miner.* **5** 98–109

[11] Drahansky M, Paridah M ., Moradbak A, Mohamed A ., Owolabi F abdulwahab taiwo, Asniza M and Abdul Khalid S H . 2016 Investigation of Electrochemical Pitting Corrosion by Linear Sweep Voltammetry: A Fast and Robust Approach *Intech* **i** 13

IOP Conf. Series: Materials Science and Engineering **694** (2019) 012032 doi:10.1088/1757-899X/694/1/012032

Influence of Heat Treatment on Structure and Corrosion Resistance of 8090 Aluminium Alloy for Ethanol Fuel Tank Application

Garimahdi Astaghfari and Bambang Soegijono

Department of Physics, Faculty of Mathematics and Natural Sciences, University of Indonesia

E-mail: naufal@ui.ac.id

Abstract Investigation was carried out to understand the influence of heat treatment on structure and corrosion resistance of 8090 aluminium alloy in ethanol solution for ethanol fuel tank application. Heat treatments were subjected to solutionizing at temperature 525°C for 4 hours to infiltrate the aluminium crystal by copper atoms and aging at temperature 190°C with different time to bring copper and lithium atoms out of aluminium crystal during aging process. X-ray diffraction was performed to investigate the phase of each sample with different aging time. Electrochemical experiments were conducted to investigate the corrosion resistance in 3,5% and 7% NaCl solution and 70% and 96% of ethanol solution. The results show different corrosion behaviour of this material. Heat treatment on this material shows significant different results.

1. Introduction

The demand of petroleum-based fuel for vehicles use keeps on rising each year and that makes the supply of petroleum-based fuel is decreasing time after time. The alternative energy is needed to replace petroleum-based fuel and also it is needed to reduce carbon emission caused by petroleum-based fuel in the future [1–3]. One of the alternative energies had been developed for vehicle use is ethanol-based fuel.

The problem of using ethanol as fuel for commercial vehicles lies on corrosion of each vehicle component. Ethanol-based fuel tends to attract water from air because of hygroscopic effect of ethanol, causing water contamination in ethanol-based fuel and the contamination can causes corrosion on each vehicle component, such as fuel tank [4,5]. 8090 aluminium alloy (Al–94.61wt%, Si–2.42wt%, Cu–1.14wt%, Mg–0.83wt%, Mn–0.70wt%, Zr–0.01wt%, Ti–0.20wt%, Zn–0.02%, Fe–0.05wt%, and Si–0.02wt%.) is used in this research to investigate the corrosion resistance in ethanol-based fuel and take advantage from its lightweight property with small value of mass density and also this type of alloy has high Young's modulus [6]. The output of this research is being able to develop fuel tank that is able to resist corrosion caused by ethanol-based fuel and also being able to develop a lightweight ethanol fuel tank for commercial vehicles.

In general, almost all type of aluminium alloys excel in mechanical properties, such as high strength and good corrosion resistance [7,8]. 8090 aluminium alloy has several mechanical disadvantages. Main cause of 8090 aluminium alloy's mechanical properties disadvantages in corrosion resistance is different applied thermomechanical treatment to reach another mechanical property advantage besides corrosion resistance [9]. Due to low content of magnesium and zirconium, which both have good corrosion resistance if the alloy has sufficient weight percent inside, 8090 aluminium alloy has poor corrosion resistance. Also, 8090 aluminium alloy has several disadvantages in mechanical properties such as poor toughness and high anisotropy [10].

In this research, each 8090 aluminium alloys with 1.7 cm × 1.5 cm surface area is heat treated differently. 5 different heat treatments conducted in this research, solutionizing and aging process. Solutionizing was done for copper and lithium to infiltrate the aluminium crystal in 8090 aluminium alloy. Only one sample is treated in this way. Solutionizing and aging process also conducted to bring out copper and lithium slowly from the aluminium crystal during aging process occurred. Variations of aging time for each sample that are conducted in this research are 24 hours, 48 hours, 72 hours, and 96 hours. Different heat treatment process will determine the corrosion behaviour for each 8090 aluminium alloy sample. Corrosion behaviour difference for each sample can be seen on the corrosion rate and shift of phase in XRD diffraction graph. Content of water in NaCl and ethanol solution determines corrosion rate value (observed in mm/year). Corrosion rate determines how much thickness loss on a material in a certain environment. Metals tend to have faster corrosion in an environment with high water content because metal alloys are easily reactive to oxygen.

2. Experiment Methods and Material

2.1. Sample Preparation

8090 aluminium alloy was selected in this research. This type of aluminium alloy was selected due to lightweight property with mass density of 2.54 gr/cm^3 and barely good corrosion resistance, which is good for application of lightweight fuel tank for commercial vehicles. In Table 1, composition of 8090 aluminium alloy can be seen. 8090 aluminium alloy was cut into 5 square-shaped pieces with 1.7cm x 1.5cm dimension and each piece of sample was heat treated differently. Each aluminium alloy surface was ground with 500 and 2000 grit SiC paper to clean the surface before heat treatment began. First aluminium alloy sample is only solutionized, the second one was gone through solutionizing and aging process for 24 hours, the third one was gone through solutionizing process and aging process for 48 hours, the fourth one was gone through solutionizing process and aging process for 72 hours, and the final one was gone through solutionizing process and aging process for 96 hours. Solutionizing process took 4 hours in furnace with temperature of 525°C and then all 8090 aluminium alloy samples were quenched in water for 5 hours.

2.2. Corrosion Testing

There were two corrosion tests conducted in this research, dipping test in 96% ethanol solution for two weeks and electrochemical test using Digi-Ivy© DY2300 series potentiostat. Electrolytes used in corrosion test were 3.5% NaCl solution, 7% NaCl solution, 70% ethanol solution, and 96% ethanol solution. Two weeks of dipping test using 96% ethanol solution was conducted to investigate changes in mass of each sample with different heat treatment. For electrochemical test, 1cm × 1cm surface area for each 8090 aluminium alloy sample was used.

Electrochemical test was also conducted to investigate different heat treatment effect on corrosion rate of 8090 aluminium alloy samples. To determine the corrosion rate of each sample, first Faraday's law could be applied in this case [11]. Mathematically, first Faraday's law defined as [11]:

$$\text{Corrosion rate of alloy, } r = \Sigma C \frac{Mi}{n\rho} = \Sigma C \frac{MI}{n\rho A} \quad (1)$$

where C is corrosion rate constant (3.27 × 10^{-3} mm^3/year was used in this research), M is atomic weight of each element in the alloy (g/mol), i is current density in Ampere/cm^2, I is current passed the samples (Ampere), A is area exposed to NaCl solutions and ethanol solutions (cm^2), n is number of electrons involved in corrosion process, ρ is mass density of each element in the alloy (g/cm^3). Faraday's law corrosion rate equation is in sums form because 8090 aluminium alloy consists of 10 elements (aluminium, lithium, copper, magnesium, manganese, zirconium, titanium, zinc, iron, and silicon).

Table 1. Compositions of 8090 aluminium alloy

Elements	Composition (% wt)
Al	94.61
Li	2.42
Cu	1.14
Mg	0.83
Mn	0.70
Zr	0.01
Ti	0.20
Zn	0.02
Fe	0.05
Si	0.02

Table 2. Electrochemical test results of 8090 aluminium alloys in different solutions

Solution	Treatment	E_{corr} (V)	I_{corr} (A)	r (mm/year)
3.5% NaCl	24 hours of aging	-0.488	4.044×10^{-6}	4.735×10^{-2}
	48 hours of aging	-0.401	1.154×10^{-5}	1.351×10^{-1}
	72 hours of aging	-0.722	8.888×10^{-5}	1.041
	96 hours of aging	-0.507	5.285×10^{-5}	6.188×10^{-1}
	Without aging	-0.673	1.078×10^{-4}	1.262
7% NaCl	24 hours of aging	-1.231	1.796×10^{-5}	2.103×10^{-1}
	48 hours of aging	-1.477	1.295×10^{-4}	1.516
	72 hours of aging	-1.433	4.465×10^{-5}	5.227×10^{-1}
	96 hours of aging	-1.49	6.629×10^{-5}	7.761×10^{-1}
	Without aging	-0.786	1.980×10^{-5}	2.318×10^{-1}
70% Ethanol	24 hours of aging	-0.341	1.604×10^{-6}	1.878×10^{-2}
	48 hours of aging	-0.471	3.514×10^{-6}	4.114×10^{-2}
	72 hours of aging	-0.681	2.261×10^{-6}	2.647×10^{-2}
	96 hours of aging	-0.893	1.62×10^{-6}	1.90×10^{-2}
	Without aging	-0.295	1.41×10^{-6}	1.66×10^{-2}
96% Ethanol	24 hours of aging	-0.25	1.206×10^{-6}	1.475×10^{-2}
	48 hours of aging	-0.411	2.072×10^{-6}	2.426×10^{-2}
	72 hours of aging	-0.58	2.460×10^{-6}	2.880×10^{-2}
	96 hours of aging	-0.645	1.548×10^{-6}	1.812×10^{-2}
	Without aging	-0.204	1.203×10^{-6}	1.408×10^{-2}

3. Results and Discussion

3.1. Corrosion Testing: Dipping Test

The results of dipping test in 96% ethanol solution can be seen on Table 2. 14 days dipping test for each sample resulted in mass reduction. For 24 hours of aging time, there is 0.002 grams mass loss between before and after dipping test. For 48 hours of aging time, there is 0.0038 grams mass loss. For 72 hours of aging time, there is 0.0029 grams mass loss. For 96 hours of aging time, there is 0.0038 grams mass loss. And for 8090 aluminium alloy sample without aging treatment, there 0.0037 grams mass loss. 8090 aluminium alloy with 24 hours of aging time has least mass loss compared to the rest of samples of 8090 aluminium alloy. It was assumed that dipping test in 96% of ethanol solution resulted in mass loss because of oxidation. Oxidation that occurred during this test attacked almost all surface of 8090 aluminium alloy caused some elements in the alloy to dissolve, resulted in mass loss of the alloy. The mass loss is not significant, but the corrosion attacked the structure of 8090 aluminium alloy and it will decrease another mechanical property advantage that the alloy used to have [9].

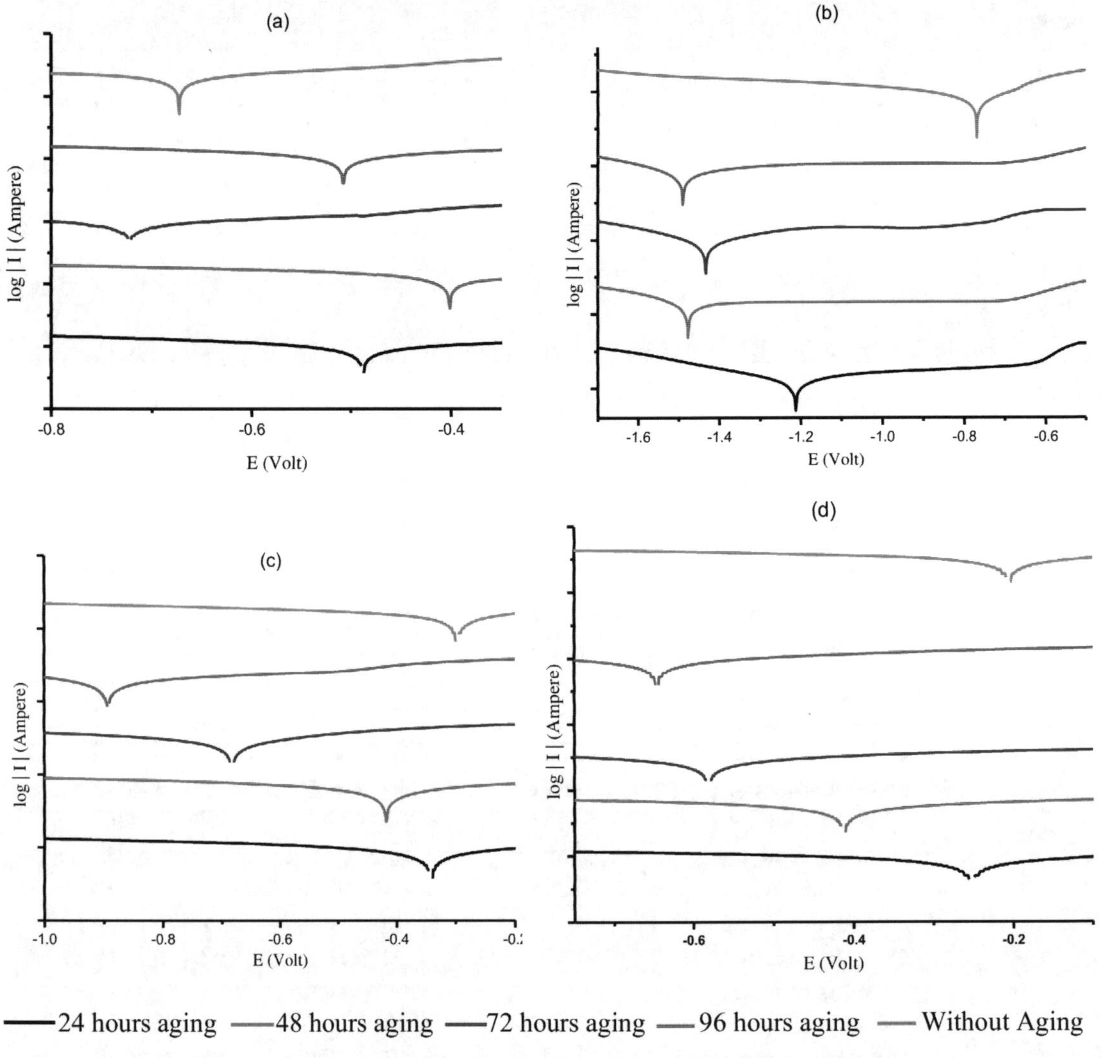

Figure 1. Tafel plot of electric potential *E* vs current *I* in logarithm form in (a) 3.5% NaCl solution, (b) in 7% NaCl solution, (c) in 70% ethanol solution, and (d) in 96% ethanol solution

3.2. Corrosion Testing: Electrochemical Test

The electrochemical behaviour of 8090 aluminium alloy in ethanol solution can be observed from the chemical reaction formula. When ethanol reacts with oxygen, the reaction will produce acetic acid and water. The water and acetic acid content in oxidized ethanol solution are the oxidation agent and also the main cause of the aluminium alloy oxidation. The reaction of ethanol oxidation can be written as:

$$C_2H_5OH + O_2 \rightarrow CH_3COOH + H_2O \qquad (2)$$

The aluminium alloy will react to acetic acid and water. When the aluminium alloy reacts to acetic acid, product of the reaction is aluminium acetate and hydrogen in form of gas. The reaction will cause corrosion of aluminium alloy [12].

$$Al + 3CH_3COOH \rightarrow Al(CH_3COO)_3 + \frac{3}{2}H_2 \qquad (3)$$

Water content also contributes to aqueous corrosion of 8090 aluminium alloy. This type of corrosion will there is enough water content in the ethanol to cause phase separation [1]. The corrosion will create bayerite from reaction of aluminium ion with water or boehmite from the same reaction, which those two products are protective hydrous oxide surface film [12].

$$Al^{3+} + 3H_2O + 3e^- \rightarrow Al(OH)_3 + \frac{3}{2}H_2 \qquad (4)$$

$$Al^{3+} + 2H_2O + 3e^- \rightarrow AlOOH + \frac{3}{2}H_2 \qquad (5)$$

The results of electrochemical test in 3.5% NaCl solution, 7% NaCl solution, 70% ethanol solution, and 96% ethanol solution can be seen on Figure 1 and the detail of electrochemical test is in Table 2. Electrochemical potential or corrosion potential of 8090 aluminium alloy samples are obviously different in different corrosive environment. Electrochemical test in 3.5% NaCl solution makes the 8090 aluminium alloy sample with 48 hours aging time treatment has least negative electrochemical potential with value of -0.401V. The alloy without aging treatment in 7% NaCl solution has least negative potential with value of -0.786V. The samples without aging process also has least negative potential in 70% ethanol solution and 96% ethanol solution, with value of -0.295V and -0.204V. It can be concluded that corrosion resistance of 8090 aluminium alloy is decreasing due to increment of aging time and the sample with most negative potential value defines the sample is prone to corrode than the others with bigger potential value, which is less prone to corrode in ethanol fuel.

Difference in electrochemical potential for each sample will leads to different corrosion rate. With using mathematical formula from Faraday's law, corrosion rate for each sample can be calculated. In 3.5% NaCl solution, the sample with 24 hours of aging time has the slowest corrosion rate with value of 4.735×10^{-2} mm/year. In 7% NaCl solution, the sample with 24 hours of aging time has the slowest corrosion rate with value of 2.103×10^{-1} mm/year. In 70% and 96% ethanol solution, the sample without aging process is the slowest with value of 1.66×10^{-2} mm/year and 1.408×10^{-2} mm/year. From all electrochemical test that has been done, it can be concluded that water content makes corrosion rate of the alloy faster. Not just water that has big contribution for corrosion of the alloy, acidic substances and solubility of oxygen in the solutions also enhance corrosion rate of the alloys. Acidic substances and solubility of oxygen increase when the concentration of ethanol has small value [13]. It is proven in the experiment that corrosion rate of the alloy in 70% ethanol solution is faster than the alloy in 96% ethanol solution.

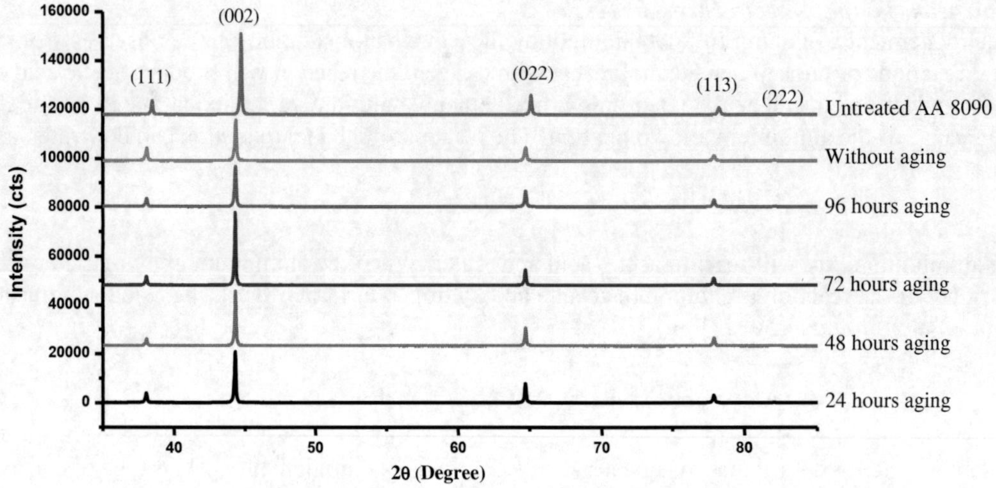

Figure 2. XRD plot of treated and untreated 8090 aluminium alloy samples

3.3. XRD Analysis

The XRD data plot of treated and untreated samples is showed in Figure 2 and the detail of the XRD data can be seen in Table 3. Each sample, treated and untreated 8090 aluminium alloy sample, had 5 peaks identified at (111), (002), (022), (113), and (222). The peaks of 24 hours aging time sample are detected at 38.03°, 44.29°, 64.68°, 77.83°, and at 82.04° angle position. The peaks of 48 hours aging time sample are detected at 38.06°, 44.31°, 64.70°, 77.84°, and at 82.06° angle position. The peaks of 72 hours aging time are detected at 38.04°, 44.29°, 64.68°, 77.83°, and 82.04° angle position. The peaks of 96 hours aging time sample are detected at 38.06°, 44.31°, 64.70°, 77.84°, and at 82.05° angle position. The peaks of sample without aging treatment are detected at 38.08°, 44.33°, 64.72°, 77.86°, and at 82.08° angle position. The peaks of untreated sample are detected at 38.46°, 44.71°, 65.09°, 78.22°, and at 82.43°.

Table 3. Crystallographic parameters of treated and untreated 8090 aluminium alloy samples

Sample	24 hours aging time	48 hours aging time	72 hours aging time	96 hours aging time	Without aging	Untreated AA 8090
Highest peak height (cts)	18765	23846	27884.06	15428.39	15659.75	33346
d-spacing (Å)	2.04345	2.04248	2.04312	2.04239	2.04158	2.02510
Density (g/cm^3)			2.7			
Volume (Å^3)	66.26342	66.28945	66.29299	66.30686	66.29393	66.38882
Lattice Parameter	a = b = c = 4.0466	a = b = c = 4.0471	a = b = c = 4.0472	a = b = c = 4.0474	a = b = c = 4.0472	a = b = c = 4.0491
FWHM (°)	0.1434	0.09327	0.11925	0.16022	0.14195	0.12065
Crystallite size (Å)	1086.158	1202.718	1585.86	1633.425	974.8201	2185.19

From the XRD data, the phase that is detected in all samples are only aluminium phase. The aluminium content in the samples have cubic crystal structure because lattice parameter of the unit cell has same value in a, b, and c. Even each sample has cubic crystal structure, the value of a, b, and c is different to each other. The longer aging duration makes the volume of the unit cell is increasing and unit cell volume will shrink if the alloy sample is going through solutionizing treatment only.

Solutionizing and aging treatment reduces height of the highest peak, which means that aluminium content on a certain grain orientation is reducing compared to the untreated 8090 aluminium alloy. The highest aluminium content when solutionizing and aging treatment conducted is the alloy with 72 hours aging time. From the highest peak height data, it can be concluded that the longer aging time does not increase the aluminium content and the alloy will have maximum aluminium content when aging time is reaching 72 hours.

Solutionizing and aging treatment also affect the grain size of 8090 aluminium alloy. When the alloy is only solutionized, the grain size is shrinking from 2185.19 Å to 974.8201 Å. If the sample is aged in 190°C for 24 to 96 hours, the grain size will increase due to duration of aging time from 1086.158 Å to 1633.425 Å. The XRD data of 8090 aluminium alloy defines that solutionizing and aging process will decrease and increase 8090 aluminium alloy's grain size, all grain size of the samples will shrink when solutionizing process is done and the grain size will increase slowly when the sample is going through aging process for a certain period of time.

4. Conclusion

Corrosion rate varies with solutionizing and aging treatment of 8090 aluminium alloy. In ethanol solutions, 8090 aluminium alloy sample without aging process has the slowest corrosion rate and also the alloy without aging process also has good corrosion resistance with the highest corrosion potential. From XRD data, aluminium phase in the alloy has cubic crystal structure because value of lattice parameter is equal between a, b, and c. solutionizing and aging treatment will affect aluminium content in certain grain orientation and maximum aluminium content in alloy will reach its maximum at 72 hours aging time. Grain size will be affected as well, long aging time will increase the grain size of 8090 aluminium alloy after grain size shrinkage in solutionizing process.

Acknowledgement

We acknowledge financial support provided by HIBAH PITTA B 2019 (NKB-0616/UN2.R3.1/HKP.05.00/2019). undergraduate Program of Physics, Department of Physics, Faculty of Mathematics and Natural Science (FMIPA) Universitas Indonesia.

References

[1] Krüger L, Tuchscheerer F, Mandel M, Müller S and Liebsch S 2012 Corrosion behaviour of aluminium alloys in ethanol fuels *J. Mater. Sci.* **47** 2798–806

[2] Chansauria P and Mandloi R K 2018 Effects of Ethanol Blends on Performance of Spark Ignition Engine-A Review *Mater. Today Proc.* **5** 4066–77

[3] Luciano M A, Castro M das M R de and Lins V de F C 2014 Corrosion Resistance of Organometallic Coating Aplicated in Fuel Tanks Using Electrochemical Impedance Spectroscopy in Biofuel Part I *Tecnol. em Metal. Mater. e Mineração* **11** 244–50

[4] Baena L M, Gómez M and Calderón J A 2012 Aggressiveness of a 20% bioethanol-80% gasoline mixture on autoparts: I behavior of metallic materials and evaluation of their electrochemical properties *Fuel* **95** 320–8

[5] Thomson J K, Pawel S J and Wilson D F 2013 Susceptibility of aluminum alloys to corrosion in simulated fuel blends containing ethanol *Fuel* **111** 592–7

[6] De Damborenea J J and Conde A 2000 Intergranular corrosion of 8090 Al-Li: Interpretation by electrochemical impedance spectroscopy *Br. Corros. J.* **35** 48–53

[7] Fitting Kourkoutis L, Lee J H, Song J H, Hwang H Y and Schlom D G 2010 2D Mapping of

IOP Conf. Series: Materials Science and Engineering **694** (2019) 012032 doi:10.1088/1757-899X/694/1/012032

Bonding Changes at Peroskite Oxide Interfaces and Around Defects *Microsc. Microanal.* **16** 1400–1

[8] Bland L G, Scully L C and Scully J R 2017 Assessing the corrosion of multi-phase Mg-Al alloys with high al content by electrochemical impedance, mass loss, hydrogen collection, and inductively coupled plasma optical emission spectrometry solution analysis *Corrosion* **73** 526–43

[9] Davó B, Conde A and De Damborenea J J 2005 Inhibition of stress corrosion cracking of alloy AA8090 T-8171 by addition of rare earth salts *Corros. Sci.* **47** 1227–37

[10] Prasad K S, Prasad N E and Gokhale A A 2013 *Microstructure and Precipitate Characteristics of Aluminum-Lithium Alloys* (Elsevier Inc.)

[11] Ahmad Z 2006 Corrosion Kinetics *Principles of Corrosion Engineering and Corrosion Control* vol 3 (Oxford, United Kingdom: Elsevier Science & Technology) pp 58–61

[12] Park I J, Yoo Y H, Kim J G, Kwak D H and Ji W S 2011 Corrosion characteristics of aluminum alloy in bio-ethanol blended gasoline fuel: Part 2. the effects of dissolved oxygen in the fuel *Fuel* **90** 633–9

[13] Matějovský L, Macák J, Pospíšil M, Baroš P, Staš M and Krausová A 2017 Study of Corrosion of Metallic Materials in Ethanol-Gasoline Blends: Application of Electrochemical Methods *Energy and Fuels* **31** 10880–9

IOP Conf. Series: Materials Science and Engineering **694** (2019) 012033 doi:10.1088/1757-899X/694/1/012033

The Effect of Vacuum Quenching on Corrosion and Hardness of the Surface of SKD61 Steel

Syifa Luthfiyah[1], Ahmad Faridh[1,2] and Bambang Soegijono[1]

[1]Department of Physics, Faculty of Mathematic and Natural Science, Universitas Indonesia, Depok, Indonesia
[2]PT. Adaro Services (Saptaindra Sejati), South Jakarta, Indonesia

E-mail: naufal@ui.ac.id

Abstract. JIS SKD61 Steel is chromium-molybdenum hot work steel which has composition 5.58 wt% Cr – 2.51 wt% Mo. This steel is a well-known material used as a die material to get good mechanical and hardening properties. Besides this material should have good corrosion resistance. Samples made of SKD61 steel were subjected to vacuum heat treatment at 1030^{0}C for 3 hours with step quenching in different rates of quenching. The difference in the cooling speed of each sample causes properties on the surface to vary. Micro Hardness Vickers is a tool used to determine micro-scale hardness. While to test corrosion resistance, electrochemical testing was carried out using NaCl solution with variations in concentrations of 1%, 3.5%, and 7%. X-Ray Diffraction was used to characterize the crystal structure and the crystallite size. The results show that the speed of quenching affects the hardness of the surface of the material, structure and corrosion resistance.

1. Introduction

JIS SKD61 steel is a chromium-based material that is included in three main groups in hot tool steel. This type of tool steel with medium carbon content, which can be used for hot work applications, such as aluminum extrusion die, forging molds, plastics, and nonferrous injection mold cavities and all types of molds for heat work applications, involving shocks.[1] JIS SKD61 steel is widely used in industry because it has good hardenability, high toughness, and excellent heat resistance. In addition, SKD61 steel has moderate wear resistance, so it can be carburatted or nitrided with the aim of increasing surface hardness. However has the effect of decreasing heat resistance. [1,2]

SKD61 steel is a type of steel that is combined with a number of chromium as a substitute for iron to form carbides, which can prevent decarburization by establishing carbon as carbide. Research shows the addition of 1% chromium produces 10.5 MPa stronger than normal steel with the same carbon content with a slight increase in toughness.[1] Although chromium content increases decarburization resistance, significant decarburization can still be found on heated SKD61 steel if the heat treatment is not well controlled.[3]

One way to increase material hardness is by vacuum heat treatment. Vacuum heat treatment is the process of heating material in a vacuum furnace. When in this condition, the material is heated in a furnace surrounded by a vacuum. This is done to prevent oxidation in the material because when heating at high temperatures allows oxidation quickly. After the heat treatment steps are completed,

IOP Conf. Series: Materials Science and Engineering **694** (2019) 012033 doi:10.1088/1757-899X/694/1/012033

the next process that is carried out is vacuum quenching. In the vacuum quenching process, there are 2 main factors that influence it, namely gas pressure and gas velocity. The gas that is usually used for cooling is nitrogen, argon, helium, or hydrogen.[4] At present, the vacuum heat treatment has been used widely, one of which is widely used in the automotive industry. Vacuum heat treatment is widely used in industry because it has due to its dimensional stability and low induced stresses in treated parts due to uniform heating and cooling.[5]

Steel SKD61, which belongs to the medium carbon steel, has fairly good corrosion resistance, but the resistance to corrosion is still lower than stainless steel. This is because the medium carbon steel only has about 5 wt% Chromium which does not have enough chromium to form the chromium oxide layer. This allows oxygen to bind with iron to produce iron oxide or rust. However, increased resistance to corrosion can be done by means of heat treatment. Therefore, it is very important to pay attention to the various stages of heat treatment carried out.[6]

2. Experimental Procedure

2.1. Material

This experiment uses medium carbon steel type SKD61 Steel, whis is divided into four different sized samples and each sample is given a different treatment. The first sample is Steel A LQ (Low Quenching) measures 5 cm x 2.6 cm x 0.6 cm, the second sample is Steel B MQ (Medium Quenching) measures 5 cm x 3.5 cm x 1 cm, the third sample is Steel C OQ (Optimum Quenching) measures 5 cm x 3 cm x 1 cm in length, width, and their thickness. Table.1 lists the chemical composition of the specimens, analyzed by a X – Ray Fluorescence (XRF).

2.2 Heat Treatment.

Heat treatment atmospheric conditions are vacuum heat treatments. At the stage of vacuum quenching using nitrogen gas and fan to accelerate cooling. To learn how each process parameter that is applied in influencing the hardness produced and corrosion resistance, the experiment is carried out in Table 2 and Table 3.

Table 1. The compositions of medium carbon steel type SKD61 Steel

Elements	**Fe**	**C**	**Si**	**Mn**	**P**	**S**	**Cr**	**Mo**	**V**
Weight %	89.411	0.353	0.32	1.17	0.004	0.003	5.58	2.62	0.539

Table 2. Heating Condition Stage Experimental Planning

Heating Condition					
Preheat 1		Preheat 2		Austinitizing	
Temperature	Time	Temperature	Time	Temperature	Time
650° C	2 H	850° C	2H	1030° C	3H

Table 3. Cooling Condition Stage Experimental Planning

Pressure		**Fan Speed**	**Time**	
Initial	2 bar	15 Hz	20 minute	
2nd	6 bar	50 Hz	Steel A LQ	30 minute
			Steel B MQ	40 minute
			Steel C OQ	50 minute
3rd	4 bar	35 Hz	20 minute	

2.3 EDM
After the vacuum heat treatment process an EDM (Electrical Discharge Machining) process is performed. The samples that were carried out by the EDM process were Steel B MQ and Steel C OQ, whereas for the A LQ samples there was no EDM process. This process was carried out in this research because the EDM process is a machining process that is often used in making dies and molds. Electro spark on EDM produces high heat on the surface of the material so that it can potentially change the nature of the material on the surface.

2.4 Microhardness Test

The microhardness testing method used in this experiment is Vicker's Hardness Test. The load applied to this test is 200 gf. Hardness testing of micro vickers is carried out in order to determine the depth of hardening softening. As for testing the surface hardness using Rockwell Hardness with a load of 150 KPressure. Hardness testing of micro vickers is carried out in order to determine the depth of hardening softening.

2.5 Corrosion Test
Corrosion test using a potentiostat. Potentiostat uses 3 electrodes namely, working electrode, auxiliary electrode, and a reference electrode. The reference electrode used in this experiment is Ag / AgCl.[4] The reference electrode serves to keep the voltage applied to the electrode stable, therefore the reference electrode is only used as a conductor of voltage without current. The electrode is connected to a wire that has been glued to the sample, then the four types of sample are dipped in 1%, 3.5% and 7% NaCl solution. Data obtained from the potentiostat are electrochemical potential and currents. From these data, the corrosion rate can be calculated at material using a formula obtained from Faraday's Law in Determining the Corrosion Rate of Metals and Alloys. [7,8]

$$r = C\frac{Mi}{n\rho} \tag{1}$$

where
M = atomic weight ($g.mol^{-1}$)
i = current density (A/cm^2)
n = number of electron involved
ρ = density (g/cm^2)
C = constant (0,129 in mpy; 3,27 in mm/year; 0,00327 in mm^3/year)

3. Results and Discussion

3.1. XRD Analysis
The results of x-ray diffraction patterns in the sample used in this study can be seen in Figure 1, and data collected from the characterization of this sample are attached in Table 4. From Figure 1 it appears that each sample tested, there are 3 peaks identified. For the three peaks of Steel A LQ has an angle position 43,98°, 64,47°, 81,71°. For the three peaks of Steel B MQ has an angle position 44,56°, 64,83°, 821,80°. For the three peaks of Steel C OQ has an angle position 44,1973°, 64,7334°, 81,82°. In the second peak and third peak, each peak has 2 different phases, namely, iron-alpha and ferchromide. This is because the two phases have diffraction reflections with almost the same d-spacing. The highest peak phase can also be obtained from the XRD test results, as shown in Table 4. Steel A LQ has the highest peak of the ferchromide phase 58,92 cps, Steel B QM 82,69 cps, Steel C QO 80,44 cps.

Table 4. Crystallografic parameters of the Untretaed Steel, Steel A QL No EDM, Steel B QM EDM, and Steel C QO EDM

Sample		Steel A QL	Steel B QM	Steel C QO
First peak's height (cps)	Iron-Alpha	43,2	59,21	60,21
	Ferchromide	58,92	82,69	80,44
First peak's d (Å)	Iron-Alpha	2,0621	2,0228	2,0613
	Ferchromide	2,0498	2,0364	2,0475
Lattice Parameter (Å)	Iron-Alpha	a = 2,8852 b = 2,8852 c = 2,8852	a = 2,8638 b = 2,8638 c = 2,8638	a = 2,8648 b = 2,8648 c = 2,8648
	Ferchromide	a = 2,8683 b = 2,8683 c = 2,8683	a = 2,8830 b = 2,8830 c = 2,8830	a = 2,8460 b = 2,8460 c = 2,8460
Volume (Å^3)	Iron-Alpha	24,0198	23,4876	23,5126
	Ferchromide	23,5988	23,9641	23,0522
Density (g/cm^3)	Iron-Alpha	7,72	7,9	7,89
	Ferchromide	7,48	7,36	7,66
Concentration (%)	Iron-Alpha	41,7	43,7	45,6
	Ferchromide	58,3	56,3	54,4

From Figure 1 the XRD plot results show that the first peak of each sample has a different FWHM. From the peak width can find out the relationship between FWHM and crystallite size. In Steel A LQ has FWHM 0.5995 °2Th and crystallite size 129.961700 Å. In Steel B MQ has FWHM 0.6246 °2Th and crystallite size 122.119800 Å. In Steel C OQ has 0.6176 °2Th FWHM and crystallite size 124.914100 Å. From these data, it can be concluded that the wider the FWHM, the smaller the crystallite size. The smallest crystallite size is owned by Steel B MQ. The ferchromide crystal system is cubic, so the lattice parameter values a, b, and c are the same. However, because each sample has a different peak position, the lattice parameter value is different for each sample.

3.2. Microhardness Analysis

After the austenitization temperature, the sample quenched using nitrogen to accelerate cooling. The cooling rate used at the core is at least 3°C / minute and more than 5 °C / minute is recommended. This is done with the aim of obtaining high toughness.[9] In this experiment, the cooling rate used was 8.5 °C / minute and used the step quenching process. The reason for using step quenching is to minimize the occurrence of induced cooling deformation and simplify controlling the cooling process in accordance with the CCT diagram to achieve the desired quality. In this process, the process is controlled by varying 2 parameters, gas pressure and fan speed.[10]

In the first step, the cooling process initially has a low gas pressure. In the second step, the gas pressure is increased significantly to prevent cooling from approaching the nasal passages of the CCT chart. This is done with the aim to avoid the formation of the pearlite phase and the bainite phase as much as possible. Then in the third stage, the gas pressure is lowered when it enters the martensite phase formation phase.[10]

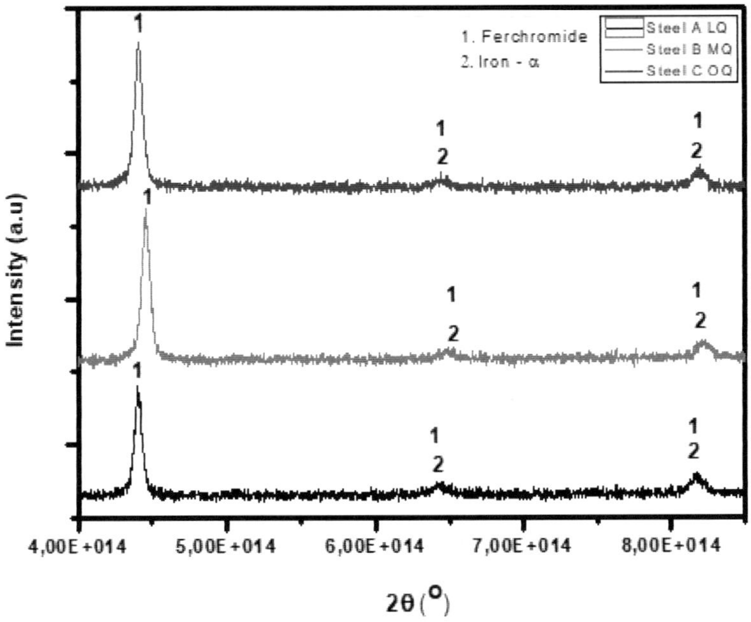

Figure 1. XRD's plot of the Steel A LQ, Steel B MQ and Steel C OQ.

From the results of vickers micro hardness test showed in Table 5 and Figure 2, the sample that has the highest hardness is Steel A LQ. This is because Steel A LQ does not accept the EDM process, so there is no decrease in hardness obtained from the vacuum heat treatment process. Then the sample that has the second highest hardness is Steel C OQ, but the core has decreased violence. This is because the time taken to cool the surface to the core is different. So that the core allows there are still residual phases of austenite, bainite, and pearlite in amounts greater than the martensite phase. The sample that has the third highest hardness is Steel B MQ. The results of the hardness obtained in Steel B MQ and Steel C OQ can be seen that there is a softening of hardness due to the EDM process of around 0.02 mm. Results of Rockwell hardness test showed in Table 6 and Figure 3. Whereas the highest hardness rockwell hardness testing is Steel A LQ, which is 43 HRC, followed by Steel B MQ 42.1 HRC, and Steel C OQ 41.5 HRC. From the data obtained reflecting the surface after EDM corrected an average decrease of 1 HRC.

Table 5. Depth of Hardness Softening (Hardness Micro Vickers)

	SKD61 Steel		
Depth	Steel A LQ No EDM	Steel B MQ EDM	Steel C OQ EDM
0.01	372.5	352.8	371.7
0.02	420.3	380.1	411.5
0.03	429.8	406.6	414.7
0.04	425.9	419.7	426.6
core	431.7	426	423.6

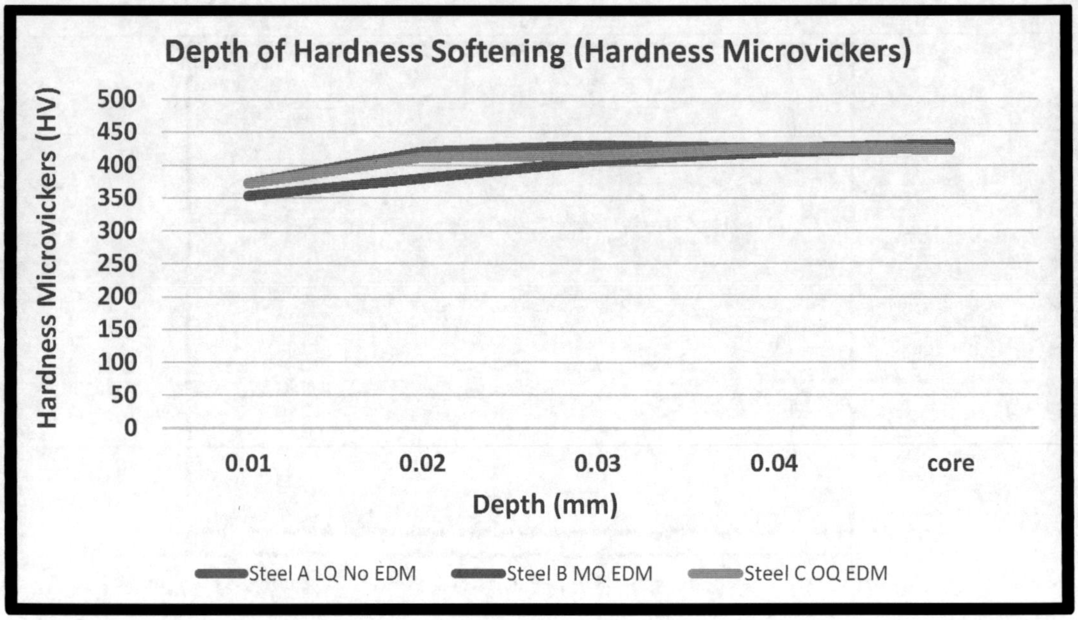

Figure 2. Hardness Micro Vickers plot of the Steel A LQ No EDM, Steel B MQ EDM and Steel C OQ EDM

3.3. Potentiodynamic Polarization Curves Analysis

Corrosion testing is carried out to produce a potentiodynamic polarization curve that describes the electrochemical behavior of SKD61 steel material. The electrochemical behavior of this sample can be seen in Figure 4 and the data obtained in Table 7. The process of corrosion can occur if there are 4 factors[11]:

- Anode
  At the anode, there is a process of oxygen binding and electron release which causes corrosion. Anodic reaction is:

$$M^+ + Cl^- + H_2O \rightarrow MOH + H^+ + Cl^- \quad (2)$$

- Cathode
  At the cathode, there is a process of releasing oxygen and binding of electrons. Cathodic reaction is:

$$2H_2O \rightarrow O_2 + 4H^+ + 4e^- \quad (3)$$

- Metallic Pathway
  A conductor that connects the anode and the cathode.

- Electrolyte
  A solution containing which can conduct electricity which is in this experiment uses a solution of NaCl electrolytes.

Table 6. Surface Hardness (Hardness Rockwell)

Item	SKD61 Steel
Steel A LQ No EDM	43 HRC
Steel B MQ EDM	42.1 HRC
Steel C OQ EDM	41.5 HRC

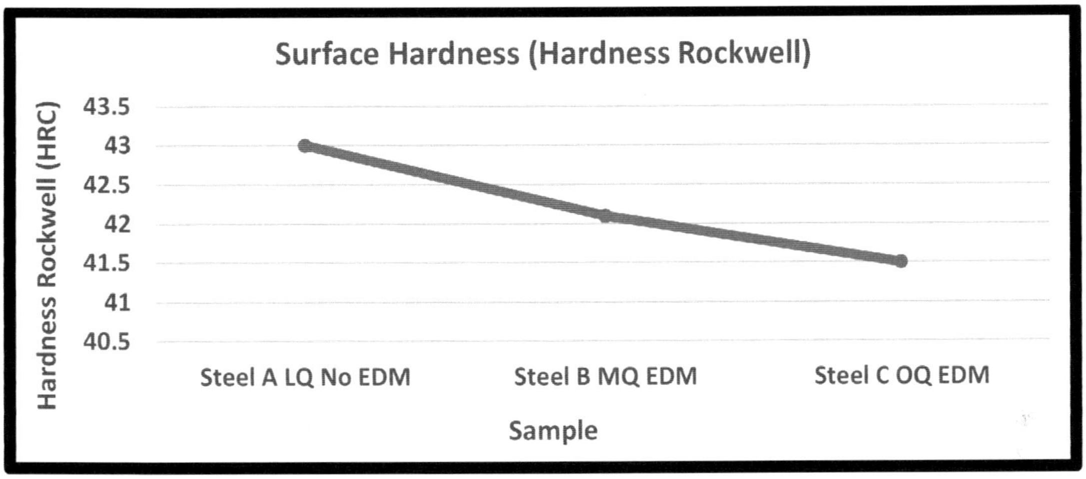

Figure 3. Hardness Rockwell plot of the Steel A LQ No EDM, Steel B MQ EDM and Steel C OQ EDM

From the potentiodynamic polarization curve, the corrosion current density can be determined by extrapolating the Tafel line. From the data on Table 7 it can be seen that in general, the corrosion rate will increase when the concentration is higher. The highest corrosion rate at 7 wt% NaCl solution. NaCl which functions as a strong oxidizing agent has successfully oxidized Fe. The greater the concentration of NaCl, the more atoms released from iron, the faster the corrosion.[11] From the curve, it can be seen as the lower the current density, causing corrosion rates slower. It happens because the lower the current density then the more electrons increase.

Table 7. Data from the corrosion test of the Steel A LQ, Steel B MQ, and Steel C OQ in the 1 wt%, 3,5 wt% and 7 wt% NaCl solution

Wt % NaCl	Sample	R (ohm)	$E_{corrosion}$ (V)	$I_{corrosion}$ (A)	Corrosion Rate (mm/year)
1 wt%	Steel A LQ	2,844 x 10^{2}	-1,088	9.036 x 10^{-5}	0.0011
	Steel B MQ	3,477 x 10^{2}	-1,094	7,39 x 10^{-5}	0.0009
	Steel C OQ	3,794 x 10^{2}	-1,097	6,772 x 10^{-5}	0.0008
3.5 wt%	Steel A LQ	1,97 x 10^{2}	-1,098	1,304 x 10^{-4}	0.0016
	Steel B MQ	1,939 x 10^{2}	-1,097	1,325 x 10^{-4}	0.0016
	Steel C OQ	2,316 x 10^{2}	-1,089	1.109 x 10^{-4}	0.0014
7 wt%	Steel A LQ	2.019 x 10^{2}	-1,086	1,272 x 10^{-4}	0.0016
	Steel B MQ	1,959 x 10^{2}	-1,085	1,312 x 10^{-4}	0.0016
	Steel C OQ	1,627 x 10^{2}	-1,109	1,579 x 10^{-4}	0.0020

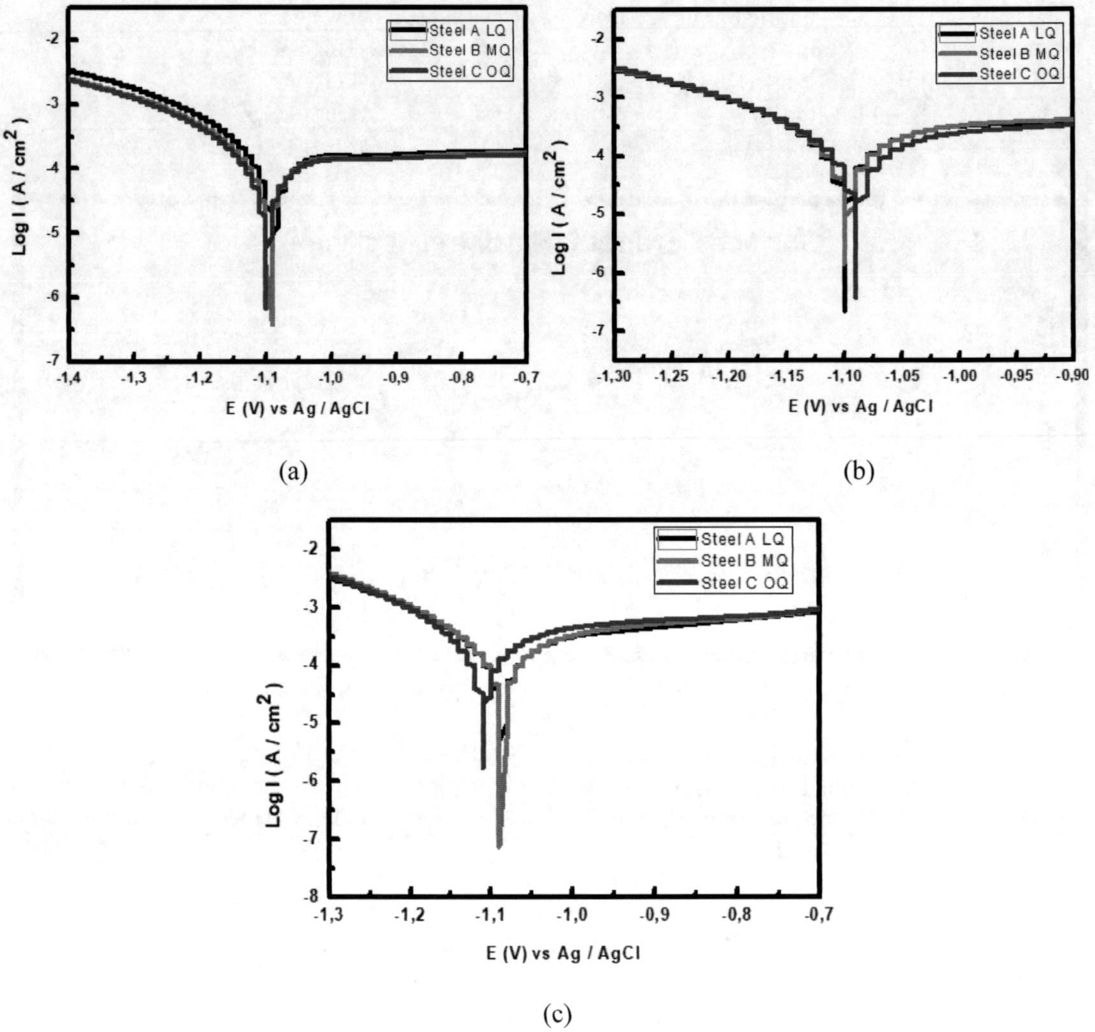

Figure 4. Potentiodynamic polarization curves of the samples in (a) 1 wt% NaCl (b) 3,5 wt% NaCl (c) 7 wt% NaCl

Different quenching speeds in each sample turned out to affect the corrosion rate. In general, the low corrosion rate is at the optimum quenching conditions and the highest corrosion rate is at the low quenching conditions. The speed of quenching will affect the volume of carbides produced in the martensite phase. Carbide volume can increase if the number of C atoms in the austenite phase decreases, which causes corrosion resistance to decrease. The low corrosion rate results in the entire sample show that vacuum heat treatment is very helpful to prevent corrosion. Because from the start of the heating process to cooling air conditions in a vacuum, thus preventing oxidation of the material because when heating at high temperatures allows oxidation quickly.

4. Conclusion

One that affects the hardness of SKD61 steel is the cooling method used and the cooling rate. Because these two things will determine the phase that is formed. So it's important to pay attention to these two things so that when the material goes through the EDM process, the reduction in hardness is not drastic.

SKD61 steel corrosion resistance is determined in the entire vacuum heat treatment process. Because the final phase formed and air condition will be the determining factor of how resistant the material is to corrosion. Another factor that determines the corrosion rate other than those mentioned above is the concentration of the electrolyte solution used. The higher the concentration of the solution used, the higher the corrosion rate of the material.

Acknowledgments
The authors would like to thanks to Universitas Indonesia for supporting the research activity for providing data XRD and SEM-EDX. This study also funded by universitas Indonesia through Hibah PITTA B No. NKB-0617/UN2.R3.1/HKP.05.00/2019

References

[1] Pasang T, Chen Z, Ramezani M, Neitzert T and Au D 2013 *The Effect of Heat Treatment Atmosphere on Hardening of Surface Region of H13 Tool Steel* **2013** 20–9
[2] Chandler H *Heat Treating Processes and Related Technology Introduction*
[3] Yong-an M I N, Rong W and Lin L I 2006 *Prediction and Analysis on Oxidation of H13 Hot Work Steel H13*
[4] Materials T and Company I 1991 *Heat Treating Volume 4 of the ASM Handbook,* p 1080
[5] Yeh S, Chiu L, Lo W and Huang C 2011 *Influence of Vacuum Heat Treatment and Salt-Bath Nitrocarburization on the Corrosion Behavior of JIS SKD61 and DH31S Hot Work Steels* **155** 1165–9
[6] Song D, Hao J, Yang F, Chen H, Liang N, Wu Y, Zhang J, Ma H, Eyram E, Gao B, Qiao Y, Sun J and Jiang J 2019 *Corrosion behavior and mechanism of Cr e Mo alloyed steel : Role of ferrite / bainite duplex microstructure J. Alloys Compd.* **809** 151787
[7] Walsh F C and Walsh F C 2016 *The Overall Rates of Electrode Reactions : Faraday ' s Laws of Electrolysis*
[8] Ahmad Z 2006 *Principles of Corrosion Engineering and Corrosion Cont* (Elsevier Science & Technology Books)
[9] International Mold Steel,Inc *DHA World Heat Treat*
[10] Since J J and Irretier O *Vacuum heat treatment and high pressure gas quenching – Aspects in Distortion Control* 1–6
[11] Rieger P H 2010 *Electrochemistry* (Chapman & Hall) pp 412–21

Characterization of Calcium Ferrite Phase from Fe_2O_3 and $CaCO_3$ Based on Iron Sand and Limestone Using XRD and SEM-EDX Analysis

Mastuki, P U Gatut, A W Brata, A G Istiawan, B Aditya and H Masrufi

Mechanical Engineering, Faculty of Engineering, University of 17 Agustus 1945 Surabaya, Indonesia

E-mail: mastuki@untag-sby.ac.id

Abstract. Synthesis of Ca-Fe-O using coprecipitation method employing $CaCO_3$ and Fe_2O_3 has been conducted. Extraction of limestone as the raw material of precipitated calcium carbonate (PCC) and iron sands as that of Fe_2O_3 was prepared to explore various compound of Ca-Fe-O. PCC and Fe_2O_3 are dissolved in HCl then mixed into homogeneous and precipitated using NH4OH. Mixing is resolved by the mass ratio of PCC and Fe_2O_3 with a ratio of 1/4, 1/6, 1/8, 1/12. The results of synthesis are sintering at temperature of 700^0C. The sintered samples were characterized by XRD and SEM-EDX. The results of XRD is indicated the formation of Ca-Fe-O phases that is dominated by $Ca_2Fe_2O_5$ and $Ca_2Fe_9O_{13}$ phases, and results of SEM/EDX indicate nanoscale particle size that is composed of Ca, Fe, O, and Si elements.

1. Introduction

Iron sand is one of the natural resources which is quite abudant in Indonesia. Almost all of the big islands in Indonesia havw iron sand wint different qualities. There are three types of iron sand ore, they are hematite magnetic iron ore, lateritic iron ore, and titan iron ore. Generally, iron sand is widely applied as a concrete aggregate in civil construction field[1], as a basic material in steel industrial field, and as a basic material in the manufacturing magnetic materials in developing advanced materials[2]. Research on the development of advanced materials that mostly use iron sand is the manufacture of soft dan permanent magnets. One of permanent magnet that utilization of iron sand is magnet type Barium Ferrite and Strontium Ferrite.

Barium Ferrite can be synthesized by some methods, they are powder metallurgy (milling, anneling, and sintering mechanisms)[3] and coprecipitation method (acid-base technique) by extracting iron sand to be pure and small dan be easily diffused with $BaCO_3$ or $SrCO_3$. Iron sand was extracted by copresipitation method to getting Fe_3O_4 that was then sintered into Fe_2O_3[4]. Barium or Strontium that using to synthesis Barium Ferrite is Barium or Strontium in the form of carbonate[5].

Besides abundant iron sand, there is abundant natural resources else, that is limestone. Limestone is a sedimentary rock which has high $CaCO_3$ phase content. In Indonesia, limestone is in various regions wiht a total 2,156 billion tons, which are spread in Aceh Darussalam (131.12 billion tons (Mt)), North Sumatra (3.24 Mt), West Sumatra (68.1 Mt), Riau (53.2 Million tons (Million)), Bengkulu (137.1 Million), Jambi (157 Million), South Sumatra (294 Million), Lampung (2 Million), Banten (61.6 Million), West Java (660.3 Million) Mt), Central Java (6 Mt), DI Yogya (10M), East Java (3,069Mt), Bali (154.64Mt), NTB (1.2Mt), NTT (132.82Mt), Central Kalimantan (449Mt), South Kalimantan

(8.33Mt), East Kalimantan (57 Mt), North Sulawesi (18.8M), Gorontalo (18.5Mt), Central Sulawesi (696Mt), South Sulawesi (31.33Mt), Southeast Sulawesi (1.527Mt), North Maluku (8.87Mt), and Papua (2.6 Mt). Generally, limestone is applied as a basic material for construction materials, the main material for portland cement production, and as an additive in the metal smelting industry[6]. Limestone (dominated by $CaCO_3$ phase) is one of the potential that has not been considered as a alternate Ba and Sr that is both form the carbonate phase.

Calcium is one of alkaline earth elements and so is same group with Barium and Strontium. The physical and chemical properties of Calcium aren't much different than Barium and Strontium. This allow a Calcium Ferrite formed. Research on Calcium Ferrite has been carried out that is a type material p-Type thermoelectric properties of the oxygen-defined $Ca_2Fe_2O_5$ perovskite[7]. Considering the abundant of iron sand dan limestone that not yet be developed maximally, and so, this study is conducted to analyze the types of Calcium Ferrite phases formed by synthesizing them based on local natural materials of iron sand and limestone.

2. Experimental

2.1. Synthesis

Calcium Ferrite samples were synthesized by a fusion of Fe_2O_3 and $CaCO_3$ with mass proportions ratio 1/4, 1/6, 1/8, and 1/12 for $CaCO_3/Fe_2O_3$. Calcium Ferrite is synthesized by coprecipitation method. Fe_2O_3 is obtained by extraction from iron sand and than be synthesized by coprecipitation method. Iron sand is dissolved in HCl 37% and the resulting solution was precipitated by NH_4OH. The precipitation is sintered at 800^0C for 2 hours. And $CaCO_3$ is obtained by extraction from limestone and than be syntesized by carbonation method. Limestone is calcined at 900^0C for 6 hours to produce CaO. CaO is dissolved into distilled water and than be carbonated by CO_2 gas. Calcium Ferrite samples were synthesized using the coprecipitation method by dissolving in HCl 37% each Fe_2O_3 and $CaCO_3$. The each resulting solution are mixed till homogeneous and than be precipitated by NH_4OH. The resulting mixture sample is then sintered at 700^0C for 1 hour.

2.2. Phase Characterization

X-ray diffractometer (XRD) is used to obtain X-ray diffraction patterns. The XRD measurement results are used to analyze the phase content of the resulting sample. The phase identification process is carried out by the search and match process using the Xpert Identify Graph and Match! Software. XRD testing for Calcium Ferrite samples was carried out in the Material Engineering laboratory of the ITS campus in the range of 2theta : 5^0 - 90^0.

2.3. SEM-EDX

Scanning Electron Microscope (SEM) is a type of electron microscope that depicts samples by scanning electrons in a raster scanning pattern. The electrons that interact with the sample-forming atoms produce a signal that gives information about the sample's topography, composition, and other properties. SEM combined with EDX informs the elements and their distribution from the sample. SEM-EDX testing was performed on Calcium Ferrite samples which had more Calcium Ferrite phase composition of XRD test data. SEM-EDX testing is carried out in the ITS Mechanical Engineering laboratory.

3. Results and Discussion

3.1. Phase and structural analysis

Calcium Ferrite phase formation occurs at temperatures above 500^0C[8]. Fe_2O_3, which is the basic ingredient of Calcium Ferrite synthesis and also the result of extraction from iron sand, is analyzed for purity of the phase. This is important as confirmation that the extraction was successful and that Fe_2O_3

was actually obtained, as well as for $CaCO_3$. XRD test results of Fe_2O_3 and $CaCO_3$ are shown in figure 1 and figure 2.

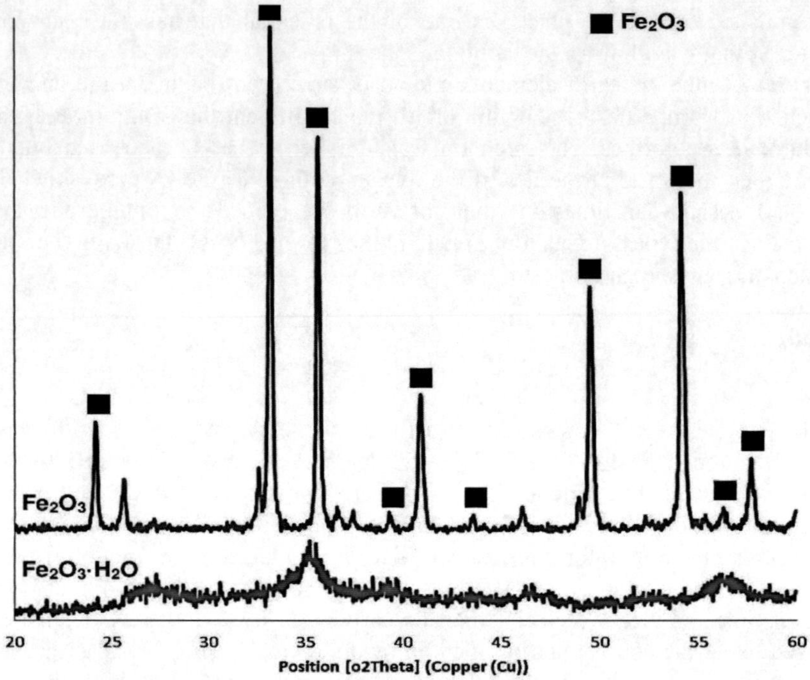

Figure 1. X-ray diffraction pattern of Fe_2O_3 synthesized from iron sand

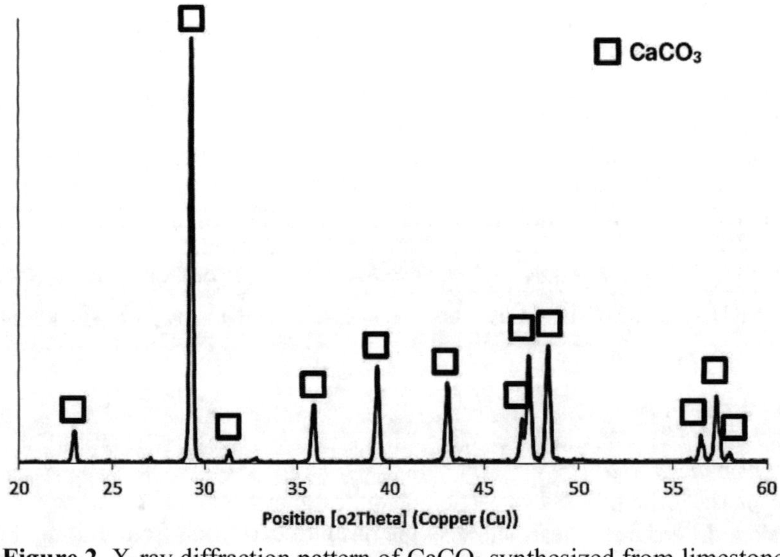

Figure 2. X-ray diffraction pattern of $CaCO_3$ synthesized from limestone

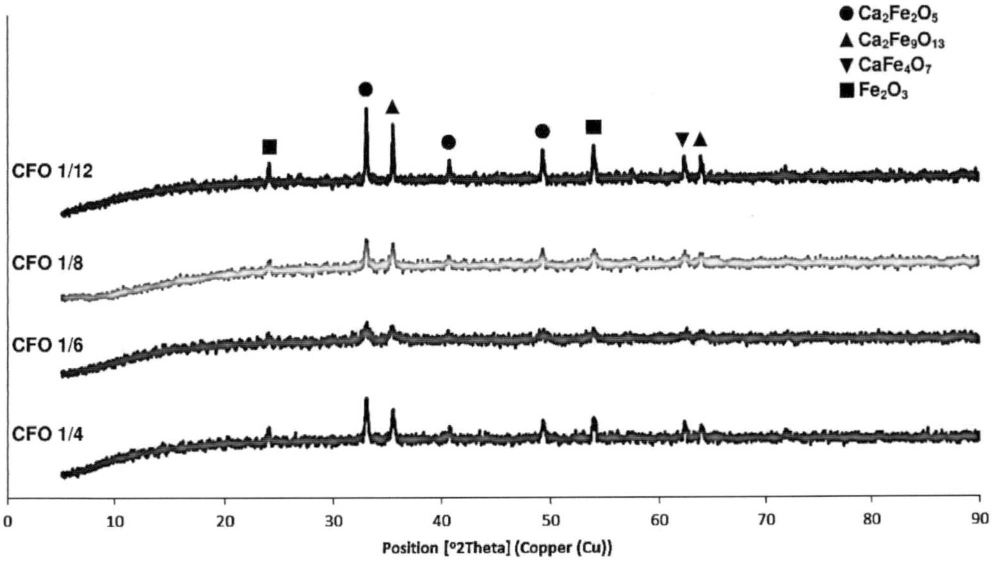

Figure 3. X-ray diffraction pattern of Calcium Ferrite synthesized by coprecipitation method

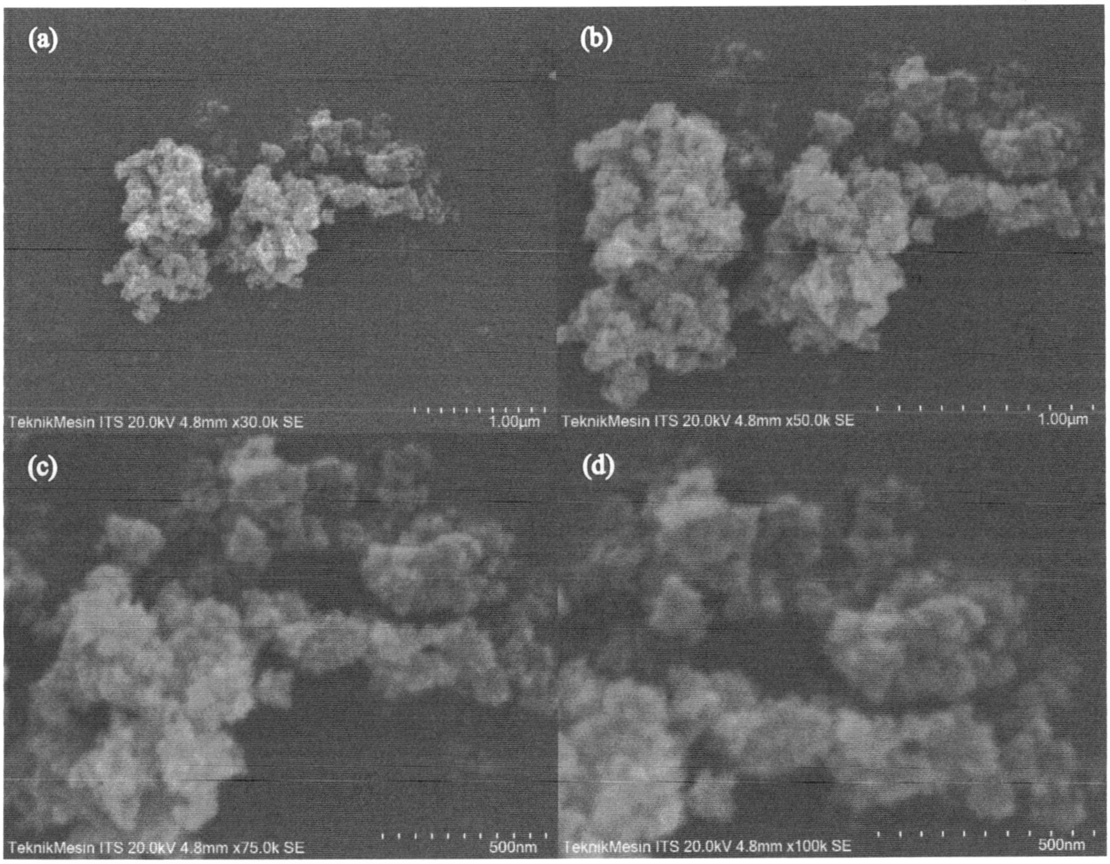

Figure 4. SEM of Calcium Ferrite samples with a ratio of 1/6

Figure 3 shows the X-ray diffraction pattern for Calcium Ferrite samples in the ratio of 4, 6, 8, and 12. The XRD test data obtained provides information that three types are formed Calcium Ferrite phase, they are $CaFe_4O_7$, $Ca_2Fe_2O_5$, and $Ca_2Fe_9O_{13}$. Each type of phase formed has different characteristics for each variation given. For the $CaFe_4O_7$ phase, this phase increases with increasing ratio of Fe_2O_3 given. That is because the $CaFe_4O_7$ phase requires more Ca elements and is formed at high temperatures. As for the other phases, namely $Ca_2Fe_2O_5$ and $Ca_2Fe_9O_{13}$, the pattern of large values of the phases formed is up and down. That is due to the growth of Fe_2O_3 phases which are indeed formed in the temperature range of 600^0C to 800^0C. From the overall data, there is still a lot of research to be done for other variations so that later can get information about properties of the phase formation and proportions in accordance with the desired phase needs.

3.2. Element composition, particle size, and sample shape analysis
Information on elemental composition, particle size, and sample shape of Calcium Ferrite was done by SEM-EDX testing. SEM-EDX testing is carried out on samples with a ratio of 1/6 by considering the volume fraction of the $Ca_2Fe_2O_5$ phase as a type of Brownmillerite Material which has attractive electrical properties. SEM results with various types of magnifications are shown in figure 4 and EDX results are shown in table 1.

Table 1. EDX of Calcium Ferrite samples with a ratio of 1/6

Element	Atomic %
Ca	2
Fe	34
O	58
Si	6

Figure 4 shows an indication of the size of nano-sized particles. Besides, particles also appear agglomerated which indicate the magnetic properties of the particles. This is reasonable considering the basic material of the magnet itself is Fe_3O_4. Table 2 also shows that the constituent elements of the sample are Ca, Fe, O, and Si. These results confirm or explain the roughness formed in the background of the XRD test data.

4. Conclusion
Calcium Ferrite from the results of mixing of Fe_2O_3 extracted from natural iron sand and $CaCO_3$ extracted from natural limestone were successfully synthesized. The XRD test results showed three types of Calcium Ferrite phases namely $CaFe_4O_7$, $Ca_2Fe_2O_5$, and $Ca_2Fe_9O_{13}$ with $Ca_2Fe_2O_5$ phases, and $Ca_2Fe_9O_{13}$ as dominant phases. SEM test results show indications of nano-sized particle size and also the particles look agglomerated which indicates the magnetic properties of these particles. The EDAX results show that the elements making up the phase are Ca, Fe, O, and there is little impurity, Si. The EDAX results confirm or explain the background roughness level from the XRD test results.

5. Acknowledgments
This research was partially supported by "Hibah Internal PT 2018" research grant provided by University of 17 Agustus 1945 Surabaya, Indonesia.

References
[1] H. Putra, I. Satyarno, A. B. Wijatna, and J. T. Fisika, "Penggunaan Pasir Besi Dari Kulon Progo Dengan Berat Jenis 4 , 311 Untuk Mortar Perisai," *Forum Teknik Sipil*, no. XVIII. pp. 909–920, 2008.
[2] Mastuki, M. A. Baqiya, and Darminto, "Explorative Synthesis of Magnetic Ca-Fe-O Compounds by Mixing Precipitated CaCO3 from Limestone and Iron Sand Extracted – Fe2O3.H2O," *Mater. Sci. Forum*, vol. 827, pp. 203–207, 2015.

[3] R. Hayati, A. Budiman, and D. Puryanti, "Karakterisasi Suseptibilitas Magnet Barium Ferit Yang Disintesis Dari Pasir Besi Dan Barium Karbonat Menggunakan Metode Metalurgi Serbuk," *J. Fis. Unand*, vol. 5, no. 2, pp. 187–192, 2016.

[4] Darminto, M. N. Cholishoh, F. A. Perdana, M. A. Baqiya, M. Y. Cahyono, and Triwikantoro, "Preparing Fe3O4 nanoparticles from Fe2+ ions source by co-precipitation process in various pH," *AIP Conf. Proc.*, vol. 1415, no. September, pp. 234–237, 2011.

[5] H. I. Hsiang and R. Q. Yao, "Hexagonal ferrite powder synthesis using chemical coprecipitation," *Mater. Chem. Phys.*, vol. 104, no. 1, pp. 1–4, 2007.

[6] M. Aziz, "Batu kapur dan peningkatan nilai tambah serta spesifikasi untuk industri," *J. Teknol. Miner. dan Batubara*, vol. 3, no. 6, pp. 116–131, 2010.

[7] E. Asenath-Smith, I. N. Lokuhewa, S. T. Misture, and D. D. Edwards, "p-Type thermoelectric properties of the oxygen-deficient perovskite Ca2Fe2O5 in the brownmillerite structure," *Journal of Solid State Chemistry*, vol. 183, no. 7. pp. 1670–1677, 2010.

[8] S. P. Mastuki, "Analisa Termal Kalsium Ferit Berbasis Bahan Alam Fe2O3•H2O Ekstraksi Pasir Besi," *Mek. J. Tek. Mesin*, vol. 4, no. 2, pp. 7–11, 2018.

IOP Conf. Series: Materials Science and Engineering **694** (2019) 012035 doi:10.1088/1757-899X/694/1/012035

Effect of water and seawater on mechanical properties of fiber reinforced polymer composites: a review for amphibious aircraft float development

K Abdurohman and M Adhitya

RCAVe (Research Center for Advanced Vehicle Universitas Indonesia)

E-mail: madhitya@eng.ui.ac.id

Abstract. One of the composite applications is its utilize in the water and seawater environments like boats, ships and float device of amphibious aircraft. This environment can affect the mechanical properties of composites since water can diffuse into the composites. Many researches have discussed about the effect of water and seawater on mechanical properties of composites like tensile, compressive, shear, and impact properties. This review is conducted on the field of durability of composites in water and seawater environment. Most studies are carried out by immersing composites in water and seawater for a certain period of time. Almost all studies exhibit the degradation of the mechanical properties of composites with ageing time of the immersion process in water or seawater.

1. Introduction

Composites are widely used in engineering practice like aeronautics, automotive, marine, construction, and packaging because they are lightweight; resistant to corrosion, high temperature, and fatigue; high specific stiffness and strength [1–4]. Composites with vinyl ester and polyester resins are often used in marine applications [2]. Other composite applications are as float devices material of amphibious aircraft which mostly utilize fiber-reinforced polymer composites. In addition, composites are also used in the development of ocean energy conversion systems such as tidal turbine [5]. This review will be used as a basis for float devices development of amphibious aircraft operating in freshwater and seawater environment.

The amphibious aircraft can take-off and landing on land and water making it suitable for missions in sea and wet areas. It has landing gear and hull or floats which must bear the impact loads of water and assure the stability when landing on sea or water [6]. The floats initially used aluminum material for applications in the water environment and will be developed using composites due to the operation of aircraft in corrosive environment of seawater. Thus, need to know the effect of water and seawater on the composite properties by reviewing papers that discuss about that.

Many researches have discussed the effect of water and seawater on the fiber-reinforced polymer composite properties, however, none has been specifically for amphibious aircraft applications. Some research is conducted on synthetic fiber composites such as carbon and glass, and others utilize natural fiber composites such as flax [3, 7–11]. The researchers used different types of composites in their research such as conventional composites, hybrid composites [1, 12-13], pultruded composites [14–17], bar composites [18-19], and sandwich composites [20-21]. Composites are made with several fabrication processes such as hand lay-up, resin vacuum infusion, RTM (Resin Transfer Molding), prepreg autoclave, and pultrusion process. The specimens is cut from the composite panel and tested for water absorption, tensile, compressive, shear, impact, etc. Researchers utilize different types of moisture as immersion media like tap water, natural seawater, and artificial seawater.

2. Water Absorption

One thing that needs to be known in the development of amphibious aircraft that will operate in the water and sea water environment is the behavior of water absorption of the float material which can affect its mechanical properties. The researchers conducted a water absorption studies using the ASTM D 570 and ASTM D 5529 standard by immersing composites on liquid media and then taking and weighing them at each specified time period. Visco, et.al., 2011 [2], showed that the water absorption of polyester and vinylester resin, and glass reinforced poliester and vinylester composites follows Fick's law with the initial linear portion and the balance plateau. Glass fiber fabric does not sufficiently influence the diffusion of water into its polymer network. Maximum water absorption value was 1.2% for polyester resin and glass-polyester composites, while for vinylester and glass-vinylester composites the value is lower by around 0.7%. Vinylester is better than polyester in avoiding diffusion of sea water so that its performance is superior than polyester in water environment. Similar to Visco et al., Boisseau, et. al. 2012 [22], showed that the diffusion of epoxy resin follows the Fick's law with the the saturation of mass gain at 2.5%. The initial weight gain of E-glass, Advantex, amd HiPer-tex composite at 60 ^{0}C also follows Fickian, then increases sharply which is probably caused by a decrease in the glass transition temperature (Tg) of the resin.

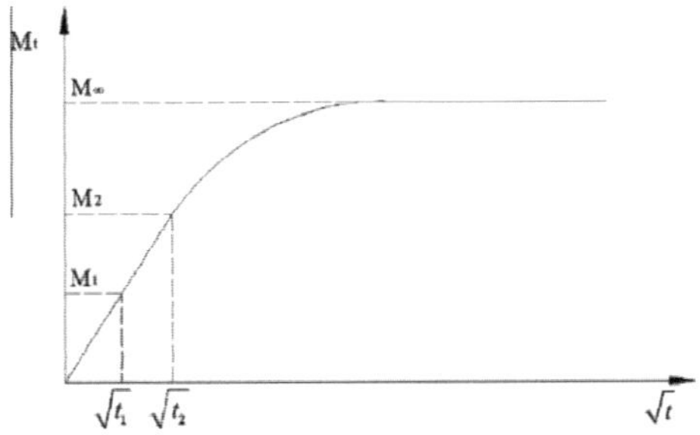

Figure 1. Fickian diffusion process [23]

In addition to the percentage of weight gain, water absorption studies also determine the diffusion coefficient of the composite. Alia, et. al., 2013 [12], showed that the diffusion coefficient of vinylester (5.14 x 10-13 m2s-1) is smaller than polyurethane (6.39 x 10-13 m2s-1) where the maximum water uptake value is also smaller (around 0.55% for vinyl ester and 0.6% for polyurethane).

The water absorption test was carried out by researchers both on GFRP and CFRP composites. The study of glass-vinyl ester composite pipes produced by filament winding conducted by Michael Berges et al. [9] stated that weight gain with water immersion was higher than humid conditions and acid immersion. Jiang, et. al., 2013 [23], showed that FRP (e-glass/polyester) experiments carried out in water produced higher saturation levels of moisture (3%). The experimental results showed that moisture uptake is more sensitive to temperature. For adhesive polyurethanes that are aging in 40 ^{0}C-water and 40 ^{0}C-96% RH, there was deviation from the Fickian diffusion that could be associated to the relaxation of the polymer at elevated temperature. Other study of Yuan Fang et al [24] stated that absorption of water and seawater in the e-glass/polyester composite mixed with cadmium sulfide (CdS) QDs did not follow the fick's law. Mass loss occurs when the immersion time is increased, which is associated with leaching of unreacted monomers and hydrolysis of ester bonds. A study of e-glass/DER-31 epoxy composite pipes conducted by Fitriah et al. [25] showed the mechanism of non-fickian water absorption. The results showed an increase in degradation in the form of cracked matrix and debonding of the fiber/matrix. An increase in the quantity of water absorption caused matrix plasticization. Yian Zhao et al. [26] state that an increase in temperature accelerates the diffusion of

IOP Conf. Series: Materials Science and Engineering **694** (2019) 012035 doi:10.1088/1757-899X/694/1/012035

sea water into the e-glass/BMI composite at an early stage, then tends to decrease in the long term immersion. Increasing the immersion temperature strengthens the effect of matrix plasticization which can increase the ductility of the material and expand the area of damage. Yinghui Hu et al. [27] states that the weight gain in epoxy samples is much faster and higher than polydicyclopentadiene and is nearing saturation after 12 months. The water absorption of epoxy is much greater when immersed in deionized water than salt water, and the difference is even greater in glass-epoxy composites, due to water diffusion and capillary diffusion along the fiber/matrix interface debonding.

The researches were also carried out by adding nano clay to the GFRP composite. Study of the additional of nano TiO2 to e-glass/epoxy composites conducted by R. K. Nayak and B. C. Ray [28] shows that seawater absorption increases with increasing nano TiO2 content, whereas 0.1 wt% nano TiO2 content causes sea water absorption to decrease by 15% compared to neat composites. Other study of GFRP/quasi isotropic epoxy composites exposed to artificial seawater conducted by Acharya Pavan et al. [29] showed that at ambient temperatures the composites absorbed water 13.22% to equilibrium, whereas for conditions under 0 degrees the maximum water absorption at equilibrium was 2.61%. Water absorption follows Ficks' law at the beginning but is gradual increment and decrements after saturation. Study of pultruded GFRP e-glass and polyester with 3% nano clay and 10% calcium carbonate conducted by A. V. Oskoueia et al. [16] showed that sea water diffusion increases with increasing immersion time and temperature, while high NaCl and chlorine content reduce seawater absorption. This is similar to Jiang et al. 2013 and Alessi et al. 2014 about the effect of elevated temperature on water diffusion.

The researchers also conducted a study of sandwich composites. Jiang, et. al., 2014 [30], conducted a study of glass-polyester sandwich wich was manufactured by vacuum infusion process exposed to variation in temperature of water and relative humidity. The results showed that the hot/wet environment can accelerate the water absorption process into the composite, as evidenced by the saturation level of the specimens which are conditioned at 40 ^{0}C-water six times the condition of 20 ^{0}C-50% RH. There was a decrease in the diffusion moisture curve from equilibrium which indicates physical and chemical degradation such as matrix hydrolysis, chain breaks, formation of small molecules and the release of these molecules from the composite. Study by Anxin Ding et al [20] at vinyl ester sandwich composites with PVC foam core showed that water absorption from composites immersed in pure water was higher than that in seawater. Water absorption for composites without void is caused by free volume between polymer molecular chains, infiltration of the polymer molecular structure which increases swelling, and capillarity in the fiber/matrix interface.

Many researchers conducted studies on carbon composites. Study of nanophased CFRP composites exposed to seawater conducted by Hossain, et. al., 2014 showed that after 12 months of immersion the sample with 2% nanophased absorbs moisture smaller (0.39%) than the conventional sample (0.67%) [31]. Alessi, et.al., 2014 [32], showed that for epoxy resins, water uptake increases gradually at an early stage then reaches a plateau as an equilibrium condition according to the Fickian mechanism. For CFRP composites, the water uptake is smaller than pure resin and showed a non-Fickian mechanism. Water absorption mechanicm in CFRP can be related to diffusion of water into meso-sclae free volume, capillarity of fiber, and microcracks due to fiber-matrix debonding. Water absorption was significantly higher and faster at increasing aging temperatures from 30 ^{0}C to 70 ^{0}C. Water absorption at CFRP can also be related to water absorption at the fiber/matrix interface. Tual, et. al., 2015 [33], conducted a study of the effect of seawater on carbon/epoxy composites fabricated using prepreg-autoclave, RTM, and vacuum infusion. The results showed that the weight saturation level under high temperature conditions is higher than the lower conditions, due to the evolution of glass transition temperature. Water diffusion is very small influenced by the thickness and orientation of the composite, and not influenced by the geometry of the specimen. However, Garcia-Espinel et al. [34] state that the time to reach the saturation level of water absorption depends on the thickness and ambient temperature. Saturation levels of moisture are 0.58% (30 days), 0.44% (210 days), and 0.4% ((30 days) in glass/polyester, glass/vinyl ester, and glass/epoxy composites, respectively. Kafodya et al. [14] state that diffusion coefficient of pultruded CFRP-epoxy submerged in water is higher than that submerged in seawater. This is caused by the dissolved particles of seawater which obstruct osmotic water diffusion. Water absorption and diffusion coefficient are reduced with increasing sustained bending due to reduced free volume fraction and evolution of swelling stresses. Arhant et al. [35] state that carbon/polyamide composites absorb water higher (3%) than carbon/polyetheretherketone (0.17%). Water absorption in composites can cause the phenomenon of palsticising, which is reversible and reduces the glass transition temperature. Study of PU resin and pultruded CFRPU immersed to seawater conducted by Bin Hong and Guijun Xian, 2018, [17] sshowed that the water absoption follow the Fick's law. Water absorption of PU (~2.65%) and CFRPU (~0.70%) higher than epoxy (~0.44%) and epoxy-based CFRP (~0.58%) due to more hydrophilic groups. However, at high temperature of immersion, water absorption of CFRP epoxy-based (1.65%) higher than CFRPU (0.83%), due to hydrolisis of epoxy. Water absorption of CFRPU are caused from voids, mocro cracks, matrix, and interface of fiber/matrix. Zike Wang, et al, [36] states that water absorption of untread interface of CFRP exposed to distilled water higher than treated CFRP due to debonding or crack water molecules attack. Other study of Yucheng Zhong et al., 2019 [37], states that the water absoption of CFRP epoxy-prepreg immersed with tap water was higher than GFRP, due to higher resin volume fraction and the pattern of weave on CFRP. Another study of glass/carbon hybrid composites immersed to seawater conducted by Jesthi and Nayak, 2019 [13], states that the water absoption of hybrid composites was lower than plain glass and carbon composites, due to good bonding of interface between glass-carbon plies and epoxy matrix.

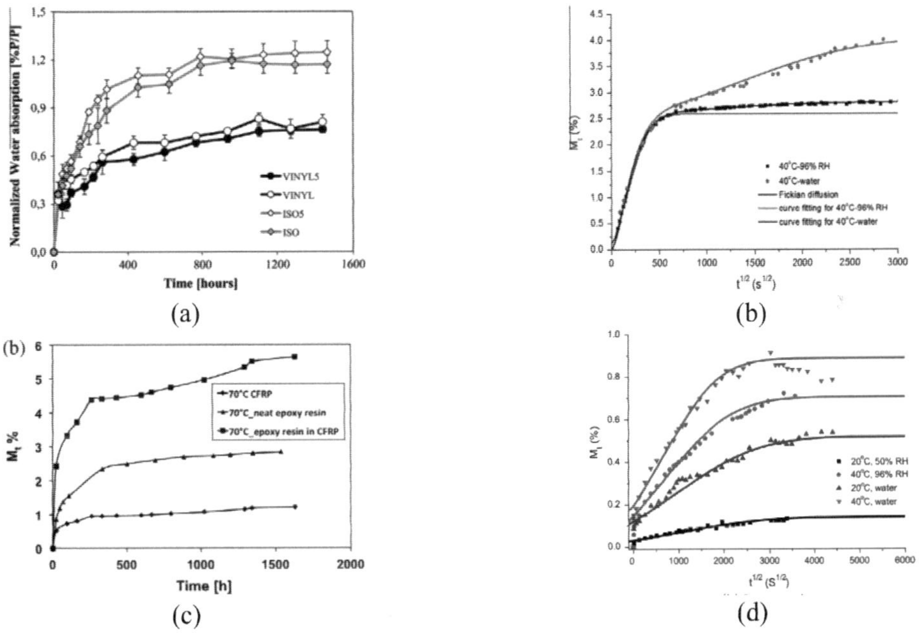

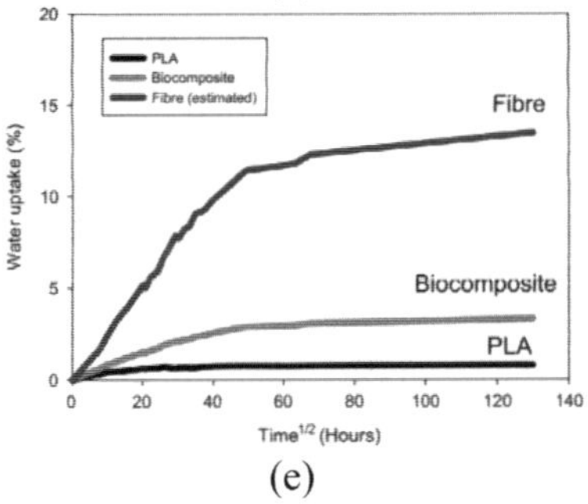

Figure 2. Water absorption curves of (a) vinyl ester, polyester, and composites [2], (b) polyurethane adhesive [23], (c) epoxy and CFRP at 70 ^{0}C [32], (d) sandwich FRP composite [30], (e) biocomposite and fiber [7]

Water absorption studies were also carried out on natural fiber composites. Assarar, et. al., 2011 [3], states that the water absorption of the flax–epoxy composite is establish to follow Fickian law. Saturation of weight gain was reached at 13.5% for flax-composites and 1.05% for glass composites. Study of Duigou, et. al., 2014 [7], showed that the diffusion of natural sea water in the flax/PLA biocomposite reached saturation after 2 months of immersion of 3.3% where the weight of the saturation of the flax fiber was 12%. Libo Yan and Nawawi Chouw [8] state that water uptake for flax/epoxy submerged in water is higher than that submerged in sea water, due to the accumulation of salt particles which inhibits the diffusion of water. Weight gain on the flax-epoxy composite is more significant than synthetic fiber composites, that is related to the hydrophilic properties of flax fibers, and poor interfaces and poor fiber resistance to water absorption due to hydroxyl groups. Similar to Assarar et al., study of flax reinforced thermoplastic and thermosetting exposed to tap water conducted by Abderrazak Chilali et al. [10] showed that the diffusion behaviour follow Fickian diffusion mechanism. Mohamed Habibi et al. 2019 [11], states that maximum water diffusion of flax-epoxy composites manufactured by combined compressing molding and hand lay-up is 12.8%. The decrease of water diffsuion after saturation level can be related to hydrophilic of flax, cause fiber swells, which present microcracks of epoxy matrix. This is similar to Libo Yan and Nawawi Chouw about the hydrophilic of flax fibre.

3. Mechanical Properties

Some mechanical properties that need to be known in the development of amphibious aircraft include tensile, compressive, shear, and impact properties.

3.1. Tensile properties

To obtain the tensile properties of composites, researchers conducted tensile testing using a universal testing machine with a constant crosshead speed. The standard that is usually used is ASTM D3039, although other standards are also used. The study of Assarar et al. [3] stated that the tensile strength of the e-glass/epoxy composite was reduced by 25% in saturated conditions (more than 40 days), whereas the flax/epoxy composite was reduced by 15% in the 20 day immersion. Furthermore, at saturation level, young modulus and maximum strain was reduced by 9% and 10% for glass/epoxy, while the reduction of young modulus was 39% for flax/epoxy. Otherwise, the maximum strain increases 63% at the saturation level.

Study of pure epoxy resin infusion by Boisseau et al. [22] show that strength and young modulus are reduced during immersion in seawater at 60 °C, but maximum strain is less affected. The tensile strength of vinyl ester and polyurethane resins after being immersed in seawater for 9 months was reduced by 12% and 21%, respectively [12].

Glass/polyester composites immersed in natural seawater for 2 years showed the tensile stiffness reduced by 10%, while the tensile strength of flax/PLA composites immersed in natural seawater for 18 months was reduced by 38% [7]. The study of Garcia Espinel et al. [34] exhibit that the tensile strength of glass/epoxy composites was reduced by 23.81% after 90 days of sea water immersion. Other study of glass/vinyl ester composites treated and untretaed MMT clay powder by Garima Mittal et al. [38] showed that the tensile strength of the composite after sea water immersion was reduced by 8% and 7% respectively, while the modulus of elasticity is reduced by 4%. Another study of glass composites using polydicyclopentadiene resin showed that the tensile strength was significantly lost, with only an average strength remaining 42% after one year water immersion [27]. Another study of glass/vinyl ester composite pipes by Essi Sarlin et al. [39] showed a reduction in tensile strength up to 63% after water immersion. Study on glass polyester modified cadmium sufide show a decrease in tensile strength of 15.5% and 13.8% after water and sea water immersion, while the tensile modulus is slightly affected by the immersion process [24]. Water absorption results in degradation of the fiber, matrix, and debonding between the fiber and the matrix.

Research on quasi isotropic e-glass/epoxy composites [29] exposed to artificial sea water shows a 27% reduction in tensile stress due to plasticization, while the failure that occurs in the composite is mainly caused by fiber failure and cracking of the reinforcement/resin interface. A study conducted by Mourada et al. [40] stated that the tensile strength was reduced by 47% for e-glass / epoxy composites, 64% for eglass/polyuretane, after soaking sea water 65 ^{0}C for 60 months to 90 months.

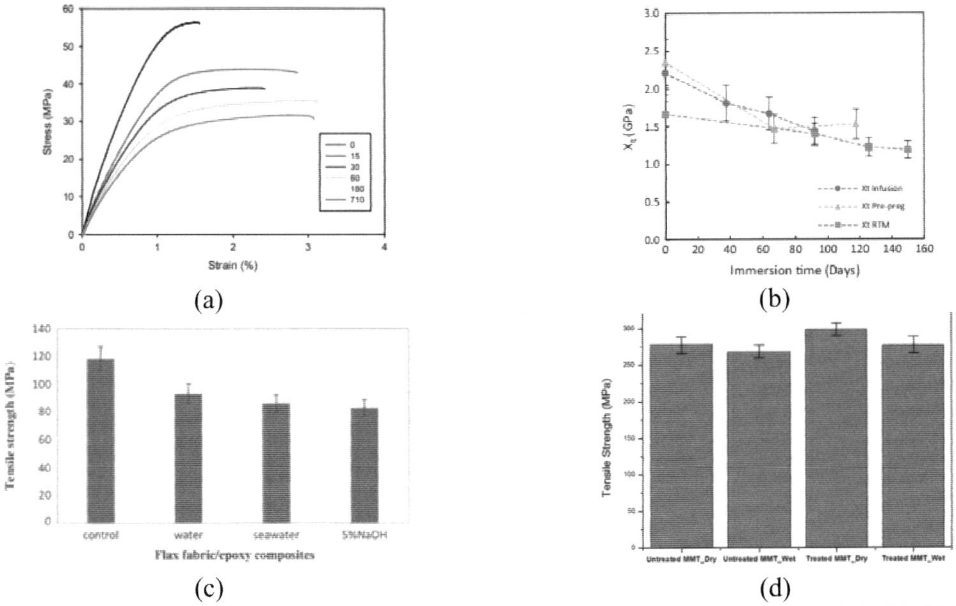

Figure 3. Effect of moisture on tensile strength of (a) flax/PLA biocomposite [7], (b) CFRP composites [33], (c) flax/epoxy composites [8], and (d) glass/vinylester composites [38]

N. Tual et al. [33] stated that water absorption affects the tensile strength of the CFRP composite due to matrix plasticization and changes in the fiber/matrix interface. However, there is no significant effect on Young Modulus of composites. The degradation of composites is more influenced by the water than temperature. Another study of pultruded epoxy-based CFRP composites immersed in water and seawater by Kafodya et al. [14] shows anomalous trends of tensile strength. Tensile strength decreases after the first 2 weeks of immersion which is associated to residual stress due to pultrusion which causes specimen failure. The increase of tensile strength after 2 weeks can be attributed to the water ingress which causes relaxation of residual stresses, and accelerate the epoxy postcuring. The tensile strength reduction in unstrained specimens after 20 weeks of sea water immersion was dominated by plasticization. Tensile modulus increases at an early stage due to postcuring progression, and decreases after 12 weeks due to the effect of plasticization. Another CFRP study using polyamide resins by Arhant et al. [35] showed that the decrease in tensile strength occurred when the water content was more than 1%, but the tensile modulus value did not change significantly.

A study of CFRP-vinyl esters by Gargano et al. [41] states that composite degradation is affected by hydrolysis, cracks caused by swelling, reduction in strength of reinforcement and resin interface, and plasticization. Another study of vinyl ester sandwich composites showed that the tensile strength dropped by 1.28% and 7.35% after water and sea water immersion, respectively [20]. A study conducted by Bin Hong [17] shows that PU and CFRPU are more resistant to seawater than epoxy and CFRP-epoxy. Yucheng Zhong et al's [37] study of GFRP and CFRP showed a reduction in strength and tensile modulus after hygrothermal conditioning. However, after drying, the mechanical properties return to similar as the specimens without conditioning. Study on hybrid glass-carbon/epoxy composites manufactured with hand lay-up [13] show a decrease in tensile strength after immersion in seawater. Meanwhile, the modulus decreases for $[GCG2C]_S$ type and increases for $[G3C2]_S$ and $[G2C2G]_S$ types.

Study on flax/epoxy composites [8] show that tensile strength decreases after 12 months immersion by 28.3% for seawater, and 22.6% for water. Whereas the tensile modulus is reduced by 27.1% for sea water, and 24% for water. This reduction is associated with degradation of fiber, epoxy, and fiber / resin interface bonding. Another study of flax composites showed a decrease in tensile strength and tensile modulus by 42% after 30 days immersion in tap water [10]. Study on flax-epoxy immersed in water at 75 ^{0}C can reduce tensile strength and modulus by 68% and 61%, respectively [11].

3.2. Compressive properties

From the references there are still few papers that discuss the effect of water or sea water environment on compressive behavior of composites. One that discusses the influence of seawater on the compressive properties of composites is the paper of Mael Arhant et al. about CFRP composites with polyamide (PA) and polyetheretherketone (PEEK) matrices [35]. This paper states that the seawater diffusion on composites reduces the compressive strength of CFRPA to 50%, whereas in compressive modulus the effect is low. Otherwise, sea water does not reduce the strength and modulus of CFRPEEK composites. The compressive strength of the composite after sea water soaking can return to its original strength after the drying process.

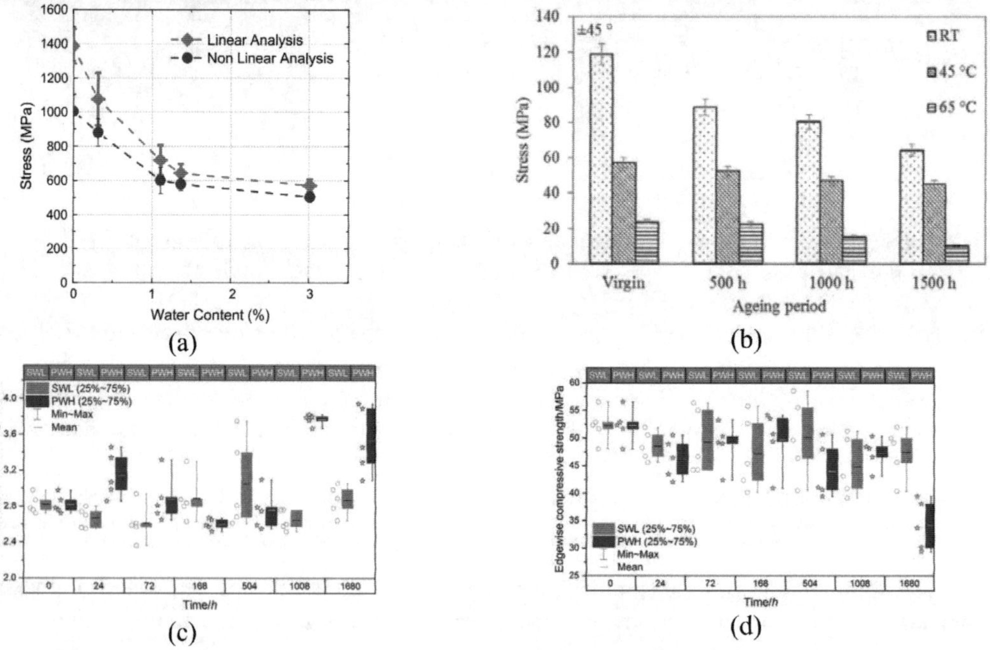

Figure 4. Effect of moisture on compressive strength of (a) CFRPA [35], (b) GFRP pipe [25], (c) flatwise sandwich composite [20], and (d) edgewise sandwich composites [20]

A study of glass/epoxy composite pipes conducted by S.N. Fitriah et al. [25] showed a deterioration in compressive strength up to 84% as a result of the longterm water immersion and temperature rises. In line with previous references, Gargano et al. [41] stated that the compressive and other mechanical properties of CFRP-vinyl ester composites dropped in seawater as a result of plasticization. Meanwhile, a study of glass/vinyl ester sandwich composites with PVC foam core showed that the decline in edgewise compressive strength values in pure water (35%) was higher than seawater (9%), but this flatwise compressive strength increased [20]. Similar to water and sea water exposure, composite sandwiches exposed to solar radiation combined with water vapor, salt spray, and hygrothermal atmosphere alternating high low temperature also show an increase in the value of flatwise compressive strength and deterioration in edgewise compressive strength [21].

3.3. Shear properties

Several methods are used by researchers to obtain the interlaminar shear strength (ILSS) of composites exposed to water and seawater. The methods used include short beam shear test (SBS) (ASTM D2344), in-plane shear test (ASTM D3518), and transverse shear test (ASTM D761). ILSS studies of e-glass/silk/epoxy (medium viscosity resin) composites, show that seawater causes swelling, plasticization, and debonding of the resin/reinforcement interphase which results in decreased of ILSS composites [1]. Other ILSS studies of glass/epoxy composites (quasi unidirectional fabric, low viscosity epoxy) [22] show a small reduction in ILSS after sea water immersion. Specimens dried in the air for 6 months show that ILSS are irreversible.

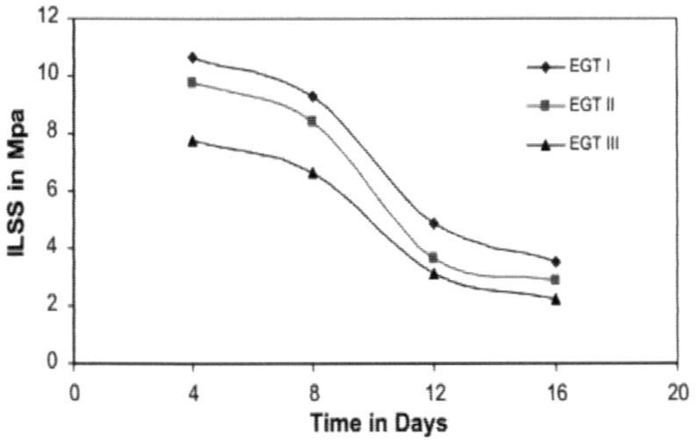

(a)

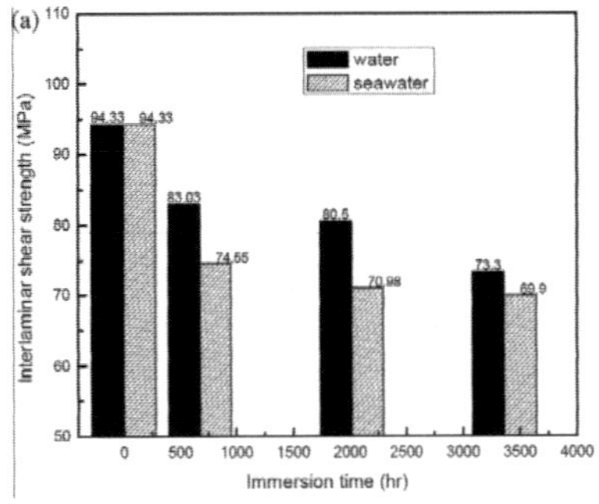

(b)

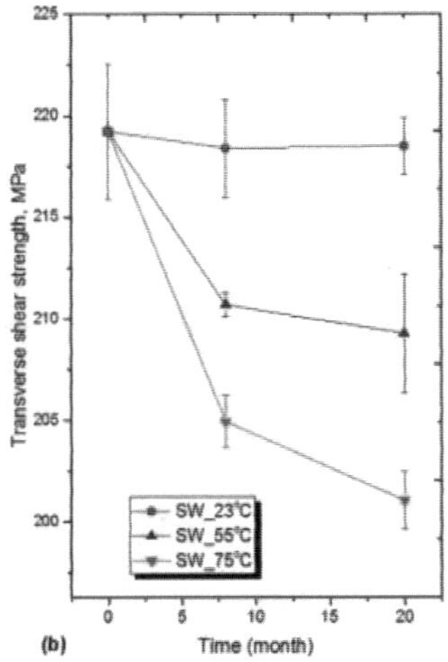

(c)

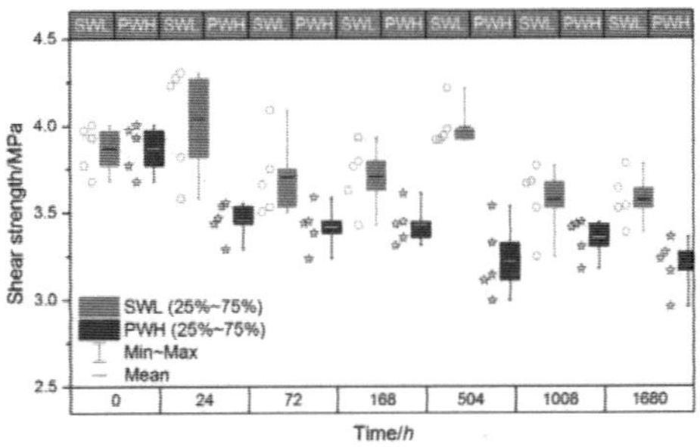

(d)

Figure 5. Effect of water and seawater on shear strength of (a) GFRP (e-glass/silk/epoxy) [1], (b) pultruded CFRP [14], (c) GFRP rebar (glass/vinyl ester) [19], and (d) sandwich composites (PVC foam) [20]

Study on CFRP composites (UD carbon, epoxy) exposed to seawater show that ILSS is reduced by 20-30% [33]. Another study of CFRP (carbon UD, epoxy) [14] showed a significant decrease in ILSS in the early stages of immersion, due to debonding and plasticization. In line with previous studies, other research [42] states that sea water immersion causes damage of reinforcement and matrix interphase, as evidenced by a reduction in ILSS by 30%, 10%, and 20% in glass/epoxy, carbon/epoxy, and glass/polyester composites, respectively. In-plane shear test of carbon-polyamide composites showed a reduction in shear strength of 41% and reduction in shear modulus of 46% [35]. The short beam shear test showed that after a 1-year immersion, the shear strength retention of polydicyclopentadiene composite was 85%, whereas for epoxy composites it was reduced by 40% [27]. Other studies have shown an increase in ILSS with the addition of nano-TiO2 to both dry and submerged GFRP composites [28]. Another study on CFRP showed a rapid decrease in ILSS at 30 days immersion, SBS strength retention and in-plane shear after 4 months immersion were 27.2% and 39.6%, respectively [36]. Other studies of GFRP rebar composites with vinyl ester resins [19] show that the trasversal shear strength after immersion in ambient temperature sea water reduced very low, while at high temperatures it drops to 95.4% due to plasticization. Short beam shear strength GFRP rebar rise 2.6% in seawater 55 ^{0}C which is associated with anti-plasticization, whereas at high temperatures the value drops 4.5% after soaking 20 months.

Shear strength studies on sandwich composites show significant changes in water immersed composites, where failure on PVC cores shows shear strength of it [20]. Sandwich composite studies on water, seawater, salt spray, hygrothermal atmosphere alternating high low temperature, and water vapor treatments showed an increase in in-plane shear strength of 3.68%, 1.57%, -6.08%, 11.3% and 7.87% [21].

3.4. Impact properties

Float devices of amphibious aircraft receive impact loads when landing in water or sea water, thus, it is necessary to know the effect of them on the impact properties of float material. Standards used by researchers to determine the composite impact behavior based on existing references include ASTM D256 (izod test), ASTM D7136 (low-velocity impact), and ASTM D3763 (drop weight impact test). One study conducted by K.V. Arun et al. [1] discussed the effect of sea water on the impact properties of hybrid glass/silk/epoxy composites. Studies conducted by Arun et al. show that the impact toughness of hybrid composites increases with increasing glass fiber volume. Meanwhile, soaking time and sea water temperature accelerate the reduction in impact toughness of hybrid composites due to fusing of the resin. Another impact study conducted by D. K. Jesthi and R. K. Nayak [13] on hybrid glass/carbon/epoxy composites showed that hybrid composites type $[GCG_2C]_S$ had the highest impact strength values compared to other hybrid types. Seawater immersion causes a decrease in the impact strength of hybrid composites where the value of the decline in composites $[GCG_2C]_S$ type (9.6%) is smaller than other types of hybrids which reach 12% in $[G_3C_2]_S$ type. Another impact study conducted by Yucheng Zhong et al. [37] showed that after exposed to hygrothermal from tap water, the performance of the prepreg epoxy CFRP composite at a low speed impact was better, while for GFRP composites the performance had decreased.

The researchers not only did an impact test on conventional composites but also on other shapes such as composite pipes. The study of GFRP filament winding composite pipes subject to variations in impact energy conducted by M. E. Deniz and R. Karakuzu [4] showed that the absorbed energy reached a minimum after three months of sea water immersion, then continue to increase until the time of immersion for 12 months. Study by M. K. Hossain et al. [31] show that the energy absorbed by nanoclay-reinforced CFRP composites is higher than conventional CFRP, as well as after prolonged immersion in seawater. Increasing sea water immersion time decreases delamination energy, absorbed energy, and damage areas. Failures resulting from impact loads are usually resin cracks, delamination, and layer splitting.

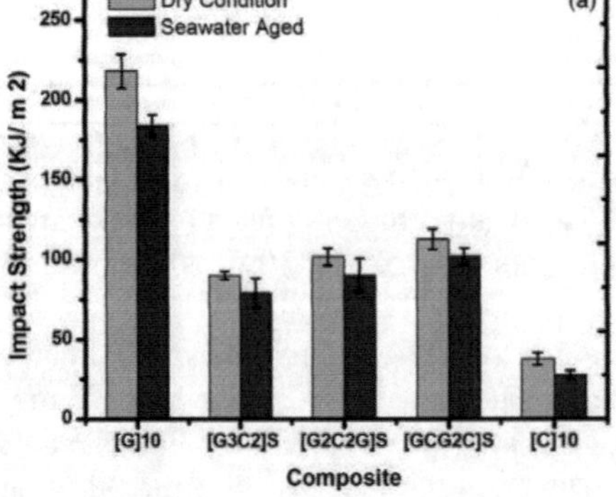

Figure 6. Impact strength of GFRP, CFRP, and hybrid glass/carbon composites [13]

4. Conlusions

Research on the influence of water and seawater environment on mechanical properties of composite was conducted by researchers on conventional, pultruded, pipes, bars, hybrid, and sandwiches composites for both synthetic and natural fiber. Most studies show that the water absorption process in composites follows the Fickian mechanism, even several studies show a non-Fickian mechanism. The longtime duration of the immersion process and the elevated temperature increases the diffusion of water into the composite. Water absorption into composites is related to water diffusion into meso-sclae free volume, capillarity of fiber, and microcracks due to debonding of resin-fibers. Sea water diffusion into the composites is lower than water due to the content of salt molecules in seawater that prevent water diffusion. Diffusion of water into the composite causes a decrease on the tensile, compressive, shear, and impact properties of the composite except the flatwise compression properties of the composite sandwich. This decrease is due to degradation of the fiber, matrix, debonding between the fiber and the matrix interface, and matrix plasticization resulted by water difffusion. Several studies have shown that the mechanical properties of composites exposed to water or seawater will return to normal after drying.

Acknowledgement

The authors gratefully acknowledge the financial support of the PITTA Grant 2019 Universitas Indonesia for this publication. Many thanks and gratitude also go to Saintek Ristekdikti scholarship for the financial support of this study.

References

[1] K. V Arun, S. Basavarajappa, and B. S. Sherigara, "Damage characterisation of glass / textile fabric polymer hybrid composites in sea water environment," *Mater. Des.*, vol. 31, pp. 930–939, 2010.

[2] A. M. Visco, N. Campo, and P. Cianciafara, "Comparison of seawater absorption properties of thermoset resins based composites," *Compos. Part A*, vol. 42, pp. 123–130, 2011.

[3] M. Assarar, D. Scida, A. El Mahi, C. Poilâne, and R. Ayad, "Influence of water ageing on mechanical properties and damage events of two reinforced composite materials : Flax – fibres and glass – fibres," *Mater. Des.*, vol. 32, pp. 788–795, 2011.

[4] M. E. Deniz and R. Karakuzu, "Seawater effect on impact behavior of glass – epoxy composite pipes," *Compos. Part B*, vol. 43, pp. 1130–1138, 2012.

[5] A. Boisseau, P. Davies, and F. Thiebaud, "Fatigue Behaviour of Glass Fibre Reinforced Composites for Ocean Energy Conversion Systems," *Appl. Compos. Mater.*, vol. 20, pp. 145–155, 2013.

[6] W. U. Dawei, W. U. Zheng, Z. Landing, and H. U. Jizhong, "Safety and Airworthiness Design of Ultra-Light and Very Light Amphibious Aircrafts," *Procedia Eng.*, vol. 17, pp. 212–225, 2011.

[7] A. Le Duigou, A. Bourmaud, P. Davies, and C. Baley, "Long term immersion in natural seawater of Flax / PLA biocomposite," *Ocean Eng.*, vol. 90, pp. 140–148, 2014.

[8] L. Yan and N. Chouw, "Effect of water , seawater and alkaline solution ageing on mechanical properties of flax fabric / epoxy composites used for civil engineering applications," *Constr. Build. Mater.*, vol. 99, pp. 118–127, 2015.

[9] M. Berges *et al.*, "Influence of moisture uptake on the static , cyclic and dynamic behaviour of unidirectional flax fibre-reinforced epoxy laminates," *Compos. Part A*, vol. 88, pp. 165–177, 2016.

[10] A. Chilali, W. Zouari, M. Assarar, H. Kebir, and R. Ayad, "Effect of water ageing on the load-unload cyclic behaviour of flax fibre-reinforced thermoplastic and thermosetting composites," *Compos. Struct.*, vol. 183, pp. 309–319, 2018.

[11] M. Habibi, L. Laperrière, and H. Mahi, "Effect of moisture absorption and temperature on quasi-static and fatigue behavior of nonwoven fl ax epoxy composite," *Compos. Part B*, vol.

166, pp. 31–40, 2019.

[12] C. Alia, M. V Biezma, P. Pinilla, J. M. Arenas, and J. C. Suárez, "Degradation in Seawater of Structural Adhesives for Hybrid Fibre-Metal Laminated Materials," *Hindawi*, vol. 2013, 2013.

[13] D. Kumar and R. Kumar, "Improvement of mechanical properties of hybrid composites through interply rearrangement of glass and carbon woven fabrics for marine application," *Compos. Part B*, vol. 168, pp. 467–475, 2019.

[14] I. Kafodya, G. Xian, and H. Li, "Durability study of pultruded CFRP plates immersed in water and seawater under sustained bending : Water uptake and effects on the mechanical properties," *Compos. Part B*, vol. 70, pp. 138–148, 2015.

[15] H. Xin, Y. Liu, A. Mosallam, Y. Zhang, and C. Wang, "Hygrothermal aging effects on flexural behavior of pultruded glass fiber reinforced polymer laminates in bridge applications," *Constr. Build. Mater.*, vol. 127, pp. 237–247, 2016.

[16] A. V. Oskouei, M. Bazli, H. Ashrafi, and M. Imani, "Flexural and web crippling properties of GFRP pultruded profiles subjected to wetting and drying cycles in different sea water conditions," *Polym. Test.*, vol. 69, pp. 417–430, 2018.

[17] B. Hong and G. Xian, "Ageing of a thermosetting polyurethane and its pultruded carbon fiber plates subjected to seawater immersion," *Constr. Build. Mater.*, vol. 165, pp. 514–522, 2018.

[18] H. El-Hassan, T. El-Maaddawy, A. Al-Sallamin, and A. Al-Saidy, "Durability of glass fiber-reinforced polymer bars conditioned in moist seawater-contaminated concrete under sustained load," *Constr. Build. Mater.*, vol. 175, pp. 1–13, 2018.

[19] M. Ahmad Sawpan, "Shear properties and durability of GFRP reinforcement bar aged in seawater," *Polym. Test.*, vol. 75, pp. 312–320, 2019.

[20] A. Ding, J. Wang, A. Ni, and S. Li, "Hygroscopic ageing of nonstandard size sandwich composites with vinylester-based composite faces and PVC foam core," *Compos. Struct.*, vol. 206, pp. 194–201, 2018.

[21] A. Ding, J. Wang, A. Ni, and S. Li, "Assessment on the ageing of sandwich composites with vinylester-based composite faces and PVC foam core in various harsh environments," *Compos. Struct.*, vol. 213, pp. 71–81, 2019.

[22] A. Boisseau, P. Davies, and F. Thiebaud, "Sea Water Ageing of Composites for Ocean Energy Conversion Systems : Influence of Glass Fibre Type on Static Behaviour," *Appl. Compos. Mater.*, vol. 19, pp. 459–473, 2012.

[23] X. Jiang, H. Kolstein, and F. S. K. Bijlaard, "Moisture diffusion in glass – fiber-reinforced polymer composite bridge under hot / wet environment," *Compos. Part B*, vol. 45, no. 1, pp. 407–416, 2013.

[24] Y. Fang *et al.*, "Monitoring of seawater immersion degradation in glass fi bre reinforced polymer composites using quantum dots," *Compos. Part B Eng.*, vol. 112, pp. 93–102, 2017.

[25] S. N. Fitriah, M. S. A. Majid, M. J. M. Ridzuan, R. Daud, A. G. Gibson, and T. A. Assaleh, "Influence of hydrothermal ageing on the compressive behaviour of glass fibre / epoxy composite pipes," *Compos. Struct.*, vol. 159, pp. 350–360, 2017.

[26] Y. Zhao, W. Liu, L. K. Seah, and G. B. Chai, "Delamination growth behavior of a woven E-glass/bismaleimide composite in seawater environment," *Compos. Part B Eng.*, vol. 106, pp. 332–343, 2016.

[27] Y. Hu, X. Li, A. W. Lang, Y. Zhang, and S. R. Nutt, "Water immersion aging of polydicyclopentadiene resin and glass fi ber composites," *Polym. Degrad. Stab.*, vol. 124, pp. 35–42, 2016.

[28] R. Kumar and B. Chandra, "Influence of seawater absorption on retention of mechanical properties of nano-TiO 2 embedded glass fiber reinforced epoxy polymer matrix composites," *Arch. Civ. Mech. Eng.*, vol. 18, no. 4, pp. 1597–1607, 2018.

[29] A. Pavan, P. Dayananda, K. M. Vijaya, S. Hegde, and P. N. Hosagade, "Influence of seawater absorption on vibrationaland tensile characteristics of quasi-isotropicglass/epoxy

composites," *J. Mater. Res. Technol.*, pp. 1–7, 2018.
[30] X. Jiang, H. Kolstein, F. Bijlaard, and X. Qiang, "Effects of hygrothermal aging on glass-fibre reinforced polymer laminates and adhesive of FRP composite bridge : Moisture diffusion characteristics," *Compos. Part A*, vol. 57, pp. 49–58, 2014.
[31] M. K. Hossain, M. M. R. Chowdhury, K. A. Imran, M. B. Salam, M. Hosur, and S. Jeelani, "Effect of low velocity impact responses on durability of conventional and nanophased CFRP composites exposed to seawater," *ASME Int. Mech. Eng. Congr. Expo. Proc.*, vol. 9, pp. 180–189, 2013.
[32] S. Alessi, G. Pitarresi, and G. Spadaro, "Effect of hydrothermal ageing on the thermal and delamination fracture behaviour of CFRP composites," *Compos. Part B*, vol. 67, pp. 145–153, 2014.
[33] N. Tual, N. Carrere, P. Davies, T. Bonnemains, and E. Lolive, "Characterization of sea water ageing effects on mechanical properties of carbon/epoxy composites for tidal turbine blades," *Compos. Part A Appl. Sci. Manuf.*, vol. 78, pp. 380–389, 2015.
[34] J. D. Garcia-Espinel, D. Castro-Fresno, P. Parbole Gayo, and F. Ballester-Muñoz, "Effects of sea water environment on glass fiber reinforced plastic materials used for marine civil engineering constructions," *Mater. Des.*, vol. 66, pp. 46–50, 2015.
[35] M. Arhant, P. Y. Le Gac, M. Le Gall, C. Burtin, C. Briançon, and P. Davies, "Effect of sea water and humidity on the tensile and compressive properties of carbon-polyamide 6 laminates," *Compos. Part A Appl. Sci. Manuf.*, vol. 91, pp. 250–261, 2016.
[36] Z. Wang, G. Xian, and X. Zhao, "Effects of hydrothermal aging on carbon fibre / epoxy composites with different interfacial bonding strength," *Constr. Build. Mater.*, vol. 161, pp. 634–648, 2018.
[37] Y. Zhong, M. Cheng, X. Zhang, H. Hu, D. Cao, and S. Li, "Hygrothermal durability of glass and carbon fi ber reinforced composites – A comparative study," *Compos. Struct.*, vol. 211, pp. 134–143, 2019.
[38] S. J. Park, H.-J. Kim, D. H. Jung, K. Y. Rhee, V. Dhand, and G. Mittal, "Investigation of seawater effects on the mechanical properties of untreated and treated MMT-based glass fiber/vinylester composites," *Ocean Eng.*, vol. 108, pp. 393–401, 2015.
[39] E. Sarlin, R. Sironen, T. Pärnänen, M. Lindgren, M. Kanerva, and J. Vuorinen, "The effect of matrix type on ageing of thick vinyl ester glass-fibre-reinforced laminates," *Compos. Struct.*, vol. 168, pp. 840–850, 2017.
[40] B. M. A.-M. Abdel-Hamid Ismail Mourad, Amir Hussain Idrisi, Maria Christina Wrage, "Long-term durability of thermoset composites in seawater environment," *Compos. Part B*, vol. 168, pp. 243–253, 2019.
[41] A. Gargano, K. Pingkarawat, V. Pickerd, T. Delaney, R. Das, and A. P. Mouritz, "Effect of seawater immersion on the explosive blast response of a carbon fi bre-polymer laminate," *Compos. Part A*, vol. 109, pp. 382–391, 2018.
[42] A. Boisseau and C. Peyrac, "Long term durability of composites in marine environment: comparative study of fatigue behavior," *Procedia Eng.*, vol. 133, pp. 535–544, 2015.

IOP Conf. Series: Materials Science and Engineering **694** (2019) 012036 doi:10.1088/1757-899X/694/1/012036

Effect of the number of layers of fiber in COPV PVC-Cotton on burst strength test

Dedi Lazuardi, Sidik Ally and Hery Sonawan

Jurusan Teknik Mesin Universitas Pasundan
Jl. Setiabudi 193 Bandung 40

E-mail: dedi.lazuardi@unpas.ac.id

Abstract. Composite overwrapped pressure vessel (COPV) is a pressure vessel wrapped or coated by composite fibers. COPV is commonly used to reduce the weight of pressure vessel. Usually COPV combine the metal liner and wrapped by the glass or the carbon fiber. How the influence of layers number to burst strength test is studied. The PVC tube is wrapped by the natural fiber (cotton) is used, and the burst test is applied. The specimens consist of four tube, a tube without wrap, a tube with one layer wrap, a tube with two layers wrap and a tube with three layers wrap. The experiment indicate that the layer effective to augment the strength of burst test until 50%, but the addition of the number of layers, did not really affect the increase in COPV burst strength test.

1. Introduction

The CNG pressure vessels which are made entirely of metal are relatively cheap, but due to the high pressure the weight of the tubes will be very heavy. This risk can be reduced by selecting composite overwrapped pressure vessels (COPV).

Composite overwrapped pressure vessel (COPV) is a pressure vessel that is wrapped or coated by a composite fiber. COPV is commonly used as a place or container of fluid that has a high pressure, such as oxygen, compressed natural gas, helium gas, and so forth. COPV consists of two layers, where the first layer serves to prevent leakage, while the second layer serves to withstand the pressure received by pressure vessels. The first layer is usually made of metal, while the second layer is made of composite.[1]

The first layer is usually made of metal but in this study will use PVC which serves to prevent leakage, while the second layer is made of a composite that serves to withstand the pressure received by the pressure vessel. COPV is designed with the hope of reducing the weight of pressure vessels without reducing resistance to high pressure.

The thickness of the composite fiber layer at COPV can determine the pressure of the pressure vessel. Pressure vessels for CNG need to be designed in such a way as to be able to withstand the enormous internal pressure of CNG. Operating pressure for CNG pressure vessels ranges from 100-200 bar. For this reason, it is necessary to test the thickness of the composite fiber in the COPV tube due to high pressure.

The thickness of the composite fiber layer at COPV can determine the pressure of the pressure vessel. Based onwall dimensions, pressure vessels can be divided into two, namely the thick pressure vessel has a wall wall thickness (shell) more than 1/20 sheel diameter and thin wall pressure vessels have thickness the wall (shell) is less than 1/20 the diameter of the shell.[2]

Difference between thin wall pressure vessel and wall thickness is in the stress distribution that occurs on the wall of the pressure vessel, on the pressure vessel thin walls, stress distribution can be ignored because of the difference in outside diameter and diameter inside is so thin that the stress distribution is occurs very small, whereas in pressure vessels thick wall stress distribution must be taken into account.

When the thin cylinder is given internal pressure, there are three principal stresses which are perpendicular to each other ie, circular stress (hoop), longitudinal tress, and radial stress that reacts at cylindrical wall.[3-5]

2. Manufacturing and Preparation of Test Specimens

The specimen is made from PVC pipe and wrapped by the cotton fiber, for this study, four specimen is used. The effect of reinforcing fibers on COPV with PVC liners is obtained by testing the specimens in the form of PVC 4 inch in diameter, 3 mm thick, and 400 mm long which is then carried out by Hoop Winding. The Each test specimen consisted of zero, one, two, and three wrapped layer. The specimen thickness are presented in Table 1.

Tabel 1. Specimen characteristic

Specimen no.	Composite Layers number	Thickness (mm)
1	n/a	3
2	1	4.1
3	2	5.3
4	3	6

Figure 1. Specimen

3. Experimental Procedure

The objective was to characterize the burst test strength of COPV PVC-cotton fibers, and to know the posibility of this COPV used for gas tank. Testing on COPV (PVC-cotton Fiber) to burst (Burst Test) is done by the Hydrotest method. This test is carried out to get at what pressure the COPV is exploding. The test set-up is presented in figure 2. The pressure is generated from a manual piston pump, which is then channeled into the tube.

IOP Conf. Series: Materials Science and Engineering **694** (2019) 012036 doi:10.1088/1757-899X/694/1/012036

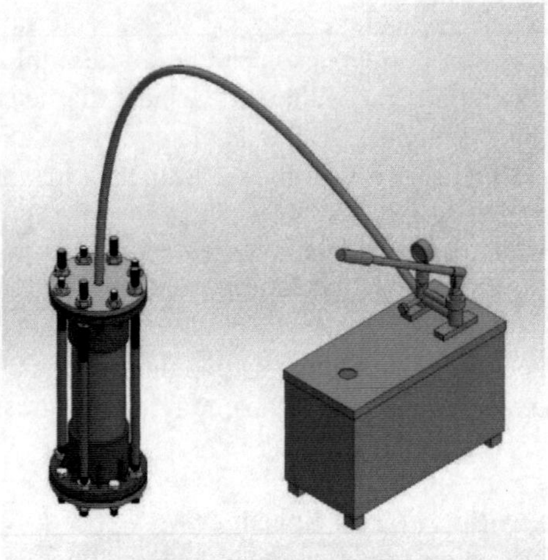

Figure 2. Test Set-up

4. Discussion of Results

The test results obtained as shown in table 2, while the specimen images of the test results can be seen in Figure 3.

Tabel 2. Experiment results

Specimen Number	Number of layers	Pressure (bar)	Remark
1	N/A	20	explode
2	1 Lapisan Komposit	25	Explode, fiber fail
3	2 Lapisan Komposit	30	Explode, fibel fail
4	3 Lapisan Komposit	30	Explode, fiber no fail

a. b. c. d.

Figure 3. Types of tubes failures; a, b, c, and d, respectively, show no layers, one layer, two layers, and three layers of composite fibers

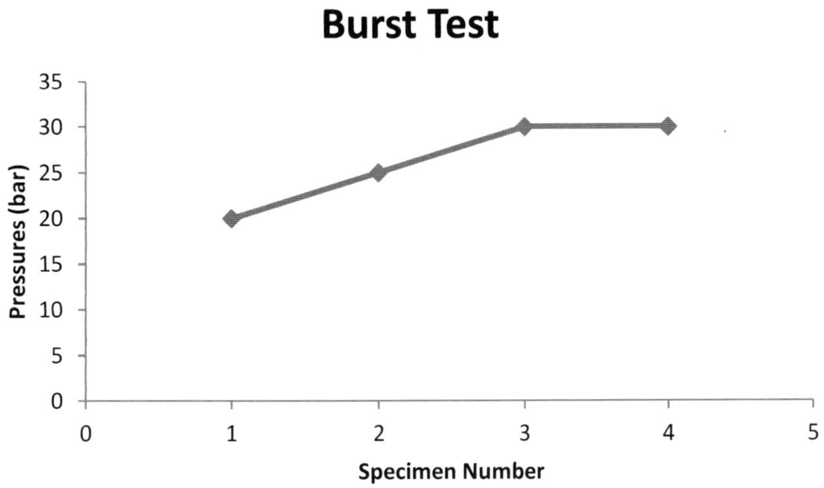

Figure 4. Burst test result of pressure as the function of numbers of layers

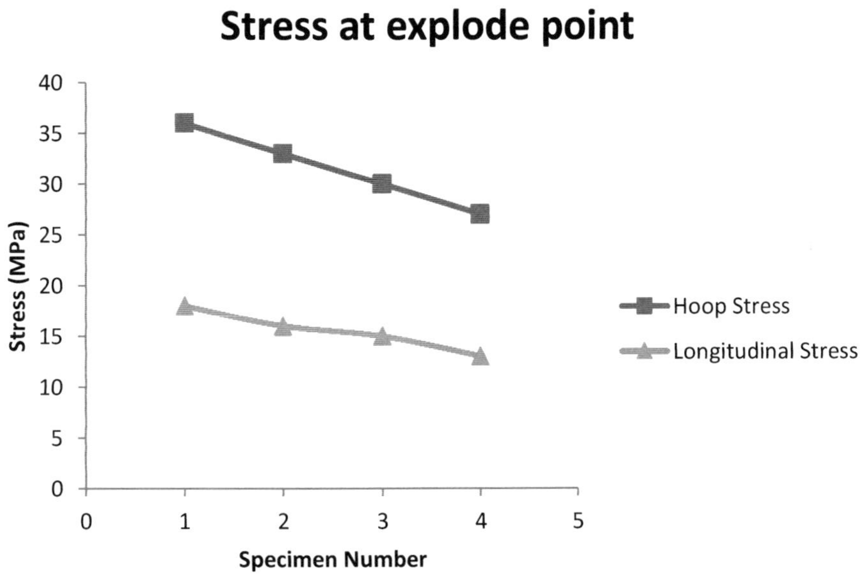

Figure 5. Stress at explode point as the function of number of layers.

From the data obtained, it appears that the pressure that can be held shows an increase with increasing layers. But the increase is not linear with the stress experienced by the fiber when the tube breaks. Assuming that the breaking strength of the fiber is fixed, the stress value, both hoop and longitudinal, should increase linearly.

If observed from the type of tube and fiber failure when the tube breaks (see figure 3), for tubes without fiber or fiber-wrapped, have the same fracture shape, while fibers do not have the same fracture characteristics, even for tubes wrapped by three layers of fiber shows that the fiber does not break.

Although the fiber does not break, but the fiber has a fairly large extension. this is what causes the stress experienced by the fiber to decrease when the tube experiences a maximum load.

5. Conclusions

From the analysis of the results, several conclusions were drawn. Namely, an increase in the number of layers of fiber used, can increase the breaking strength of COPV up to 50%. It became clear that the fiber elasticity doesn't accommodate to strengthen the tube, this is also caused by the tension of the fiber when the filament winding process is not optimal.

Although the strength of the fiber is high, but the percentage of elongation is higher than the percentage of the linear strain, it is not recommended to be used as a COPV reinforcing fiber.

Acknowledgements

The authors acknowledge the Indonesian Higher Education for funding the researche project within which this study was conducted.

References

[1] Pat B. McLaughlan, P. E., Scott C. Forth, Ph. D. and Lorie R. Grimes-Ledesma, Ph. D. Composite Overwrapped Pressure Vessels, A Primer. Houston : National Aeronautics and Space Administration, 2011. SP-2011-573.

[2] Moss, Dennis R. Pressure Vessel Design Manual. s.l. : Elsevier Sciense, 2004. 0570677406/9780750677400.

[3] Kaminski, Dr. Clemens. Stress Analysis & Pressure Vessel. s.l. : University Of Cambridge, 2005.

[4] Ryder, G. H. Strengt of Materials. London : The Macmillan Press LTD. 03331444430.

[5] M. Nebe et al., "Experimental Characterization of the Structural Deformation of Type IV Pressure Vessels Subjected to Internal Pressure", Key Engineering Materials, Vol. 809, pp. 47-52, 2019

Water absorption and tensile properties of the hybrid of fly ash-$CaCO_3$/epoxy composites

P D Setyawan, W Martadinata, S Sugiman*, S Salman, A D Catur and A Zainuri

Department of Mechanical Engineering, Faculty of Engineering, University of Mataram, Mataram, Indonesia

*E-mail: s.sugiman@unram.ac.id

Abstract. The paper presents water absorption and its effect on the tensile properties of hybrid of fly ash-$CaCO_3$-filled epoxy. The total filler content was 20wt%, with the fly ash to $CaCO_3$ ratios were 20/0, 15/5, 10/10, 5/15 and 0/20. Adding fillers into epoxy decreased the water uptake in the ranges of 15 - 22%, where the lowest and the highest water uptake of the hybrid systems belonged to the fly ash/$CaCO_3$ ratios of 15/5 and 20/0, respectively. The tensile strength and elastic modulus continued to decrease with aging time for all epoxy systems. At the saturation level, the tensile strength and elastic modulus of filled epoxies were improved compared to the unfilled epoxy, except for the hybrid of fly ash/$CaCO_3$ of 15/5. The fracture strain as well as the tensile toughness were considerably increased at the equilibrium. A scanning electron microscopy revealed that in dry condition the fly ash was well bonded with the matrix, but not for the $CaCO_3$ particles. In the semi equilibrium and equilibrium states, the fly ash and $CaCO_3$ were mostly detached from the epoxy. The hybrid of fly ash/$CaCO_3$ of 15/5 was not recommended as it decreased the tensile properties more than the other hybrid compositions.

1. Introduction

The increasing use of epoxy in coating, polymeric matrix composites and structural adhesives [1] has attracted the material engineers to improve the property in order to fulfill a specific requirement such as high stiffness and water resistance. Hard inorganic fillers, such as mica, silica, and glass bead have been extensively used to fill the thermoset such as unsaturated polyester and epoxy resin [2,3] and often they acted individually to improve the stiffness and/or toughness. Filler can effectively improve a specific property such as elastic modulus, while other fillers can improve the other aspects, such as water resistance. Therefore, hybridization of fillers may improve many aspects required for increasing the performance of thermosetting polymers.

Hybridization can include nano dan micro-fillers, soft and hard particles and different type of fillers. Imai et al. [4] investigated hybridization of micro-silica and layered silicate to improve the thermal expansion dan insulation breakdown properties of epoxy. The results indicated that the thermal expansion was low and the insulation properties were improved. Javni et al. [5] used micro- and nano-silica to fill the rigid and flexible polyurethane. They found that the nano-silica improved the hardness and compressive strength but decreased the rebound resilience. In the contrary, micro-silica decreased the hardness and compressive strength and increased the rebound resilience. Carolan et al. [6] investigated the hybrid of soft rubber particles and nano silica to improve the toughness as well as to maintain the stiffness of epoxy. They found that at the hybrid content of 8% rubber particles and 8%

nano-silica increased the toughness by about 900% compared to unmodified epoxy, but the elastic modulus decreased by about 15%.

Among the micro-fillers, fly ash and $CaCO_3$ are low cost fillers. Fly ash is a spherical particle as waste of coal combustion in a thermal power plant. Fly ash has been extensively studied to fill the epoxy resin [7,8], the unsaturated polyester resin [9], while micro- and nano-$CaCO_3$ particles have been utilized to modify the polyurethane [10], polylactic acid [11] and epoxy [8,12,13]. Our previous study [8] revealed that the fly ash improved the tensile strength and the toughness of epoxy better than the $CaCO_3$ particles. On the other hand, $CaCO_3$ particles reduced the water absorption and increased the elastic modulus of epoxy more than that of the fly ash particles. Therefore, it is interesting to hybridize both particles to improve the mechanical properties as well as to reduce water absorption of epoxy. The paper investigates the water absorption and its effect on the tensile properties of epoxy as well as the morphologies of the fracture surfaces.

2. Methods

2.1. Materials

Epoxy was based on a diglycidyl ether bisphenol-A with a hardener of polyamideamino. As filler, fly ash and calcium carbonate ($CaCO_3$) were used. Fly ash particles were obtained from Paiton's thermal power plant with the particle size of 200 mesh and less, while $CaCO_3$ was commercially purchased from a local store with the particles size of 800 mesh.

2.2. Specimen preparation.

For fabricating the tensile specimens, epoxy resin was mixed with fillers with total filler content of 20wt%. The ratios of fly ash (FA) to $CaCO_3$ (CC) were 20/0, 15/5, 10/10, 5/15, 0/20 (by weight), which were named as FA20/CC0, FA15/CC5, FA10/CC10, FA5/CC15 and FA0/CC20, respectively. The mixture was stirred using a hand mixer for 30 min. The mixture then was degassed in a vacuum chamber to release the air bubble trapped during mixing. Hardener was added to the mixture (with epoxy resin to hardener ratio of 1/1), stirred slowly to avoid significant air bubbles and then degassed again. The mixture then was transferred into a dog bone shaped-silicon rubber mould with dimension according to ASTM D638 [14]. The mould was left to cure for at least 24 hours.

2.3. Conditioning and testing

Conditioning was carried out by immersing the specimens in distilled water at a temperature of 50°C. The water uptake studies were performed for each specimen using the dog-bone specimen directly. Weighing of the specimens were conducted at every three hours during the first day and once a day for the following days. Conditioning was grouped as dry, semi-equilibrium and equilibrium. For the aged specimens, after the water content of the specimen reached approximately the semi equilibrium (144 hours) and the equilibrium states (2479 hours), they were taken out for tensile testing. The tensile test for the dry, semi equilibrium and the equilibrium specimens was conducted using a Tensilon testing machine with displacement rate of 5 mm/min. Scanning electron microscopy was performed to examine the fracture surface morphologies using a JEOL 6510 LA machine operating at 10 kV.

3. Results and Discussion

3.1. Water uptake

Figure 1 shows the water uptake vs. square root of aging time for the epoxy systems. It can be seen that the water uptakes were initially linear and then slowing down when reached close to the equilibrium (saturation). The non-linier water uptake at the very beginning stage might be caused by the difficulty of water diffused into the very smooth surfaces, as the specimen surface was not abraded before aging. It was also observed by another researcher [15]. The water uptakes seemed to follow the Fickian diffusion law as also reported by Sugiman and Salman [8].

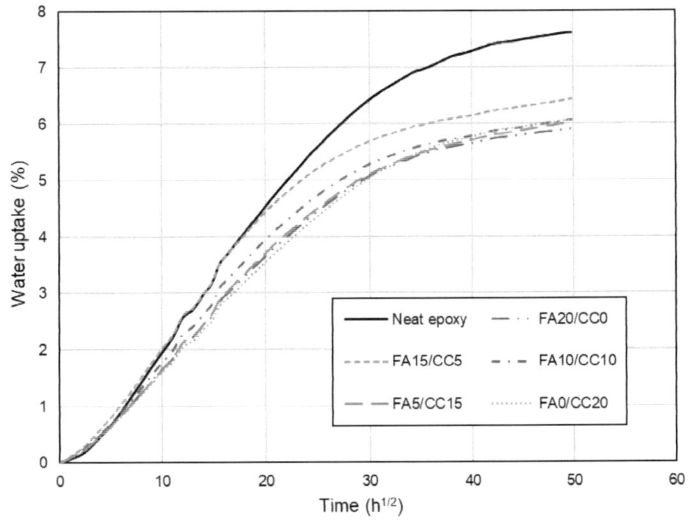

Figure 1. Water uptake vs. square root of aging time for epoxy systems.

Table 1. The diffusion properties of the epoxy systems.

Specimen	Water uptake ± SD (%)		Diffusion rate ± SD (mm^2/h) ($\times 10^{-3}$)
	At semi-equilibrium	At equilibrium	
Neat epoxy	2.51 ± 0.13	7.60 ± 0.07	3.60 ± 0.41
FA20/CC0	2.05 ± 0.03	5.90 ± 0.01	3.03 ± 0.03
FA15/CC5	2.41 ± 0.34	6.42 ± 0.09	3.13 ± 0.05
FA10/CC10	2.09 ± 0.14	6.05 ± 0.04	3.14 ± 0.13
FA5/CC15	1.87 ± 0.14	6.01 ± 0.10	3.34 ± 0.52
FA0/CC20	2.15 ± 0.16	6.06 ± 0.01	3.10 ± 0.02

As seen in Figure 1 and Table 1, the equilibrium water uptakes and the diffusion rate of the filled epoxies decreased compared to the neat epoxy. Compared to the neat epoxy, the decreases of the equilibrium water uptake were 22, 15, 20, 21 and 20% for FA20/CC0, FA15/CC5, FA10/CC10, FA5/CC15 and FA0/CC20, respectively. Meanwhile for the diffusion rate, the decreases were in the range of 12-16%, compared to the neat epoxy. For the filled epoxy systems, the diffusion of water took longer route due to the existence of particles. Individually (not hybridized), fly ash and $CaCO_3$ fillers acted the similar way as the equilibrium and the diffusion rate was about the same. The hybridization with the composition of FA10/CC10 and FA5/CC15 also did not significantly affect the diffusion properties compared to the fly ash- dan $CaCO_3$-filled epoxies. However, at the composition of FA15/CC5, the effectivity in reducing the equilibrium and the diffusion rate of neat epoxy was less than the other compositions. A fly ash particle contained some elements such as silicone (Si), aluminium (Al), magnesium (Mg), iron (Fe), potassium (K), and calcium (Ca) [8]. The elements of Na and Ca are soluble to water forming NaOH and $Ca(OH)_2$ respectively [16], while SiO_2 reacted to water forming silica acid [17]. The existence of a small amount of $CaCO_3$ together with the hydrolysed fly ash might be favourable for forming the calcium silicate hydrates that more consumed water than the other compositions [18].

3.2. Tensile properties

3.2.1. Tensile strength. Figure 2 shows the tensile strength of the epoxy systems in the dry, semi equilibrium and the equilibrium states. In the dry condition, compared to the neat epoxy the significant decrease of the tensile strength, by about 15% was observed for the FA15/CC15 composition, while the tensile strength of other compositions did not differ significantly (based on the analysis of variance). After being aged until the semi equilibrium state, the tensile strength for all epoxy systems decreased significantly. The highest decrease was also observed for the FA15/CC5 compositions, while for the other compositions, the tensile strength did not significantly differ to that of the neat epoxy. The decrease of the tensile strength for the neat epoxy, FA20/CC0-, FA15/CC5-, FA10/CC10-, FA5/CC15- and FA0/CC20-filled epoxies were about 28, 32, 54, 33, 34 and 33%, respectively. Further aging to the equilibrium state, the tensile strength significantly decreased and the decreases were 83, 76, 85, 79, 80, and 79% compared to those in the dry condition for each epoxy system. Again, the highest decrease was observed for the FA15/CC5-filled epoxy. In this case, plasticization of the matrix was the main cause in decreasing the tensile strength for the neat epoxy [8,15]. For the filled epoxies, in addition to the plasticization, particles debonding also contributed to the decrease of the tensile strength [8]. This effect was clearly seen in the FA15/CC5 filled epoxy, while for the other compositions, it was not observed as the tensile strength did not significantly differ with that of the neat epoxy. Nevertheless, the hybrid composition of FA5/CC15 was preferred as it has the highest tensile strength in the dry condition and has relatively the modest reduction in the wet condition.

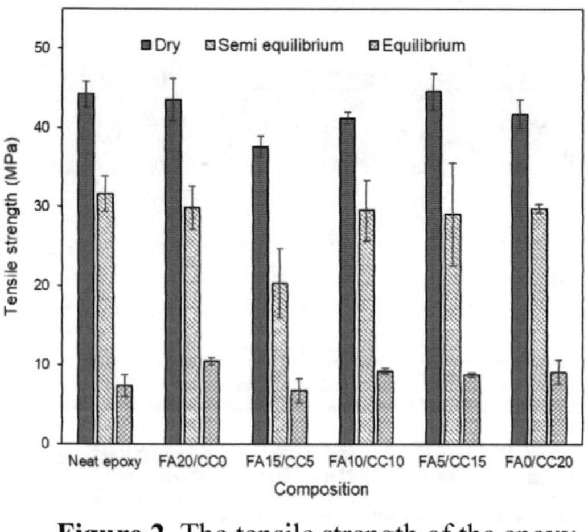

Figure 2. The tensile strength of the epoxy systems in various conditions.

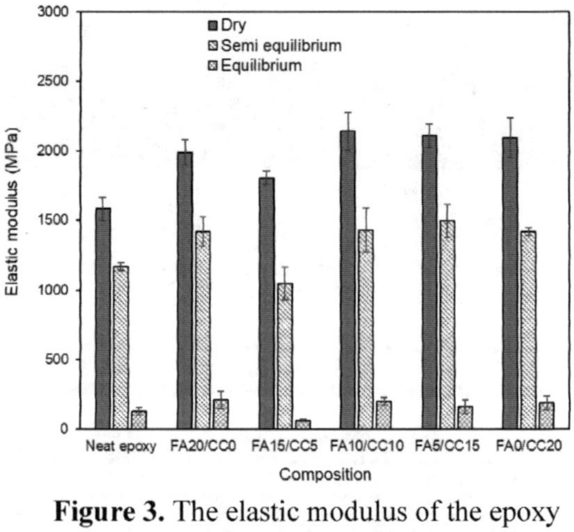

Figure 3. The elastic modulus of the epoxy systems in various conditions.

3.2.2. Elastic modulus. Figure 3 shows the elastic modulus of the epoxy systems in various conditions. In the dry condition, the elastic modulus increased compared to the neat epoxy. This was expected as the elastic modulus of fillers was much higher than the epoxy matrix. The increase of the elastic modulus for the FA20/CC0-, FA15/CC5-, FA10/CC10-, FA5/CC15- and FA0/CC20-filled epoxies were 26, 14, 35, 33 and 33%, respectively. It was seen that when the content of $CaCO_3$ was higher than 10% in the total filler content, the elastic moduli were about the same. The elastic modulus for the FA was about 17 GPa [19], while the elastic modulus of $CaCO_3$ was about 35 GPa [20], so it was expected that high content of $CaCO_3$ increased the elastic modulus more than that of the fly ash. Similar to that in the tensile strength, compared to the neat epoxy, the increase of the elastic modulus for FA15/CC5 was the lowest among the other compositions.

For the specimens aged until the semi equilibrium state, compared to the dry condition, the elastic moduli decreased and the reductions were 26, 10, 34, 10, 6 and 10% for the neat epoxy, FA20/CC0-, FA15/CC5-, FA10/CC10-, FA5/CC15- and FA0/CC20-filled epoxies, respectively. The elastic moduli for the filled epoxies were still higher than that of the neat epoxy; except the elastic modulus of the FA15/CC5-filled epoxy, which was lower than that of the neat epoxy. After aging until the equilibrium, the elastic moduli of the epoxy systems had the similar the trend to those of the semi equilibrium. The elastic modulus at the equilibrium state were dropped by 92, 87, 96, 87, 90 and 88% for the neat epoxy, FA20/CC0-, FA15/CC5-, FA10/CC10-, FA5/CC15- and FA0/CC20-filled epoxies, respectively, compared to those in the dry condition. This was caused by plasticization effect that increased with the increase of aging time (or water content). For the filled epoxies, the effect of plasticization was compensated by the particles by hindering plastic deformation, so the decrease of the elastic moduli was not as high as that for the neat epoxy. It seemed that the effect of plasticization on the elastic modulus was more pronounced than the effect on the tensile strength [15].

3.2.3. Fracture strain. Figure 4 shows the fracture strains of the epoxy systems under various conditions. In contrast to the tensile strength and the elastic modulus, in the dry condition, the fracture strains of the filled epoxies were slightly lower than that of the neat epoxy. However, after being aged until the semi equilibrium state, compared to those in the dry condition, the fracture strains increased by about 64, 61, 110, 32, 9, and 19% for the neat epoxy, FA20/CC0-, FA15/CC5-, FA10/CC10-, FA5/CC15- and FA0/CC20-filled epoxies, respectively. Further aging to the equilibrium state, the increases of fracture strains were remarkably, by about 588, 602, 531, 557, 544 and 623% for the neat epoxy, FA20/CC0-, FA15/CC5-, FA10/CC10-, FA5/CC15-, and the FA0/CC20-filled epoxies, respectively, compared to those in the dry condition. The remarkable increase of the fracture strain indicated that the aged epoxy systems became very ductile due to the plasticization effect.

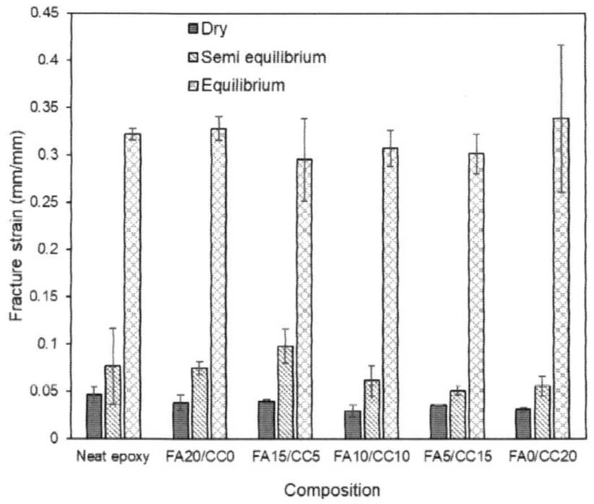

Figure 4. The fracture strain of the epoxy systems in various conditions.

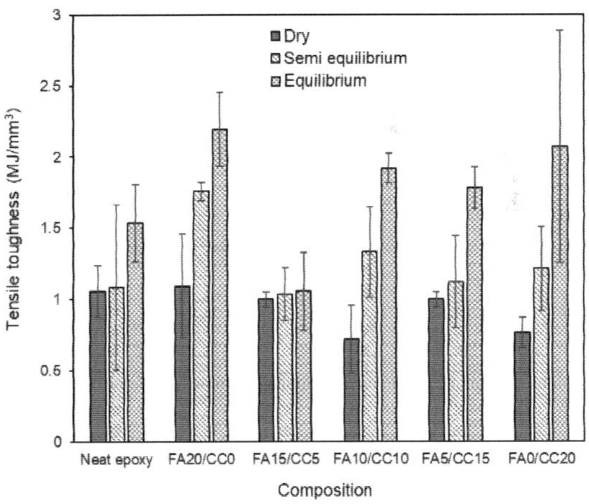

Figure 5. The tensile toughness of the epoxy systems in various conditions.

3.2.4. Tensile toughness. Figure 5 shows the tensile toughness of the epoxy systems in various conditions. The toughness was obtained by measuring the area underneath the tensile stress-strain curve, which indicated the amount of energy absorbed for breaking [21]. Referring to Figures 3 and 4, it seemed that the toughness was the opposite to the elastic modulus, and it was proportional to the fracture strain. As seen in Figure 5, in the dry condition, the toughness did not differ significantly among the epoxy systems (based on the analysis of variance); however, they tended to increase after being aged until the semi equilibrium and further increased with increasing aging time up to the equilibrium state. The highest toughness belonged to FA20/CC0 composition, which was the fly ash-filled epoxy. Sugiman and Salman [8] also reported that the fly ash-filled epoxy had higher fracture toughness than the $CaCO_3$-filled epoxy. Nonetheless, the other compositions except FA15/CC5 also indicated high toughness after aging and they were higher than that of the neat epoxy.

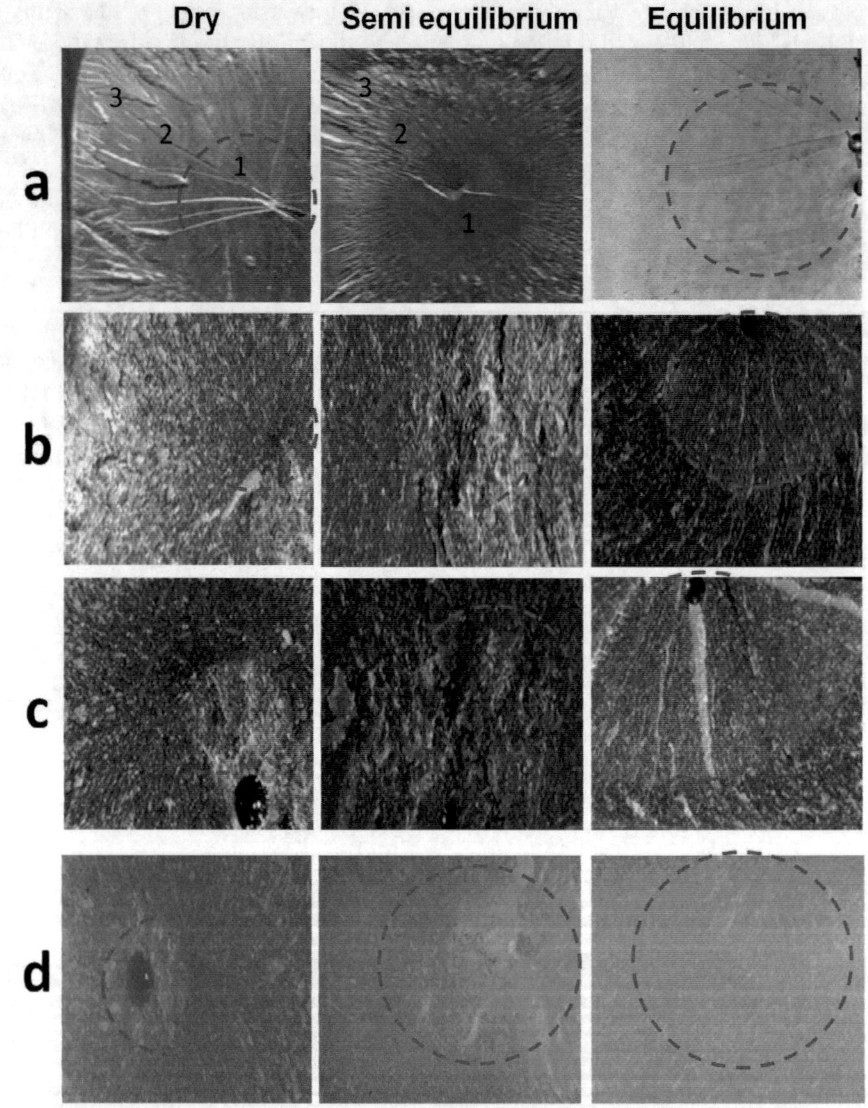

Figure 6. Optical micrograph of the fracture surfaces of (a) neat epoxy, (b) FA20/CC0, (c) FA15/CC5, and (d) FA0/CC20. The magnification is 50×.

3.3. Fracture morphologies

Figure 6 shows the optical micrographs of the fracture surfaces of the selected compositions in the dry, semi equilibrium and the equilibrium states. For brittle fracture like epoxy, the fracture surfaces consisted of three regions; mirror, mist and final fracture regions [22,23]. The mirror region is an area surrounding the crack source. The mist region is the transition between the mirror and the final fracture regions. The surface roughness is smooth like mirror in the mirror region and gradually increases into the transition region and very rough in the final fracture region [22,23]. It was seen in Figure 6a that for the neat epoxy, the mirror, mist and the final fracture regions were distinguishable, indicated by marks 1, 2 and 3, respectively. As seen in Figure 6a, in the dry condition, the mirror region (in the red dashed line, indicated by mark 1) was clearly seen, but the mist region was not. In the semi equilibrium state, the mirror and the mist regions were obvious. In this condition, the mechanical properties between the inner and outer of the bulk specimen were different, as at the outer layer the water content was higher than the inner part of the bulk. Due to plasticization, the outer layer was more ductile than the inner; therefore, this affected the fracture mechanism leading to the different appearances of the fracture surfaces. In the equilibrium state, the fracture surface was smooth, where the mirror, mist and the final fracture regions were hardly distinguished. In this condition, the bulk of epoxy has reached the saturation level, so the difference in the mechanical properties between the outer dan the inner of bulk might be small, similar to that in dry condition. The area of mirror region increased with the increase of aging time (water content), due to increasing the plasticization effect [24].

For the filled-epoxy systems, the fracture surfaces also showed the three regions, although they were not as obvious as that for the neat epoxy. The fracture surfaces were rougher than those of neat epoxy due to the fillers that acted as loci of microcrack initiation, crack branching, crack pinning and void growth [2,3,8]. As seen in Figures 6(b-d), the fracture surfaces for the filled epoxies in the same conditions (dry, semi equilibrium and equilibrium) seemed similar. The area of mirror region increased with aging time, which was consistent with that in the neat epoxy.

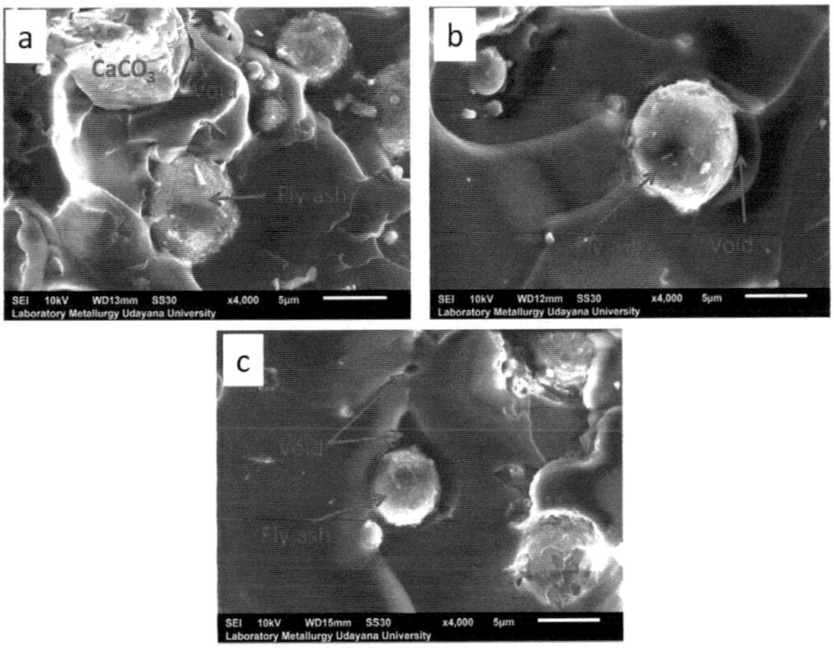

Figure 7. The scanning electron microscopy (SEM) micrograph of the fracture surfaces of FA15/CC5-filled epoxy in (a) dry, (b) semi equilibrium, and (c) equilibrium states.

The scanning electron microscopy (SEM) was conducted on the selected composition, which was FA15/CC5-filled epoxy, as this composition have the lowest properties in all conditions. As seen in Figure 7, the fracture surface in the dry condition was rougher than the other conditions. In dry condition (Figure 7a), the fly ash particles were well bonded by the epoxy matrix; however, the $CaCO_3$ particle appeared to debond, as void was observed around the $CaCO_3$ particle. Similar results on this behaviour were reported in Ref. [8]. The debonded $CaCO_3$ particle might initiate the microcrack that could propagate causing the failure. This might cause the lower properties of the FA15/CC5-filled epoxy. After being aged in the semi equilibrium and the equilibrium states, the matrix plasticized, and water might attack the interface filler-matrix. When the filler has a weak interface bonding with epoxy matrix, water may displace the particle from the matrix and upon loading, and then the particles debonded from the matrix. This decreased further the tensile strength. The debonded fly ash particles after being aged in the semi-equilibrium and equilibrium are shown in Figures 7b and 7c, respectively. It can be seen that the fly ash particle debonded from the matrix and the void grew upon loading. The matrix void growth became wider after aging to the equilibrium state, where the matrix plasticization was massive.

4. Conclusions

Investigation of water absorption and its effect on the tensile properties of the neat and the hybrid of fly ash (FA)-$CaCO_3$ (CC)-filled epoxies have been undertaken. The total filler content was 20wt%, and the hybrid compositions were FA20/CC0, FA15/CC5, FA10/CC10, FA5/CC15, FA0/CC20. It can be concluded as follows:

- The hybrid of fly ash-$CaCO_3$-filled epoxy decreased the equilibrium water uptake and the diffusion rate of epoxy resin. The decrease of the water uptake was about the same (20-22%), but for the composition of FA15/CC5, it was about 15%. While the decrease of the diffusion rate was about 12-16%.
- The tensile strength of the hybrid of fly ash-$CaCO_3$-filled epoxy did not differ significantly to that of the neat epoxy, except the composition of FA15/CC5, where the tensile strength decreased by about 15%. Meanwhile, the hybrid of fly ash-$CaCO_3$ significantly increased the elastic modulus of epoxy.
- Aging in distilled water at temperature of 50°C caused deleterious effect on the tensile strength and the elastic modulus for all of the studied epoxy systems; however, the fracture strain and the toughness increased considerably. Again, the effect of aging has the highest impact on the composition of FA15/CC5, so it is not recommended for hybridizing the fly ash and $CaCO_3$ with the fly ash to $CaCO_3$ ratio was close to 15/5, but the hybrid composition of FA5/CC15 was recommended.

5. Acknowledgements

This work was funded by University of Mataram through PNBP research grant and partly by Fundamental Research Scheme (DRPM, Kemenristekdikti), Republic of Indonesia under contract number 1836/UN18.L1/PP/ 2019.

References

[1] Kong J, Ning R, Tang Y 2006 *J Mater Sci.* **41** 1639–41.
[2] Adachi T, Osaki M, Araki W, et al. 2008 *Acta Mater.* **56** 2101–09.
[3] Lee J, Yee AF 2000 *Polymer* **41** 8375–85.
[4] Imai T, Sawa F, Nakano T, Ozaki T, Shimizu T, Kozako M, and Tanaka T 2006 *IEEE Transactions on Dielectrics and Electrical Insulation* **13(**1).
[5] Javni I, Zhang W, Karajkov V, Petrovic ZS, and Divjakovic V 2002 *Journal of Cellular Plastics* 38 229.
[6] Carolan D, Ivankovic A, Kinloch AJ, Sprenger S, Taylor AC 2016 *Polymer* **97** 179-190.
[7] Baheti V, Militky J, Mishra R, Behera BK 2016 *Composites Part B* **85** 268-276.
[8] Sugiman S & Salman S 2019 *Journal of Adhesion Science and Technology* 33(7) 691-714.
[9] Sugiman and Setyawan PD 2013 *Makara Technology Series* **17**(3) 128-132.
[10] Donate-Robles J, Martın-Martınez JM 2011 *International Journal of Adhesion & Adhesives* **31** 795–804.
[11] Kumar V, Dev A, Gupta AP 2014 *Composites: Part B* **56** 184–8.
[12] Eskizeybeka V, Ulus H, Kaybal HB, Şahin OS, Avcı A 2018 *Composites Part B* **140** 223–231.

[13] Atta AM, Al-Lohedan HA, Ezzat AO, Al-Hussain SA 2016 *Progress in Organic Coatings* **101** 577–586.
[14] ASTM D638-02a 2007 *Standard test method for tensile properties of plastics* (Pennsylvania: ASTM International).
[15] Sugiman, Sularjaka 2016 *International Journal of Technology* **3** 438-446.
[16] Mattigod SV, Rai D, Eary LE, Ainsworth CC 1990 *J. Environ. Qual.* **19** 188-201.
[17] Iyer R 2002 *Journal of Hazardous Materials* **B93** 321–329.
[18] Capiel G, Uicich J, Fasce D, Montemartini PE 2018 *Polymer Degradation and Stability* **153** 165-171.
[19] Matsunaga T, Kim JK, Hardcastle S, Rohatgi PK 2002 *Materials Science and Engineering A* **325** 333–343.
[20] Xanthos M 2010 Modification of Polymer Properties with Functional Fillers *Functional Fillers for Plastics,* ed. M Xanthos (Weinhem: WILEY-VCH Verlag GmbH & Co. KGaA).
[21] Askeland DR, Fulay PP 2009 *Essentials of Materials Science and Engineering* 2nd (Toronto: Cengage Learning).
[22] Kinloch AJ, Young RJ, 1983 *Fracture Behaviour of Polymers* (London: Applied Science Publishers).
[23] Greenhalgh ES, 2009 *Failure Analysis and Fractography of Polymer Composites* (Cambridge: Woodhead Publishing Limited and CRC Press LLC).
[24] Sugiman S, Putra IKP, Setyawan PD 2016 *Polymer Degradation and Stability* **134** (2016) 311-321.

IOP Conf. Series: Materials Science and Engineering **694** (2019) 012038 doi:10.1088/1757-899X/694/1/012038

Effect of Ramie Fibers on Fatique Strength Automotive Product

I Komang Astana Widi, Gerald Pohan, Wayan Sujana and Tutut Nani Prihatmi

National Institute of Technology of Malang (ITN Malang)

E-mail: aswidi@yahoo.com

Abstract. The automotive industry is already utilizing many composite materials for its products since composite materials are very environmentally friendly, have low economic costs and able to reduce weight. Ramie fiber reinforced by natural composites have a great opportunity to be used as materials for automatic components. The mechanical bonding of the ramie fiber reinforced composite pad can be improved by increasing the compatibility between the matrices and good product handling technique. In addition, the mechanical behavior is also greatly influenced by the orientation of the fibers in different matrices. This paper will discuss the effect of fiber orientation (unidirectional, woven and random) on several resin matrices (epoxy and polypropilene). The purpose of this study is to observe the effect of ramie fiber orientation on fatigue resistance properties in the form of handle components. The observations used SEM (scanning electron microscope) for observing of the failure mechanism. Fatigue test of the specimens using 80s carry-type handles (in situ). Handling fatigue testing equipment is specially designed in accordance with the application of the load. The strength of the fatigue observed in the handle component shows that the optimum value is produced in the handle with the orientation variation of fiber in both the epoxy matrix and the polypropylene matrix, namely 17141 and 8414 cycles. The presence of ramie reinforcement fibers in general can increase the fatigue strength of handle components. Whereas the lowest fatigue strength is owned by fibers which have a random orientation. However, when compared with no fiber, the component handles with randomly oriented fibers still have better fatigue strength.

1. Introduction

The use of natural fiber composites is one solution to support automotive development, as Indonesia is a big producer of fiber. However, the fiber potential has not been utilized optimally for bigger industrial needs. One of the fibrous plants that is currently becoming a trend and is being developed is ramie fiber [1]. This annual plant in the form of clumps is easy to grow and develop in the tropics; the ramie plant known by its Latin name Boehmeria Nivea (L) Goud produces natural vegetable fibers from ribbons on its very hard and shiny wood bark. In certain cases, ramie fiber has advantages over other fibers such as tensile strength, water absorption, and moisture and bacteria resistance.

Some reasons that underlie this research include:

(1) Conserve the environment (government programs, global warming) ultimate disposability a sustainability resources: natural fiber (FRNP)

(2) Natural fibers in Indonesia have very many types and quantities but have not been utilized optimally, especially for technical applications
(3) Effect of natural ramie fiber orientation as reinforcement material on polymer composites can increase the product's life
(4) Natural ramie fiber composites can be used to improve the performance of materials made of plastic materials
(5) Demand for automotive interior products, especially the old type car door panels

Ramie plants that have existed since the Japanese era during World War II are annual plants in the form of clumps that are easily grown and developed in the tropics, which are resistant to diseases and pests, and can support the preservation of nature and the environment. The ramie plant, known by its Latin name Boehmeria nivea (L) Goud, are able to produce natural vegetable fibers from ribbons on its very hard and shiny wooden bark. Ramie fiber has the characteristics of cotton fiber that is both spun and mixed with other fibers to be used as raw material for textiles. In certain cases ramie fiber has advantages over other fibers such as tensile strength, water absorption, moisture and bacterial resistance, heat resistance, ranked number 2 after silk compared to other natural fibers, lighter than sentimental and friendly fibers environment (does not pollute the environment so it is good for health). Ramie fiber can be used as a raw material for textiles, by mixing together cotton or polyester fibers, (as a substitute for cotton / cotton fiber). The development of ramie plants has a very bright prospect, the world's need for ramie fiber is 400,000 tons per year, so far the supply is 270,000 tons per year, with a total supply of 130,000 tons. From the results of the study, the quality of ramie fiber in Indonesia is able to compete with ramie fiber from China, Brazil, the Philippines, Taiwan, Korea, Cambodia, Thailand and Vietnam. Thus the development of this plant has a very bright prospect, because until now Indonesia has great potential to drive the people's economy through the rural economy, farmers' income and non-oil export commodities [2].

Kompas data [3] stated that polypropylene is a safe plastic material in sequence number 5. Safe means that the material can be used as packaging material for both food and drinks. Therefore Polypropylene (PP) is widely used to package soup, tomato sauce and margarine.

Table 1. Polypropylene and polypropylene with recycled rice husk fiber, a) melt flow index value, b) mechanical properties [4]

Samples	Tensile Strength (Mpa)	Flexural Strength (Mpa)	Impact Strength (Mpa)
Polypropilene	30.5	52.6	2.76
Polypropilene + Rice husk	21.5	48.7	3.17
Polypropilene + Rice husk (1st Recycled)	20.5	47.0	3.04
Polypropilene + Rice husk (2nd Recycled)	20.5	46.8	2.88
Polypropilene + Rice husk (3rd Recycled)	20.6	45.9	2.82
Polypropilene + Rice husk (4th Recycled)	20.3	43.3	2.61

Thermoplastic resins heated to liquid (plastic) are continuously extruded through the mold hole. The extrusion machine consists of a mold and a pressure unit. The mold provides the shape and pressure unit to push material out of the mold hole, cools and controls the shape and dimensions of the extruded product. This process is widely used to make rods, pipes, thin sheets and so on.

IOP Conf. Series: Materials Science and Engineering **694** (2019) 012038 doi:10.1088/1757-899X/694/1/012038

2. Methodology

The methodology for the research in this paper is shown in Figure 1.

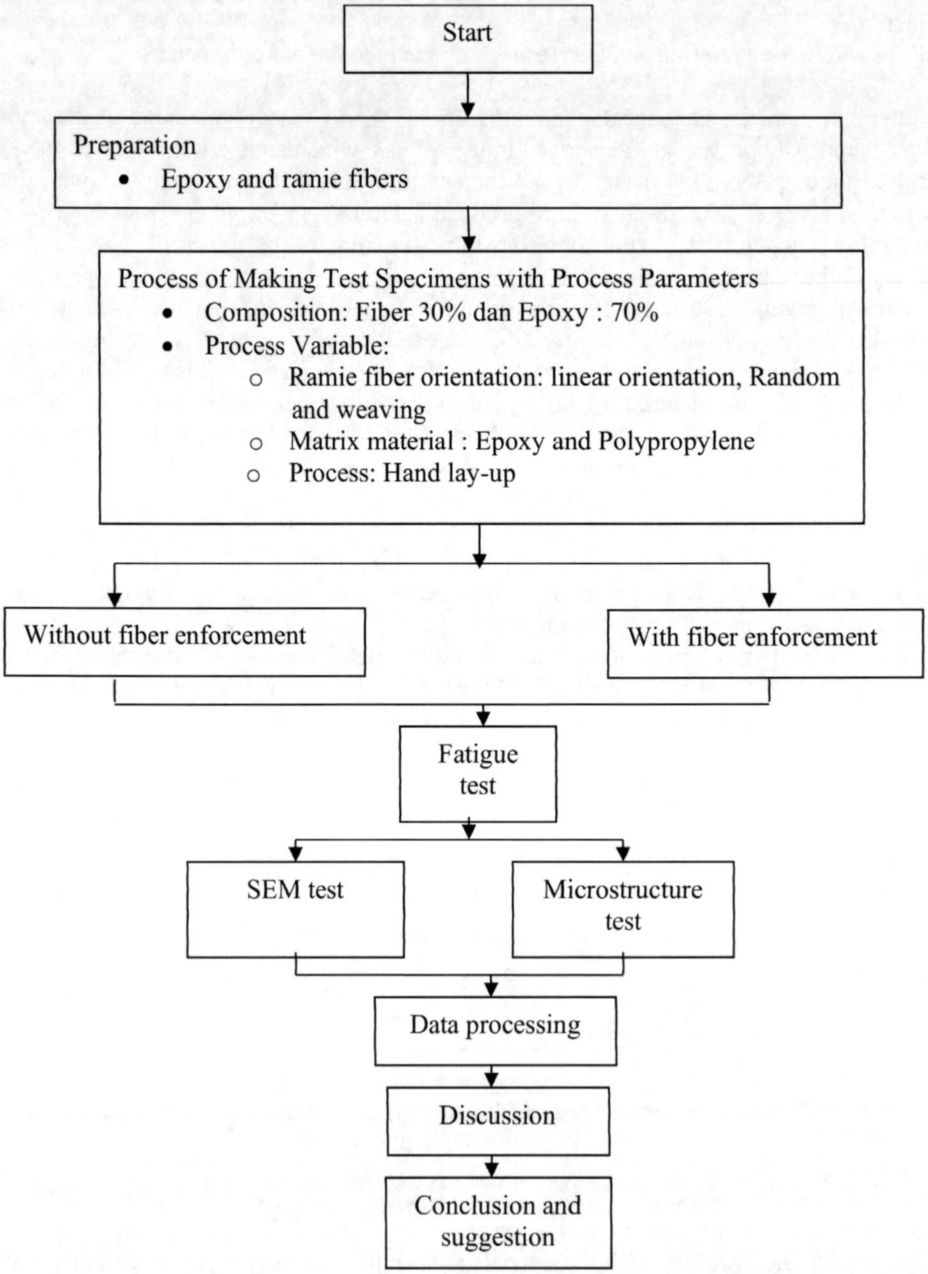

Figure 1. Flowchart of the methodology

IOP Conf. Series: Materials Science and Engineering **694** (2019) 012038 doi:10.1088/1757-899X/694/1/012038

3. Result and Discussion

In general, the orientation of the fiber is very influential to the change in fatigue resistance of car door panels. This is indicated by significant data which sequentially the orientation increases the fatigue resistance of the handle product from random orientation with the lowest fatigue strength followed by weaving orientation while the best is at unidirectional orientation. The analysis shows that fatigue strength is greatly influenced by the stress distribution contained in the component due to changes in the position of the fiber (fiber orientation). This stress distribution is also influenced by the design of the handle components [5].

Table 2. Data on Testing Results of epoxy matrix products with Ramie Fiber Orientation and Without Ramie Fiber

No	Fiber Type	Speed (rpm)	Cycle to Broken Specimens (N)				
			1	2	3	4	Average (∑)
1	Without fiber	800	6896	5287	6495	7479	6540
2	Random	800	9495	8479	10676	8297	9236
3	Weaving	800	13622	14246	11275	11387	12632
4	Linear	800	19721	17174	16557	15113	17141

Table 3. Data on Testing Results of polypropilene matrix products with Ramie Fiber Orientation and Without Ramie Fiber

No	Fiber Type	Speed (rpm)	Cycle to Broken Specimens (N)				
			1	2	3	4	Average (∑)
1	Without fiber	800	8522	7089	7255	6692	7389
2	Random	800	5485	7634	3795	4314	5307
3	Weaving	800	7233	8101	5244	3805	6095
4	Linear	800	9324	10112	6787	7436	8414

Table 4. Data on Test Results of fatigue matrix products and polypropylene matrices with Ramie Fiber Orientation and Without Ramie Fiber

No	Fiber Type	Cycle to Broken Specimens (N)	
		Epoxy Matrix	Polypropylene Matrix
1	Without fiber	6540	7389
2	Random	9236	5307
3	Weaving	12632	6095
4	Linear	17141	8414

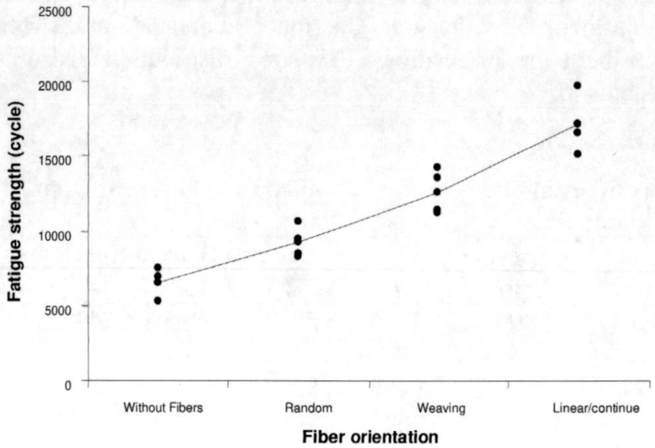

Figure 2. Test results of Epoxy matrics product fatigue with Ramie Fiber Orientation and Without Ramie Fiber Orientation

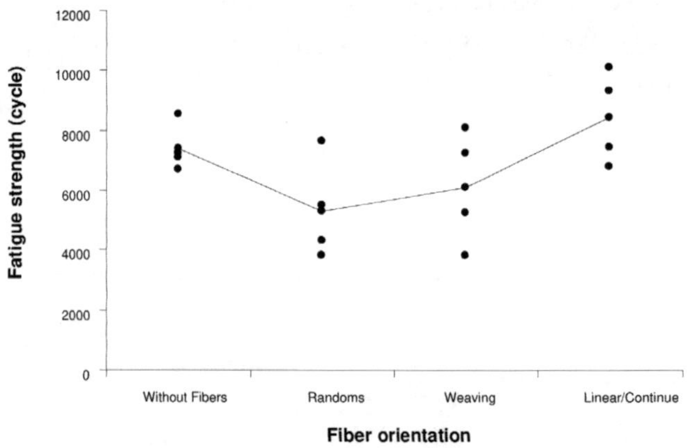

Figure 3. Test results of polypropilene matrics product fatigue with Ramie Fiber Orientation and Without Ramie Fiber

The initial fatigue resistance before the use of ramie fiber as a reinforcing element in both matrices, shows the fatigue resistance of the polypropilene matrix is better than the epoxy matrix. Meanwhile, if viewed from the results of fatigue strength testing by using fiber as a reinforcement shows that compared to polypropylene composites, epoxy matrix that is not too strong in this composite has two important roles, namely as a medium for transferring loads to the fiber, and the fiber-matrix interface deflects and stops small cracks, so that the composite can withstand cracks better than the single component/material forming. This is what causes the composite with the ramie-epoxy combination to be more effective than the ramie-polypropilene combination. Therefore, it can be concluded that the effectiveness of the epoxy matrix is better than polypropilene.

The fatigue test data shows that the presence of fiber as a reinforcement in the car door panel components can increase the service life of the car door panel components. However, fiber orientation can also affect the decreasing fatigue life of the component, ie from PP matrix without fiber with a fatique value of 7389 with the use of fiber in the direction of random orientation showing a decrease in the value of the average cycle endurance to 5307 cycles (N) and this fatigue strength will increase in the orientation of woven and unidirectional fibers. The influence of orientation of the fiber with random directions can increase the concentration of stress on the component, as evidenced by the number of initial cracks which are more numerous than other orientations. In addition, other factors that can increase concentration are increased formation of porosity due to uneven bonding between the fiber and the matrix, which with random orientation or irregularity, so that more air is trapped because the matrix or epoxy resin cannot unite the overlapping fibers. Then the bond between the fiber and the matrix cannot be so fused. The surface area of the trapped air (porosity) in the area of the fiber that is not or is less integrated is the beginning of the crack (Figure 4).

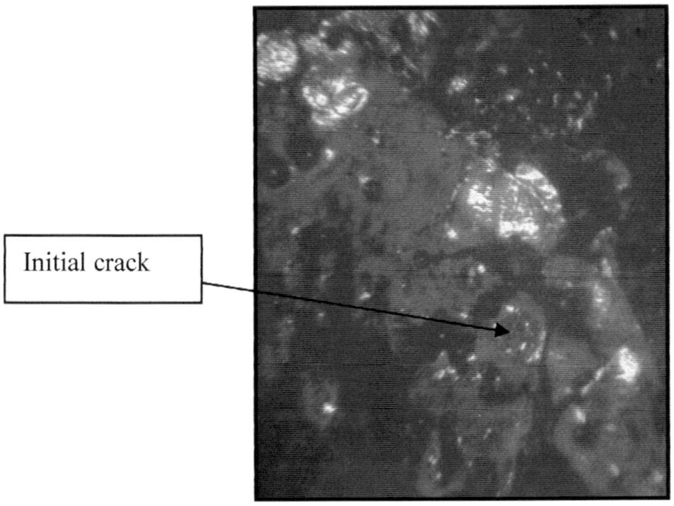

Figure 4. Random orientation microstructure shows a number of initial cracks.

The fatigue strength of the same direction orientation is the highest compared to other orientations namely, an average of 9124 cycles (N), this is due to the arrangement of the position or placement of fibers that are evenly distributed (unidirectional), in every part of the component and gives an influence on dynamic loads. Where the stress concentration of the components can be reduced, it can be made more evenly as indicated by the broken position of the car door panel, the same as the initial design simulation (Ansys).

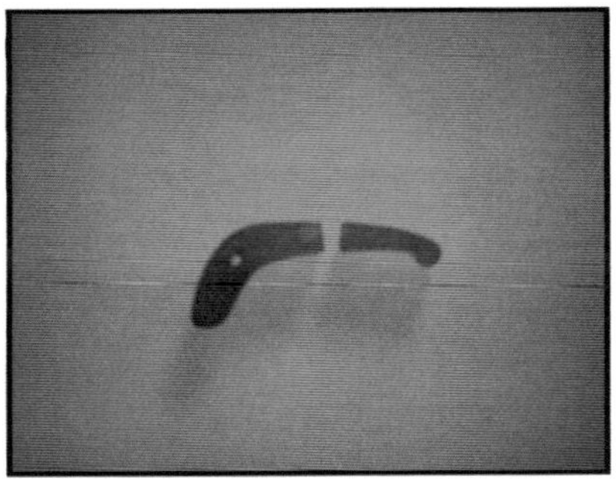

Figure 5. There is an unequal stress concentration due to the influence of the handle design

The lower resistance of ramie-polypropilene fatigue is also caused by the fabrication factor of making composites where the fabrication process that utilizes the injection process with a process temperature of around 120 0C will cause some ramie fibers to burn so that it will cause combustion gases trapped inside the fibers to form porosity. This is due to the topography of the ramie fiber surface which has a variety of surface textures and ramie fiber has pores ranging from 20 - 500 Angstroms (mesoporous type). With the size of these pores, the temperature of oxygenation and oxygen trapped inside the fiber pore can cause fiber burning and an increase in the size of porosity in the composite (Figure 6).

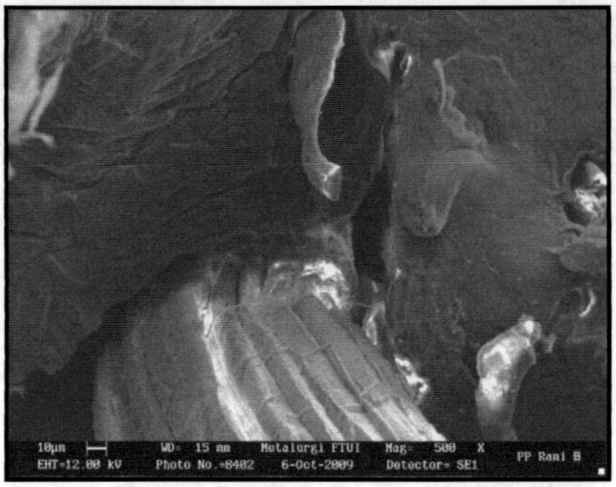

Figure 6. SEM analysis photo forming porosity around the burning fiber.

In addition, the injection of this matrix will also change the composition of the ramie reinforcement fibers inside the matrix so that the resulting stress becomes uneven in each handle component, thus resulting in lower fatigue resistance as shown in the ideal geometry model of fiber-packing by Gibson (Figure 7) following the equation below:

Volume of fiber fraction in the square model: $V_f = \pi/4 \, . \, (d/s)^2$ (1)

The maximum fiber fraction volume if s = d: $v_{fmak} = v_{fmak} = \pi/4 = 0.785$ (2)

Volume of fiber fraction in the triangle model: $V_f = \pi/2\sqrt{3} \, . \, (d/s)^2$ (3)

The maximum fiber fraction volume if s = d: $V_f = \pi/2\sqrt{3} = 0.907$ (4)

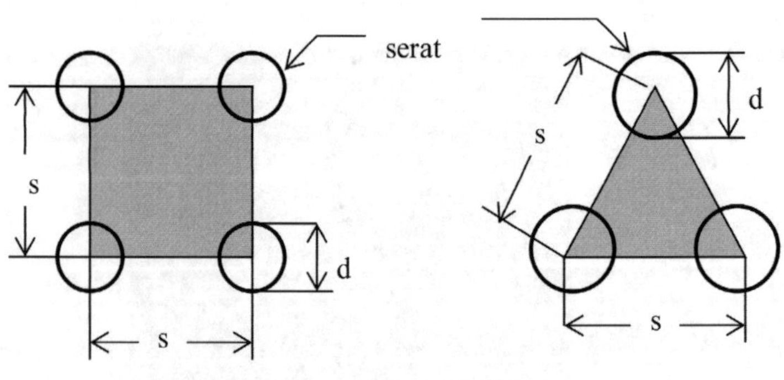

Figure 7. Fiber geometry bound by a matrix

The equation is in accordance with the simulation model ansys which shows the maximum voltage tendency with the variation of fiber orientation will form a packing system in which the ramie-polypropilene changes in the volume fraction from a square model to a triangle model which means the distance d/s will increase due to injection pressure so that the strength packing will be lower which means the strength of fatigue will also be reduced. this points to an increase in the maximum stress value in the simulation results for the composite ramie-polypropilene (Table 4).

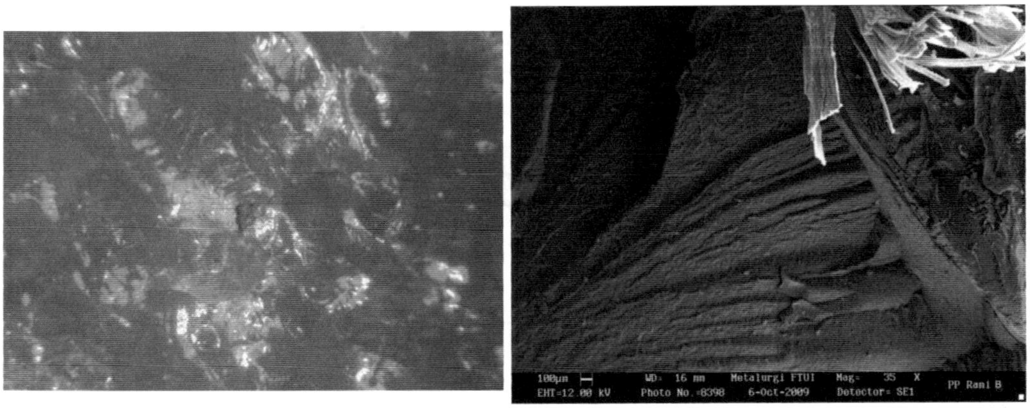

(a) **(b)**

Figure 8. Photograph of (a) macro structure and (b) SEM Analysis of grooved fractures on the surface of the handle components

Based on the analysis of failure due to fatigue on the handle products on natural fiber fiber composite materials can be correlated with the theory of failure due to fluctuating loads, generally there are two possible mechanisms, namely due to fiber breaking (matrix breaking) and matrix cracking.

The mechanism of failure of broken fibers generally occurs in epoxy matrix composites. Because this epoxy matrix has quite resilient properties which is shown by the strain value which is quite high when compared with the polypropylene matrix. What is analyzed is based on the modulus of elasticity which shows the stress value formed on the epoxy matrix is lower than the polypropylene matrix (as Ansys analysis shows). So that the fibers at the crack ends will be broken and the cracks on the matrix will propagate. If the cracks that occur are long enough, the shear stress that occurs will be able to damage the bond between the fiber and the matrix so as to allow parallel cracks with the fiber. This mechanism occurs due to the high loading level so that it exceeds the weakest fiber strength of the composite and will cause a shear stress which is concentrated at the fiber end. This shear stress will cause the fiber to be separated from the matrix and the length of the area released from the matrix depends on the magnitude of the shear stress that occurs. In the area of fiber that is separated from the matrix (debanded area) stress concentrations will occur and eventually cause failure in the composite material. This mechanism can be shown from the results of photo analysis on the following epoxy matrix composites.

IOP Conf. Series: Materials Science and Engineering **694** (2019) 012038 doi:10.1088/1757-899X/694/1/012038

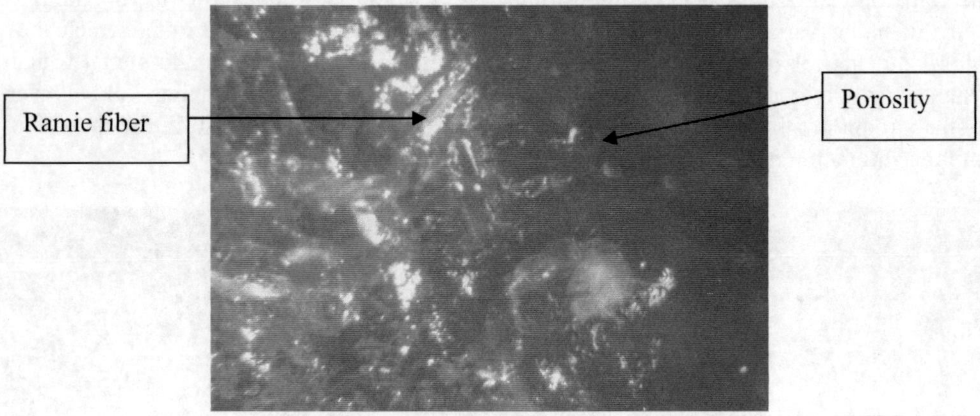

Figure 9. Random fiber fatigue test result

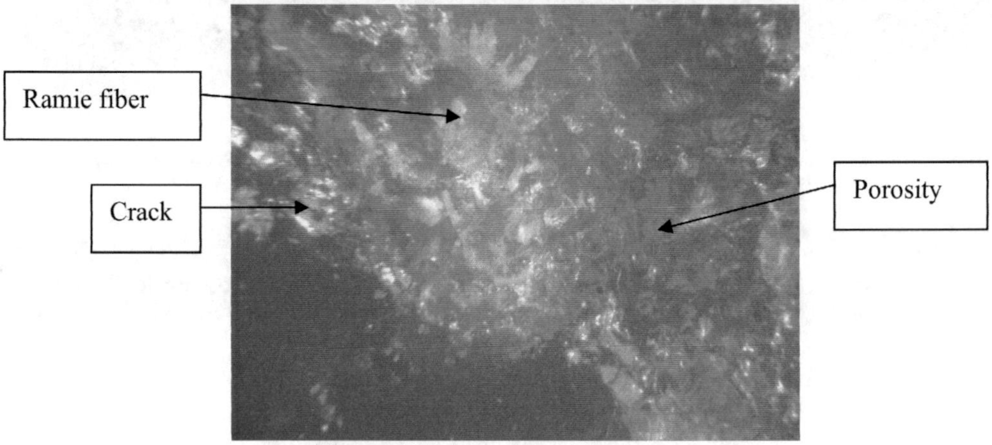

Figure 10. Linear fiber fatigue test result

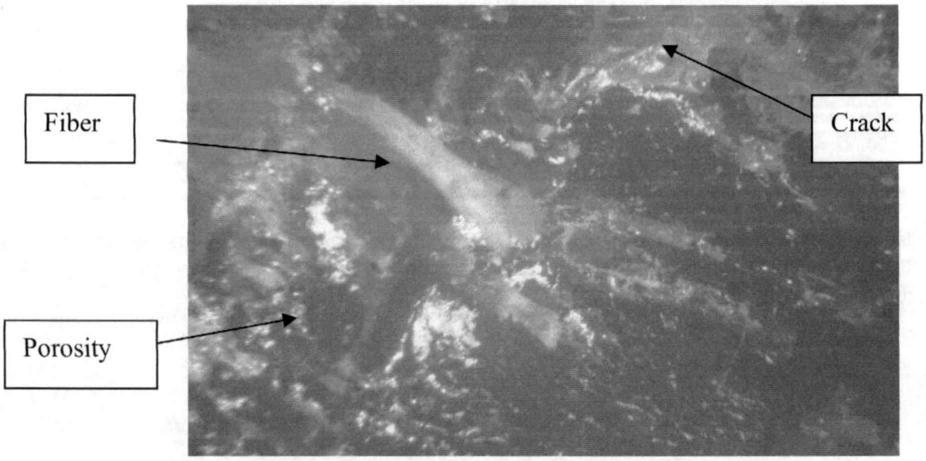

Figure 11. Testing results for woven fiber fatigue

IOP Conf. Series: Materials Science and Engineering **694** (2019) 012038 doi:10.1088/1757-899X/694/1/012038

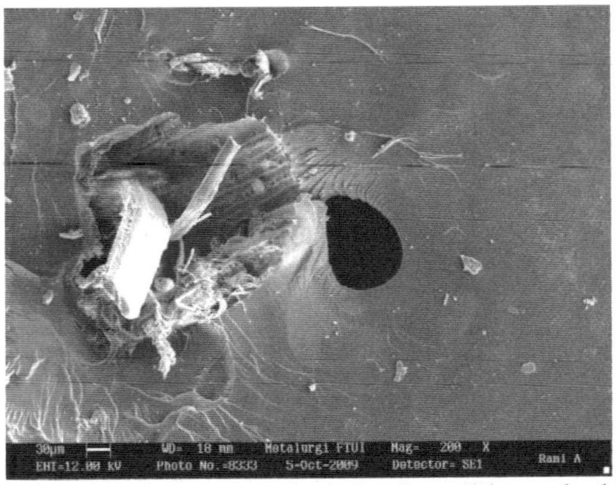

Figure 12. SEM image showing fracture analysis begins with cracked ramie fibers (fibers released from the matrix) on the surface of the epoxy composite matrix

Mechanism of damage due to fatigue in the matrix (matrix damage) generally occurs in polypropilene bermatrik composite where the results of the analysis of the following photo shows it begins with a crack initiation which is then followed by crack propagation. As explained earlier that the initial cracking in the matrix is generally caused by the porosity formed in the PP matrix due to the gas resulting from the burning of trapped fibers. The formation of cracks in this matrix will propagate to areas that have the highest stress and in general at the interface area of the fiber matrix and these cracks will stop propagating if the stress at the crack end is unable to break the fiber. This generally occurs in matrices that have a fairly low strain so that cracking only occurs in matrices with a sufficient number of cracks.

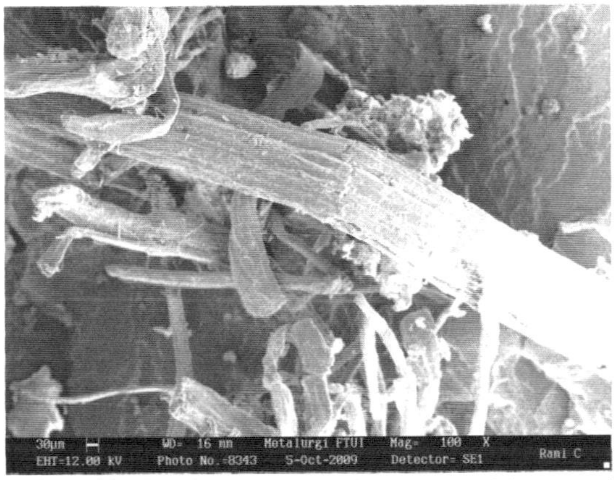

Figure 13. SEM Photo Fracture analysis begins with a matrix crack on the surface of the Polypropilene matrix composite

From the results of the fatigue test analysis, it can be concluded that some conditions can reduce the failure rate due to fluctuating loads on these handle products, including:

1) The function of the fiber, which is the reinforcing element, determines the mechanical properties of the composite because it continues the load distributed by the matrix. Orientation, size and shape and fiber matrial are factors that influence stress concentration.
2) There is a good placement and position of the fiber so that the bond between the fiber and the matrix is very good. In unidirectional fibers have a good level of direction arrangement of fiber and the surface area can bind entirely between the fiber and the matrix.

IOP Conf. Series: Materials Science and Engineering **694** (2019) 012038 doi:10.1088/1757-899X/694/1/012038

3) Reduced porosity, since porosity is the initial occurrence of cracks or cracks where the shape and size greatly affect the amount of stress concentration. In the same direction the porosity of the fiber is very little compared to the others.

4. Conclusion

From the results of all the mechanical tests, it can be concluded that the presence of ramie reinforcing fibers in both the epoxy and polypropylene-patterned handle products is strongly influenced by the orientation of the fiber. Some randomly oriented specimens even exhibit lower mechanical properties compared to handle specimens without reinforcing fibers. On the results of fatigue testing, for epoxy matrix has a significant increase of 162% while the PP matrix only has a low fatigue strength value of only 13%

Based on observations of microstructure and SEM for all tests in general, it shows that the mechanical properties of the composite are reduced due to several things, including porosity, adhesion between fiber-matrices and between matrices, overlapping fibers and broken fibers. Natural composites with PP matrix given an injection process tend to produce poor fiber orientation because the pressure will cause the fibers to accumulate in a given area which results in uneven distribution of fibers so that the concentration formed is the beginning of cracking in the handle product.

The utilization of technology will greatly assist in producing more optimal mechanical properties both in terms of quantity of processing and product quality. With a consistent result it is hoped that later it can be easily applied en masse both for SMEs and industries who want to develop the use of these composite materials as their products.

References

[1] Jang, B., 1994, *Polymer Composites for Automotive Aplications*, Advanced Polymer Composite, pen. ASM International, London

[2] Gibson, R.F., 1994. *Principles of Composite Material Mechanics*, McGraw Hill Int., Singapore

[3] Kompas Newspaper, Sunday, March 19, 2006 p. 19

[4] Prachayawarakorn, J. dan Yaembunying, N, 1995, *Characterization of husk rice with recycle Polyprilene*, Composite Journal.

[5] Anang S, K. A. Widi, W. Sujana, Sibut, The Rule of Rice Husk Ash in PET Type Polymer Composites, Flywheel Journal, Vol 8 No 1, February 2017

IOP Conf. Series: Materials Science and Engineering **694** (2019) 012039 doi:10.1088/1757-899X/694/1/012039

Using Geopolymer Composites for Heat and Corrosion Resistant Pipes

A Su'udi, J Akmal, N Tanti and Arizon

Department of Mechanical Engineering, Universitas Lampung, Prof. Soemantri Brodjonegoro St No.1 Bandar Lampung, 35145, Indonesia

E-mail: jamiatulakmal@gmail.com

Abstract. This research studied the use of carbon fiber reinforced geopolymer composites as corrosion resistant pipe material. Geopolymer material is a general term for material synthesized from fly ash. Fly ash is an industrial by-product (coal combustion residue) with high silica content (SiO_2) and Alumina content (Al_2O_3). In addition there are also other elements in a relatively small percentage. This material will compound like cement if activated with Sodium Silicate. This study aims to obtain the optimum composition for high quality mechanical properties. The method used is to synthesize material with various elemental compositions. Furthermore, a mechanical test (bending test) is performed to see the effect of each element to be optimized. Besides that, a physical test in the form of SEM-EDX was also conducted to see the microstructure. Materials with optimum composition have been fabricated as pipes with centrifugal casting process. Mechanical testing of pipe samples is done by hoop tensile strength test. The test results show that the strength of geopolymer composites can achieve 58 MPa.

1. Introduction

Pipeline is transportation mean to drain fluid that is commonly used in industry. Nowadays, steel pipe is mostly used in indutry field. However, in a certain condition, it is needed a pipe that is resistant to corrosion and heat temperature, expecially it is in geothermal industry. One of the problems that appears while using steel pipe is corrosion rate and deposits of crust in the pipes used. This case can make the company get lossed. In Indonesia, PERTAMINA company have to spend USD 6-7 for steel pipe treatment used [1].

Geopolymer is one of the materials that is resistant to heat temperature and corrosion. Geopolymer is the synthetic result from basic material that contains a lot of bonding elements of alumina (Al), silica (Si) and oxide. Geopolymer is made of coal industry waste thish are fly-ash dust and others materials that contain silica. The material will become strong material if it is activated with activator like sodium Hidroksida and sodium Silicate [2].

The aim of this research is to know the mechanic characteristic of geopolymer pipe as an alternative pipe. The success of this research is hoped which can answer the problem that is faced in pipe, like the level of corrosion and resistant to heat temperature. Therefore, it can impact to eficiency and low treatment cost. This reasearch is also hoped can facilitate in resolving fly-ash waste and give point plus to the waste indutry.

IOP Conf. Series: Materials Science and Engineering **694** (2019) 012039 doi:10.1088/1757-899X/694/1/012039

The composition of geopolymer composite that is added the element of silicon, dioxide and sodium silicate that will be molded as the standard of ASTM C 1161 [3], and then it will be tested by using Bending Three Point method. After getting the best compotition, geopolymer composite will be molded on a pipe mould. The pipe mould will be powered by mover bike. The mover bike will give centrifugal force, so geopolymer material that is in pipe mould will be formed like the mould. After becoming geopolymer pipe [4-7].

The optimization of geopolymer composite by adding the element of silicon dioxide and sodium silicate in making geopolymer pipe by using centrifugal casting molding, it is hoped that it can increase the power of geopolymer composite pipe which is resistant to corrosion and heat temperature.

2. Materials and Experiment

Materials used in the manufacture of geopolymers are Fly Ash (FA), Kaolin (K), Calcium Oxide (CaO) and Carbon Fiber (CF). This material mixture is then added with SiO_2, NaOH and Na_2SiO_3 which functions as activator. Fly ash comes from PLTU Tarahan Sector III Lampung Selatan Indonesia and kaolin used commercial type. The chemical composition of FA and K is investigated by XRF (X-Ray Flourences) testing and the respective content is shown in Table 1. Silica fume used is a commercial type from PT. NORMET Indonesia and Carbon Fiber medium type (1.56 kg / m3).

Table 1. Chemical Composition of FA dan Kaolin (%) [1]

Materials	Si	Al	Fe	Ti	Zr	Mn	Sb	Zn	Sn	Ni	Pb	Ga
Fly-ash	48,2	26,0	22,1	2,53	0,49	0,25	0,20	0,10	0,07	0,06	0,02	0,02
Kaolin	57,6	37,2	3,82	1,23	0,1	-	-	0,04	-	0,04	-	0,02

2.1 Design of Matrix Composition by Taguchi Method

Composite Geopolymer is made with 2 stages: activator preparation and mixing of solid binder. In preparation for making the activator to be done first is to dissolve the crystalline NaOH into a solution of 14M by adding a solution of Aquades. The ratio of Na_2SiO_3 and NaOH ratio is 0.3. The carbon fiber is cut 1,5 cm and then weigh all the material according to the predetermined variation.

There are 4 materials that will be varied by the taguchi method [8]: Silicon dioxide, kaolin, calcium oxide (CaO), and carbon fiber. Silicon dioxide, kaolin, and CaO use the best composition performed by Jamiatul Akmal [2], SiO_2 27% wt, Caolin 51% wt, and CaO 0,5 wt%, where as the carbon fiber is 13%.

Composition are respectively: A, B, C, D,E and F , as shown in Table 2. Table 3 is an orthogonal table showing the mixing of factors with varying levels. There are 9 mixtures to be tested to obtain the optimum mixture.

Table 2. Mixture composition and relevansi factor

Level	Faktor					
	SiO_2 (gr) A	K (gr) B	CaO (gr) C	CF (%) D	Na_2SiO (gr) E	NaOH (gr) F
1	27	55	5	13	15	15
2	30	53	5	13	20	10
3	33	51	5	13	22,5	7,5

Table 3. Taguchi orthogonal matriks L9(3^4)

No	SiO_2 (gr)	K (gr)	Na_2SiO (gr)	NaOH (gr)
1	27	55	15	15
2	27	53	20	10
3	27	51	22,5	7,5
4	30	55	20	7,5
5	30	53	22,5	15
6	30	51	15	20
7	33	55	22,5	20
8	33	53	15	7,5
9	33	51	20	15

The next step is mixing the powder materials: Silicon dioxide, kaolin, CaO and CF until evenly distributed. Then add activator solution, with the ratio of powder to the activator solution is 0.7. After stirring evenly, put the mixture into a 90x6x8 mm mold and vibrate at a frequency of 54000 Hz for about 30 minutes. The sample is then stored in an incubator with a temperature of 30-35 °C for ≥ 28 days.

2.2 Manufactures pipes with centrifugal casting machines

Geopolymer composites with the best composition are then molded into pipes with a centrifugal casting machine, as shown in Figure 1a. Figure 1b shows the molded pipe.

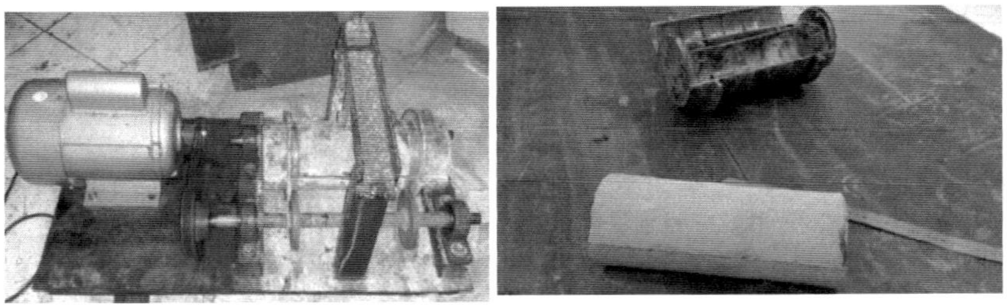

(a) (b)

Figure 1. (a) Centrifugal machine testing, (b) The molded pipe

2.3 Experiment Testing

Rectangular bars with size (90x6x8) mm subjected to the three point bending test to determine the strength and flexural modulus (based on ASTM C 1161), as shown in Fig. 2. The test machine used is MTS Landmark (capacity 50 kN) with loading speed 1.0 mm / min. Flexural strength can be determined by Eq. (1).

$$\sigma_f = \frac{3P_m S}{2WD^2} \quad (1)$$

where σ_f is flexural strength (MPa), P_m is ultimate load (N), S is the span of the sample, W is the specimen with and D is the specimen thickness. The flexural modulus was computed using the initial slope of the load-displacement curve, $\left(\frac{\Delta P}{\Delta x}\right)$, using the Eq. (2)

$$E_f = \frac{S3}{4WD^3}\left(\frac{\Delta P}{\Delta x}\right) \quad (2)$$

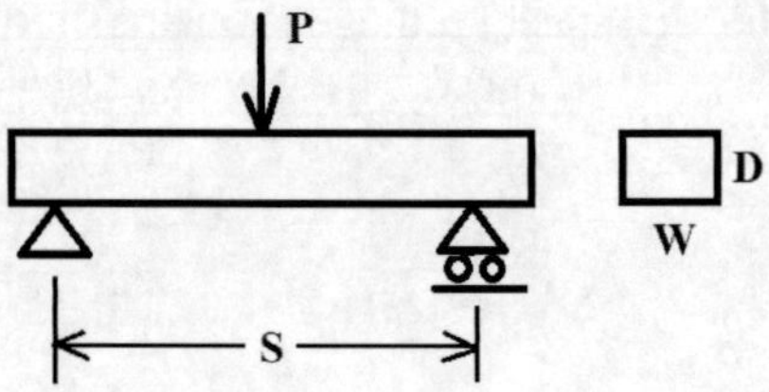

Figure 2. Schematic of the three point bend test

3. Result and discussion

There were 9 samples tested and each sample consisted of 3 specimens. Figure 3 shows the tested sample and the cross-sectional view broken. Flexural strength of each specimen is shown in Table 4.

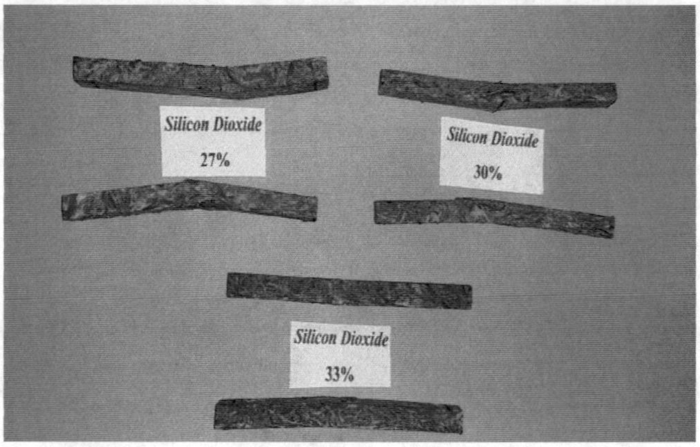

Figure 3. Deformation specimens after testing

The influence of factor A,B,E and factor F to SN Ratio are shown in Figures 4 and 5, respectively. Factor E and F shows the effect of Na_2SiO_3 and NaOH addition on flexural strength, to optimum at S3 level, with 27%wt SiO_2, 51%wt kaolin, 5%wt CaO and 13%wt CF addition. Compositions are shown in Table 4.

Table 4. Flexural strength of the samples (MPa)

Mix	A	B	C	D	E	F	Flexural Strength (MPa)			Average
							Spesimen 1	Spesimen 2	Spesimen 3	
S1	27	55	5	13	15	15	15,26	9,34	4,38	9,66
S2	27	53	5	13	20	10	7,83	31,57	9,15	16,18
S3	27	51	5	13	22,5	7,5	64,87	49,60	60,00	58,16
S4	30	55	5	13	20	7,5	19,84	24,51	12,11	18,82
S5	30	53	5	13	22,5	15	7,63	20,89	12,02	13,51
S6	30	51	5	13	15	20	23,56	6,48	26,61	18,88
S7	33	55	5	13	22,5	20	18,03	17,45	30,24	21,91
S8	33	53	5	13	15	7,5	34,34	44,74	43,40	40,83
S9	33	51	5	13	20	15	34,72	25,75	32,34	30,94

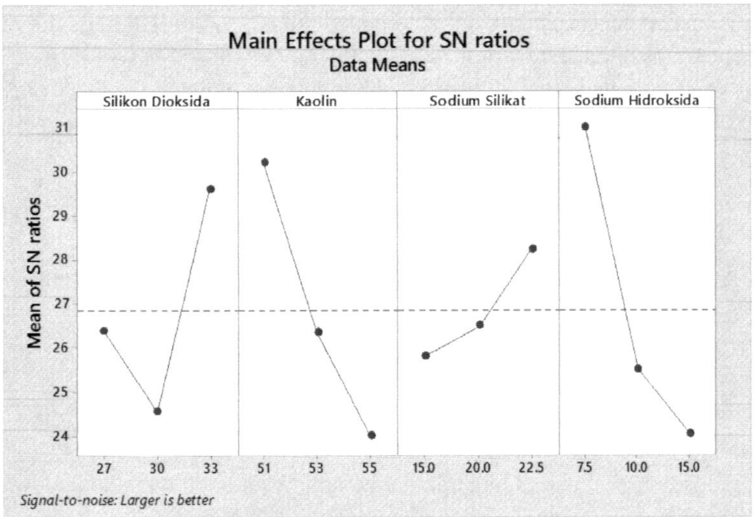

Figure 4. Effect plot for SN ratios

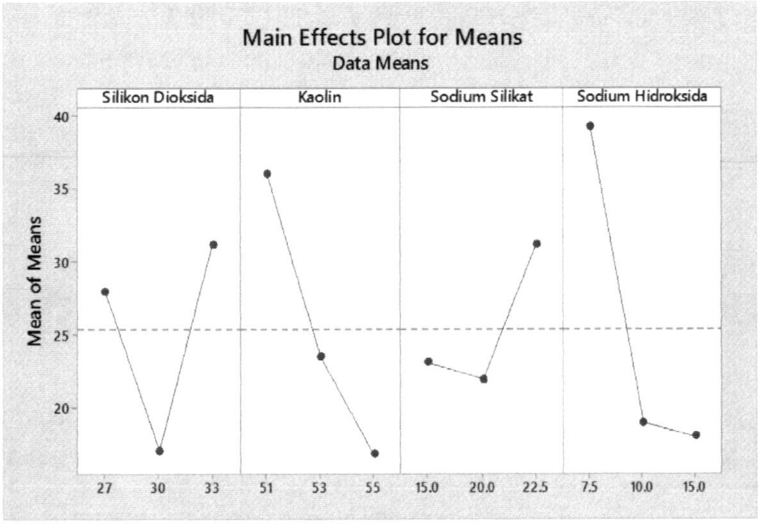

Figure 5. Effect plot for Means

Table 5. The 5 optimum samples and their compositions

Mix Number	SiO_2 (gr)	K (gr)	CaO (gr)	CF (%)	Na_2SiO_3 (gr)	NaOH (gr)
S3	27	51	5	13	22,5	7,5
S6	30	51	5	13	15	20
S7	33	55	5	13	22,5	20
S8	33	53	5	13	15	7,5
S9	33	51	5	13	20	15

Mix S3 is the best specimen with composition: Silicon dioxide (27% wt), kaolin (51% wt), CaO (5% wt), and carbon fiber (13% wt), with flexural strength 58,16 MPa . Figure 6 shows the stress-displacement curve on the geopolymer composite for 5 optimum samples. The curves appear to be 2 segments: linear segments and non-linear segments. This shows a similar trend with steel performance. In the first segment the curve is relatively straight and no plastic damage has occurred. In the second segment, after reaching the maximum strength the curve drops suddenly and then horizontally until the specimen breaks out debonding between the matrix and fiber. This curve also explains that the addition of carbon fiber (CF) can significantly increase flexural strength and modulus. Figures 7 and 8 respectively show flexural strength and modulus strength.

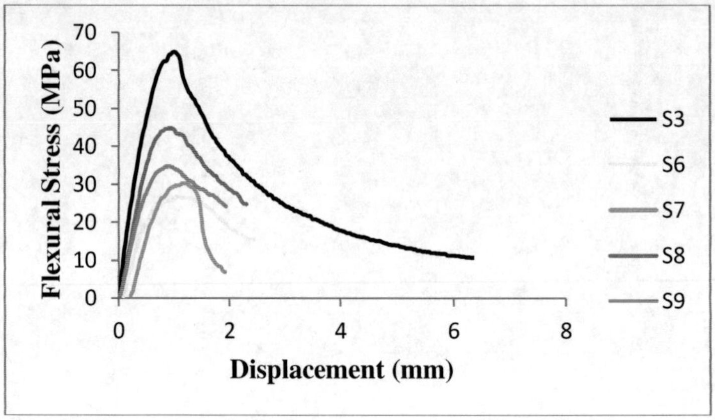

Figure 6. The stress and displacement curves of the 5 optimum samples

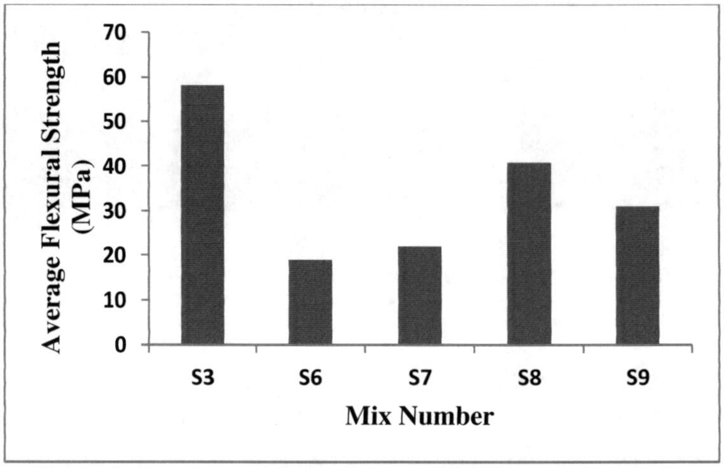

Figure 7. The average flexural strength of 5 optimum samples

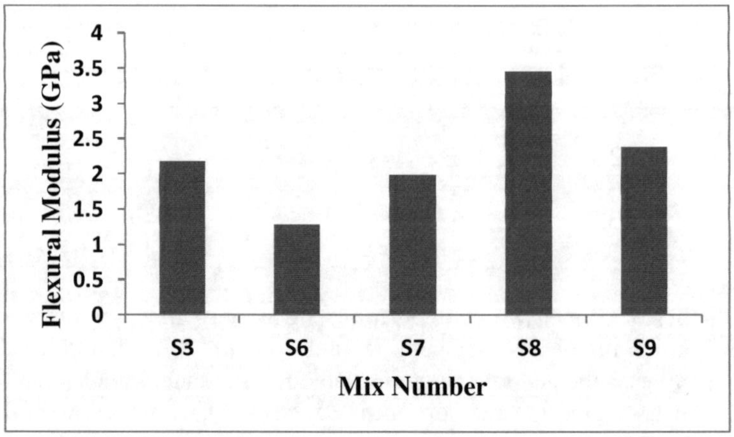

Figure 8. The flexural modulus of 5 optimum samples

Back Scatter Electron (BSE) Images for samples with different compositions of Na_2SiO_3 and NaOH are shown in Figure 9. Surface morphology for all samples shows a low density and there are holes with even distribution. The low density level is caused by the process of compacting that is not optimal when molding samples, only by the vibration process. This becomes the input for further research so that compaction is done with pressurized molding process.

The composition of the elements for each sample is shown in Table 6. In this picture is shown spectrum 1 and spectrum 2 for each sample, indicating the points analyzed. In Figure 9a, the points shown are matrix (without fiber), there is an element of carbon with a small percentage (15% -22%). This carbon comes from carbon tape which is added as an electron intermediate during the scaning process Figure 9b shows the analyzed points on carbon fiber (CF) with relatively high carbon content (about 81-95%). Likewise, Figure 9c and Figure 9d, each showing the same analysis form as Figure 9a and Figure 9b.

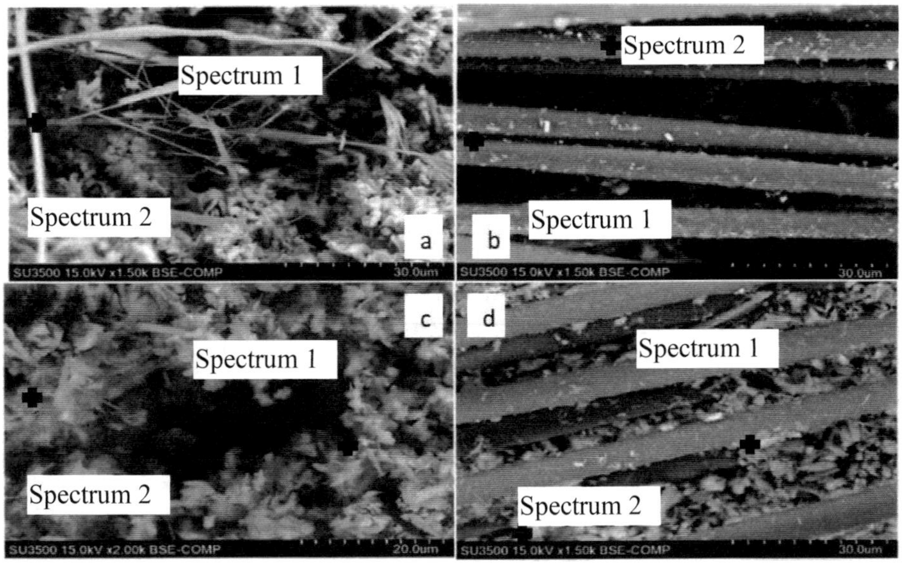

Figure 9. BSE image for samples with different compositions of Na_2SiO_3 and NaOH (a) Na_2SiO_3 0 gr and NaOH 30 gr (b) Na_2SiO_3 15 gr and NaOH 15 gr (c) Na_2SiO_3 20 gr and NaOH 10 gr, (d) Na_2SiO_3 22,5 gr and NaOH 7,5 gr.

Table 6. Result Point Analysis Sample

	CaO=0 CF=0 (% Atom)		CaO=0 CF= 15% (% Atom)		CaO=4% CF=0 (% Atom)		CaO=4% CF=15% (% Atom)	
Elemen	1	2	1	2	1	2	1	2
C K	15.50	22.83	81.66	94.51	18.51	25.72	75.80	90.25
O K	59.08	52.98	16.75	5.38	62.81	52.76	13.86	9.09
Na K	17.71	18.99	1.15	0.08	10.26	17.53	0.57	0.62
Mg K	-	-	-	-	0.11	-	-	-
Al K	2.58	0.37	0.11	-	0.74	0.90	3.49	-
Si K	13.96	0.88	0.28	0.04	1.75	3.09	6.05	0.04
S K	-	0.88	-	-	-	-	-	-
Ca K	0.78	1.93	0.05	-	5.77	-	0.17	-
Fe K	0.40	1.15	-	-	0.04	-	0.07	-
Totals	100.00	100.00	100.00	100.00	100.00	100.00	100.00	100.00

IOP Conf. Series: Materials Science and Engineering **694** (2019) 012039 doi:10.1088/1757-899X/694/1/012039

4. Conclusion

Geopolymers with the variation of Na_2SiO_3 and NaOH have a significant effect on increasing the strength of the geopolymer and changing the brittle properties to be more ductile. The relatively good geopolymer composite can achieve flexural strength of 58,16 MPa with the content: Silicon dioxide (27% wt), kaolin (51% wt), CaO (5% wt), and carbon fiber (13% wt), with flexural strength 58,16 MPa. The Sodium hidroksida and Sodium silicate content still has the potential to be increased because point F3 is not the optimum level.

Acknowledgements
This work was supported by the DIPA-BLU, LPPM Universitas Lampung, 2019.

References

[1] Akmal, J., Badaruddin M., Ismoyo M.K., Yuwono S.D., (2019) Optimization Of Matrix Composition Al_2O_3, Caolin and CaO on the mechanical properties of a geopolymer composite with short carbon fiber. *IOP Conference Series: Journal Materials Science and Engineering* **602**.

[2] Suwanto E.K., (2018) Effect of Addition of Calcium Oxide and Carbon Fiber to the Mechanical Properties of Geopolymer Composites. Thesis. Faculty of Engineering. University of Lampung. Bandar Lampung.

[3] ASTM C 1161 (1990) Standard Test Method for Flexural Strength of Advanced Ceramics at Ambient Temperature. Annual Book of ASTM Standard 15. pp 333-339

[4] Wahid, M., (2018). Design of the Centrifugal Casting. Faculty of Engineering. Departement of Mechanical Engineering. University of Lampung. Bandar Lampung. pp 30-43

[5] Okoye, F.N., Durgaprasad, J., Singh, N.B., (2015), Mechanical Properties of Alkali Activated flyash/Kaolin Based Geopolymer Concrete. Journal Construction and Buildings Materials **98**. pp 685-691

[6] Okoye, F.N., Durgaprasad, J., Singh, N.B., (2015), Flyash/Kaolin Based Geopolyner Green Concretes and Their Mechanical Properties. Journal Data in Brief **15**. pp 739-744

[7] Callister, W.D., (1940), Materials Science and Engineering. Department of Metallurgical Engineering. 7th Ed. University Of Utah

[8] Soejanto, I., (2009), Experiment design with taguchi Method. Graha Ilmu, Yogyakarta.

IOP Conf. Series: Materials Science and Engineering **694** (2019) 012040 doi:10.1088/1757-899X/694/1/012040

Stirring Effect on Surface Morphology, Structure, and Electrochemical Behavior of Electrodeposited Nickel Film on Copper Substrates

Ferry Budhi Susetyo[1,2], Ahmad Faridh[1] and Bambang Soegijono[1,*]

[1]Dept. Physic, Universitas Indonesia, Depok 16424, Indonesia
[2]Mechanical Engineering, Universitas Negeri Jakarta, Jakarta 13220, Indonesia

[*]E-mail: naufal@ui.ac.id

Abstract. Nickel as engineering material has good corrosion resistance properties. Nickel has been deposited on copper surface by electrochemical deposition with various stirring speed. Objective on this research are to observe the effect of stirring on morphology, structure and electrochemical behavior. Electrodeposited nickel film on copper substrates was conducted with various stirring speeds (0 rpm, 75 rpm, and 150 rpm). Surface morphology was observed with Field Emission Scanning Electron Microscope (FE-SEM), structure was identified with X-ray diffraction (XRD), and electrochemical behavior was investigated by potentiodynamic polarization and cyclic voltammetry. Stirring speed effect clearly on the surface morphology, structure and electrochemical behavior. Corrosion potential and corrosion current are changes while the stirring speed is increase.

1. Introduction

Electrodeposited of nickel films has been used widely in many automotive component manufacturer to improve corrosion protection of metal coated [1-3]. Nickel films a applied at automotive component such as handle comp, fuel tank cap, and door handle etc. Chemical properties and Physical properties of metallic materials are strongly influences by surface morphology [4]. Corrosion resistance of material influences by microstructure behavior, such as surface morphology, grain size, and texture [5-7]. These are closely related to the electrodeposition parameters, such as stirring electrolyte during deposition, electrolyte pH, temperature of the electrolyte solution, current density, additive and texture of the substrate [5,8,9]. The texture crystallographic of substrate such as copper influence significantly properties of deposited nanocrystalline nickel [10]. Nickel as engineering material has good corrosion resistance in aqueous aggressive environments [11]. When nickel a deposited on copper surface, it can protect copper surface from uniform corrosion [12]. Therefore, the aims of the present work were to prepare the nickel films on copper substrates using electrodeposited technique at different magnetic stirring speeds. The Nickel films specimens was observed of by Field Emission Scanning Electron Microscope (FE-SEM), X-ray diffraction (XRD), potentiodynamic polarization and cyclic voltammetry.

2. Experimental Methods

Nickel film was prepared from a watts solution consisting of $NiSO_4.6H_2O$ (250 g/L), $NiCl_2.6H_2O$ (40 g/L), and H_3BO_3 (30g/L). The electrodeposition was carried out using current density of 2mA/cm² for 120 minutes and the solution was stirring with various speed 0 rpm, 75 rpm and 150 rpm. The samples were designated as Ni-0, Ni-75 and Ni-150. Nickel sheet used as anode and copper were used as cathode. Surface of the substrates were polished using silicon carbide (SiC) abrasive papers 500 up to 2000 prior electrodeposition. Observed of the nickel films morphology was using the FE-SEM. The nickel films crystal structure was identified by XRD. The XRD pattern scans with the resolution of 0.01° from 40° up to 80°. To refine the XRD pattern, software Highscore plus were used. For Electrochemical investigation (potentiodynamic polarization, and cyclic voltammetry), Digi-Ivy DY2311 were used, and Ag/AgCl as Reference Electrode (RE) and Platinum wire (Pt) as Counter Electrode (CE). The working area of test samples is 1 cm².

3. Results and Discussion

3.1. Surface Morphology

Surface morphology of samples Ni-0, Ni-75 and Ni-150 was shows in Figure 1. Those images were capture from FE-SEM with magnification 10000 X. The grain size with perform stirring speed 75 rpm and 150 rpm (figure 1 (b) and 1 (c)) are bigger than without stirring. The samples without stirring show smooth surface compared to samples with stirring during electrodeposition.

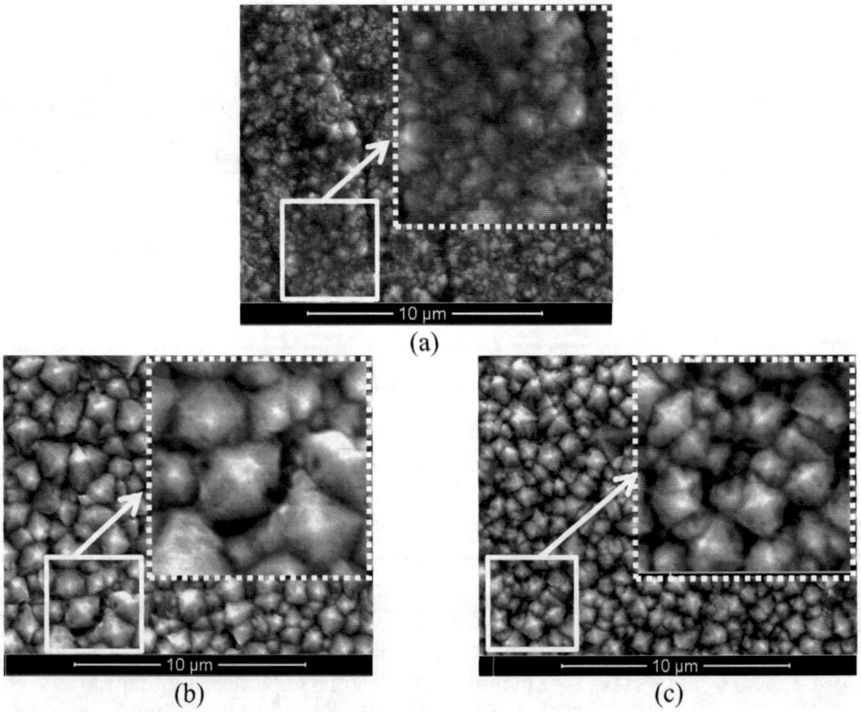

Figure 1. FE-SEM showing comparison of various nickel film deposited on copper substrate at different samples Ni-0 (a), Ni-75 (b), and Ni-150 (c)

3.2. Structure

The XRD patterns of nickel films with various stirring speeds are shown in figure 2. The entire sample shows peaks of (111), (002), and (022) that represents the nickel film crystallographic orientation. Increased stirring speed during electrodeposition clearly affected the reduction peak of (002). However, the increase intensity peaks of (111) are revealed on all samples. The nickel films with stirring speed 75 rpm show the intensity of (111) peak higher than the other samples. But, it decrease when stirring changes to 150 rpm.

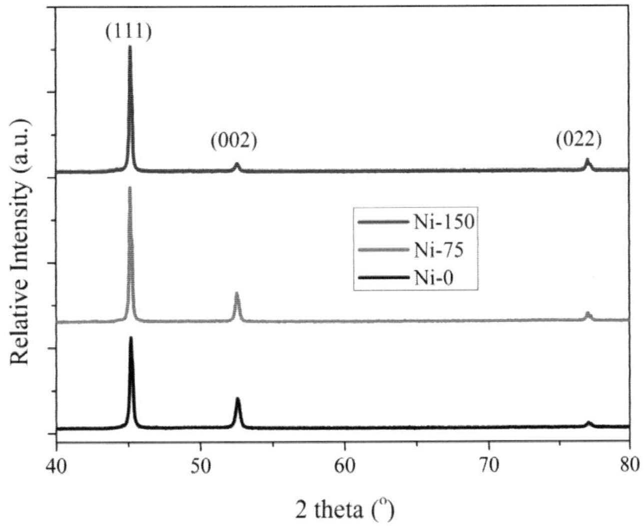

Figure 2. XRD pattern of electrodeposited nickel film samples with various stirring speed

Refinement results with highscore plus software of XRD pattern show that the crystal systems of nickel films are Cubic FCC with Space Group Fm-3m. Sample Ni-75 has smaller density than the other samples. The Ni-75 sample also has micro strain 0.097 % and difference chemical formula with the other samples. The refinement results with Highscore software are shown in Table 1.

Table 1. Crystals parameter of the nickel films with various stirring speed

Parameter	Samples		
	Ni-0	**Ni-75**	**Ni-150**
Crystal System	Cubic FCC	Cubic FCC	Cubic FCC
Space Group	Fm-3m	Fm-3m	Fm-3m
a=b=c (Å)	3.5284	3.5281	3.5286
V (Å^3)	43.928	43.917	43.935
density (g cm^{-3})	8.87	8.06	8.87
Crystallite size (nm)	2791.34	2791.34	2791.34
Micro strain (%)	0.000	0.097	0.000
Chemical formula	Ni 4.00	Ni 3.99	Ni 4.00
R_{wp} (%)	8.54	8.06	8.43
GOF	1.59	1.38	1.49

3.3. Electrochemical Behavior

To determine the corrosion behavior of nickel coated to copper, potentiodynamic polarization test is used for examination. In order to deep explore corrosion behavior, potentiodynamic polarization measurements was employed to observe nickel film in 3.5 % NaCl solution. The results of potentiodynamic polarization test are shown in Table 2.

Table 2. Potentiodynamic polarization measurement at various stirring speeds

Sample	$I_{Corrosion}$ (A)	$E_{Corrosion}$(V)	Corrosion Rate (mmpy)
Ni-0	1.22E-05	-0.898	0.131
Ni-75	1.55E-05	-0.851	0.167
Ni-150	1.78E-05	-0.850	0.192

Figure 3 shows of the potentiodynamic polarization behavior of nickel film deposited on copper substrat with various stirring speed in a 3.5% NaCl. Nickel film of all samples has showed behavior an active to passive and trans passive. The passive layer was form on the nickel surface is one of the key that affected the nickel films corrosion behavior [7]. Passive region shown on figure 3 are difference to each sample. Nickel film with no stirring has higher passive region than nickel film with stirring. Figure 3 also shows that nickel film with stirring has shifted to positive corrosion potential ($E_{corrosion}$) than nickel film with no stirring. Sample with no stirring (Ni-0) resulting smaller corrosion current and corrosion rate than the other samples (table 2).

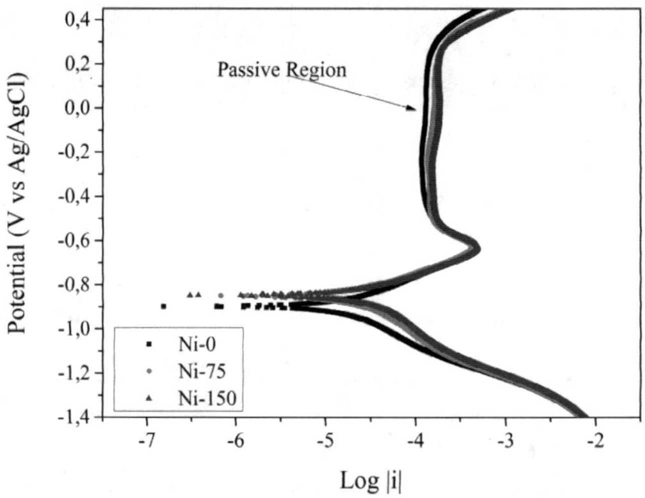

Figure 3. Potentiodynamic polarization curve in 3.5% NaCl solution at a scan rate 0.05 mV/s.

Cyclic voltammetry characteristic of the samples were performed with scan rate 0.05mV/S at 100 mL 3.5% NaCl solution and showed in Figure 4. Cyclic voltammetry are characterized generally by the presence of anodic and cathodic peaks in the respective area, which are related to the dissolution and reduction of metallic ion [13]. Cyclic voltammetry of nickel film are shown in Figure 4, and it can be seen that swept of the potential is forward direction increase in the cathodic current and resulting difference cathodic peaks of samples. This current increment are due to reduction nickel film (Ni^{2+}/Ni) with simultaneous evolution of hydrogen as shown in equation (1).

$$Ni^{2+} + 2e^{-} \rightarrow Ni \quad (1)$$

On reverse sweep in direction of anodic region resulting potential around -0.65 V for oxidation peak attributed to the nickel film dissolution of as shown in equation (2).

$$Ni \rightarrow Ni^{2+} + 2e^{-} \quad (2)$$

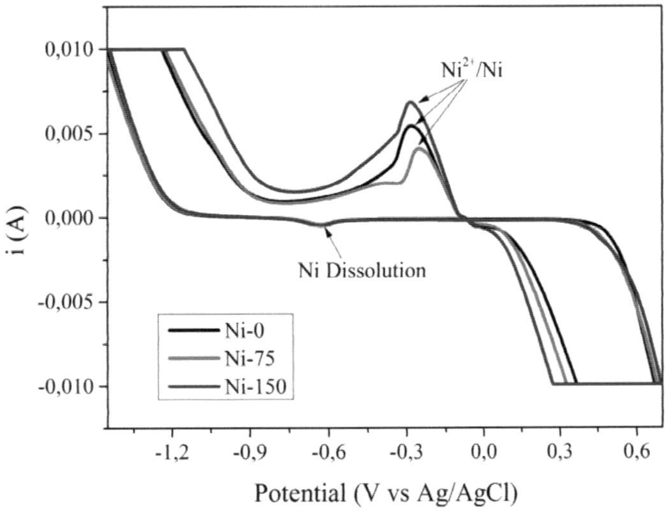

Figure 4. Cyclic voltammetry curve investigated in 3.5% NaCl

4. Conclusions

The Nickel film deposit on copper substrate has been carried out successfully. The stirring speed influenced on the grain size, structure parameter and electrochemical behavior. Increasing the stirring speed clearly affected the reduction peak of (002). The electrochemical investigation shows that Ni-0 sample possessed the best corrosion resistance. The corrosion potential for samples Ni-75 and Ni-150 samples shift to more positive potential value than Ni-0 sample.

Acknowledgement

The authors gratefully acknowledge with the financial supports from the Universitas Indonesia under grant PITTA, No. 2239/UN2.R3.1/HKPO5.00/2018.

References

[1] Bazzaoui M, Martins JI, Bazzaoui EA, and Albourine A 2012 *App. Surf. Sci.* **258** 7968
[2] Lampke T, Dietrich D, Leopold A, Alisch G, and Wielage B 2008 *Surf. Coat. Technol.* **202** 3967
[3] Snyder DL 1997 *Metal Finishing* **95** 29
[4] Tan L, Ren X, Sridharan K, and Allen TR 2008 *Corros. Sci.* **50** 3056
[5] Zhao H, Liu L, Zhu J, Tang Y, Hu W 2007 *Mater. Lett.* **61** 1605
[6] Zamanzad-Ghavidel MR, Raeissi K, and Saatchi A 2009 *Mater. Lett.* **63** 1807
[7] Nasirpouri F, Sanaeian MR, Samardak AS, Sukovatitsina EV, Ognev AV, Chebotkevich LA, Hosseini MG, and Abdolmaleki M 2014 *App. Surf. Sci.* **292** 795
[8] Jinlong L, Tongxiang L, and Chen W 2016 *Journal of Solid State Chemistry* **240** 109
[9] Wasekar NP, Haridoss P, Seshadri SK, and Sundararajan G 2016 *Surf. Coat. Technol.* **291** 130
[10] Ebrahimi F, and Ahmed Z 2002 *Mater. Charact.* **49** 373
[11] Guiñón-Pina V, Igual-Muñoz A, García-Antón J 2009 *Corros. Sci.* **51** 2406
[12] Zhang W, Yu Z, Chen Z, and Li M 2012 *Mater. Lett.* **67** 327
[13] Nath P, Sahu DK, and Mallik A 2016 *Surf. Coat. Technol.* **307** 772

The Effect of Tempering Temperature on Medium Carbon Steel Plate of Surface Hardening Result Using Induction Heating as Ballistic Resistant Material Study

Helmy Purwanto[1*], Muhammad Dzulfikar[1], Mohammad Tauviqirrahman[2], Rifky Ismail[2] and Nyoto Lestari[1]

[1]Department of Mechanical Engineering, Engineering Faculty, Universitas Wahid Hasyim, Semarang – Indonesia
[2]Department of Mechanical Engineering, Engineering Faculty, Universitas Diponegoro, Semarang – Indonesia

*E-mail: helmypurwanto@unwahas.ac.id

Abstract. Steel that has been heat treatment using quenching is hard but brittle. Tempering was done to remove the agility caused by the residual stress. The objective of this paper writing is to know and analyze the effect of tempering temperature on the mechanic property of steel plate of the surface hardening result. Steel plate with thickness of 8 mm of surface hardening result quenched by oil media was tempered at temperature of 100, 200, 300 and 400°C. Simulation based on finite element method was conducted to know the ballistic resistance. Experiment result showed that heating through induction and quenching on oil resulted in surface hardening. Tempering treatment decreased the hardness but at low temperature, the tensile strength increases. The simulation result showed that quenching and tempering material at temperature of 100°C was more able to resist the ballistic rate compared to other various temperatures.

1. Introduction

Medium carbon steel's physical and mechanical nature in solid condition can be changed through heat treatment. Heat treatment through quenching produces very hard martensitic structure but brittle. The presence of residual stress or stress relieves during rapid cooling causes weak ductility and toughness. Removing the stress relieved as well as improving the ductility and toughness was conducted through tempering heat treatment. Tempering was conducted by heating the martensitic steel below eutectoid temperature in certain period of time and cooled slowly.

There are many factors affecting quench-temper process result on medium carbon steel. Chemical material composition takes role on transformation phase. The additional of alloying elements on low carbon steel will affect the size of austenite particle so that it produces various physical and mechanical properties after the quenching [1]. Heating method in quenching-tempering process also affects the result. Hardening through pre-heating was continued by repeated heating until it reached austenite temperature and decreases until it reaches temperature of $\alpha + \gamma$ and quenched to produce high strength and toughness

[2]. In addition to the composition elements, heating method and temperature, cooling media also affects the hardening result. Quenching marula oil media was able to produce hard sample without any cracked and distortion compared to quenching water media [3].

Tempering temperature affects the mechanical properties, in which the higher the tempering temperature, then change the broken-brittle properties into broken-tough [4]. However, tempering at low temperature can increase the steel's mechanical properties [5]. Cooling rate in tempering process affects the formed structure transformation. Increasing the cooling rate or decreasing the temperature on-line tempering contributes to resist the austenite transformation into ferrite, and simultaneously maintain more austenite to transform into perlite [6].

Ballistic resistant material is a material which is able to resist ballistic rate. In army, ballistic resistant material is a material which is able to resist projectile rate. The ultra-high velocity impact of projectile must be resisted by target material of special properties. Combination between hardness, force and toughness is needed in ballistic resistant material [7]. Target hardness affected the hole of projectile impact shape [8], since it was affected by different deformation ability [9]. Difference in hardness value due to tempering temperature is studied on the depth of projectile penetration on high strength armor steel [10].

Simulation method based on finite element was used in ballistic testing was more efficient. Finite element modeling along with optimization probabilistic hole distribution modeling is a fast , robust and effective tool used in parametric study to maximize the efficiency of perforated armor systems [11]. Through the simulation, the ballistic hole morphology and energy absorbed in the test can be predicted [12]. Good agreement between the experiment test and simulation was conducted on steel plate shot using blunt ended projectile on velocity above ballistic limit [13].

There were many reviews regarding ballistic resistant steel making using quench temper. However, the use of surface hardening using induction heating is still rarely found. The objective of this research is to know and analyze the effect of tempering temperature on medium carbon steel plate done through quenching treatment on oil media towards the hardening distribution and ballistic resistant through simulation based on finite element.

2. Method

Composition of medium carbon steel plate on Table 1 with the size of Ø150 x 8 mm was austenitized at temperature of 900°C through induction heating in one of its surface for 3 seconds. After the temperature and time was fulfilled, plat was quenched on 15 liters oil media. The quenching result was treated using tempering heating by putting the sample on the furnace. Tempering temperature variety chosen were 100, 200, 300 and 400°C, while the holding time was for 15 minutes. The quench result and quench-temper result was made into specimen for micro-Vickers hardness and tensile. Micro-Vickers hardness test was conducted on cross-section as many as 20 points from the surface to the bottom. Tensile test specimen was in accordance with ASTM E8/E8M. The experiment test result data was used as reference in implementing simulation of ballistic test. Simulation based on finite element was validated using the previous research Børvik et al. [13]. The simulation was conducted on the plate based on experiment test data using blunt deformable projectile with diameter of 20 mm, length of 80 mm, mass of 0.197 kg and initial velocity of 303.5 m/s. Other boundary condition in the simulation was presented in Table 2 and Table 3. Fine meshing size was 0.5 mm as seen in Figure 1.

Table 1. Chemical Composition (%wt.)

Material	Chemical Compositions
Medium carbon steel	0.4667 %C; 0.5584 %Mn; 0.0556 %Cr; 0.0411 %Ni; 0.2441 %Si; 0.0002 %Nb; 0.0163 %Al; 0.001 %V; 0.0106 %S; 0.0013 %Mo; 0.0016 %W; 0.0195 %P; 0.0577 %Cu; 0.0044 %Ti; 0.0078 %N; 0.0002 %B; 0.0064 %Sb; 0.0003 %Ca; 0.0009 %Mg; 0.0004 %Zn; 0.0108 %Co; Fe bal.

Table 2. Projectile material properties [13]

Description	Notation	Nominal
Modulus young	E (MPa)	200000
Poisson ratio	v	0.33
Density	ρ (Kg/m^3)	7850
Yield stress	A (MPa)	1900
Plastic Strain	$\varepsilon_r(\%)$	1

Table 3. Target material properties [13]

Description	Notation	Nominal
Modulus young	E (MPa)	200000
Poisson ratio	v	0.33
Density	ρ (Kg/m^3)	7850
Yield stress	A (MPa)	ET
Strain Hardening	B (MPa)	807
Constant	n	0.73
Viscous effect	C	0.012
Thermal Softening Constanta	m	0.94
Strain rate hardening	$p_{0,}r_0(S^{-1})$	5.10^{-4}
Specific Heat	Cp (J/Kg.K)	452
Melt temperature	Tm (K)	1800
Transition temperature	To (K)	293
Fracture strain constrain	$D1$	0.0705
	$D2$	1.732
	$D3$	-0.54
	$D4$	-0.0123
	$D5$	0
User Defined reference strain rate	Po	1

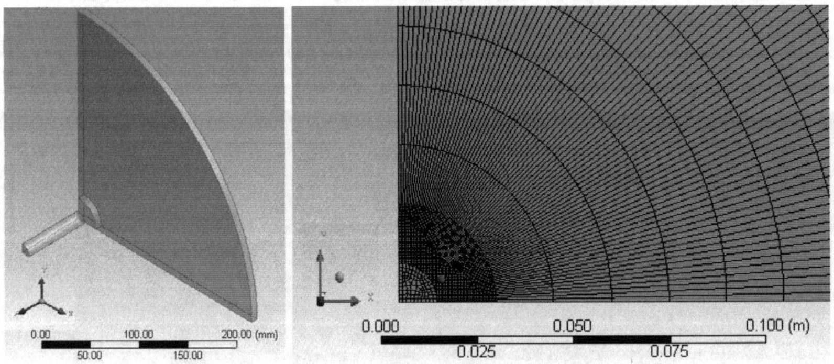

Figure 1. Modeling and meshing scheme

3. Result and Discussion

The test result of hardness distribution using micro-Vickers method is presented in Figure 2.

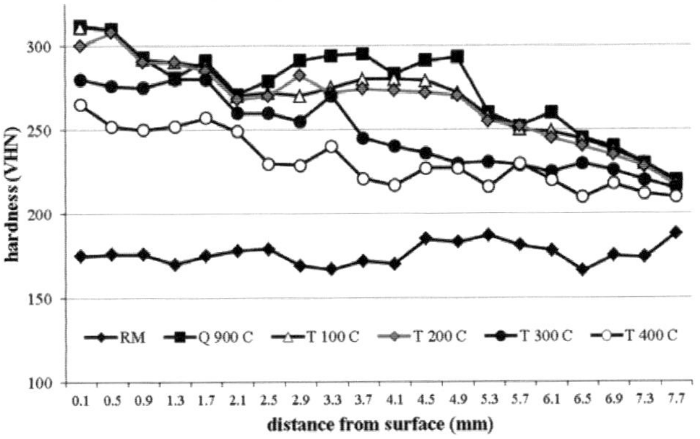

Figure 2. Distribution of cross-section plate hardness using various tempering temperature

Figure 2 is graph of hardness distribution presenting the mean hardness of raw material which is 176.2 VHN. From the experiment with various austenization temperatures, the optimal result was obtained at temperature of 900°C using quench oil media. The mean hardness on the surface was 312 VHN, in the middle was about 290 VHN, while in the bottom area was about 220 VHN. This proves that by using induction heating, hardness could occur in one of parts of the surface. While on the other surfaces, the hardness was maintained in order to maintain its ductility. Hard material due to quenching produces remaining tension. A very hard material which is exposed to ballistic weight can experience failure. The increase of hardness on steel plate does not affect linearly on the ballistic penetration and resistance [10].

Treatment of tempering at the temperature of 100, 200, 300 and 400°C decreases the hardness. The hardness does not decrease significantly at the tempering temperature of 100 and 200°C. Meanwhile, the tempering at temperature of 300 and 400°C, the hardness decrease more. The hardness on the surface at tempering temperature of 100°C was 311 VHN and its hardness decreases into 219 VHN at the bottom. The biggest decrease of the hardness occurred at tempering temperature of 400°C, in which on the surface, it decreased into 265 VHN, while at the bottom, it decreased into 210. However, the hardness value was still above raw material hardness which is 175 VHN. This proves that the higher the tempering temperature, the bigger the decrease occur on the hardness value, as reported by Mishra et al. [14] on tempered high strength steel.

Tempering temperature also affected the tensile strength. Figure 3 presents the effect of tempering temperature variety on tensile strength.

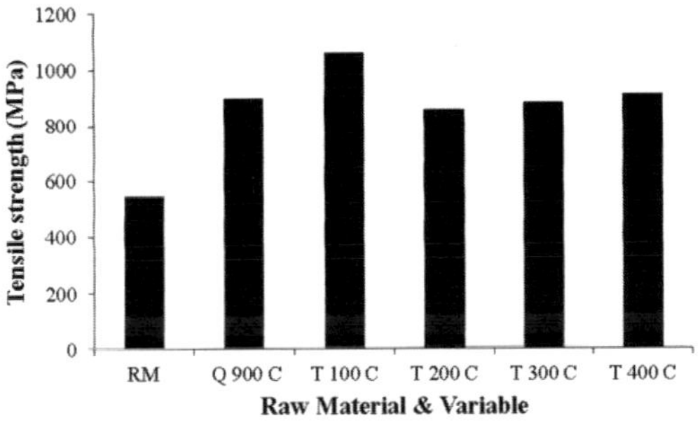

Figure 3. Graph of Tensile Strength on Various Tempering Temperature

The tensile strength of quenched steel plate was 900 MPa, which significantly increased to 548 MPa. The tensile strength of a steel plate occupied at 100°C increase to 1063 MPa from the quenched steel plate which was 900 MPa. Meanwhile, the tensile strength of tempered steel plate at temperature of 200 and 300°C experienced decrease into 848 and 883 MPa. The tensile strength of steel plate at the tempering temperature of 400°C experienced small increase from the quenched steel plate which became 912 MPa.
The hardness value increased causes the tensile strength also increased, while the strain decreased. Hard material has high tensile strength but tends to fragile and decrease ductility. In this case, samples tempered at temperature of 100°C generally experienced increase in its tensile strength although there was small decrease in the hardness. This is different from the cases reported before. However, Xiao et al. [5] stated that tempering at low temperature can increase the steel mechanical properties. Such increase of hardness may be caused by low tempering temperature and samples' homogeneity, although there is sufficient samples amount.
Ballistic test through simulation based on finite element presents that all target plates with various temperature experienced perforation. Projectile with velocity of 302 m/s succeed in penetrating the plate with thickness of 8 mm. Therefore, the ballistic resistant analysis was conducted by measuring the projectile velocity after passing through the target plate. The result of ballistic test of the effect of various tempering temperature on remaining projectile velocity V_r and plugging velocity V_{rpl} is presented in Table 4 and Figure 4.

Table 4. Residual velocity projectile and plugging on various tempering temperature

No	*Variable*	projectile velocity **V_r (m/s)**	plugging velocity **V_{rpl} (m/s)**
1	RAW	244.36	272.15
2	Q 900°C	234.46	240.39
3	T 100°C	185.9	230.03
4	T 200°C	193.89	236.12
5	T 300°C	191.28	267.98
6	T 400°C	191.02	267.98

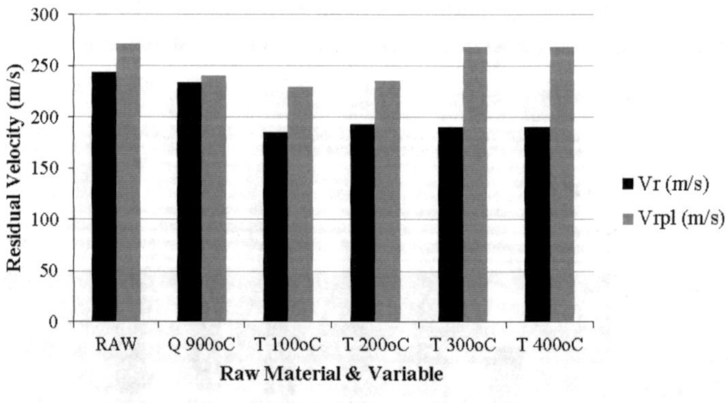

Figure 4. Residual velocity projectiles and plugging on various tempering temperature

Steel plate that was tempered at temperature of 100°C has better ballistic resistant compared to other treatment. It was seen that the residual velocity projectile and plugging of plate that was the lowest was V_r 185.90 m/s and V_{rpl} 230.03 m/s. Ballistic resistant of a material is directly proportional with its tensile strength. The higher the tensile strength, the more resistant it is towards the projectile impact. Simulation of ballistic test on target plate that was heated at austenization temperature of 900°C, quenched using oil media and tempered at temperature of 100°C is seen in Figure 5.

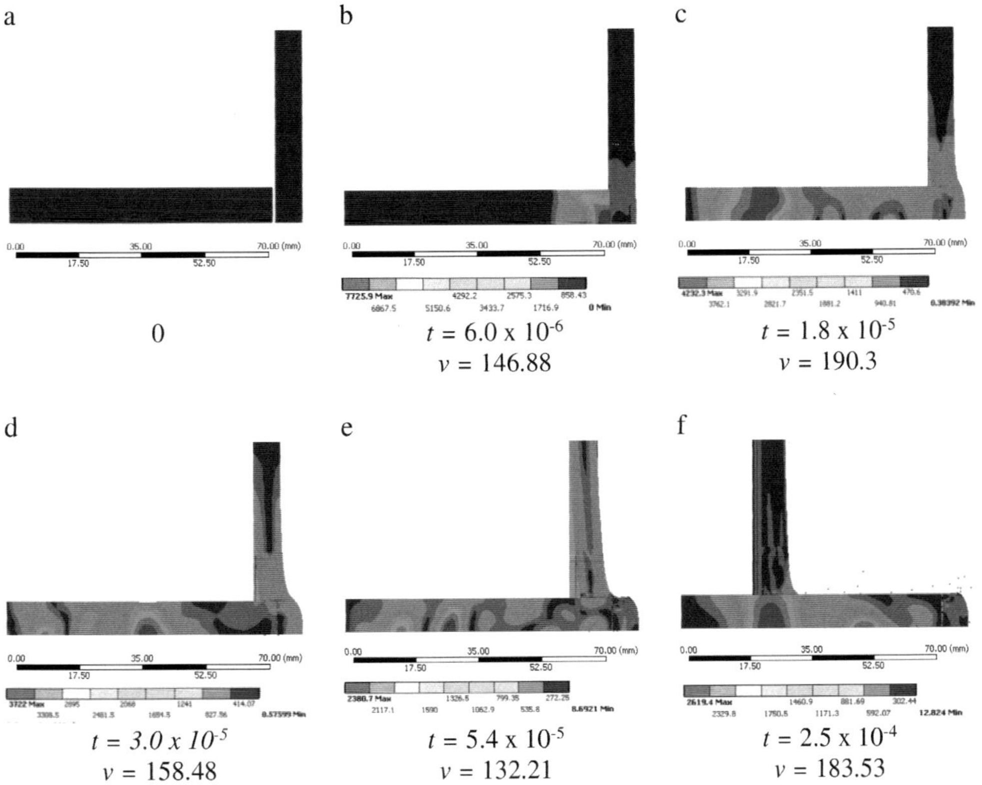

Figure 5. Residual velocity projectile and plugging at various tempering temperature

Projectile was given initial velocity of 301.2 m/s. Wend of the projectile touches the plate surface, the projectile velocity decreased to 146.88 m/s (Figure 5b). Maximum tensile stress occurred at the end of projectile and plate which were directly touched by the end of the projectile. However, when the projectile is succeed in broken the target plate or penetrating half of plat thickness, then the projectile velocity increases to 190.30 m/s (Figure 5c). This is due to quite big projectile's kinetic energy. Maximum tensile stress shifted to the back of projectile. Furthermore, the projectile's velocity decreased after perforation on the target plate. The velocity slowly decreased due to the fraction between the projectile's outer surface and hole's wall (Figure 5d, e and f). After the projectile passed through the target plate, the velocity became 185.90 m/s.

4. Conclusions

Based on the experiment and simulation result on low carbon steel plate through tempering heating, it can be concluded that:

1. Heating through induction and dipping on oil media results in higher surface's hardness compared to other surfaces.
2. Tempering temperature can decrease the hardness, but the tempering temperature at 100°C can increase its tensile strength.

3. Through the simulation of finite element, all variation can be penetrated by the projectile. However, the quenched and tempered material at low temperature is more able to resist the projectile velocity compared to other various tempering temperature.

Acknowledgement
Gratitude is expressed towards the Research and Service Community Institution of Universitas Wahid Hasyim and the Ministry of Research Technology and Higher Education, which have funded this research and publication.

References

[1] Abbasi E, Luo Q and Owens D 2018 A comparison of microstructure and mechanical properties of low-alloy-medium-carbon steels after quench-hardening *Materials Science and Engineering: A* **725** 65–75
[2] Miernik K, Bogucki R and Pytel S 2010 Effect of quenching techniques on the mechanical properties of low carbon structural steel **10** 6
[3] Johnson O T, Ogunmuyiwa E N, Ude A U, Gwangwava N and Addo-Tenkorang R 2019 Mechanical Properties of Heat-treated Medium Carbon Steel in Renewable and Biodegradable Oil *Procedia Manufacturing* **35** 229–35
[4] Chen K, Jiang Z, Liu F, Yu J, Li Y, Gong W and Chen C 2019 Effect of quenching and tempering temperature on microstructure and tensile properties of microalloyed ultra-high strength suspension spring steel *Materials Science and Engineering: A* **766** 138272
[5] Xiao Y, Li W, Zhao H S, Lu X W and Jin X J 2016 Investigation of carbon segregation during low temperature tempering in a medium carbon steel *Materials Characterization* **117** 84–90
[6] Zhang H, Wu Y, Fu J, Xu J and Zhai Q 2017 Influence of on-line tempering parameters on microstructure of medium-carbon steel *Journal of Iron and Steel Research, International* **24** 59–66
[7] Børvik T, Dey S and Clausen A H 2009 Perforation resistance of five different high-strength steel plates subjected to small-arms projectiles *Int. J. Impact Eng.* **36** 948–64
[8] Karagoz S, Atapek H and Yilmaz A 2008 A Fractographical Study on Boron Added Armor Steel Developed By Alloying And Heat Treatment to Understand its Ballistic Performance *13th International Conference on Applied Mechanics and Mechanical Engineering* (Cairo, Egypt)
[9] Purwanto H, Soenoko R, Purnowidodo A and Suprapto A 2018 Effect of Soft-hard Plate and Rubber Sandwich Against on 5.56 mm Deformable Projectile *Journal of Engineering Science and Technology Review* **11** 44–54
[10] Jena P K, Mishra B, RameshBabu M, Babu A, Singh A K, SivaKumar K and Bhat T B 2010 Effect of heat treatment on mechanical and ballistic properties of a high strength armour steel *International Journal of Impact Engineering* **37** 242–9
[11] Burian W, Żochowski P, Gmitrzuk M, Marcisz J, Starczewski L, Juszczyk B and Magier M 2019 Protection effectiveness of perforated plates made of high strength steel *International Journal of Impact Engineering* **126** 27–39
[12] Purwanto H, Soenoko R, Purnowidodo A and Suprapto A 2018 Energy absorbers on the steel plate – rubber laminate after deformable projectile impact *Eastern-European Journal of Enterprise Technologies* **4** 6–12
[13] Børvik T, Langseth M, Hopperstad O S and Malo K A 1999 Ballistic penetration of steel plates *International Journal of Impact Engineering* **22** 855–86
[14] Mishra B, Jena P K, Ramakrishna B, Madhu V, Bhat T B and Gupta N K 2012 Effect of tempering temperature, plate thickness and presence of holes on ballistic impact behavior and ASB formation of a high strength steel *International Journal of Impact Engineering* **44** 17–28

IOP Conf. Series: Materials Science and Engineering **694** (2019) 012042 doi:10.1088/1757-899X/694/1/012042

Electrochemical Behavior of Full Annealed Austenitic Stainless Steel

Victoria Henniwuriyama[1], Evi Ulina Margaretha Situmorang[1,2] and Bambang Soegijono[1,a)]

[1] Department of Physics, University of Indonesia, Depok 16424, Indonesia
[2] School of Medicine and Health Sciences, Atmajaya Catholic University of Indonesia, Jakarta 12930, Indonesia
[a)]Corresponding author: naufal@ui.ac.id

E-mail: victoria.henniwuriyama@ui.ac.id

Abstract. Austenitic Stainless Steel has been being used in many fields. Austenitic Stainless Steel which has composition 19 wt%Cr–7wt%Ni has been subjected to heat treatment. One of the many heat treatments applicable to improve the corrosion resistance is the full annealing method. This material subject to full annealing method with rapid quenching in various annealing temperatures; 1010°C, 1065°C, and 1120°C. Potentiodynamic curves before and after heat treatment were acquired to elaborate the electrochemical behavior of this austenitic stainless steel. Besides the corrosion rate, the structure of the austenitic stainless steel has been characterized using x-ray diffraction (XRD) to study the structure. The results showed the full annealing modified the structure and the electrochemical behavior of the austenitic stainless steel.

1. Introduction

Stainless steel is one type of ferrous-alloy that contains a minimum 10,5% Cr and can be divided into four major groups, as one of them is austenitic stainless steel. This type of stainless steel is the most produced stainless steel and also the most common type of stainless steel used in the world. Some good properties of the austenitic stainless steel are weldable, formable, high strength, ductility, and toughness, and also resistant to corrosion in many conditions. These excellent characteristics beneficially used in manufacturing industries, chemical plants, and other corrosive and severe environment [1]–[5].

The good corrosion resistance of austenitic stainless steel is due to the spontaneous appearance of the chromium oxide layer of its surface when the austenitic stainless steel exposed by the corrosive solution or environment [2]. Austenitic stainless steel is said to have the highest corrosion resistance caused by the higher amount of chromium and the addition of nickel, so the chromium is the foremost element of the austenitic stainless steel [6].

The type of austenitic stainless steel in this experiment is type 302 that is categorized in the normal unstabilized composition. By the American Society of Metals, one in three heat treatment which applicable for the austenitic stainless steel is full annealing with rapid cooling to make sure the dissolved carbides reside in the water. The full annealing temperature range for the type 302 is from 1010°C - 1120°C [3]. This experiment intends to see the sample's corrosion resistance when different temperatures applied with full annealing method because this still being unclear.

2. Experimental Procedure

The austenitic stainless steel type 302 has been cut into approximately 3 cm x 2 cm x 0,2 cm in length, width, and thickness. Three samples used in this experiment were fully annealed with different temperatures for an hour a half. The first sample has been through full annealing treatment with temperature 1010°C (hereinafter referred as 1010ASS), the second one was fully annealed with temperature 1065°C (hereinafter referred as 1065ASS), and the last sample was fully annealed with temperature 1120°C (hereinafter referred as 1120ASS). The full annealing method for austenitic stainless steel is heat treatment with rapid quenching, which is different when it applied to the low-alloy or carbon steels [7].

To see the characterization before and after heat treatment, these were analyzed at UPP IPD Research Laboratory, Faculty of Mathematics and Natural Sciences, Universitas Indonesia, and the compositions of the austenitic stainless steel type 302 shown in TABLE 1. For the corrosion tests, all of the samples were dipped into 3,5 wt% NaCl solution and the data were collected using Potentiostat. The corrosion rate values were obtained from Faraday's Law in Determination Corrosion Rates of Metals and Alloys shown below [8].

$$r = C\frac{\mathrm{Mi}}{\mathrm{n}\rho} \quad (1)$$

where
M = atomic weight ($g.mol^{-1}$)
i = current density (A/cm^2)
n = number of electron involved
ρ = density (g/cm^2)
C = constant (0,129 in mpy; 3,27 in mm/year; 0,00327 in mm^3/year)

The value of i has SI unit A/cm^2, where the cm^2 means the cross-sectional area of the samples which will be tested for the corrosion resistance. And for the experiment is 0,785 cm^2 which obtained from the calculation of area of a circle with a diameter of 1 cm.

3. Results and Discussion

3.1. XRD Analysis

In Figure 1 we can see the XRD plot and the data of characterization shown in TABLE 2. For the 1010ASS, angle positions of the six peaks are 43,680°, 44,543°, 50,754°, 64,559°, 74,669°, and 82,072°. For the 1065ASS, angle positions of the six peaks are 43,776°, 44,594°, 50,755°, 64,636°, 74,767°, and 81,997°. For the 1120ASS, angle positions of the six peaks are 43,778°, 44,599°, 50,671°, 64,983°, 74,862°, and 81,928°. The highest first iron-alpha's peak is on the 1010ASS which value is 116,92 count per second, then followed by the 1065ASS which value is 105,67 count per second and 97,99 count per second for the 1120ASS.

Table 1. The compositions of austenitic stainless steels type 302 identified by XRF

Fe	**Cr**	**Ni**	**Mn**
71 wt%	19 wt%	7 wt%	2 wt%

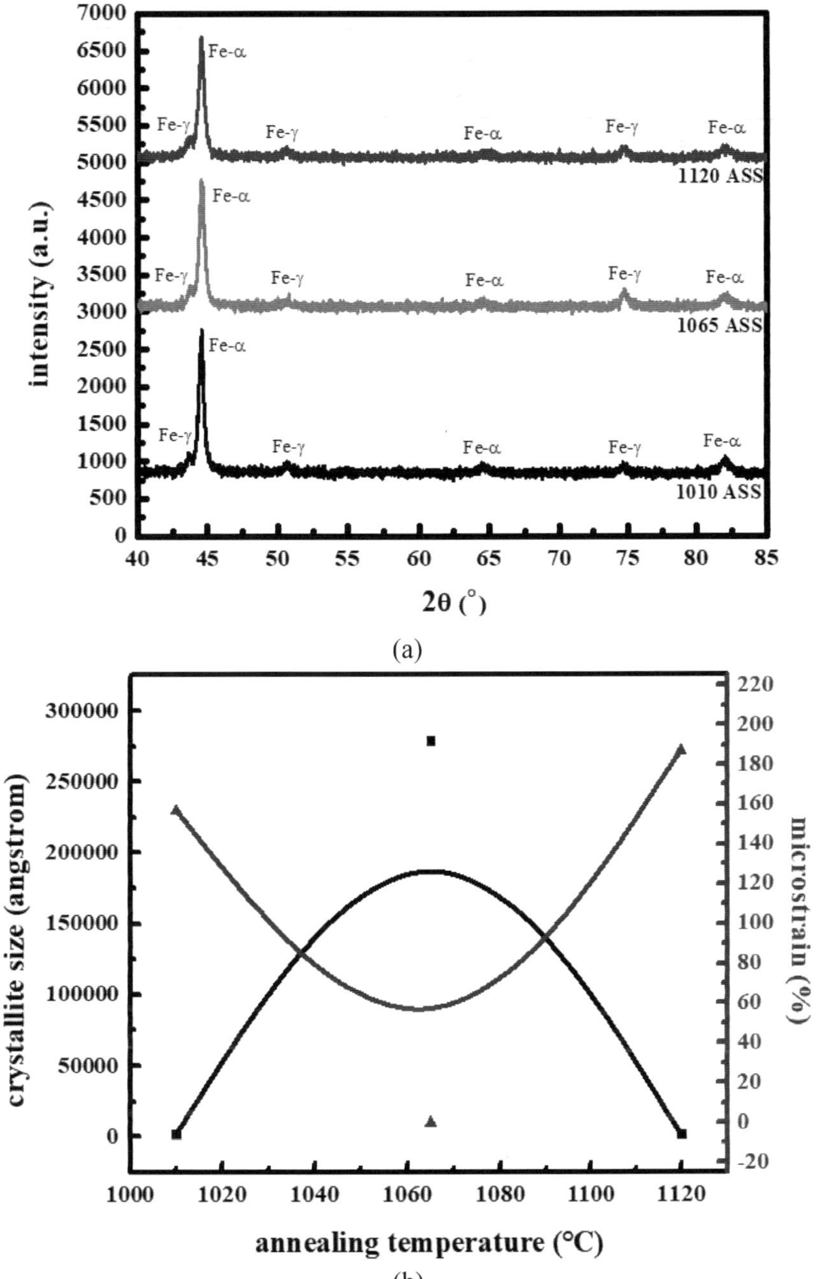

(a)

(b)

Figure 1. (a) XRD's plot of the 1010ASS, 1065ASS, 1120ASS;
(b) Relation between crystallite size and microstrain on the 1010ASS, 1065ASS, 1120ASS

The largest d value for the first iron-alpha's peak is on the 1010ASS which value is 2,03424 Å, then followed by the 1120ASS which value is 2,03311 Å for the 1065ASS. The lattice parameters have the same value for a, b, and c because the crystal system of these samples is cubic. For the iron-alpha, the lattice parameter change in each sample, where the biggest value is on the 1120ASS which value is 2,90586 Å, then followed by the 1010ASS which value is 2,89876 Å, and 2,89262 Å for the 1065ASS. For the iron-gamma, the biggest value of the lattice parameter is on the 1120ASS which value is 3,61667 Å and 3,61 Å for the 1010ASS and 1065ASS. The volume of the iron-alpha and iron-gamma are also

Table 2. Crystallographic parameters of the 1010ASS, 1065ASS, and 1120ASS.

Sample		1010ASS	1065ASS	1120ASS
First peak's height (cps)	Iron-Alpha	116,92	105,67	97,99
	Iron-Gamma	20,94	31,71	16,30
First peak's d (Å)	Iron-Alpha	2,0342	2,0328	2,0331
	Iron-Gamma	2,0681	2,0711	2,0656
Lattice Parameter (Å)	Iron-Alpha	a = 2,8987 b = 2,8987 c = 2,8987	a = 2,8926 b = 2,8926 c = 2,8926	a = 2,9058 b = 2,9058 c = 2,9058
	Iron-Gamma	a = 3,6100 b = 3,6100 c = 3,6100	a = 3,6100 b = 3,6100 c = 3,6100	a = 3,6166 b = 3,6166 c = 3,6166
Volume (Å^3)	Iron-Alpha	24,3576	24,2033	24,5371
	Iron-Gamma	47,0458	47,0458	47,3072
Density (g/cm^3)	Iron-Alpha	7,61	7,66	7,56
	Iron-Gamma	7,88	7,88	7,84
Concentration (%)	Iron-Alpha	77,8	91,9	72,8
	Iron-Gamma	22,2	8,1	27,2
Microstrain (%)	Iron-Alpha	0,157	0,000	0,187
	Iron-Gamma	0,189	0,000	0,315
Crystallite Size (Å)	Iron-Alpha	208,1	27913,4	175,4
	Iron-Gamma	173,2	27913,4	103,8

changed following their lattice parameter's value. The highest concentration of iron-alpha phase is on the 1065ASS which value is 91,9%, followed by 1010ASS which value is 77,8%, and 72,8% for the 1120ASS.

From the XRD's plot in FIGURE 1(a), the only changes can be seen slightly is the peak's height of iron-alpha phase, where the highest is on the 1010ASS, then followed by the 1065ASS, and the 1120ASS. This explains that increase in annealing temperature will decrease the peak's height of iron-alpha phase.

It can be seen in FIGURE 1(b) that the highest crystallite size value is on the 1065ASS, while the other two annealing temperature has the lower value. However, 1065ASS has the lowest microstrain value compared by the other two samples. This explains that crystalite size value inversely proportional to the microstrain value [9].

Table 3. Data from the corrosion test of the 1010ASS, 1065ASS, 1120ASS in the NaCl solution 7wt%

Wt % NaCl	Sample	R (ohm)	$E_{corrosion}$ (V)	$I_{corrosion}$ (A)	Corrosion Rate (mm/year)
7 wt%	1010ASS	3,566 x 10^2	-1,099	7,206 x 10^{-5}	0,0587
	1065ASS	2,789 x 10^2	-1,080	9,211 x 10^{-5}	0,0751
	1120ASS	3,894 x 10^2	-1,091	6,599 x 10^{-5}	0,0538

IOP Conf. Series: Materials Science and Engineering **694** (2019) 012042 doi:10.1088/1757-899X/694/1/012042

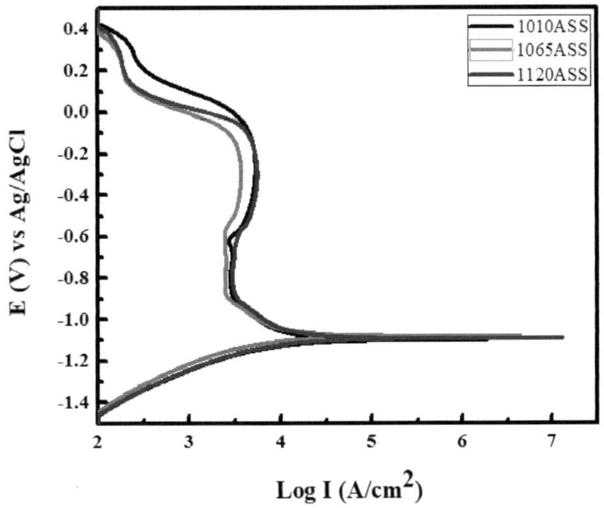

(a)

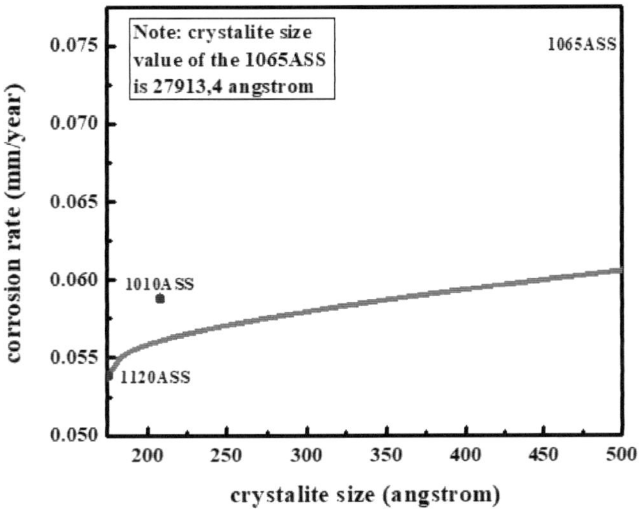

(b)

Figure 2. (a) Potentiodynamic polarization curves of the samples in 7 wt% NaCl; (b) Relation between crystallite size and corrosion rate of the 1010ASS, 1065ASS, 1120ASS

3.2. Potentiodynamic Polarization Curves Analysis

The mechanism of corrosion of these materials is the occurrence of anodic dissolution on their surfaces and then followed by ionization and hydrolysis of the samples [2,10]. These phenomena are called the anodic and cathodic reaction, which its potentiodynamic curves shown in FIGURE 2(a).

Figure 2(b) summarizes the data from TABLE 2 and TABLE 3, which is display the relation between corrosion rate and crystallite size into a plot. The results obtained are convinced by another experiment which explains that an increase in grain size (or crystallite size) is proportional to the increase of the corrosion rate value (i.e corrosion resistance reduced) [11].

From Table 3, it can be seen that 1120ASS has the lowest corrosion rate, which its value is 0,0538 mm/year. It means the 1120°C annealing temperature is better for improving corrosion resistance from the other annealing temperatures used in this experiment. This is due to the nearness to the ideal temperature which is limit the grain growth is 1095°C [3]. The 1010ASS has the next best annealing temperature for improving corrosion resistance, which its value is 0,0587, even though the corrosion rate difference between 1120ASS and 1010ASS is not significant. This temperature (1010°C) is close to 1030°C, which is used to some manufacturing industry to annealing their material. And the last annealing temperature which has the highest corrosion rate value is 1065°C which its value is 0,0751. This explains that the increase in annealing temperature is not directly proportional to the corrosion rate value. It might be related to the carbide precipitation and need a further experiment.

4. Conclusion

The three different annealing temperatures used in this experiment affect the crystallographic parameters of the samples, and also affect the corrosion rate value, in which the best annealing temperature to improve corrosion resistance is 1120°C, followed by the 1010°C, and then 1065°C. The corrosion rate also depends on the grain or crystallite size, in which the decrease of grain size is directly proportional to the corrosion rate reduction.

Acknowledgements

The authors acknowledge with gratitude the financial supports from University of Indonesia, under grant Hibah Publikasi Internasional Terindeks Tugas Akhir (Hibah PITTA B), No. NKB-0617/UN2.R3.1/HKP.05.00/2019.

References

[1] R. T. Loto, "Corrosion resistance and morphological deterioration of 316Ti austenitic, *GX4CrNiMo16-5-1 martensitic and 444 ferritic stainless steels in aqueous corrosive environments,"* in Results in Physics 14(2019)102423.

[2] R. T. Loto, C. A. Loto, and I. Ohijeagbon, "*Effect of heat treatment processes on the localized corrosion resistance of austenitic stainless steel type 301 in chloride/sulphate solution*," inResults in Physics (Elsevier B. V., 2018)

[3] T. L. Gall, R. L. Stedfeld, and W. H. Cubberly, "Stainless Steels," and "Austenitic Stainless Steels," in Heat Treater's Guide Standard Practices and Procedures for Steel, edited by P. M. Underweiser, H. E. Boyer, and J. J. Kubbs (American Society for Metals, United States of America, 1984), pp. 401-417.

[4] A. F. Padilha, R. L. Plaut, and P. R. Rios, "*Annealing of Cold-worked Austenitic Stainless Steels*," ISIJ International, Vol.43 (2003), No.2, pp. 135-143.

[5] M. F. McGuire, "Austenitic Stainless Steels," in *Stainless Steels for Design Engineers* (ASM International, United States of America, 2008), pp. 69-71.

[6] W. D. Callister, Jr., "Stainless Steels," in *Materials Science and Engineering An Introduction* (John Wiley & Sons, Inc., United States of America, 2007), pp. 364-365.

[7] R. A. Covert and A. H. Turthill, "Stainless Steels: An Introduction to Their Metallurgy and Corrosion Resistance," in Dairy, Food, and Environmental Sanitation (Association for Food Protection, Iowa, 2000), pp. 506-517.

[8] Z. Ahmad, "Application of Faraday's Laws In Determination of Corrosion Rates of Metals and Alloys," in Principle of Corrosion Engineering and Corrosion Control (Butterworth-Heinemann, United Kingdom, 2006), pp. 58-61.

[9] K. Maniammal, G. Madhu, and V. Biju, "X-ray diffraction line profile analysis of nanostructured *nickel oxide: Shape factor and convolution of crystallite size and microestrain,"* in Physica E 85 (2017) 214-222.

[10] P. Marcus and V. Maurice, "Fundamental Aspects of Corrosion of Metallic Materials," in Materials Science and Engineering-Vol. II.

[11] K. D. Ralston, N. Birbilis, and C. H. J. Davies, "Revealling the relationship between grain size and corrosion rate of metals," in Scripta Materialia 63 (2010) 1201-1204.

Effect of Solution Heat Treatment of Aluminum Alloy 4032 on the Structure and Corrosion Resistance in 3,5% and 10,5% NaCl Solution

Fadli Nur Cahyo and Bambang Soegijono

Department of Physics, Universitas Indonesia, Depok, Indonesia

E-mail: naufal@ui.ac.id

Abstract. Aluminum Alloy 4032 is usually used as a material for making engine pistons, which has a composition of Aluminum (Al), 21%wt Silicon (Si), 0.82%wt Nickel (Ni) and 0.67%wt Copper (Cu). The alloy is known to have mild properties and should have good corrosion resistance. Heat treatment was carried out with the aim of the enhancement of the corrosion resistance. The solution heat treatment process was performed at 510°C for 2 hours and then quenching to room temperature in water media. Some of the material were also performed an aging process at 171°C with varying duration of time and then cooling to room temperature. Corrosion resistance testing was carried out by electrochemical testing using 3.5% wt and 10.5% wt NaCl solution. The structure before and after heat treatment were characterized by using X-Ray Diffraction (XRD) were analyzed. The results of the characterization indicate that the heat treatment on the Alloy is affected by solution heat treatment and the aging process.

1. Introduction

Aluminum Alloy 4032 is a 4000 series aluminum alloy, which has a very good surface, without hot ductility, good for wrought and cast alloys. The aluminum-silicon alloy 4000 series is identical with good durability, lightweight, very small expansion coefficient and as a good conductor of heat. Aluminum Alloy 4032 has the main alloy is silicon, followed by Nickel and Copper. Aluminum Alloy 4032 is very widely used because it has a striking advantage. In the automotive sector and industry, it is used for several engine components such as pistons, transmission valves, master brake cylinders, bushings for rack and pinion steering systems, hydraulic applications, and many more [1].

Aluminum Alloy 4032 has a lower melting point than other compositions made from the same material or what is often referred to as having eutectic properties. Based on the silicon content, this alloy is divided into three, namely Hypoeutectic alloys which have a silicon content of between 5% - 10%, Eutectic alloys that have a silicon content of between 11% -13%, and Hypereutectic alloys which have a silicon content of between 14% -25% [2]. The presence of silicon alloys into aluminum will increase fluidity, heat resistance. And the addition of silicon can also reduce the value of specific gravity and thermal coefficient of aluminum. The addition of Nickel can improve the properties of aluminum alloys at high temperatures and can also reduce the coefficient of thermal expansion of the alloy. As for copper which is very influential in increasing the strength and stiffness of aluminum both as-cast and also after the heat treatment process, as well as the addition of copper can reduce heat cracking resistance. The presence of Cu and Ni elements, causing this alloy is resistant to rust [3].

Based on the purity of the base material, Al-Si alloys consist of various impurities such as iron, manganese, silver, and copper. The presence of copper and magnesium often added as an alloy of elements to increase the strength and ability of the material to be cast. Impurities and alloying elements will dissolve into solid solutions in the matrix and some form intermetallic particles during the freezing process.

Solution Heat Treatment in Aluminum alloy 4032 aims to produce solid solutions with the number of main alloying elements such as Al, Si, and Cu in aluminum alloys so that their mechanical properties will increase. Solution heat treatment consists of three steps, namely solution, quenching, and aging. Microstructural changes occur because the solution treatment stages include the process of quenching and aging [4]. In the process of heat treatment solution that needs to be considered is the use of temperature does not exceed the eutectic temperature because it will cause overheating, the tensile strength will be ductile and fracture toughness will be reduced, because when overheating the grain boundaries melt.

2. Experimental Procedure

Aluminum-Alloy 4032 samples with a size of approximately 2.6 cm x 2.1 cm x 0.5 cm were used for this experiment. Each of the samples were heat treated differently. For all samples treated were polished with sandpaper. In the first sample only treated normally or not treated (hereinafter referred to as *AA 4032 NH*). In the second sample (hereinafter referred to as *AA 4032 HT*) through the Solution Heat Treatment. In the third sample (hereinafter referred to as *AA 4032 A5*) through the Solution Heat Treatment and the Artificial Aging Treatment. In the fourth sample (hereinafter referred to as *AA 4032 A10*) through the Solution Heat Treatment and the Artificial Aging Treatment. In the fifth sample (hereinafter referred to as *AA 4032 A15*) through the Solution Heat Treatment and the Artificial Aging Treatment.

All samples were analyzed at the UPP IPD Research Laboratory, Faculty of Mathematics and Natural Sciences, University of Indonesia. The composition of Aluminum-Alloy 4032 used in this experiment can be seen in TABLE 1. List of compositions of Aluminum Alloy 4032 in weight%.

Table 1. List of composition of Aluminum Alloy 4032 in weight %

Elements	Weight %
Al	76.73
Si	21.42
Ni	0.82
Cu	0.67
Fe	0.22
Ag	0.10
Mn	0.04

2.1. Solution Heat Treatment

Solution Heat Treatment was carried out by samples of *AA 4032 HT*, *AA 4032 A5*, *AA 4032 A10*, and *AA 4032 A15*. Solution Heat treatment is carried out at a temperature of 510 ° C with a duration of 2 hours and then quenching using water at room temperature. The process is intended to stabilize physical or mechanical properties or corrosion resistance and is intended to quench the structure of the atoms in solution.

2.2. Artificial Aging Treatment
Artificial Aging Treatment is a heat treatment process carried out on a material to improve its hardness by combining heating above room temperature with heating time [5]. The Artificial Aging Treatment process was carried out at 171 ° C at 5 hours, 10 hours and 15 hours in three samples, namely *AA 4032 A5*, *AA 4032 A10*, and *AA 4032 A15*. This treatment is carried out to obtain deposition hardening, which is hardening due to deposition of the second phase which is evenly distributed. Precipitate hardening aims to enhance the metal's resistance to deformation by inhibiting the movement of dislocations by the precipitate.

2.3. Corrosion Test
Corrosion test was carried out using the Digi-Ivy © DY2300 series potentiometer with two different concentration solutions, 3.5% wt and 10.5% wt NaCl solution. Also for the sample, we have prepared by connecting the sample with the wire together, so that the sample can be entered into the solution before the wire collects data. It aims to investigate the change in mass of each sample by heat treatment. The surface area of each sample of Aluminum Alloy 4032 is used to test corrosion with a size of 1 cm x 1 cm. To collect data in the corrosion test process using a Potentiostat, the formula used to calculate the value of the corrosion rate is obtained from Faraday's Law in Determining the Corrosion Rate of Metals and Alloys [6].

$$\text{Corrosion rate of alloy, } r = C\frac{Mi}{n\rho} \tag{1}$$

where C is the constant corrosion rate (3.27 × 10-3 mm3 / year used in this study), M is the atomic weight of each element in the alloy (gr/mol), i is the current density in Amperes / cm2, where cm2 defines the cross-sectional area of the sample tested for corrosion resistance., n is the number of electrons involved in the corrosion process, ρ is the mass of each element in the alloy (gr / cm3).

3. Results and Discussion

3.1. XRD Analysis
In Figure 1, there are results from the XRD Plot sample of Aluminum Alloy 4032 and there are also data from the characterization of the Aluminum Alloy 4032 sample in TABLE 2. Seen from FIGURE 2, each of them identified seven peaks in the graph of each sample. For *AA 4032 NH* samples, there are seven peaks identified with angle positions of 28,616 °, 38,626 °, 44,867 °, 47,458 °, 56,269 °, 65,214 ° and 78,325 °. For *AA 4032 HT* samples, there are seven peaks identified with angle positions of 28,616 °, 38,617 °, 44,859 °, 47,473 °, 56,290 °, 65,213 ° and 78,328 °. For *AA 4032 A5* samples, there are seven peaks identified with angle positions of 28,481 °, 38,470 °, 44,713 °, 47,352 °, 56,177 °, 65,071 ° and 78,189 °. For the *AA 4032 A10* sample, there are seven peaks identified with sample positions of 28,463 °, 38,469 °, 44,713 °, 47,323 °, 56,142 °, 65,073 °, and 78,193 °. For *AA 4032 A15* samples, there are seven peaks identified with sample positions of 28,471 °, 38.47 °, 44,715 °, 47.34 °, 56,165 °, 65,079 °, and 78,202 °. In FIGURE 1, there is an X-Ray diffraction pattern that there are seven peaks of the high intensity, namely, (111), (111), (002), (022), (113), (022), and (113) seen clear in the aluminum phase. which value is 3,132 Å, then followed by the *AA 4032 A5* which value is 3,131 Å, and 3,116 Å for the *AA 4032 NH* & *AA 4032 HT*.

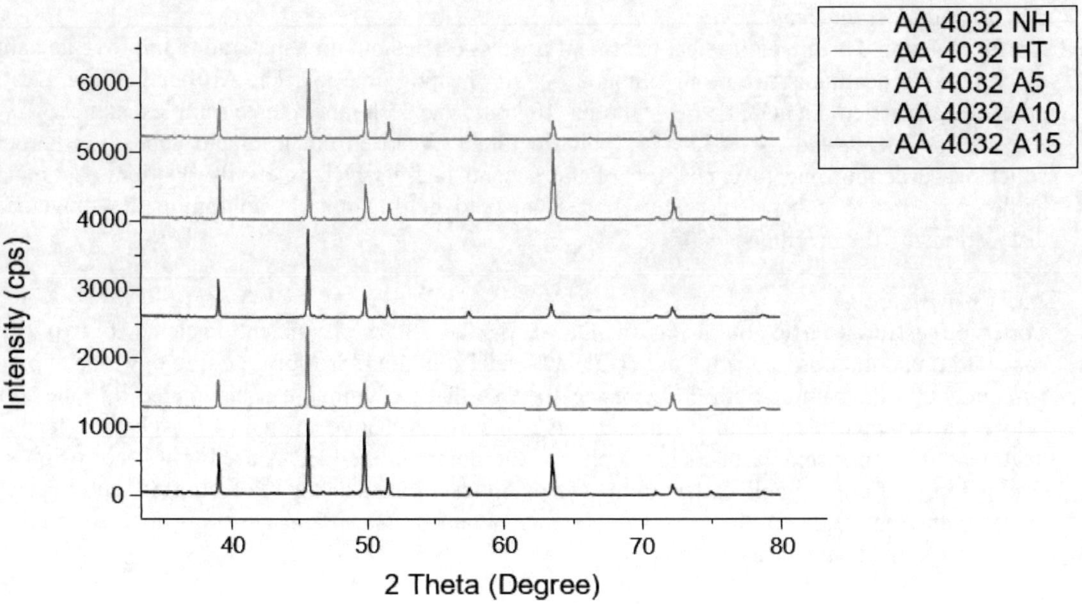

Figure 1. XRD plot of Aluminum Alloy 4032 Non Heat Treatment, Heat Treatment, 5 hours of aging, 10 hours of aging, and 15 hours of aging.

Table 2. Crystallografic parameters of *AA 4032 NH*, *AA 4032 HT*, *AA 4032 A5*, *AA 4032 A10*, and the *AA 4032 A15*

Sample		AA 4032 NH	AA 4032 HT	AA 4032 A5	AA 4032 A10	AA 4032 A15
Highest peak height (cps)	Alumunium	247,64	247,06	236,97	215,73	211,3
	Sillicon	107,72	147,70	103,70	94,12	81,6
Highest peak d (Å)	Alumunium	2,329	2,329	2,338	2,338	2,338
	Sillicon	3,116	3,116	3,131	3,133	3,132
Lattice Parameter (Å)	Alumunium	a = 4,054	a = 4,053	a = 4,052	a = 4,052	a = 4,051
		b = 4,054	b = 4,053	b = 4,052	b = 4,052	b = 4,051
		c = 4,054	c = 4,053	c = 4,052	c = 4,052	c = 4,051
	Sillicon	a = 5,435	a = 5,431	a = 5,428	a = 5,431	a = 5,428
		b = 5,435	b = 5,431	b = 5,428	b = 5,431	b = 5,428
		c = 5,435	c = 5,431	c = 5,428	c = 5,431	c = 5,428
Volume (Å^3)	Alumunium	66,630	66,581	66,55	66,535	66,501
	Sillicon	160,592	160,267	159,958	160,202	159,973
Density (g/cm^3)	Alumunium	2,69	2,69	2,69	2,69	2,69
	Sillicon	2,32	2,33	2,33	2,33	2,33
Concentration (%)	Alumunium	74,8	74,5	76	82,5	76,5
	Sillicon	25,2	25,5	24	17,5	23,5
Microstrain (%)	Almunium	0,073	0,088	0,306	0	0
	Sillicon	0	0,116	0	0	0
Crystalite Size (Å)	Alumunium	1527	748,2	173,3	27913,4	27913,4
	Sillicon	27913,4	27913,4	27913,4	27913,4	27913,4

Referring to Table 2, the lattice parameter value has the same value in a, b, and c which means that the aluminum and silicon content has a cubic crystal structure. The highest aluminum peak (111) is at AA 4032 A10, then followed by AA 4032 A15, AA 4032 A5, AA 4032 NH, and AA 4032 HT. This explains that the artificial aging heat treatment and coolant temperature increase the aluminum phase. Based on X-Ray diffraction analysis (XRD) which can explain that Aluminum has more concentration and is resistant to corrosion. [7].

Table 3. Data from the corrosion test of *AA 4032 NH*, *AA 4032 HT*, *AA 4032 A5*, *AA 4032 A10*, and the *AA 4032 A15* in the 3,5 wt% and 10,5 wt% NaCl solution

Wt % NaCl	Sample	R (ohm)	$E_{corrosion}$ (V)	$I_{corrosion}$ (A)	Corrosion Rate (mm/year)
3,5 wt%	*AA 4032 NH*	3,568 x 10^1	-1,403	7,201 x 10^{-4}	0,170637
	AA 4032 HT	3,462 x 10^1	-1,426	7,421 x 10^{-4}	0,17585
	AA 4032 A5	6,523 x 10^1	-1.450	3,939 x 10^{-4}	0,09334
	AA 4032 A10	8,180 x 10^1	-1,430	3,141 x 10^{-4}	0,07443
	AA 4032 A15	4,783 x 10^1	-1,405	5,372 x 10^{-4}	0,127297
10.5 wt%	*AA 4032 NH*	3,474 x 10^1	-1,330	7,396 x 10^{-4}	0,175258
	AA 4032 HT	2,380 x 10^1	-1,374	1,080 x 10^{-3}	0,25592
	AA 4032 A5	3,939 x 10^1	-1,365	6,522 x 10^{-4}	0,154547
	AA 4032 A10	2,960 x 10^1	-1,380	8,679 x 10^{-4}	0,20566
	AA 4032 A15	3,416 x 10^1	-1,354	7,523 x 10^{-4}	0,178267

3.2. Corrosion Testing Analysis: Electrochemical Test

Electrochemical behavior of the samples were carried out by using Potentiodynamic. The data collected were listed TABLE 3 and potentiodynamic curve are shown FIGURE 2. The corrosion mechanism of aluminum alloy 4032 can be explained by anodic and cathodic reactions that occur when the sample is dipped in a solution of NaCl. It is generally stated that a 10.5 wt% NaCl solution gives a higher corrosion rate in these five samples, compared to a 3.5 wt% NaCl solution. It states that there is a trend of decreasing the value of the corrosion rate along with the decreasing concentration of NaCl solution and also it appears that the higher the concentration of NaCl solution which causes the corrosion rate to increase. In a 3.5% NaCl solution, it was seen that *AA 4032 A10* which underwent an aging process for 10 hours had the lowest corrosion rate with a value of 0.07443 mm / year. In the 10.5% NaCl solution, it was seen that *AA 4032 A5* which underwent an aging process for 5 hours had the lowest corrosion rate with a value of 0.154547 mm / year. The highest corrosion rate among these samples is the corrosion rate of *AA 4032 HT*, this happens because this sample does not go through an artificial aging process, because it will occur in the form of a second phase precipitation which begins with the nucleation process and the onset of the atomic cluster that becomes the beginning from precipitate and precipitate can increase the strength and hardness of the material. With increasing strength and hardness, it is also due to the dominant aluminum concentration [8]. The corrosion rate in each sample has a very significant difference which is the reason for using the concentration of NaCl solution which has a considerable distance between 3.5 wt% to 10.5 wt%.

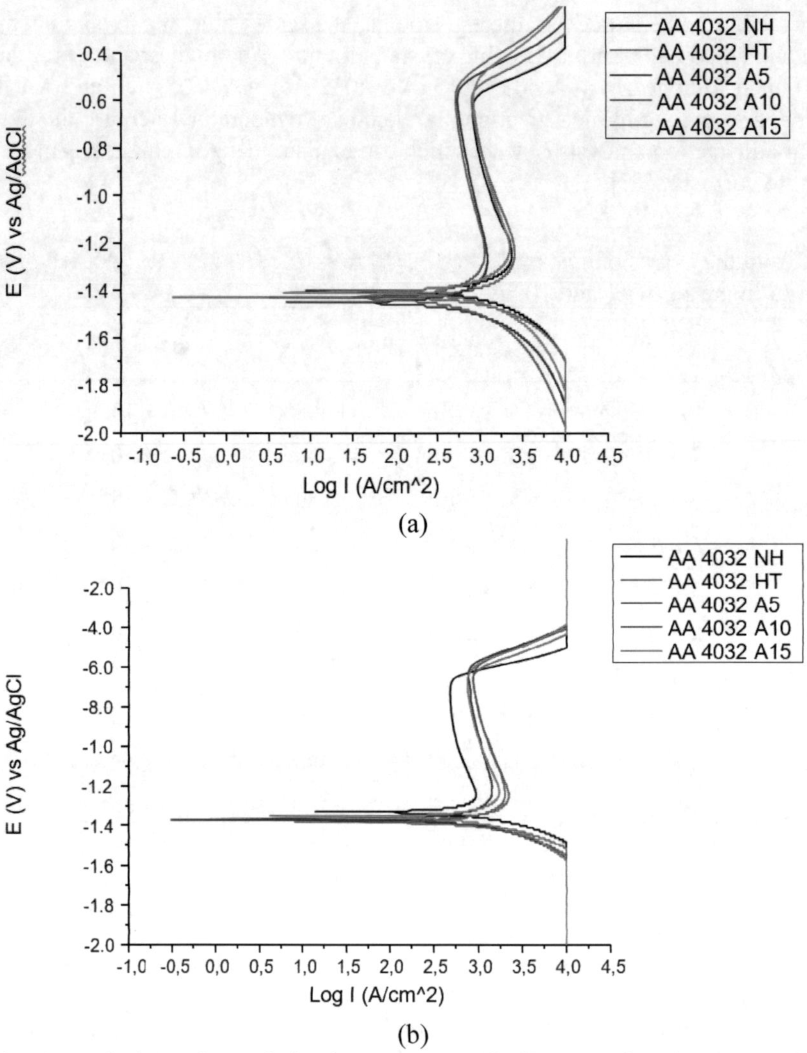

Figure 2. (a). Potentiodynamic polarization curves of the samples in 3,5 wt% NaCl (b). Potentiodynamic polarization curves of the samples in 10,5 wt% NaCl

Table 3 show the results of potentiodynamic characterization. The corrosion potential (E (V)) that has a negative value, due to the electron content that enters the metal due to anodic reaction (oxidation) which thus shifts the potential to negative, which slows the anodic reaction and speeds up the reaction cathodic. It can be concluded that aluminum alloy 4032 tends to be more or less anodic, depending on the duration and type of heat treatment that has been carried out and the phenomenon on aluminum alloy 4032 treated. With the difference in corrosion potential in each sample, it will also cause a change in the corrosion rate in each sample.
The anodic reaction is:

$$Al \rightarrow Al^{3+} + 3e^{-} \quad (2)$$

$$Al^{3+} + 6(H_2O) \rightarrow (Al(H_2O)_6)^{3+} \quad (3)$$

$$6NaCl + 6H_2O \rightarrow 6NaOH + 6Cl_6 + 3H_2 \quad (4)$$

The cathodic reaction is:

$$O_2 + 2(H_2O) + 4e^- \rightarrow 4OH^- \quad (5)$$

Main reaction is:

$$2Al^{3+} + 6Cl^- \rightarrow Al_2Cl_6 \quad (6)$$

Al_2Cl_6 works in $6H_2O$, the result is: [9]

$$Al_2Cl_6 + H_2O \rightarrow Al_2(OH)_6 + 6HCl \quad (7)$$

Based on Table 3 that has the highest current density is *AA 4032 HT*, so when the current density is lower, the corrosion rate is also lower, because there are not fewer electrons flowing in the material or the smaller electron exchange occurs, so the smaller the rate of formation the corrosion and also the presence of aluminum alloys with silicon make the current density relatively low. Also, the effect of corrosion is from the corrosion test with NaCl which has been concluded that the water content, due to acidic carrier water which makes accelerating the rate of corrosion, so that the rapid occurrence of corrosion on the alloy [10].

4. Conclusion

The dependence of the corrosion rate on the concentration of the electrolyte solution with the greater the concentration of the electrolyte solution, the greater the corrosion rate. Based on experimental results that the value of the corrosion rate of aluminum alloy 4032 which has been affected by the heat treatment process, cooling process, and aging. The samples that most resistance to corrosion is AA 4032 A10, with artificial aging and quenching. From XRD data, the unit cell of aluminium and silicon have cubic structure because the lattice parameter have equal value in a, b, and c.

Acknowledgement

The authors would like to show appreciation to the Research of Center of Material Science (RCMS) and CMPFA (Center for Material and Processing Failure Analysis) Universitas Indonesia for supporting the research activity for providing data XRD and SEM-EDX. This study also funded by Universitas Indonesia through Hibah PITTA B No. NKB-0617/UN2.R3.1/HKP.05.00/2019.

References

[1] Rogante M, Lebedev V T, Nicolaie F, Rétfalvi E and Rosta L 2005 SANS study of the precipitates microstructural evolution in Al 4032 car engine pistons *Phys. B Condens. Matter*
[2] Davis J R 2001 Aluminum and Aluminum Alloys 351–416
[3] Schwartz M 2016 *Encyclopedia and Handbook of Materials, Parts and Finishes*
[4] Polmear I J 2005 Physical metallurgy of aluminium alloys *Light Alloys*
[5] Pogatscher S, Antrekowitsch H, Leitner H, Ebner T and Uggowitzer P J 2011 Mechanisms controlling the artificial aging of Al-Mg-Si Alloys *Acta Mater.*
[6] Ahmad Z 2006 *Principles of Corrosion Engineering and Corrosion Control*
[7] Ortiz A L and Shaw L 2004 X-ray diffraction analysis of a severely plastically deformed aluminum alloy *Acta Mater.*
[8] Loto R T, Loto C A and Ohijeagbon I 2018 Effect of heat treatment processes on the localized corrosion resistance of austenitic stainless steel type 301 in chloride/sulphate solution *Results Phys.*
[9] Liu Y, Sun Q, Li W, Adair K R, Li J and Sun X 2017 ScienceDirect A comprehensive review on recent progress in aluminum e air batteries *Green Energy Environ.* **2** 246–77
[10] Melchers R E 2005 The effect of corrosion on the structural reliability of steel offshore structures *Corrosion Science*

IOP Conf. Series: Materials Science and Engineering **694** (2019) 012044 doi:10.1088/1757-899X/694/1/012044

The Effect of Manganese and Heat treatment on Structure and Corrosion Resistance of Cobalt Chrome Molybdenum

Aliya Qonita, Daniel Eka Perkasa and Bambang Soegijono*

Department of Physics, Faculty of Mathematics and Natural Science, University of Indonesia

*E-mail: naufal@ui.ac.id

Abstract. The selection of implant biomaterials must be based on the compatibility of the human body so that it has significant properties that will last long without experiencing rejection during use in the body. One of the commonly used for biomedical implants is Co-Cr-Mo alloy. Since these alloy should also fulfill implant requirements such as high corrosion resistance. To increase corrosion resistance, the heat treatment temperature and time (heat treatment parameters) were 65 hours at 700℃. Characterization was performed using X-ray diffraction and Potensio dynamic. The different X-ray diffraction patterns and corrosion resistance were obtained. The results show both structure and corrosion resistance are markedly influenced by the addition of Manganese after heat treatment.

1. Introduction

For economic reasons, as well as health issues, Co-Cr-Mo alloys have attracted the attention of researches for surgical implants. These metal implants will be implanted in the body to replace the hip joint and artificial knee joint. [1] Co-Cr-Mo alloys have high durability, high corrosion resistance, superior wear behavior, and modest biocompatibility, so they are often used for metal-to-metal hip coatings. [2] Co-Cr-Mo alloy material consists of Cobalt (Co) as the main constituent element, where this element gives elastic modulus, strength, and hardness to the alloy. Chromium (Cr) provides alloy resistance to corrosion. Molybdenum (Mo) makes the surface of the grain smoother, increasing the strength of the atom.

The problem of using Co-Cr-Mo alloys as body implants is affected by corrosion and wear resistance. [4] Co-Cr-Mo alloys are considered to have good corrosion resistance, because it has a slow oxidation rate at room temperature. However, interactions between metal materials used as implants that interact with body fluids can cause tribocorrosion. [3] Tribocorrosion is a form of degradation of materials due to the combined effects of corrosion and wear. Also, tribocorrosion increases the possibility of releasing metal ions into the body which will harm the tissues in the body and the possibility of implant failure. A new study shows that heat treatment can affect its resistance to tribocorrosion and the release of metal ions. [5]

The behavior of the material can be determined by the composition of the alloying elements and the heat treatment applied. Cobalt as an alloying element in high-temperature applications will be discussed in this paper. Through the alloy design, it will be observed the possibility that occurs if the alloy is given heat treatment. Heat treatments are used to change microstructure to improve mechanical properties. Heat treatments are also applied to release stress and to remove crystal defects, such as dislocation. The percentage of carbon present in the alloy will affect the hardness of the material. Co-Cr-Mo alloys with a low carbon percentage can be used as implants because the material is bio-compatible.

The present paper will deal with the study of the effect of manganese, heat treatment on corrosion resistance of Co-Cr-Mo based on experimental results supported by reference theory.

2. Experimental Procedure

The raw material of Co-Cr-Mo and W were provided to fabricate samples with different weight percentages (wt%) of Mn. The material used in this experiment is Co-Cr-Mo alloys type CDS 2, type G6, type G7, and type G8. All four samples received the same treatment, such as the XRD test, heat treatment test, and corrosion test. The composition of Co-Cr-Mo alloys which is used can be seen in Table 1.

Table 1. Composition of Co-Cr-Mo in weight percentage.

Type of alloys	Analyse (weight percentage)				
	Co	Cr	Mo	Mn	Si
CDS 2	Balance	20.3	6,1	0,85	1,1
G6	Balance	20,3	6,1	3	1,1
G7	Balance	20,3	6,1	5	1,1
G8	Balance	20,3	6,1	7,5	1,1

Cds2 is a material that has a manganese composition of 0.85%, G6 has a manganese composition of 3%, G7 has a manganese 5%, and G8 has manganese 7.5%.

2.1. XRD

By studying the crystal structure, phase identification is achieved by comparing the data obtained with that in the reference database. X-ray powder diffraction (XRD) analysis was carried out to see the characterization of Co-Cr-Mo alloys before and after heat treatment. [6]

2.2. Corrosion Testing

Corrosion test conducted in this study, an electrochemical test using a Digi-Ivy © DY2300 series potentiometer. The electrolyte used in the corrosion test in a 3,5 % NaCl solution. To determine the corrosion rate of each alloy sample, Faraday's first law can be applied in this experiment. Mathematically, Faraday's first law was defined as:

$$\text{Corrosion rate of alloy, } r = \Sigma C\,\frac{Mi}{n\rho} = \Sigma C\,\frac{MI}{n\rho A} \tag{1}$$

where

M = atomic weight ($g.mol^{-1}$), i = current density (A/cm^2)

n = number of electron involved

ρ = density (g/cm^2)

C = constant (0,129 in mpy; 3,27 in mm/year; 0,00327 in mm^3/year)

The value of I from the information above has SI unit A/cm^2, where the cm^2 means the cross-sectional area of the samples which will be tested for the corrosion resistance.

2.3. Solution Heat Treatment

The heat treatment solution was tested for all samples at 700 ℃ for 65 hours. The initial temperature of the heating machine is 426 ℃ with a rate of 40. After 65 hours of heat treatment, the materials were followed by water quenching at room temperature. This treatment was done so that the constituents or

IOP Conf. Series: Materials Science and Engineering **694** (2019) 012044 doi:10.1088/1757-899X/694/1/012044

alloying elements enter into a homogeneous solid solution. Followed by rapid cooling or quenching in the room temperature water to hold the microstructure of materials.

3. Result and Discussion

3.1. XRD analysis

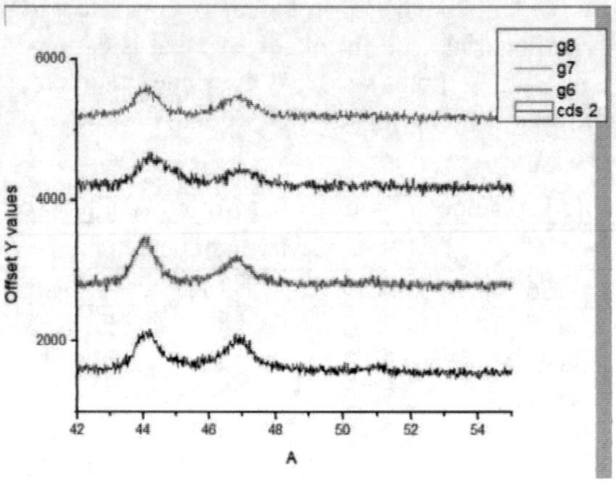

Figure 1. XRD's plot of the materials of Co-Cr-Mo alloys type CDS2, G6, G7, G8.

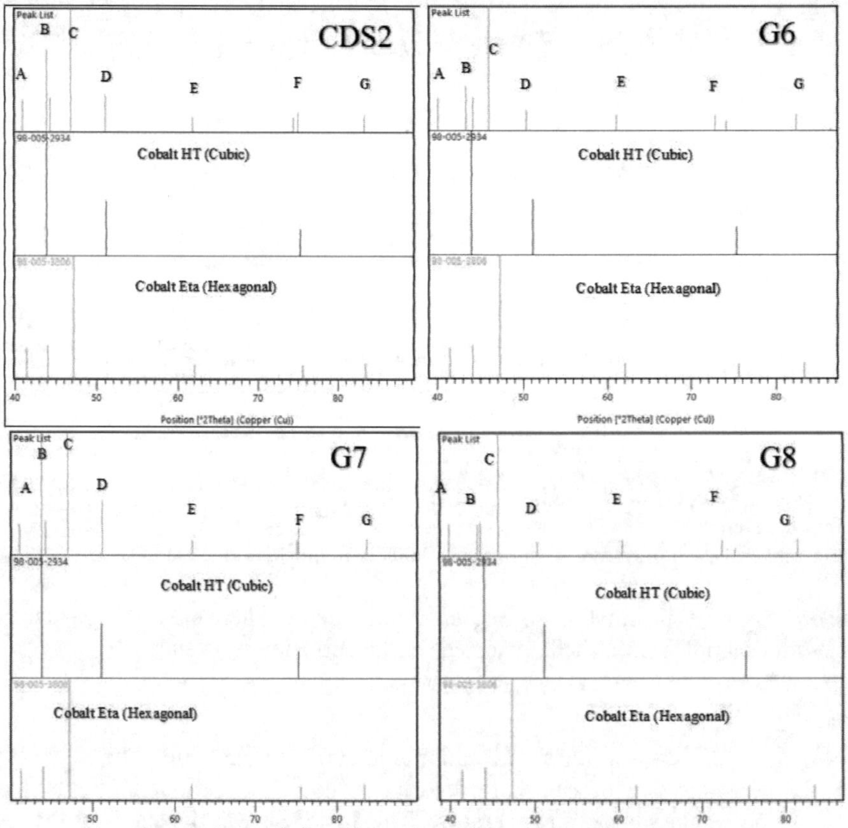

Figure 2. Peak list of Co-Cr-Mo in CDS2, G6, G7, G8

XRD identification through High Score using Cobalt HT and Cobalt Eta phase. 7 main peaks that can be identified, where 2 peaks (B,f) have 2 phases, namely Cobalt HT phase and Cobalt Eta phase in the same peak. There are a total of 9 peaks identified. For CDS2 material, 7 peaks can be identified with the following angles at 40.286 ; 43.066 ; 46.200 ; 50.412 ; 60.985 ; 74.515 ; 82.669. For G6 material has 7 peaks with each angle position as follows at 39.466 ; 42.611 ; 45.232 ; 49.775 ; 60.347 ; 71.967 ; 81.577. For G7 material has 7 peaks can be identified with the following angles at 40.520 ; 43.214 ; 46.447 ; 50.399 ; 61.356 ; 74.648 ; 82.911. And for G8 material has 7 peaks with each angle position as follows at 39.157 ; 42.805 ; 44.890 ; 49.580 59.665 ; 71.555 ; 80.675.

Table 2. Crystallographic parameters of the materials of Co-Cr-Mo alloys type CDS2, G6, G7, G8

Sample		**CDS 2**	**G6**	**G7**	**G8**
First peak's Height (cps)	Cobalt HT	270.67	389.24	147.31	487.48
	Cobalt Eta	181.85	1087.91	148.25	122.62
First peak's d (Å)	Cobalt HT	2.0905	2.0874	2.0918	2.0867
	Cobalt Eta	2.2086	2.2111	2.2058	2.2140
Lattice Parameter (Å)	Cobalt HT	a= 3.58497 / 90 b= 3.58497 / 90 c= 3.58497 / 90	a= 3.61923 / 90 b= 3.61923 / 90 c= 3.61923 / 90	a= 3.57098 / 90 b= 3.57098 / 90 c= 3.57098 / 90	a= 3.62392 / 90 b= 3.62392 / 90 c= 3.62392 / 90
	Cobalt Eta	a= 2.55147/ 90 b= 2.55147 90 c= 4.09735 / 120	a= 2.59883 / 90 b= 2.59883 / 90 c= 4.10672 / 120	a= 2.52884 / 90 b= 2.52884 / 90 c= 4.086 / 120	a= 2.61008 / 90 b= 2.61008 / 90 c= 4.15531 / 120
Volume (Å^3)	Cobalt HT	46.07398	47.40752	45.53677	47.59233
	Cobalt Eta	23.10015	24.02044	22.62928	24.51559
Density (g/cm^3)	Cobalt HT	8.49	8.26	8.59	8.22
	Cobalt Eta	8.47	8.15	8.65	7.98
Concentration (%)	Cobalt	37.8	16.7	48.4	16.3
	Cobalt Eta	62.2	83.3	51.6	83.7
Microstrain (%)	Cobalt HT	0.00	0.00	0.00	0.00
	Cobalt Eta	0.00	0.00	0.00	0.00
Crstalite Size (Å)	Cobalt HT	0.00	0.00	0.00	0.00
	Cobalt Eta	0.00	0.00	0.00	0.00

The intensity of the X-ray diffraction pattern is directly proportional to the concentration of the components that produce it. As can be seen in Figure 1 that in the Cobalt HT phase and Cobalt Eta phase have maximum intensity, so there is a change in the concentration of Cobalt HT and Cobalt Eta in the sample. Based on the crystal structure, the cobalt HT phase has a cubic shaped crystal structure with the same lattice parameters in a, b and c. The size of the crystalline volume of the cobalt HT phase changes for each of the largest samples is G8, G6, G7, and CDS2. Whereas in the Cobalt Eta has a hexagonal-shaped crystal structure with a lattice parameter an equal to b, but not equal to c. The size of the crystalline volume of the Cobalt Eta phase changes from the largest G6, G8, CDS2, and G6.

Meanwhile, if we compare the concentration of Cobalt HT with Cobalt Eta, the largest concentration of Cobalt HT is found in type G7, type G8, type CDS2, and type G6. This explains the composition of manganese contained in the Co-Cr-Mo alloys influences the cobalt composition of cubic and hexagonal structure.

3.2. Corrosion analysis

The electrochemical behavior can be known from the analysis of the corrosion mechanism and the corrosion rate values of the material of Co-Cr-Mo alloys used. Corrosion mechanism in this experiment, where Co-Cr-Mo was analyzed using electrochemical behavior in a solution of 3.5% NaCl. Since the chloride content of the environment is an important factor to consider when looking corrosion of an alloy. Chloride ions are an aggressive species that can cause local corrosion processes (attack occurs in specific sites where there are high local dissolution rates, which lead to high rates of penetration) in the form of pitting and crevice corrosion that are the most relevant types for artificial hip joints.

Figure 3 Tafel graph is presented to express the relationship between anodic reaction and a cathodic reaction. Corrosion E is the potential at which the rate of anodic reaction is proportional to the rate of cathodic reaction. The data presented in Table 3 contains the values of E corrosion (Volt), I corrosion (Ampere), and corrosion rate (mm/year).

Table 3. Data from the corrosion test of the materials of Co-Cr-Mo alloys type CDS2, G6, G7, G8 room temperature in the 3,5 wt% NaCl solution

Wt % NaCl	Sample	R (ohm)	$E_{corrosion}$ (V)	$I_{corrosion}$ (A)	Corrosion Rate (mm/year)
3,5 wt%	CDS 2	1.530×10^3	-0,649	1.679×10^{-5}	3.221×10^{-2}
	G6	2.733×10^3	-0.808	9.402×10^{-6}	1.804×10^{-2}
	G7	3.839×10^2	-0.972	6.692×10^{-5}	1.284×10^{-1}
	G8	4.007×10^3	-0.570	6.412×10^{-6}	1.230×10^{-2}

The reactions that occur during the corrosion process on Co-Cr-Mo alloys on NaCl electrolytes are shown below

Anodic reaction:
$$Co \rightarrow Co^{2+} + 2e \quad (2)$$
$$Co \rightarrow Co^{2+} + 2e \quad (3)$$

The formation of $Co(OH)^+$ is mainly responsible for the sudden increase in current due to the dissolution of the Co metal. The electrons that are produced in the oxidation reaction must be consumed in a cathodic reaction of oxygen in a solution:

Cathodic reaction:
$$O_2 + 2H_2O + 4e^- \rightarrow 4OH^- \quad (4)$$
Main reaction:
$$Co^{2+} + 2Cl^- \rightarrow CoCl_2 \quad (5)$$
$$CoCl_2 + 2H_2O \rightarrow Co(OH)_2 + 2HCl \quad (6)$$

From table 3, the corrosion rate is influenced by potential E corrosion (Volt) and I corrosion (Ampere). Table 3 shows the greatest corrosion rate if the corrosion potential of I corrosion is getting negative, the fastest corrosion rate is the material of Co-Cr-Mo alloy types G7, CDS2, G6,G8.

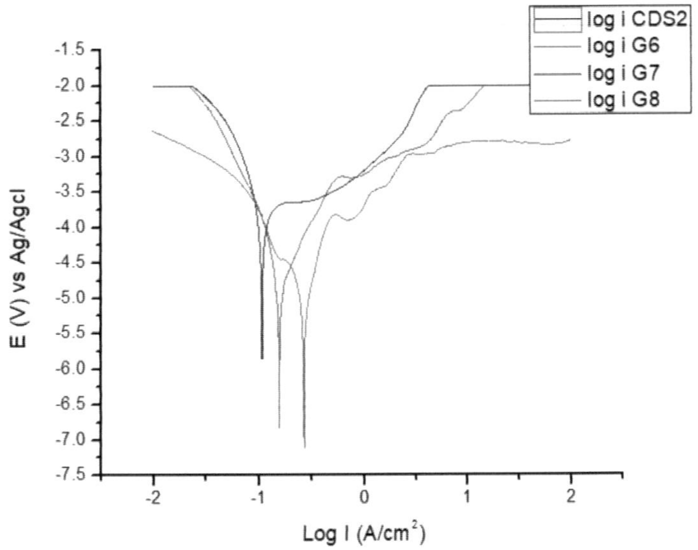

Figure 3. Potentiodynamic polarization curves of the samples in 3,5 wt% NaCl

Corrosion delays are caused by an anodic reaction which releases electrons to the metal, so that excess electrons in the metal shift the potential of the metal in a more negative direction and cause the anodic reaction to slow down and speed up the cathodic process.

3.3. Heat Treatment analysis

Table 2 can be seen if there is a difference in the first peak's high in the state of cubic and hexagonal structure, indicating a structural change that occurs in the material of Co-Cr-Mo alloy.

4. Conclusion

The variation of manganese in the material of Co-Cr-Mo alloy effects the value of the corrosion rate. Co-Cr-Mo alloy type G7 with a manganese content of 5% has the greatest corrosion resistance. The material of Co-Cr-Mo alloy type G7 has the largest phase of cubic and hexagonal structures. The heat treatment has a considerable effect on structural change.

Acknowledgment

This research fully funded by universitas Indonesia through Hibah PITTA B No. NKB-0616/UN2.R3.1/HKP.05.00/2019

References

[1] Kauser, F., *Corrosion of CoCrMo alloys for biomedical applications*, in Department of Metallurgy and Materials, School of Engineering. 2007, University of Birmingham: Birmingham. P. 4-285.

[2] Sinnett-Jones, P.E., J.A. Wharton, and R.J.K. Wood, Micro-abrasion-corrosion of a CoCrMo alloy in simulated artificial hip joint environments. Wear, 2005. 259: p. 898-909.

[3] Yan, Y., A. Neville, and D. Dowson, *Biotribocorrosion – an appraisal of the time dependence of wear and corrosion interactions*: II. *Surface analysis*. Journal of Physics D-Applied Physics, 2006. **39**(15): p.3206-3212.

[4] Bellfontaine G. *The corrosion of CoCrMo alloys for biomedical applications thesis*, 2010.

[5] Doni Z., Alves A. C., Toptan F., Gomes J.R., Ramalho A., Buciumeanu M., Palaghian L., and

Silva F.S. *Dry sliding and tribocorrosion behavior of hot-pressed CoCrMo biomedical alloy as compared with the cast CoCrMo and Ti6Al4V alloys Mater. Des.,* 2013.
[6] Artioli G. *X-ray diffraction (XRD) Encycl. Earth Sci. Ser*., 2017.
[7] Kaita W., Hagihara K., Rocha L.A., and Nakano T. *Plastic deformation mechanisms of biomedical Co-Cr-Mo alloy single crystals with hexagonal close-packed structure Scr. Mater*, 2018. **142** 111-5
[8] Yan X., Xu Y.X., Wu Y., Lin H. *Effects of heat treatment on metal-ceramic combination of selective-laser-melted cobalt-chromium alloy* J. Prosthet Dent., 2018. **120** 319. E1-319. E6
[9] Wang W., Hou Z., Lizarraga R., Tian Y., Babu R.P., Holmström E., Mao H., and Larsson H. *An experimental and theoretical study of duplex fcc+hcp cobalt-based entropic alloys Acta Mater*, 2018. **176** 11-8
[10] Mallik M.K., Rao C.S., and Kesava Rao V.V.S., *Effect of heat treatment on hardness of Co-Cr-Mo alloy deposited with laser engineered net shaping Procedia Eng*., 2014. **97** 1718-23
[11] Valero Vidal C., and Igual Muñoz A. *Electrochemical characterization of biomedical alloys for surgical implants in simulated body fluids Corros. Sci.,* 2008.

IOP Conf. Series: Materials Science and Engineering **694** (2019) 012045 doi:10.1088/1757-899X/694/1/012045

The Effect of Copper on Corrosion Resistance and Structure of Al-8%wt Si Alloys for Engine Block Application

Adina Aji Setiady and Bambang Soegijono

Department of Physics, Faculty of Mathematics and Natural Sciences, Universitas Indonesia, Depok, Indonesia

E-mail: naufal@ui.ac.id

Abstract. Al-Si Alloys are mostly used in an application for the engine block. An experimental investigation was carried out to understand the corrosion resistance of alloys with various Copper (Cu) content in Al-Si alloys. The corrosion properties of Aluminium alloys (Al-Si-0wt%Cu, Al-Si-2wt%Cu, Al-Si-5wt%Cu, and Al-Si-9wt%Cu) were investigated. X-ray diffraction and Electrochemical test were performed to investigate the structure of each sample, and corrosion resistance in 3.5%wt and 7%wt NaCl solution, and coolant water were used to understand the corrosion rate of Al-Si alloys. X-ray diffraction shows a different diffraction pattern from one sample to another. The effect of Copper addition shows marks different in corrosion resistance and structure of this alloy.

1. Introduction

Pure Aluminium does not have sufficient strength; therefore various materials are added to improve mechanical properties [1]. Recently, Aluminium-Silicon alloys are used in application for the engine block due to their good properties such as low density, good formability, high strength and excellent corrosion resistance with various condition are relative to other mixture [2]. The influence mixture of alloys is very important on microstructure.

Corrosion is main factor for the failure aluminium alloys [3]. The additional of Copper content in Aluminium-Silicon could improve their hardness and strength [4]. Generally, addition of Copper may change the corrosion resistance. Copper is needed when combining in Aluminium-Silicon to investigate the appropriate properties of these alloys. The aluminium alloys (Al-Si-0 wt%Cu, Al-Si-2wt%Cu, Al-Si-5wt%Cu, and Al-Si-9wt%Cu) were used in this research to investigate corrosion resistance. These alloys are resistant to attack from water, concentration and viscosity, making Aluminium alloys useful in a various environments and take advantage from its lightweight property. Therefore, this research attempts to investigate the characteristic of ethylene-glycol as coolant water. Ethylene Glycol is mostly used as coolant in engine block, in a pH range between 7 and 8, so it is great heat adsorption capacity [5].

To understand this behavior, various Copper content (0 wt%, 2 wt%, 5 wt%, and 9 wt%) Aluminium-Silicon were tested using electrochemical test in 3.5% wt and 7% wt NaCl solution, and coolant water. Content of water in NaCl and coolant water solution determine corrosion rate value (observed in mm/year). Corrosion rate has a major impact on structural integrity. It determines the rate of corrosion at material thickness will be loss in a specific environment.

2. Experimental Methods and Materials

2.1. Sample Preparation

This type of Aluminium-Silicon alloy with various Copper was selected in this experiment to achieve the appropriate properties of these alloy. The chemical composition of the alloys Al-Si-Cu-Mg-Zn-Fe are given in Table 1. There were four samples used in this experiment, and each of the sample was given different Copper content. These alloys were cut into four pieces with 2 cm of diameter dimensions. The size of the exposed area affects the result of the measured corrosion behaviour when potentiodynamic polarization tests. Small scale of sample area tend to better than big scale measurements on electrochemical experiments [6]. The exposed area to the corrosive solution was 0.949 cm^2. A Copper wire was provided an electrical connection. Prior each aluminium alloy surface was further ground with 120, 240, 400, and 2000 grit SiC paper to scarp of layer on the surface before the corrosion test began.

2.2. Corrosion Testing

There were two corrosion tests conducted in this experiment. Electrochemical measurements were carried out in 3.5 wt% NaCl and 7 wt% NaCl, and coolant water solution at room conditions. Different concentration of NaCl was taken for corrosion behaviour to investigate the surface degradation in different environmental condition [4]. A computer-controlled Digi-Ivy© DY2300 potentiostatic was used for the electrochemical measurements. It was used to measure the current and voltage between Working Electrodes and the Reference Electrode [1]. The polarization curves were obtained potentiodynamically over the range -2.0- + 2.0 V at a scan rate of 0.05 $Vsec^{-1}$. The formula to measure the corrosion rate of each sample, applying First Faraday's law in this case [7].

$$\text{Corrosion rate of alloy, } r = \Sigma C \frac{Mi}{n\rho} = \Sigma C \frac{MI}{n\rho A} \quad (1)$$

where r is corrosion rate, C is corrosion rate constant (3.27×10^{-3} mm^3/year was used in this research), M is atomic weight of each element in the alloy (g/mol), *I* is current density in Ampere/cm^2, *n* is number of electrons involved in corrosion process, ρ is mass density of each element in the alloy (gr/cm^3). Faraday's law corrosion rate equation in sums form because Al-8Si aluminium alloy consist of six elements (Aluminium, Silicon, Magnesium, Iron, Copper, and Zinc).

Table 1. Chemical Composition of Alloy in weight percent. A11 (Cu=0%), A12 (Cu=2%), A13 (Cu=5%) and A14 (Cu=9%)

Alloy	Mg	Si	Fe	Cu	Zn	Al
A11	4	8	0.4	0	2	Balance
A12	4	8	0.4	2.1	2	Balance
A13	4	8	0.4	5.12	2	Balance
A14	4	8	0.4	9.45	2	Balance

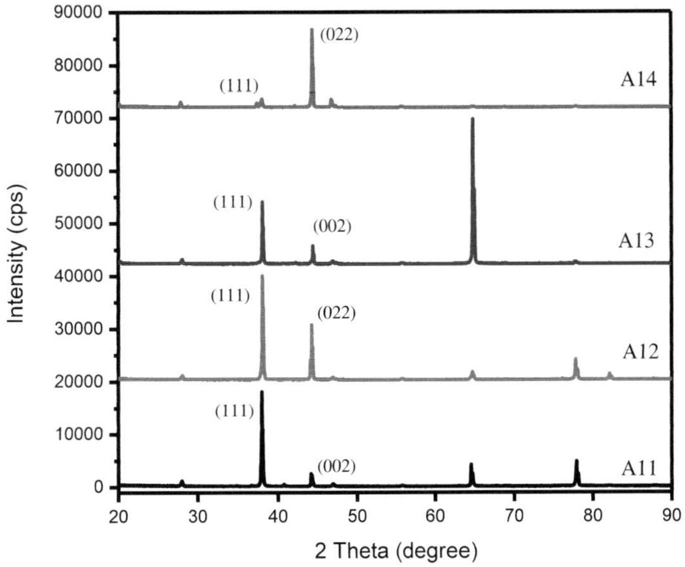

Figure 1. X-Ray Diffraction of the Aluminium-Silicon with Various Copper Content (A11, A12, A13 and A14)

3. Results and Discussion

3.1 XRD Analysis

To describe a better identification of the phase, the X-Ray Diffraction plots of Aluminium-Silicon alloy with various Copper content samples are shown in Figure 1 and the data of characterization was served in Table 2. Figure 1 shows there were four identified peaks. For A11 with 0wt% Cu, four identified peaks with angle positions 28.942°, 37.953°, 44.219° and 64.659°. For A12 with 2wt% Cu, four identified peaks with angle positions 26.415°, 38.054°, 44.129°, and 64.422°. For A13 with 5wt% Cu, four identified peaks with angle positions 29.025°, 38.063°, 44.323°, and 64.741°. And for A14 with 9wt% Cu, five identified peaks with angle positions 29.020°, 38.063°, 44.378°, and 64.741°. And from Figure 1, X-Ray Diffraction pattern shows that the highest intensity peak value on A11 with 0wt% Cu which the value is 541.97 cps. For A12 with 2wt% Cu the height peak value is 645.53 cps. For A13 with 5wt% Cu the height value is 610.12 cps. And for A14 with 9wt% Cu the height value is 82.29 cps. It seems that the Copper content significantly causes the phases in the alloys [8]. We can be concluded that the intensity value is decreasing due to Aluminium is balance. In fact, as these alloys have a small difference of shift peaks from A11 to A14 caused Copper addition.

The lattice parameters have the same value for a, b and c, so the crystal system on these alloys are cubic. From Table 2, size of volume crystal from Aluminium-Iron phase was changed along with the additional of Copper from sequentially A11, A12, A13 and A14. Meanwhile in Aluminium phase, A14 has the biggest lattice parameter then followed by A13, A12 and A11. Each volume of these alloys are also transformed following alteration in lattice parameter.

Table 2. Cystallographic Parameters of the Aluminium-Silicon with Various Copper Content (A11, A12, A13 and A14)

Sample		A11	A12	A13	A14
Highest Peak's Height (cps)	Aluminium	0.880	64.490	5.510	64.75
	Iron				
	Aluminium	541.97	645.53	610.12	82.29
d spacing (Å)	Aluminium	2.165	2.050	2.162	2.435
	Iron				
	Aluminium	2.368	2.362	2.362	2.365
	Aluminium Iron	a = 6.039 b = 6.039 c = 6.039	a = 5.745 b = 5.745 c = 5.745	a = 6.052 b = 6.052 c = 6.052	a = 5.694 b = 5.694 c = 5.694
Lattice Parameter (Å)	Aluminium	a = 4.039 b = 4.039 c = 4.039	a = 4,047 b = 4,047 c = 4,047	a = 4.043 b = 4.043 c = 4.043	a = 2.730 b = 2.730 c = 2.730
Volume (Å^3)	Aluminium	220.249	189.65	221.675	184.614
	Iron				
	Aluminium	66.003	66.300	66.087	52
Density (g/cm^3)	Aluminium	5.87	6.81	5.83	7
	Iron				
	Aluminium	2.71	2.7	2.71	2.73
Concentration (%)	Aluminium	0	2.8	0.4	19.7
	Iron				
	Aluminium	100	97.2	99.6	80.3

It is shown in Figure 1, the phase that is detected in each samples, there is a single peak which described as an Aluminium and Aluminium-Iron phase at angle points. In this case, it is considered that which peak has the highest intensity. The concentration an Aluminium greater than Alumunium-Iron, hence the height of intensity peak for Alumunium more dominant than Aluminium-Iron. Both of these phases have cubic crystal structure because of the lattice parameter has the same value in a, b, and c.

3.2. Potentiodynamic Polarization Curves Analysis

The polarization curves prove Tafel type behavior of these samples in the active condition. The potentiodynamics curves in NaCl solution is shown in Figure 2, and the data were collected in Table 3. The corrosion mechanism of these Aluminium-Silicon alloys can be depicted by the anodic and cathodic reaction when the sample dipped in NaCl and coolant water solution. It can be caused an electron transfer process which make electrochemical reaction will be occur. From Table 3, the corrosion potential of alloys shifts to more a negative value, and corrosion current density is reduced with increasing Copper content. The electrochemical reaction on electrode can be written as follows [11]:

The anodic reaction is:

$$Al \rightarrow Al^{3+} + 3e^- \quad (2)$$

$$Al^{3+} + 6(H_2O) \rightarrow (Al(H_2O)_6)^{3+} \quad (3)$$

When NaCl reacts with H_2O :

$$\mathbf{6NaCl + 6H_2O \rightarrow 6NaOH + 6Cl_6 + 3H_2} \quad (4)$$

The cathodic reaction is:

$$\mathbf{O_2 + 2(H_2O) + 4e^- \rightarrow 4OH^-} \quad (5)$$

Main reaction is:

$$2Al^{3+} + 6Cl^- \rightarrow Al_2Cl_6 \quad (6)$$

Al_2Cl_6 works in $6H_2O$, the result is:

$$Al_2Cl_6 + H_2O \rightarrow Al_2(OH)_6 + 6HCl \quad (7)$$

The result of the reaction between NaCl and H_2O generate Cl^- ion, it was absorbed into the protective film on these alloys. So, this is one of the triggering factors of the process nucleation of corrosion on the alloy surface [10].

Potentiodynamic polarization curves of Aluminium-Silicon alloys in coolant water is shown in Figure 2. The polarization curves show that the samples tendency to positive corrosion potential because of anodic polarization. The main cathodic reaction is reduction of oxygen under the condition[5][12]. Generally, the film formed during the dipping in the presence of that Silicon enhances the growth of a protective film which prevents corrosion during anodic polarization. Thus, the corrosion rate in coolant water solution is lower than NaCl solution. Mostly, coolant water contains Ethylene-Glycol for cooling applications. Ethylene-Glycol molecule is larger than water, so the adsorption of the surface is better. Thus, the aluminium alloy is more completely protected by Ethylene-Glycol, which more effective prevent aluminium alloy from the corrosion attack.
The anodic reaction is:

$$Al \rightarrow Al^{3+} + 3e^- \quad (8)$$

$$Al^{3+} + 3OH^- \rightarrow Al(OH)_3 \quad (9)$$

The cathodic reaction is:

$$\mathbf{O_2 + 2(H_2O) + 4e^- \rightarrow 4OH^-} \quad (10)$$

When Alumunium alloy in environment with low content of oxygen, the Aluminium oxide film will be formed on Aluminium surface. When the electrochemical test used in Ethylene-Glycol as the solution, it would be reduced and Aluminium-Alcohol will appeared [13].
Main reaction is:

$$CH_2CH_{2(}OH)_2 + e \rightarrow OHCH_2CH_2O^- + H \quad (11)$$

$$Al^{3+} + 3OHCH_2CH_2O^- \rightarrow Al(OHCH_2CH_2O)_3 \quad (12)$$

Where there is no oxygen in the environment, the cathodic reaction that occur are either reduction of Ethylene-Glycol or reduction of water:

$$H_2O + e^- \rightarrow H + OH^- \quad (13)$$

Hence, the protective film is formed on each alloys which are contain combination of Aluminium Oxide and Aluminium Alcohol [14].

3.3. Corrosion Testing: Electrochemical Test

The current-potential curves were obtained from each potentiodynamic polarization test. The result of electrochemical test in 3.5% NaCl solution, 7% NaCl solution, and coolant water solution can be seen on Table 3. Corrosion potential of Aluminium-Silicon with various Copper are evidently different in different medium. Electrochemical test in 3.5% NaCl solution makes the A14 alloy with 9wt% Copper content has the most negative corrosion potential (*Ecorr*) with value of -1.247 V. Meanwhile, the A11 alloy with 0wt% Copper content in 7% NaCl solution has the most negative potential with value of -1.494 V. The A14 alloy with 9wt% Copper content in Coolant water solution has the most negative potential with value of -0.744 V; therefore, it can be concluded that corrosion resistance of Aluminium-Silicon alloy is depend on Copper content and phases. In an environment with high content of NaCl, all alloys are prone to be corroded due to Hydro Chloride Acid concentration.

The differences of electrochemical current (*Icorr*) for each sample will affect the different corrosion rate. The Faraday's Law formula was used to calculate corrosion rate for each sample. In 3.5% NaCl solution, the A11 alloy with 0wt% Copper content has the slowest corrosion rate with value is 1.172 x 10^{-2} mm/year. In 7% NaCl solution, the A13 alloy with 5wt% Copper content has the slowest corrosion rate with value is 2.816 x 10^{-2} mm/year. Meanwhile, in coolant water solution, the A12 alloy with 2wt% Copper content has the slowest corrosion rate with value of 3.079 x 10^{-4} mm/year. After electrochemical test has been done, there some factors that affect the corrosion rate are effect of pH and water content. These two factors are important components of a failure analysis because a lot of the failures related to environments are linked to corrosion. Solutions with different percentages varying the NaCl concentration enhance corrosion rate of the alloy. When the concentration of NaCl is increased, corrosion rate of the alloy has great value. Meanwhile, the effect of pH between NaCl and coolant water also affect the corrosion rate. It was shown that the corrosion rate was reduced with increasing the value of pH [9].

Table 3. Data from the corrosion test of the Aluminium-Silicon with Various Copper Content (A11, A12, A13 and A14)

Solution	Sample	R (ohm)	$E_{corrosion}$ (V)	$I_{corrosion}$ (A)	Corrosion Rate (mm/year)
3,5 wt%	A11	3,837 x 10^2	-1,163	6,695 x 10^{-5}	1.172 x 10^{-2}
	A12	2,751 x 10^2	-1,181	9,339 x 10^{-5}	1.634 x 10^{-2}
	A13	3,756 x 10^2	-1,051	6,841 x 10^{-5}	1.197 x 10-^{2}
	A14	1,146 x 10^2	-1,247	2,242 x 10^{-4}	3.924 x 10^{-2}
7 wt%	A11	5,555 x 10^1	-1,494	4,625 x 10^{-4}	8.096 x 10^{-2}
	A12	3,649 x 10^1	-1,454	7,041 x 10^{-4}	1.232 x 10^{-1}
	A13	1,597 x 10^2	-0,507	1,609 x 10^{-4}	2.816 x 10^{-2}
	A14	2,624 x 10^1	-1,448	9,790 x 10^{-4}	1.713 x 10^{-1}
Coolant Water	A11	5,761 x 10^3	-0,623	4,460 x 10^{-6}	7.807 x 10^{-4}
	A12	1,461 x 10^4	-0,705	1,759 x 10^{-6}	3.079 x 10^{-4}
	A13	1,278 x 10^4	-0,673	2,010 x 10^{-6}	3.518 x 10^{-4}

IOP Conf. Series: Materials Science and Engineering **694** (2019) 012045 doi:10.1088/1757-899X/694/1/012045

A14	5,806 x 10^3	-0.744	4,425 x 10^{-6}	7.747 x 10^{-4}

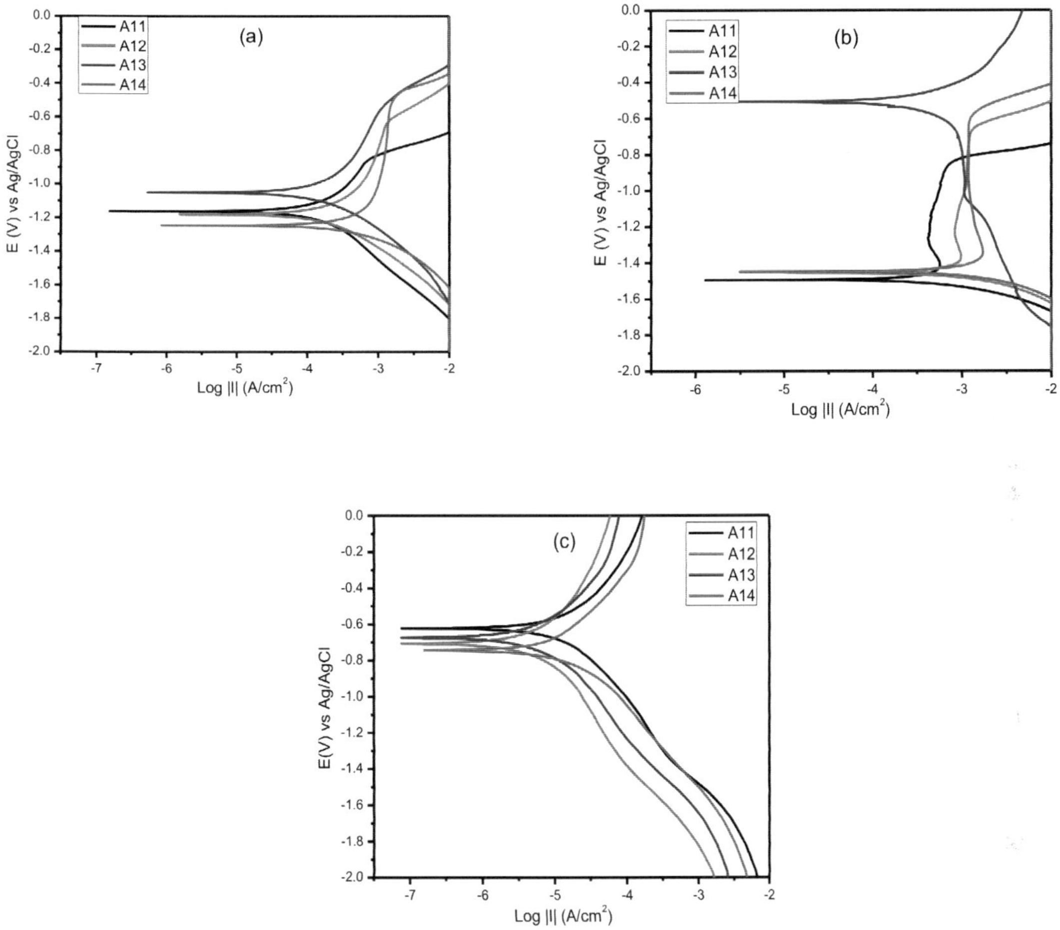

Figure 2. (a) Potentiodynamic polarization curves of the samples in 3,5 wt% NaCl;
(b) Potentiodynamic polarization curves of the samples in 7 wt% NaCl;
(c) Potentiodynamic polarization curves of the samples in Coolant Water

4. Conclusion

From this study of the corrosion behaviour of the Alumunium-Silicon alloy engine block in 3.5% NaCl solution, 7% NaCl solution, and coolant water solution to investigate the corrosion resistance. It can be concluded that various of Copper content influence the corrosion rate. The environments have a great effect on corrosion behaviour in alloy. In NaCl solution, when Copper content is increased, the corrosion potential of alloys shifts to more a negative value and corrosion current density is reduced. From the above result it is clear that the passive film became weak in corrosion caused by Cl^- ion was the reason of corrosion pit. In coolant water solution, Alumunium-Silicon alloys with various Copper content have the slowest corrosion rate than 3.5% NaCl solution and 7% NaCl because of Ethylene Glycol molecule is larger than water. Thus, it prevents alloys from the corrosion attack because the surface of these alloys more completely protective by Ethylene Glycol. The presence of the intermetallic phases in these alloys has shown that the oxide film is not consistent. However, the additional of Copper content contributes to improve hardness, strength, and corrosion behaviour. So, it can be applied in an application for engine block.

Acknowledgments
The authors would like to thanks to Universitas Indonesia for supporting the research activity for providing data XRD and SEM-EDX. This study also funded by Universitas Indonesia through Hibah PITTA B No. NKB-0616/UN2.R3.1/HKP.05.00/2019.

References

[1] Zor S, Zeren M, Ozkazanc H and Karakulak E 2010 Effect of Cu content on the corrosion of Al-Si eutectic alloys in acidic solutions *Anti-Corrosion Methods Mater.* **57** 185–91
[2] Haque M M and Maleque M A 1998 Effect of process variables on structure and properties of aluminium - Silicon piston alloy *J. Mater. Process. Technol.* **300** 122–8
[3] Mishra R K 2019 Study the effect of pre-corrosion on mechanical properties and fatigue life of aluminum alloy 8011 *Mater. Today Proc.*
[4] Šćepanović J, Asanović V, Radonjić D, Vuksanović D, Herenda S, Korać F and Bikić F 2019 Mechanical properties and corrosion behaviour of Al-Si alloys for IC engine *J. Serbian Chem. Soc.* **84** 503–16
[5] Harooni A Q A, Eskandari H, Maddahy M H, Danaee I and Nikmanesh S 2015 Corrosion behavior of 6063 aluminum alloy in ethylene glycol-water solution *Iran. J. Mater. Sci. Eng.*
[6] Perren R A, Suter T, Solenthaler C, Gullo G, Uggowitzer P J, Böhni H and Speidel M O 2001 Corrosion resistance of super duplex stainless steels in chloride ion containing environments: Investigations by means of a new microelectrochemical method. II. Influence of precipitates *Corros. Sci.* **43** 727–45
[7] Z. Ahmad, "Application of Faraday's Laws in Determination of Corrosion Rates of Metals and Alloys," in Principle of Corrosion Engineering and Corrosion Control (Butterworth-Heinemann, United Kingdom, 2006), pp. 58-61.
[8] Medjahed A, Henniche A, Derradji M, Yu T, Wang Y, Wu R, Hou L, Zhang J, Li X and Zhang M 2018 Effects of Cu/Mg ratio on the microstructure, mechanical and corrosion properties of Al-Li-Cu-Mg-X alloys *Mater. Sci. Eng. A* **718** 241–9
[9] Kadhum M Y, Bkhakh C K and Albehadili M H 2018 The Effect of Sodium Hydroxide Solutions with Different Ph on the Corrosion Of Iron Alloy (C1010) In Industrial Water *J. Adv. Chem.* **15** 6136–41
[10] Kaiser M S and Dutta S 2014 Corrosion Behaviour of Aluminium Engine Block in 3.5% NaCl Solution *J. Mater. Sci. Chem. Eng.* **02** 52–8
[11] Liu Y, Sun Q, Li W, Adair K R, Li J and Sun X 2017 A comprehensive review on recent progress in aluminum–air batteries *Green Energy Environ.* **2** 246–77
[12] Zaharieva J, Milanova M, Mitov M, Lutov L, Manev S and Todorovsky D 2009 Corrosion of aluminium and aluminium alloy in ethylene glycol-water mixtures *J. Alloys Compd.*
[13] Chen X, Tian W, Li S, Yu M and Liu J 2016 Effect of temperature on corrosion behavior of 3003 aluminum alloy in ethylene glycol–water solution *Chinese J. Aeronaut.* **29** 1142–50
[14] Liu Y and Cheng Y F 2010 Effects of coolant chemistry on corrosion of 3003 aluminum alloy in automotive cooling system *Mater. Corros.*

IOP Conf. Series: Materials Science and Engineering **694** (2019) 012046 doi:10.1088/1757-899X/694/1/012046

The Impact of Aging Time on Structure and Corrosion Resistance of Aluminum 2091 Lithium Alloy for Aerospace Application

Arya Octavian and Bambang Soegijono*

Department of Physics, University of Indonesia, Depok, Indonesia
*E-mail: naufal@ui.ac.id

Abstract. In need of aerospace and automotive industries, the aluminum lithium alloy becomes a worldwide topic because of its corrosion resistance, hardness, and categorized as a lightweight material. This alloy has high resistance to corrosion due to doped alloying element, such as copper. Addition of lithium to aluminum alloys lowers its density. This paper was meant to study the impact of aging time at 190°C for 24 hours, 42 hours, and 72 hours on aluminum alloy crystal. The materials had solutionized at 525°C for 5 hours and had quenched at water before it was artificially aged. Besides corrosion resistance, the structure of this alloy are change as the aging times rise that caused by Al_2Cu precipitation. To examine the electrochemical response of this material, potentiodynamic curves were obtained for aluminium lithium alloy 2091 during standard test in natural deaerated 3.5% wt and 7.0% wt NaCl solution after heat treatment. The structure after heat treatment were characterized by using X-ray diffraction. The results showed that aging time modified the structure and lead to change the electrochemical behaviour.

1. Introduction

Aluminum alloys are divided into two different major categories which are wrought composition and cast composition. A further distinction for each category is whether heat-treatable or non-heat-treatable [1]. Al-Li 2091 heat-treatable alloy is start to developed in 1924 by several researcher to determine the properties for aerospace application. This type of aluminum alloy has advantageous properties such as easy to fabricated, available in a wide range of strength value, lightweight, high tensile strength, and resistance to corrosion in various condition [2,3]. Its prominent properties can be applied to many part of aerospace craft. For example, fuselage, wing, empennage, and supporting structure [4]. Because of its higher modulus and lower density, Al-Li become attractive for more investigation.

The additional of lithium atom which is lighter than aluminum makes a displacement towards one of aluminum atom from crystal lattice which can helps block dislocation. Considering at high temperature, solubility of lithium in aluminum is also high. Each wt% of lithium increases the modulus of aluminum about 6% and decreases the density of aluminum about 3% [2]. The Al-Li 2091 alloy provide an increases the specific modulus of approximately 15% and offers a density reduction of approximately about 8% compared to 2xxx series [5].

The heat treatments that have been done into Al-Li 2091 may lead the corrosion to modify based on the quantity and the distribution of microstructure. As a 2xxx series, alloying element copper dissolves in α-Al matrix at 550°C which increase corrosion potential. After aging treatment, nucleation of Al_2Cu forms (i.e. α and θ phase) at grain boundaries which more cathodic than α-Al [6]. Furthermore, it can cause a galvanic action. Although, aluminum may create a thin invisible oxide layer, which protects it from corrosion, when the surfaces are exposed to any environment [3]

The Objective of this work is to study the electrochemical behavior of the Al-Li alloy 2091 after different condition of heat treatment, aging time (AT), using electrochemical corrosion test and trying to make some correlation within its structure.

2. Experimental Procedure

2.1. *Material*

The material used in this experiment is Al-Li alloy 2091, which have been formed into approximately a rectangle shaped 3 cm in length, 0.7 cm in width, 0.2 cm in thickness. There were 4 samples used in this experiment with various treatment for each samples. Before the experiments, samples were polished using silicon carbide paper grit-180. Its chemical composition is listed in the Table 1.

Table 1. Aluminum-Lithium Alloy 2091 Chemical Composition

Elements	% Weight
Al	94.72
Li	1.9
Cu	1.85
Mg	1.2
Zr	0.07
Fe	0.07
Si	0.02
Ti	0.02

2.2. *Solution Heat Treatment*

Solution heat treatment was done to all of the samples at temperature 525°C on furnace for 5 hours and then quenched in the water medium in room temperature. This treatment intended to make the constituents or alloying elements enter into the homogenous solid solution. Followed by rapid cooling or quenching in the water to hold the microstructure.

2.3. *Artificially Aging Treatment*

Afterwards, only 3 samples were artificially aged below the solvus line at temperature 190°C for 24 hours, 48 hours, and 72 hours. These treatment were done to see assorted effect of one another and also compared to solutionized only sample. XRD analysis for all samples were done at UPP IPD Research Laboratory, Faculty of Mathematics and Natural Sciences, Universitas Indonesia. This test was meant to see the characterization of each treatment of Al-Li alloy 2091.

2.4. *Corrosion Test*

Corrosion tests data were collected out using Potentiostat. The Working Electrode and the Reference Electrode are used to quantify the current and the voltage, where Ag/AgCl was used as the reference electrode in this experiment [7]. All of the samples were connected with the wire so the samples can be dipped into the solution to collect the data. These electrodes were soaked into 3.5 wt% of NaCl solution. The corrosion rate of each samples had been calculated after all data were collected from Potentiostat. The formula that used to calculated the value of corrosion rate was obtained from Law in Determination Corrosion Rates of Metals and Alloys [8].

$$r = C\frac{Mi}{n\rho} \tag{1}$$

Where C stands for the constant of corrosion rate (0.129 in mpy; 3.27 in mm/year; 3.27×10^{-3} in mm^3/year), M stands for atomic weight of every constituent in alloy ($g{\cdot}mol^{-1}$), i stands for current density (A/cm^2), n stands for number of electron involved, ρ stands for density of every constituent in alloy (g/cm^2). Corrosion rate constant was used in this experiment is 3.27×10^{-3} in mm^3/year. The value of i has SI unit A/cm^2, where the cm^2 defines the cross-sectional area of the samples that was tested for the corrosion resistance. In this experiment, all of 4 samples have the same cross-sectional area which is rectangle-shaped of 1.05 cm^2 with length for 1.5 cm and width for 0.7 cm.

3. Result and Discussion

3.1. XRD Analysis

The XRD plots of all Al-Li alloy 2091 sample can be seen in Figure 1 and the data of the characterization is served in Table 2. There were five demonstrated peaks, namely, (111), (002), (022), (113), and (222) which are clearly visible for aluminum phase. This happens because the intensity of XRD patterns are equal to the composition of constituent that produce it. For the *Solutionized Only* sample, five peaks were identified with angle positions of 37.98^{O}, 44.24^{O}, 64.5999^{O}, 77.72^{O}, and 81.96^{O}. For *24 Hours of Aging* sample, five peaks were identified with angle positions of 38.00^{O}, 44.24^{O}, 64.63^{O}, 77.74^{O}, and 81.95^{O}. For *48 Hours of Aging* sample, five peaks were identified with angle positions of 38.039^{O}, 44.26^{O}, 64.62^{O}, 77.76^{O}, and 81.99^{O}. For 72 Hours of Aging sample, five peaks were identified with angle positions of 37.98^{O}, 44.24^{O}, 64.61^{O}, 77.78^{O}, and 81.99^{O}.

The highest peak value is on *Solutionized Only* sample which the value is 2.3675 Å. This is believed to be due to the heat treatment lower the intensity of the light that diffracted by atoms in the crystal structure which proportional to the square of the amplitude of the structural factor [9]. All samples have a little difference at each peak which states that there was a slight peak shift caused by heat treatment. This will not change the phase that is already owned by the alloy, but it can affect the differences in the microstructure of the crystal system as evidenced by differences in microstrain.

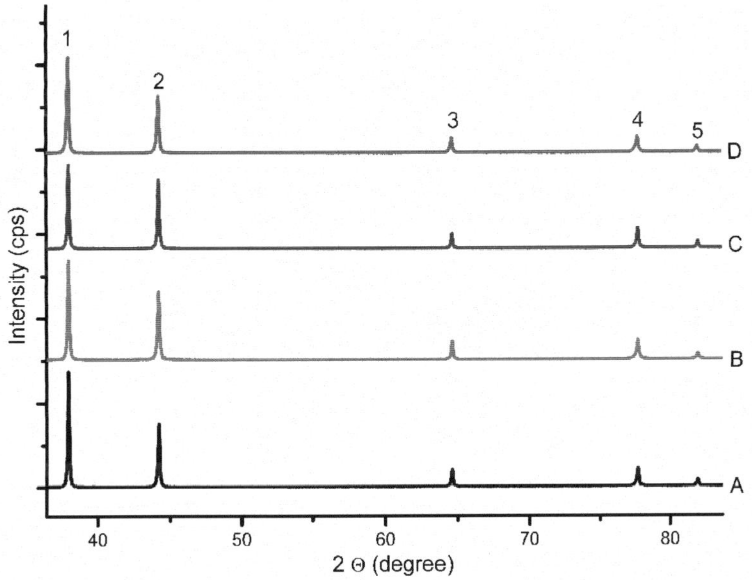

A = Solutionized Only B = 24 h, aging C = 48 h, aging D = 72 h, aging
1 = (111) 2 = (002) 3 = (022) 4 = (113) 5 = (222)

Figure 1. XRD plot of aluminum lithium alloy 2091

IOP Conf. Series: Materials Science and Engineering **694** (2019) 012046 doi:10.1088/1757-899X/694/1/012046

The crystallite system of these samples are cubic, so the lattice parameters have the same value for a, b, and c. For aluminum reference, *Solutionized Only* sample has the biggest lattice parameter, then followed by *24 Hours of Aging* sample, *48 Hours of Aging* sample, and *72 Hours of Aging* sample. The lattice parameter changes in each sample as the time of artificially aging are raised. This is considered as the impact of Al_2Cu precipitations existence in the alloy lattice. The volume are also transformed following changes in lattice parameter.

Aging heat treatment slightly reduce the height of the peaks. It can be proven by analizing the diffraction angle which is related to the peak's width. By considering the crystallite size, the various treatments change the grain of the Al-Li 2091. In certain period of time, the grain size will decrease and increase the grain size. It has commonly assumed that there are deformation and grain growth that depend on the strain.

Table 2. Crystallographic parameters of the aluminum lithium alloy 2091.

	Sample	A	B	C	D
Aluminum Magnesium (3.7/0.3)	**Crystallite Size (Å)**	27913.4	1839.5	27913.4	1134.6
	Lattice Parameter (Å)	a = 4.0494 b = 4.0494 c = 4.0494	a = 4.0486 b = 4.0486 c = 4.0486	a = 4.0483 b = 4.0483 c = 4.0483	a = 4.0472 b = 4.0472 c = 4.0472
	Volume (Å^3)	66.40	66.36	66.34	66.29
	Microstrain (%)	0.044	0.071	0.050	0.094

A = Solutionized Only B = 24 h, aging C = 48 h, aging D = 72 h, aging

3.2. Potentiodyncamic Curves Analysis

By analysing the corrosion mechanism and understanding the change in corrosion rate, electrochemical behaviour of the Al-Li alloy 2091 samples in this experiment are observed. The outcome potentiodynamic curves test in 3.5 wt% NaCl solution and 7.0 wt% NaCl solution for samples with various heat treatment are shown on Figure 2. For all samples, similar trend are displayed in potentiodynamic polarization curves. I_{corr} and V_{corr} are obtained by applying Tafel plotting exploration. The reactions that represent for corrosion in Al-Li 2091 in NaCl electrolytes are shown below.

Anodic reaction : $Al \rightarrow Al^{3+} + 3e$ (2)

Cathodic reaction : $O_2 + 2H_2O + 4e^- \rightarrow 4OH^-$ (3)

In cathodic reactions, electrons from anodic reaction are absorbed to do some reduction of oxygen in solution. The ion Cl^- plays the role in corrosion by breaking and preventing the aluminum thin oxide layer [10].

Main reaction : $2Al^{3+} + 6Cl^- \rightarrow Al_2Cl_6$ (4)

$$Al_2Cl_6 + 6H_2O \rightarrow Al_2(OH)_6 + 6HCl \quad (5)$$

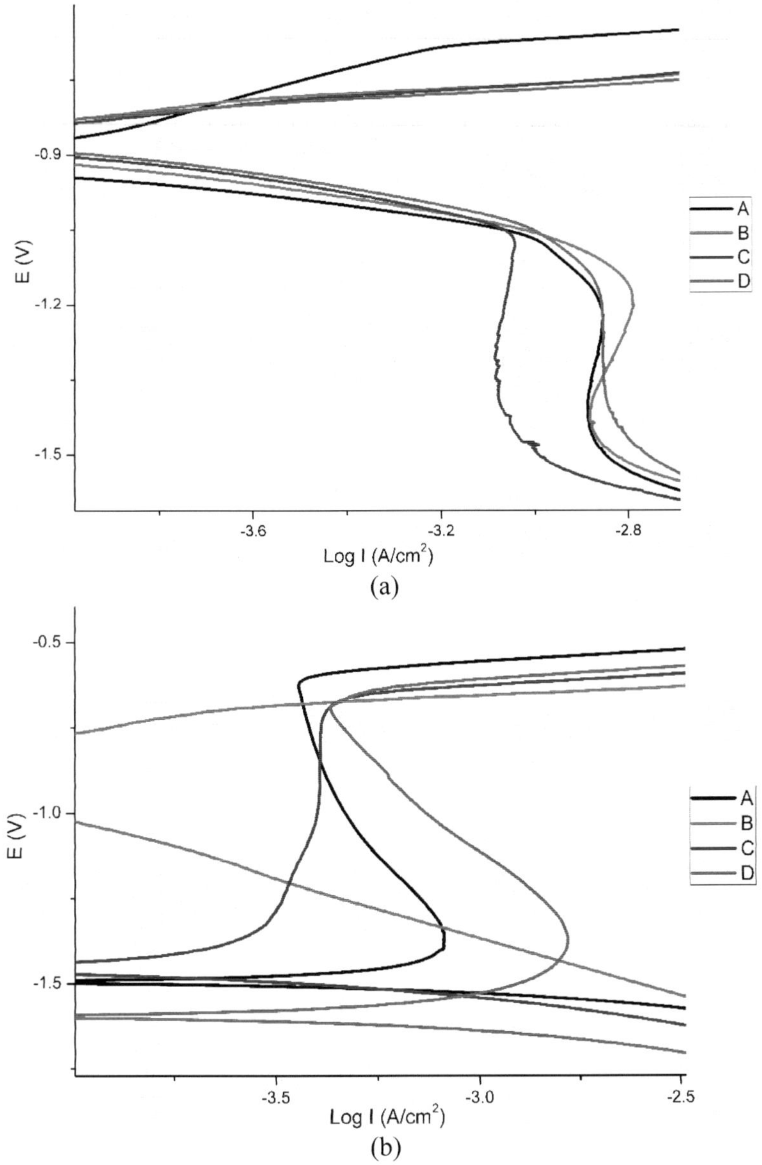

A = Solutionized Only B = 24 h, aging C = 48 h, aging D = 72 h, aging

Figure 2. (a) Potentiodynamic polarization curves of the aluminum lithium alloy 2091 in 3.5 wt% NaCl solution. (b) Potentiodynamic polarization curves of the aluminum lithium alloy 2091 in 7.0 wt% NaCl solution.

Potential corrosion E_{corr} (V) and corrosion current I_{corr} (A) affect changes in the corrosion rate. Table 3 shows *solutionized only samples* and *72 hours of aging* samples have significant values of corrosion rate. It is expected that in *solutionized sample*, the alloying elements are solutionized into the solution of aluminum which increasing the current transition to anodic dissolution of these metals exceeds the ability of oxygen. Aging time not only increase the current density in Al-Li 2091 alloy, but also increase the corrosion rate due to the precipitate size [6]. The release of electrons in an anodic reaction can make the metal potential more negative which can accelerate the cathodic reaction and decelerate anodic reaction and this can slow down the rate of corrosion [11].

Table 3. Data from the corrosion test of the aluminum lithium alloy 2091 in the 3.5 wt% and 7.0 wt% NaCl solution.

NaCl Solution	Sample	R (ohm)	$E_{corrosion}$ (V)	$I_{corrosion}$ (A)	Corrosion Rate (mm/year)
3.5%	**Solutionized Only**	3.734×10^{2}	-0.921	6.880×10^{-5}	7.623×10^{-1}
	24 Hours of Aging	4.659×10^{2}	-0.876	5.515×10^{-5}	6.111×10^{-1}
	48 Hours of Aging	3.578×10^{2}	-0.868	7.182×10^{-5}	7.958×10^{-1}
	72 Hours of Aging	3.322×10^{2}	-0.861	7.735×10^{-5}	8.571×10^{-1}
7.0%	**Solutionized Only**	4.627×10^{1}	-1.495	5.553×10^{-4}	6.153
	24 Hours of Aging	1.432×10^{3}	-0.890	1.795×10^{-5}	1.989×10^{-1}
	48 Hours of Aging	1.754×10^{2}	-1.457	1.465×10^{-4}	1.623
	72 Hours of Aging	5.070×10^{2}	-1.597	5.068×10^{-5}	5.615×10^{-1}

4. Conclusion

Variation of heat treatment in aluminum lithium alloy 2091 cause various impacts in structure and corrosion rate on its structure and corrosion rate. Structural changes in these alloys are caused by artificial aging. Solution heat treatment has great impact at corrosion rate due to the reaction of every constituent in its solution. Lower artificial aging time at 190°C resulted lower the rate of corrosion.

Acknowledgement

The authors thanks for the financial supports from the Universitas Indonesia under grant Publikasi Internasional Terindeks untuk Tugas Akhir Mahasiswa (PITTA), No. 2239/UN2.R3.1/HKPO5.00/2018.

References

[1] J. R. Davis, "Corrosion of Aluminum and Aluminum Alloys," 1999.
[2] E. A. Starke, *Historical Development and Present Status of Aluminum-Lithium Alloys*. Elsevier Inc., 2013.
[3] M. Engineering, "Effect of Aging on the Corrosion of Aluminum Alloy 6061 Effect of Aging on the Corrosion of Aluminum Alloy 6061," 2010.
[4] J. Staley and D. Lege, "Advances in aluminium alloy products for structural applications in transportation To cite this version : HAL Id : jpa-00251921," 1993.
[5] K. Welpmann *et al.*, "MECHANICAL PROPERTIES AND CORROSION BEHAVIOUR OF 2091 SHEET MATERIAL To cite this version : HAL Id : jpa-00226609," 1987.
[6] V. Gadpale, P. N. Banjare, and M. K. Manoj, "Effect of ageing time and temperature on corrosion behaviour of aluminum alloy 2014" 2018.
[7] A. Vats, "Corrosion Measursement , Friction testing and XRD Analysis of Single Layer CrN Coatings on AISI 304 Stainless Steel," no. 5, pp. 435–445, 2017.
[8] Z. Ahmad, "Application of Faraday's Laws In Determination of Corrosion Rates of Metals and Alloys," in Principle of Corrosion Engineering and Corrosion Control (Butterworth-Heinemann, United Kingdom, 2006), pp. 58-61.

IOP Conf. Series: Materials Science and Engineering **694** (2019) 012046 doi:10.1088/1757-899X/694/1/012046

[9] J. Kepler and N. Year, "X-ray crystallography," no. 1784.
[10] J. Anggono, S. Tjitro, D. F. Teknik, J. Teknik, M. Universitas, and K. Petra, "Studi Perbandingan Kinerja Anoda Korban Paduan Aluminium dengan Paduan Seng dalam Lingkungan Air Laut."
[11] G. Bellefontaine, "THE CORROSION OF CoCrMo ALLOYS FOR BIOMEDICAL," no. January, 2010.

Comparative study on stress analysis in human molar tooth between metallic and nonmetallic dental filling material

M A M Soliheen[1], A Z A Kadir[1], D Kurniawan[2] and F M Nor[3]

[1]School of Mechanical Engineering, Universiti Teknologi Malaysia, Johor Bahru 81310, Malaysia
[2]Faculty of Engineering, Universiti Teknologi Brunei, Gadong BE1410, Brunei Darussalam
[3]Department of Mechanical Engineering, Curtin University, Miri 98009, Malaysia

E-mail: fethma@curtin.edu.my

Abstract. Dental filling is regarded as proven dental restoration method in dealing with partial missing of tooth crown due to caries or trauma. This study analyses the reliability and durability of materials for dental filling using finite element analysis. Specifically, it determines the stress induced at a dental filling due to occlusal load. Two filling depths were simulated: (a) filling depth in enamel and (b) filling depth in enamel/dentine on top of molar tooth. Dental filling made of glass ionomer cement (GIC), composite, and amalgam (as a reference material) which represented metallic and non-metallic filling materials were chosen as the materials for this study. A 100 MPa pressure was applied on the dental filling to represent an occlusal load. The results obtained were the stress (magnitude and distribution) for enamel/amalgam, enamel/GIC, enamel/composite, enamel/dentine/amalgam, enamel/dentine/GIC, and enamel/dentine/composite. It was found that the maximum induced stress was higher for enamel/dentine depth while there is no significant difference in stress magnitude for all dental filling materials.

1. Introduction

Dental filling basically replaces the partially lost tooth structure due to external causes such as trauma or dental caries and is subjected to similar variety of load as intact tooth. These loads may be due to mastication, swallowing, biting, chewing, speech, clinching, bruxism and also by the action of tongue, perioral and circumoral musculature [1]. This required biologically compatible replacement of tooth crown. It is interesting to note that to date there is still no filling material that has properties similar to dental crown (enamel and dentine).

In the past years, studies have been done using finite element analysis on dental filling. Some looked at the biomechanics which is also related to compatibility, mechanical and clinical behavior of filling materials. Others discussed about the development of modelling techniques in outlining the risk of failure in filling materials. In a previous study, the authors [1] stated that improvement in computer modelling render the finite element method a very reliable and accurate approach in biomechanical applications. The advantage of finite element analysis to alter structures and material properties under controlled conditions is described in previous literature [2].

Amalgam is the commonly used dental filling material, considering its relatively low cost, mechanical strength sufficient to apply in posterior teeth, ease of preparation, and ease of applying. However, considering amalgam contains mercury and its aesthetic is low, alternative filling materials are made available. A previous study to investigate the stress analysis in amalgam and porcelain dental fillings found that amalgam outperforms porcelain in terms of stress [3]. This study intends to make comparison between amalgam and other nonmetallic filling materials which are commercially available. Composite and glass ionomer cements are the materials of interest in this study. Both are dental fillings that have visually similar look as tooth and bond well to dental tissue, suitable for posterior teeth restoration [4]. Clinically, use of composite filling for class II restoration recorded 100% retention rate after over 3.5 years, showing no recurring caries, sensitivity, or fractures [5].

Dental filling functions when the material can withstand the loads (mechanical or thermal) it is subjected to and there is proper bonding between the dental filling and the tooth it attaches on. Both can be analysed with revealing the stress induced in the dental filling. This study intends to determine the stress induced on a dental filling on a tooth, loaded with mastication load, using finite element analysis. Scenarios where dental filling material only restores the enamel and when it is restoring both enamel and dentin were simulated.

2. Materials and methods

The developed three-dimensional model represents an approximate physical model of human third molar. Two software, a computer-aided design SolidWorks (Dassault Systemes S.A.) and a finite element analysis Abaqus FEA v. 6.13 (Dassault Systemes S.A.) were used in this study.

A solid model was created [3] from the mean dimensions of a large population of sound teeth based on Asian people which represented an "average" lower third molar (posterior) using the SolidWorks. The restored molar tooth was divided into parts (enamel, dentine, adhesive and filling) which were then exported into an ACIS SAT 3D file. The three-dimensional model was filled with tetrahedral (C3D10M) elements and the meshing was created with 0.3 mm long elements using Abaqus. A convergence test was done [6]. It was indicated that convergence was achieved when using elements smaller than 0.5 mm in length. The generated model consisted of 161316 elements. All other nodes were given three degrees of freedom (X, Y and Z). The classification for dental filling shape [7] is Class I, i.e. caries affecting on occlusal surface of molar tooth.

The occlusal load during chewing and biting for molar tooth was presumed to be a pressure (P) of 100 MPa on top of the enamel/filling surface as shown in figure 1. This pressure is equivalent to a force of 380 N as reported by previous work [8, 9]. The root was made to be fully constrained in all direction.

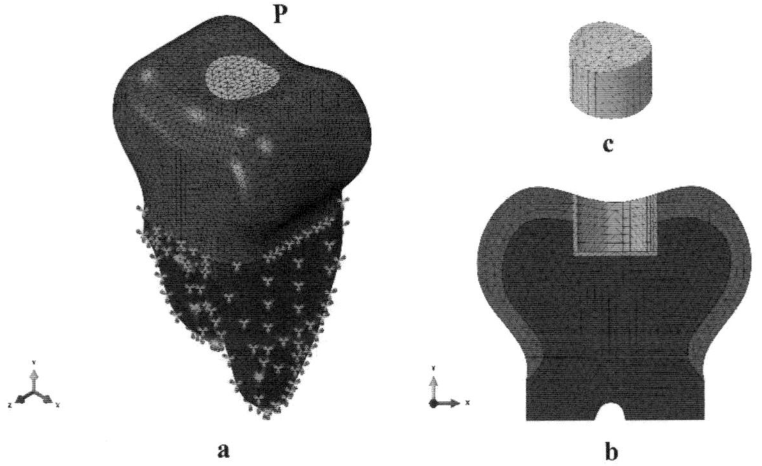

Figure 1. Geometry model used in this study: (a) Restored molar tooth with constrains fixed to the root and vertical occlusal load (P) applied to enamel/filling surface area. (b) Sectional view of restored molar tooth with adhesive layer, (c) composite filling [3].

Following assumptions were made into the model:

i. Complete bonding between enamel and dentine,
ii. Pulp was not included in the model,
iii. Root of the tooth was not included in stress analysis,
iv. There was no time-dependent behavior of the elastic modulus of the materials,
v. Only static loading was applied, and
vi. The dental filling has uniform cylindrical shape as the basis comparison.

The material types and properties of the tooth as described by [6, 10, 11] are listed in table 1. All materials, tooth structures (enamel and dentine) were modeled as isotropic, homogeneous materials. For composite filling case study which required bonding agent, the adhesive layer between the filling material, enamel and dentine has been included in the model shown in figure 1. For that purpose, a thickness for the adhesive layer is defined in accordance to [12, 13]. GIC type II is used in this study which serve as filling materials as classified by [14].

Table 1. Elastic properties of isotropic restorative materials and tooth tissue.

Material	Young's Modulus, E [GPa]	Poisson's ratio, v	References
Amalgam	50	0.29	[6]
GIC (Fuji IX GP)	10.8	0.30	[11]
Composite	15	0.30	[10]
Adhesive	4.5	0.30	[10]
Enamel	80	0.30	[6]
Dentine	20	0.31	[6]

3. Results and discussion

All the finite element analysis of restored molar tooth with amalgam, composite and GIC are performed under static loading with magnitude of 100 MPa occlusal load with two different filling depths. The results indicate that maximum value of stress (Von Mises) for enamel/amalgam, enamel/GIC, enamel/composite, enamel/dentine/amalgam, enamel/dentine/GIC and enamel/dentine/ composite fillings are 257 MPa, 259 MPa, 258 MPa, 292 MPa, 292 MPa and 293 MPa, respectively. The highest magnitude of stress is found in enamel/dentine/GIC filling and lowest magnitude of stress is observed in enamel/amalgam filling.

The overview of stress distribution and magnitude within the dental filling and restored molar are presented in figures 2 and 3. Each of the model shows different stress distribution as well as stress magnitude.

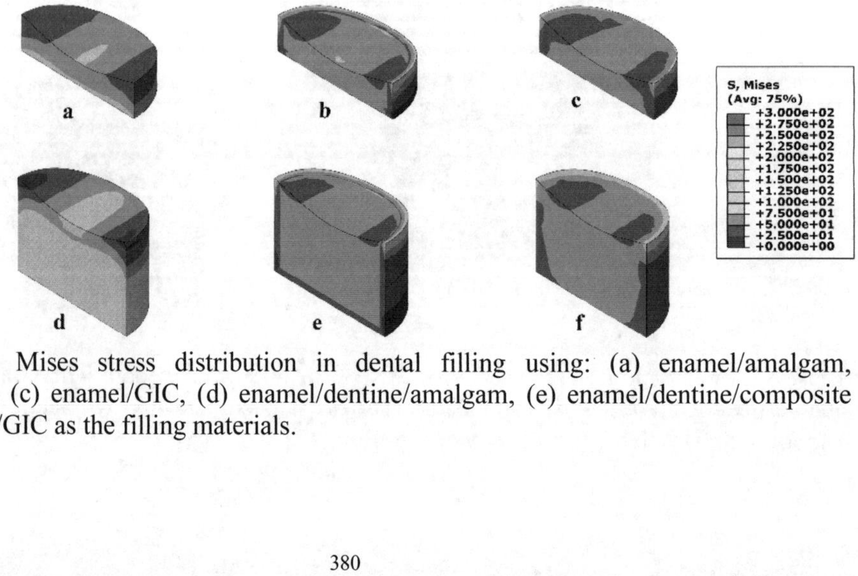

Figure 2. Von Mises stress distribution in dental filling using: (a) enamel/amalgam, (b) enamel/composite, (c) enamel/GIC, (d) enamel/dentine/amalgam, (e) enamel/dentine/composite and (f) enamel/dentine/GIC as the filling materials.

From figure 2, generally the stress is lower in enamel depth fillings compared to enamel/dentin depth fillings. This shows a good agreement with past study [15] where smaller restorations showed better survival compared with larger restorations and the more surfaces a restoration covered the greater the chance of failure. Stress distribution and magnitude in enamel/GIC filling was the lowest of all fillings. Enamel has higher stiffness compared to dentin. Hence, more stress was borne by the enamel, leaving less to be borne by the dental filling. Dentin has lower stiffness than enamel. Hence, less stress is induced at the dentin compared to at the enamel. GIC has the lowest Young's modulus of the dental filling materials. Consequently, this makes the most stress was borne by the tooth region when GIC was used. Although, it should be noted that the difference in maximum stress magnitude is not significant compared to the other two filling materials.

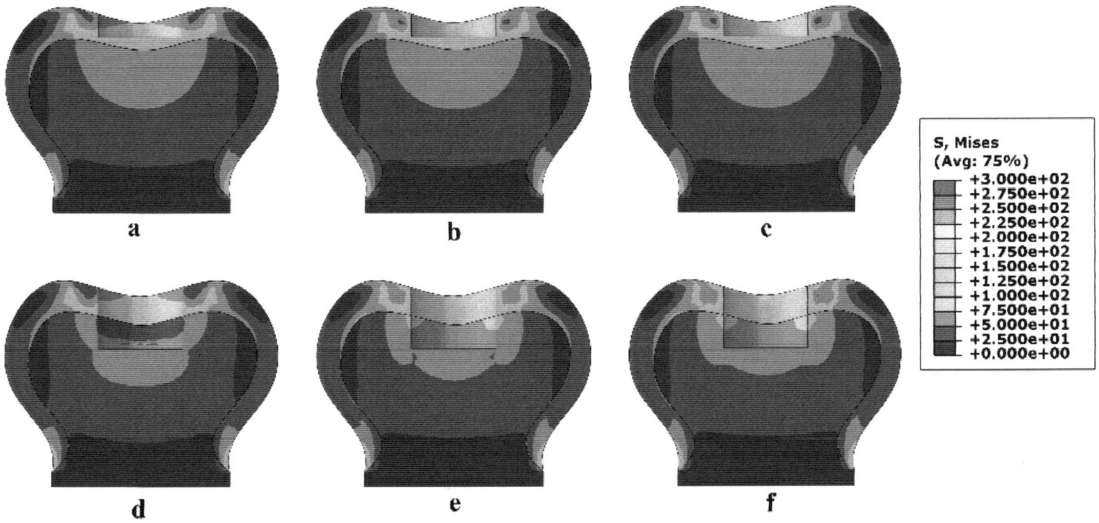

Figure 3. Von Mises stress (MPa) distribution for the Class I restored molar tooth using: (a) enamel/amalgam, (b) enamel/composite, (c) enamel/GIC, (d) enamel/dentine/amalgam, (e) enamel/dentine/composite and (f) enamel/dentine/GIC as the filling materials. For the clarity, dental fillings are not shown.

Seen through the tooth's cross section in buccal-lingual direction (figure 3), the location of the maximum stress resulting from occlusal loading of the restored tooth can be seen more clearly. For enamel depth fillings, high stress occurs predominantly at the bottom surface of enamel/filling interface, as evident in figures 3(b) and 3(c). While for enamel/dentin depth fillings, high stress is concentrated in dentine wall and occlusal surface. The former can be attributed to the lower mechanical properties of the dentin compared to enamel. It should be noted that the filling materials gave no significant effect on the maximum stress magnitude for the same depth of filling.

To have clear results on the stress magnitude on occlusal surface, the magnitude of stress can be plotted against its location. In figure 4, the path outlines are indicated with the red-coloured line. Both paths are plotted on the occlusal surface between the dental tissue (enamel) and filling material which covered the loading area. Von Mises stress was assessed along these paths (figure 5). Two paths are defined in these models, i.e. (a) following the mesial and distal wall profiles of the tooth and (b) following buccal and lingual wall profiles of the tooth.

Overall, the stress induced on occlusal surface is lower than the stress in filling-tooth interface, both for enamel depth and enamel/dentin depth. From figure 5, the enamel/dentine/GIC filling has the highest magnitude of stress which exceeds 180 MPa, while the lowest magnitude of stress is found in enamel/amalgam filling, which is close to 80 MPa. In figure 5(a), points 0.2 and 0.8 and in figure 5(b), points 0.1 and 0.9 are points of enamel/filling interface respectively. This means the highest stress occurs at the filling/tooth interface. This interface is where the expected failure to occur when a static load is applied. This finding supports previous report [6] that found the highest stress is along the filling/tooth interface.

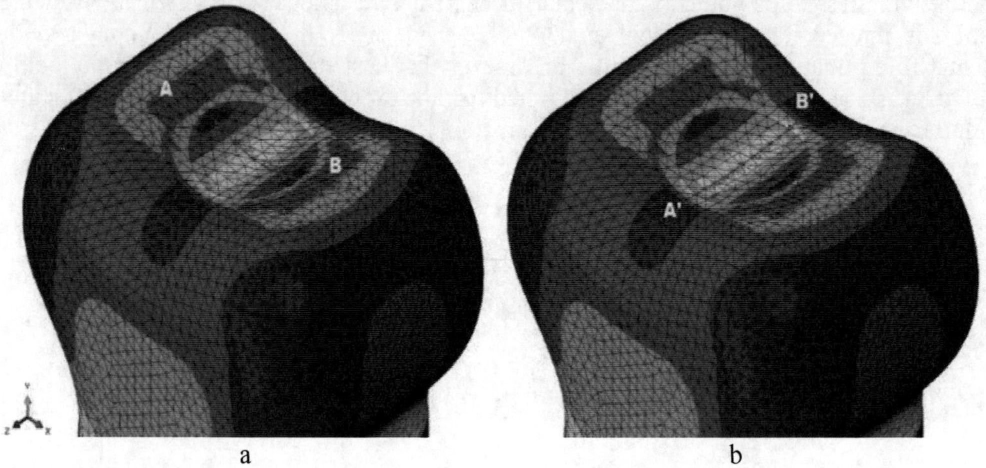

Figure 4. Paths defined in restored molar tooth considering the area of loaded enamel/filling surface. (a) Path A-B mesial-distally directed and (b) Path A'-B' buccal-lingual directed.

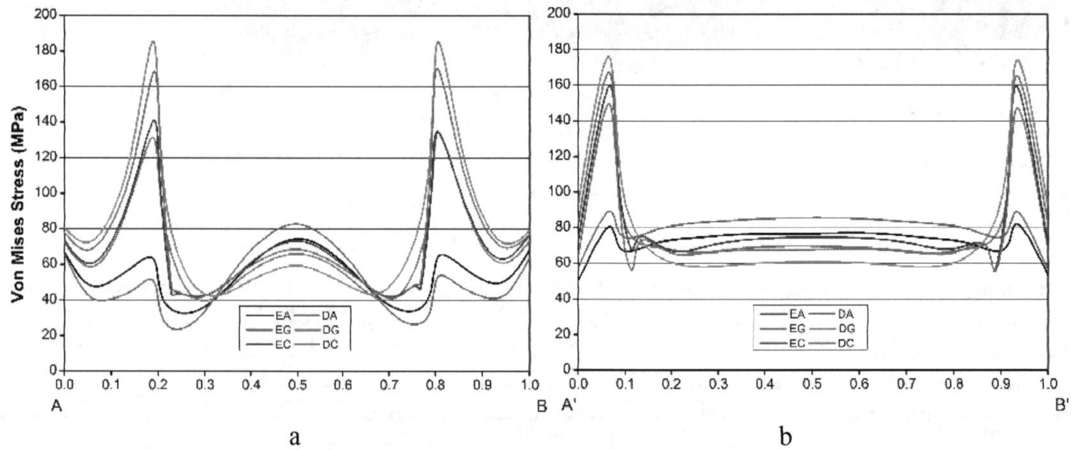

Figure 5. Von Mises stress along previously defined (a) path A-B and (b) path A'-B' for Class I dental filling. (Note: EA: Enamel/Amalgam, EG: Enamel/GIC, EC: Enamel/Composite, DA: Enamel/Dentine/Amalgam, DG: Enamel/Dentine/GIC, and DC: Enamel/Dentine/Composite filling.)

Another point to note is that a material can be a failure when the magnitude of stress exceeds their strength. The strength of amalgam is 380 – 414 MPa, GIC (Fuji IX GP) is 88 MPa [11, 16] and composite 300 MPa [17]. Considering the stress induced and the strength of filling material, it is likely that GIC will fail due to the occlusal loading of 100 MPa. This is because the maximum stress induced at occlusal surface is 180MPa, and it is even higher at internal interface of filling/enamel beneath the occlusal surface, while its strength is only 88MPa. Composite and amalgam dental fillings are likely to survive the simulated masticatory load.

4. Conclusion

Finite element analysis was performed to investigate the stress (distribution and magnitude) in a Class I restored molar tooth loaded by masticatory load. The variation parameters were filling depth (enamel depth and enamel/dentin depth) and filling materials (composite, GIC, and amalgam). Under the load, it can be concluded that

1. Stress at enamel/dentin depth filling is higher than at enamel depth filling, regardless of the filling material.
2. There is no significant difference in stress for all dental filling materials.
3. The highest stress occurs at interface between filling and tooth for all dental fillings.
4. Maximum stress can be used to determine whether dental filling fails under the load. In this case, GIC is likely to fail.

References

[1] U. Andreaus, M. Colloca and D. Iacoviello. "Damage detection in a human premolar tooth from image processing to finite element analysis," In N. Jorge et al. (Eds.). *Biodental Engineering*, pp. 29-34, 2010.

[2] R. L. Sakaguchi, E. W. Brust, M. Cross, R. DeLong and W. H. Douglas. "Independent movement of cusps during occlusal loading," *Dental Materials*, 7, pp. 186-190, 1991.

[3] M. A. M. Soliheen, D. Kurniawan, and F. M. Nor. "Stress Distribution between Bonding Surface of Dental Filling in Enamel and Dentine," *Procedia Manufacturing*, 2, pp. 212 – 217, 2015.

[4] T. Cornacchia, E. Las Casas, C. Cimini and R. Peixoto. "3D finite element analysis on esthetic indirect dental restorations under thermal and mechanical loading," *Medical & Biological Engineering & Computing*, 48, pp. 1107-1113, 2010.

[5] R. C. Spreafico, I. Krejci and D. Dietschi. "Clinical performance and marginal adaptation of class II direct and semidirect composite restorations over 3.5 years in vivo," *J. Dent.*, 33(6), pp. 499–507, 2005.

[6] G. Couegnat, S. L. Fok, J. E. Cooper and A. J. E. Qualtrough, "Structural optimization of dental restorations using the principle of adaptive growth," *Dent. Mater.*, 22, pp. 3–12, 2006.

[7] D. A. Young, B. B. Nový, G. G. Zeller, R. Hale, T. C. Hart and E. L. Truelove, "The American Dental Association caries classification system for clinical practice," *The Journal of the American Dental Association*, 146(2), pp. 79 – 86, 2015.

[8] D. Kurniawan, F. M. Nor, H. Y. Lee and J. Y. Lim. "Finite element analysis of bone – implant biomechanics : refinement through featuring various osseointegration conditions," *Int. J. Oral Maxillofac. Surg.*, 41, pp. 1090–1096, 2012.

[9] K. F. Woodmansey, M. Ayik, P. H. Buschang, C. A. White and J. He. "Differences in masticatory function in patients with endodontically treated teeth and single-implant-supported prostheses: a pilot study," *J. Endod.*, 35, pp. 10-14, 2009.

[10] K. Palka, J. Bienias and H. Debski. "Finite element analysis of thermo-mechanical loaded teeth," *Computational Materials Science*, 64, pp. 289-294, 2012.

[11] C. Sinescu, L Marsavina and A. Cernescu. "Asymptotic stress field for the interface between teeth and different restorative materials," *Computational Materials Science*, 59, pp. 57-64, 2012.

[12] P. F. Hubsch, J. Middleton and J. Knox. "A finite element analysis of the stress at the restoration-tooth interface, comparing inlays and bulk fillings," *Biomaterials*, 21, pp. 1015-1019, 2000.

[13] N. De Jager, P. Pallaz and A. J. Feilzer. "Finite element analysis model to simulate the behavior of luting cements during setting," *Dental Materials*, 21, pp. 1025-1032, 2005.

[14] G. J. Mount. "*An atlas of Glass Ionomer Cements: A Clinician's Guide*," 2nd Ed., Martin Dunitz, London, 1994.

[15] M. F. Burrow. "Composite adhesive restorative materials for dental applications," In *Non-*

metallic biomaterials for tooth repair and replacement, pp. 235-269, 2013.
[16] L. C. Pereira, M. C. Nunes, R. G. Dibb, J. M. Powers, J. F. Roulet and M. F. Navarro. "Mechanical properties and bond strength of glass-ionomer cements," *J Adhes Dent*., 4(1), pp. 73-80, 2002.
[17] P. Ausiello, S. Rengo, C. L. Davidson and D. C. Watts. "3D finite element analysis on esthetic indirect dental restorations under thermal and mechanical loading, *Dent. Mater*.", 20, pp. 862–872, 2004.

IOP Conf. Series: Materials Science and Engineering **694** (2019) 012048 doi:10.1088/1757-899X/694/1/012048

Effect of part orientation on dimensional accuracy, part strength, and surface quality of three dimensional printed part

Ahmed M. Abdelrhman[1], Wen Wei Gan[1], D. Kurniawan[2]

[1] Department of Mechanical Engineering, Curtin University, Miri 98009, Malaysia
[2] Faculty of Engineering, Universiti Teknologi Brunei, Gadong BE1410, Brunei Darussalam

E-mail: denni.kurniawan@utb.edu.bn

Abstract. Additive manufacturing such as fused deposition modelling (FDM) three dimensional printing machine can deliver complex shapes with fine details in relatively short building time using various materials (albeit mostly polymers in this case). However, the quality of the printed results might vary depending on the set input. The present study aims to investigate the effects of part orientation on dimensional accuracy, part strength, total cost, and surface quality of a three dimensional printed part. Five test specimens were printed on FDM machine with polylactic acid (PLA) as the material by varying the build orientation, i.e. X0°Y0°, X90°Y0°, X0°Y90°, X0°Y45°, and X90°Y45°. Dimensional accuracy, surface roughness, and part strength were measured by using digital vernier calliper, universal testing machine, and portable surface roughness tester, respectively. The results show that part orientation influences printed part quality, especially in accuracy and surface roughness. No part orientation studied delivered all positive quality expected, but overall, X0°Y0° orientation resulted the highest quality in terms of dimensional accuracy, mechanical properties, and cost.

1. Introduction

The first applications of three dimensional (3D) printing or additive manufacturing (AM) were in the area of rapid prototyping, which started as a fundamental procedure to construct prototype models. Rapid prototyping with 3D printing offers the adaptability essential to make this vital experimentation procedure conceivable for physical items and products. Among all of the additive manufacturing technologies, FDM is the standout amongst the regularly used technologies for manufacturing or engineering purposes. FDM is applicable for an assortment of thermoplastics resins and the process is low-cost when contrasted with other AM processes. Besides, the support material is less demanding to break away from the completed part. FDM part is perfect for shape, impeccable for fit and function testing since it can endure testing thoroughly without twisting, contracting or dampness assimilation [1].

For products with non-complex features, it is normally the best approach to emphasise on the commercial value. Targeting at orientation with the most minimal amount of deposited layers allows rising for at least two significant part parameters, with high opportunity to increase accuracy and occasionally surface quality. A final product made utilising FDM can be described by certain quantities, which are subjected to many conditions. Creating a precise and dimensionally stable part have always been a challenging goal of FDM technology. In the past, the effect of manufacturing build parameters on a single objective of 3D printed part has been broadly investigated by different theoretical and experimental methods. Raut et al. [2] prepared and fabricated ASTM standard specimens in three geometrical axes with various part orientations to perform the tensile and flexural tests. They perceived that high tensile strength FDM parts with low manufacturing cost would occur at the built up orientation of y-axis of 0°; whereas FDM parts with good flexural strength and moderate fabrication cost would take place at built up orientation of x-axis of 0°. Afrose et al. [3] studied the fatigue characteristics of FDM printed part with polylactic acid (PLA) material by deliberating the impact of various build orientations for which the samples were manufactured in ASTM D638 standard. As reported by Górski et al. [4, 5], there are two main reasons for the effect of part orientation on mechanical characteristics. The first reason is the layers in between the FDM parts appear in weak bond, which leads to the strength properties to drop. The second cause is the FDM models which consistently show volume error within.

The issue of dimensional stability has become significant when components are proposed for assembly and customary utilisation. Tak et al. [6] studied the shrinkage degrees on the produced models by examining the effect of 3D printing setting on dimensional accuracy. They found that the decrease in size rises the level of shrinkage where the geometric forms directly influence on the deviations of the measurement. Nizam et al [7] studied on dimensional accuracy by using FDM technology to manufacture a sample with four general features of parts in various magnitudes. The authors observed and declared that FDM machine they were using was having low precision, especially in fabricating a part with rounded figure as the most of them exceed the device's tolerance. Górski et al [8] utlised part orientation as the primary parameter of the FDM technology and determined the relationship with dimensional accuracy and repeatability of acquired products. They concluded that deviations are comparatively fewer for greater scales which implied that the bigger the body dimension, the more accurately it will be produced. Singh et al [9] found that the effect of parts placement on accuracy and mechanical properties has noticeable change when advancing along Y axis from the origin but not the extent in the X direction. The authors observed that there is a little abatement in mechanical properties and a small increment in dimensional accuracy error when progress along Y axis in the build. The slicing layer produced and the orientation at which the part is fabricated can directly affect the surface quality of the printed model. Badola and Vaishya [10] have conducted an experimental study and presented a couple of more parameters which could be adopted to manage the surface finish of a part, for example, build style, build envelope's temperature as well as multi-contouring. Stair case effect is inevitable because of the nature of layer manufacturing process. It is usually established on inclined planes and rounded surface, resulting in a high value of surface roughness of the part. This effect is difficult to eliminate, but it can be reduced to some extent and thus increase the surface quality. Sreedhar et al. [11] reported that due to some layers of the FDM sample involve void material, the boundaries of the lateral part could not acquire sufficient tool movement. However, they stated that the issue could be prevented by increasing the amount of travel of the contour tool in those layers and limiting the raster route to assure the value of surface roughness is reducing. The research of Bakar et al. [12] indicated that contour width and internal raster could help in the bonding substance amongst layers and result in an improved surface quality. Vijay et al. [13] performed fractional factorial design in two stages for layer thickness and three tiers for orientation aspect to evaluate the ideal surface quality of the printed part.

From previous studies, it can be summarised that most of the investigations were centred on one particular build goal or a few build objectives regarding only one process variable, for example, the part orientation. In the present work, the characteristics and the effect of build orientation are examined on several build goals, including dimensional accuracy, part strength, building time, build material, support material, manufacturing cost, and surface quality of the 3D printed part. This investigation is performed based on the subsequent methodology, in which first step is to design and create 3D model. The second step is to transfer the designed model to 3D printer and print the parts with five different orientations (X0°Y0°, X90°Y0°, X0°Y90°, X0°Y45°, and X90°Y45°). The third step is to measure and average the results of dimensional accuracy, mechanical strength and surface roughness. The last step is to analyse the experiment results and identify the relationship between accuracy, strength and surface roughness with part orientation.

IOP Conf. Series: Materials Science and Engineering **694** (2019) 012048 doi:10.1088/1757-899X/694/1/012048

2. Materials and methods

The experimental test target was designed using Solidworks software and printed as shown in figure1. The specimen used is a bracket which was chosen because it contains many of the geometrical dimensioning and tolerance features. The experiment specimens were set to be five test prints which each of the five printed specimens applied its particular unique set of printed variables with changing part orientation. For the result authentication, the experiment was repeated thrice and therefore in total there are 15 samples. The models are produced by using the STRATASYS FDM machine (MOJO) default parameters setting and the other parameters, which are shown in table 1. Post-processing was needed to remove the support material. The model material required for each part, the support material required for each part as well as the build time were estimated by Control Panel software as shown in table 2.

Table 1. FDM Process Parameters

No.	Feature	Dimension	Value (mm)
1	Hole 1	Diameter	5.71
2	Hole 2	Diameter	6.71
3	Hole 3	Diameter	8.23
4	Cylinder	Thickness	6.10

Table 2. Properties for various orientation

Sample No.	Orientation	Building Time (hr:min)	Model Material (cm^3)	Support Material (cm^3)
1	X0°Y0°	4:15	28.3	14.5
2	X90°Y0°	4:50	28.6	16.9
3	X0°Y90°	7:23	29.3	29.6
4	X0°Y45°	6:25	29.0	25.8
5	X90°Y45°	5:54	28.8	20.1

The data of building time, model material used and support material used can be obtained from the FDM machine log file. In the present case, adopted from Tagore et al. [14], the cost of each criterion and the estimated manufacturing cost for each orientation can be calculated by applying the formula in equation 1.

Total cost = (Machine hr. /min × Build up time) + (Model material × Model material cost) + (Support material × Support material cost) (1)

where

Building hr cost = $10 / hr

Model material cost = $ 0.2169 / cc

Support material cost = $ 0.1735 /cc.

The measurement of dimensional accuracy was conducted by determining the dimensions of specimen features. Measurements were conducted via digital calliper with an accuracy of ±0.02 mm and the magnitudes were contrasted with the nominal dimensions (CAD model). The full analysis and investigation of dimensional accuracy were attained which showed in result segment. According to Chua et al [15], the tolerance in the x-axis is 0.127 mm.

$$D(\%) = \left| \frac{A_{CAD} - A_{mea}}{A_{CAD}} \right| \times 100\% \quad (2)$$

where A_{CAD} is the nominal size of CAD model set by the computer, A_{mea} is the real size measured by a digital vernier calliper.

For determining the surface roughness of every specimen, a Mitutoyo Surftest SJ-301 portable surface tester is used. Depends on the size of the printed model, the contact needle of the surface tester can navigate to a range of 0.25 mm to 25 mm. Both surface roughness, Arithmetic Mean Deviation (R_a) and Average peak to valley height (R_z) was measured at four different positions. These quantities were applied in transversal path of the material surface for three times at each location of different part orientation.

Among mechanical strength tests, tensile test is the most common investigations to determine the material and part properties. Tensile tests were performed on the printed specimen three times for each part orientation where the experiment was carried out on an LR10KN Universal Materials Testing Machine. The force values were set to zero to neutralise any tensions produced in the equipment during part setting. This calibration secured an equivalent starting point over all the tests. The upper wedge grip of the machine shifted upward at a rate of 2 mm/min until the point when the test part broken. The stress and maximum fracture load were recorded continuously through Tinius Olsen software which associated with the analyser. Based on the parameters, the test times is extended to 2 to 5 minutes.

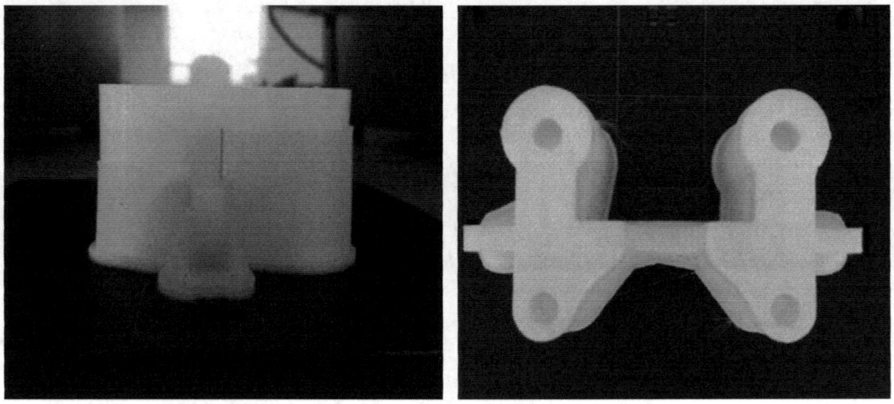

Figure 1. Specimen built at 1st orientation (X0° Y0°).

3. Results and discussion

As shown in table 3, orientation has significant influence on several main features such as the height of the part in the build direction. It determines the required manufacturing time and cost of final parts. The outcomes validate that specimen with a small degree have short building time. Besides, diminishing the contact region between the model and the support structure will reduce the building time and the cost as well. Therefore, based on the obtained results, the best option for the least manufacturing cost and fabricating time is sample 1 at the orientation of X0°Y0°.

Table 3. Manufacturing cost of different part orientation

Sample No. (SN)	Building Time (hr:min)	Model Material (cc)	Support Material (cc)	Total Cost ($)
1	4:15	28.3	14.5	51.23
2	4:50	28.6	16.9	51.71
3	7:23	29.3	29.6	54.07
4	6:25	29	25.8	53.35
5	5:54	28.8	20.1	52.31

Table 4. Dimensional accuracy of part features with different part orientation

SN	Hole 1 Diameter (mm)					Deviation (mm)	Dimensional Accuracy (%)
	Measured Dimension				Nominal Dimension		
	Test 1	Test 2	Test 3	Avg			
1	5.71	5.60	5.67	5.660	5.71	-0.048	99.16
2	5.49	5.37	5.44	5.433	5.71	-0.275	95.19
3	5.50	5.64	5.65	5.597	5.71	-0.111	98.05
4	5.42	5.43	5.44	5.430	5.71	-0.278	95.13
5	5.53	5.50	5.53	5.520	5.71	-0.188	96.71
SN	Hole 2 Diameter (mm)					Deviation (mm)	Dimensional Accuracy (%)
	Measured Dimension				Nominal Dimension		
	Test 1	Test 2	Test 3	Avg			
1	6.68	6.73	6.64	6.683	6.708	-0.025	99.63
2	6.52	6.48	6.47	6.490	6.708	-0.218	96.75
3	6.51	6.68	6.60	6.597	6.708	-0.111	98.34
4	6.57	6.63	6.53	6.577	6.708	-0.131	98.04
5	6.57	6.47	6.53	6.523	6.708	-0.185	97.25
SN	Hole 3 Diameter (mm)					Deviation (mm)	Dimensional Accuracy (%)
	Measured Dimension				Nominal Dimension		
	Test 1	Test 2	Test 3	Avg			
1	8.24	8.21	8.22	8.223	8.23	-0.009	99.89
2	8.21	8.25	8.19	8.217	8.23	-0.015	99.81
3	8.09	8.06	8.27	8.140	8.23	-0.092	98.88
4	8.22	8.20	8.23	8.217	8.23	-0.015	99.81
5	8.03	8.05	8.03	8.037	8.23	-0.195	97.63
SN	Cylinder Thickness (mm)					Deviation (mm)	Dimensional Accuracy (%)
	Measured Dimension				Nominal Dimension		
	Test 1	Test 2	Test 3	Avg			
1	6.26	6.27	6.31	6.280	6.1	0.180	97.05
2	6.13	6.18	6.17	6.160	6.1	0.060	99.02
3	6.23	6.22	6.24	6.230	6.1	0.130	97.87
4	6.31	6.30	6.35	6.320	6.1	0.220	96.39
5	6.20	6.19	6.18	6.190	6.1	0.090	98.52

From the results obtained in tables 5 and 6, it can be seen that all the three actual diameters of the hole presented negative deviation which is below their nominal values. As for the standard tolerance of the FDM device is ±0.127mm, the results showed that except for sample 1 and 3, the dimensional differences of holes for the other two samples are surpassing the range. While for the case of cylinder feature, all the measured values of the cylinder thickness are above the nominal dimension (positive deviation) and which some of them is beyond the standard tolerance except for samples 2 and 5.

Based on table 4, the accuracy of the hole is linearly increasing with the diameter of the hole. The most accurate of the hole is up to 99.89% which is sample 1, Hole 3 (Ø8.23mm) and the least is 95.13% at sample 4, Hole 1 (Ø5.71mm). The highest dimensional accuracy of the cylinder is 99.02% at sample 2 whereas the lowest is 96.39% at sample 4. Hence, it can be concluded that the best dimensional accuracy obtained for producing both hole and cylinder feature is at orientation X0°Y0°.

In general, the printed model displays a rougher surface because of the extrusion of unwinding semi-liquefied thermoplastic thin filament. Based on the table 5, the highest values of the surface roughness are recorded where the support materials are located.

Table 5. Surface roughness variation of each part orientation

Location	Part 1							
	Ra (um)				Rz (um)			
	Test 1	Test 2	Test 3	Avg	Test 1	Test 2	Test 3	Avg
1	2.32	1.88	1.18	1.793	7.24	5.94	4.87	6.017
2	2.42	2.69	1.91	2.340	7.88	7.67	7.73	7.760
3	3.22	1.06	3.08	2.453	9.40	5.79	12.06	9.083
4	15.34	14.47	12.01	13.940	58.32	49.22	44.39	50.643
	Part 2							
Location	Ra (um)				Rz (um)			
	Test 1	Test 2	Test 3	Avg	Test 1	Test 2	Test 3	Avg
1	6.51	6.60	4.80	5.970	31.09	34.33	27.27	30.897
2	5.77	9.55	8.06	7.793	34.76	42.46	38.54	38.587
3	10.17	13.04	10.34	11.183	58.34	51.58	51.65	53.857
4	1.84	0.81	1.89	1.513	5.09	3.55	5.94	4.860
	Part 3							
Location	Ra (um)				Rz (um)			
	Test 1	Test 2	Test 3	Avg	Test 1	Test 2	Test 3	Avg
1	10.46	9.40	10.37	10.077	46.23	44.22	42.27	44.240
2	10.34	9.82	10.43	10.197	50.44	47.58	48.25	48.757
3	4.73	3.44	3.90	4.023	20.45	24.54	19.47	21.487
4	10.28	10.19	13.49	11.320	44.54	49.97	55.10	49.870
	Part 4							
Location	Ra (um)				Rz (um)			
	Test 1	Test 2	Test 3	Avg	Test 1	Test 2	Test 3	Avg
1	12.57	11.18	11.92	11.890	48.87	47.35	49.75	48.657
2	12.54	10.94	11.43	11.637	53.26	52.95	57.35	54.520
3	5.05	5.61	4.26	4.973	19.58	18.71	18.05	18.780
4	14.25	13.89	14.98	14.373	61.35	59.53	61.89	60.923
	Part 5							
Location	Ra (um)				Rz (um)			
	Test 1	Test 2	Test 3	Avg	Test 1	Test 2	Test 3	Avg
1	14.79	14.95	15.02	14.920	57.14	60.70	57.42	58.420
2	13.24	16.19	15.73	15.053	59.40	65.66	63.99	63.017
3	18.11	15.74	16.93	16.927	72.48	61.00	68.30	67.260
4	10.96	10.62	10.58	10.720	48.95	46.57	44.98	46.833

In FDM technologies, part orientation influences shape quality due to fabricating the component in rises in the Z-direction and also the routing of slicing. A fine surface finish is acquired from the areas of the part that is regularly smoothened by the extrusion nozzle which is not directly in contact with the support material and build platform. Hence, the low value of surface roughness will always achieve at the top facing surface of a 3D printed part [16]. The result indicated that sample 2 is having the lowest value of surface roughness (1.51μm) among all five parts. This is because for the other samples, there is a curved surface of the part which the part mass is upright on the support. Stair-stepping would occur, thus making a rough surface structure and giving out the large value of surface roughness. Moreover, for the specimen 2, there is not much load from the part weight on the contact area between the support material and the part or the part load is not falling in the support. Therefore, it can be concluded that the best surface quality obtained for producing a part is at orientation X90°Y0°.

A general experimental study was proposed to investigate the effects of part orientation on tensile strength and maximum fracture load of printed samples. Based on table 6, the orientation of X0°Y° (sample 1) resulted in the largest values for tensile strength and load at fracture point, whereas X0°Y90° (sample 3) demonstrated the lowest values for both mechanical properties. For other samples, the amounts mainly fall in between that of 0° and 90° in the y-axis. This variance can be expounded by examining inter raster composite adhesions as well as trans-raster strength, which means the tensile force of each raster. In 0° orientation of the printed parts, the overshoot of inter-raster compound had minimal effect on the mechanical strength as every raster was moving upward of its longitudinal axis, instigating trans-raster tensile failure. However, for 90° orientation samples, the load was applied vertically to raster longitudinal axis which leads to weaker bonds. For this circumstance, layer bond alongside the shell number in the samples with 90° orientation directly influences the flexible capability, since the internal raster synthesis bonds amongst adjoining raster endured through the greater part of the exerted force.

Table 6. Experimental results for mechanical properties of different part orietation

SN	Tensile Strength (MPa)				Maximum Fracture Load (N)			
	Experimental				Experimental			
	Test 1	Test 2	Test 3	Avg	Test 1	Test 2	Test 3	Avg
1	29.01	29.98	29.08	29.36	1392.25	1439.23	1395.78	1409.09
2	23.16	24.02	24.88	24.02	1111.55	1153.02	1194.15	1152.91
3	12.49	16.22	15.43	14.71	599.65	778.62	740.56	706.28
4	19.77	17.49	20.55	19.27	948.78	839.49	986.27	924.84
5	18.92	23.11	21.76	21.26	907.96	1109.06	1044.50	1020.51

4. Conclusion

In this study, a 3D model was established to study the effects of the part orientation on dimensional accuracy, part strength, and surface quality. Results showed that X0°Y0° (sample 1) and X90°Y0° (sample 2) would be the most appropriate part orientation to produce the part with the highest dimensional accuracy and good surface quality, respectively. Orientation X0°Y0° demonstrated the highest values for tensile strength and maximum fracture load, giving 29.36 MPa and 1409.09 N, respectively. To fabricate a model with the least manufacturing cost, part orientation X0°Y0° would be the best choice.

References

[1] Calignano, Flaviana, Diego Manfredi, Elisa Paola Ambrosio, Sara Biamino, Mariangela Lombardi, Eleonora Atzeni, Alessandro Salmi, Paolo Minetola, Luca Iuliano, and Paolo Fino., 2017, *Overview on Additive Manufacturing Technologies,* Proceedings of the IEEE, **105**(4): p. 593-612.

[2] Raut, Sandeep, VijayKumar S. Jatti, Nitin K. Khedkar, and T. P. Singh., 2014, *Investigation of the Effect of Built Orientation on Mechanical Properties and Total Cost of FDM Parts,* Procedia Materials Science, **6**: p. 1625-1630.

[3] Afrose, Mst Faujiya, S. H. Masood, Pio Iovenitti, Mostafa Nikzad, and Igor Sbarski., 2015, *Effects of part build orientations on fatigue behaviour of FDM-processed PLA material,* Progress in Additive Manufacturing. **1**(1-2): p. 21-28.

[4] Górski, FILIP, WIESŁAW Kuczko, and RADOSŁAW Wichniarek., 2014, *Impact strength of ABS parts manufactured using Fused Deposition Modeling technology,* Archives of Mechanical Technology and Automation. **31**(1): p. 3-12.

[5] Górski, Filip, Radosław Wichniarek, Wiesław Kuczko, Przemysław Zawadzki, and Paweł Buń., 2015, *Strength of Abs Parts Produced by Fused Deposition Modelling Technology – a Critical Orientation Problem.* Advances in Science and Technology Research Journal, **9**: p. 12-19.

IOP Conf. Series: Materials Science and Engineering **694** (2019) 012048 doi:10.1088/1757-899X/694/1/012048

[6] Tak, Ashish, Prabhakar Ingavale, Chaitanya Khopatkar, Sumit Bhure, Vinod Barve, and Avadhoot Rajurkar, Effect of 3D Printing Parameters on Dimensional Accuracy & Shrinkage on Printed Parts.

[7] Mohd Nizam, S., S. Shamsul Anuar, and A. Mohd Azman., 2016, *Effect Of Part Features On Dimensional Accuracy Of FDM Model.* ARPN Journal Of Engineering And Applied Sciences, **11**(13): p. 8067-8072.

[8] Górski, F., W. Kuczko, and R. Wichniarek., 2013, *Influence of Process Parameters on Dimensional Accuracy of Parts Manufactured Using Fused Deposition Modelling Technology,* Advances in Science and Technology – Research Journal. **7**(19): p. 27-35.

[9] Singh, S., A. Sachdeva, and V.S. Sharma., 2013, *Investigation of dimensional accuracy/mechanical properties of part produced by selective laser sintering,* International Journal of Applied Science and Engineering 10 (1), p. 59-68.

[10] Badola, M. and R. Vaishya., 2015, *Parameters Affecting Surface Roughness of Fused Deposition Modeling.* i-Manager's Journal on Mechanical Engineering, **6**(1): p. 34.

[11] Sreedhar, P., C. Mathikumar Manikandan, and G. Jothi., 2012, *Experimental investigation of surface roughness for fused deposition modeled part with different angular orientation.* Technology, **5**(3): p. 21-28.

[12] Bakar, N.S.A., M.R. Alkahari, and H. Boejang., 2010, *Analysis on fused deposition modelling performance.* Journal of Zhejiang University-SCIENCE A, **11**(12): p. 972-977.

[13] Vijay, P., P. Danaiah, and K.V. D. Rajesh., 2012, *Critical Parameters Effecting the Rapid Prototyping Surface Finish.* Journal of Mechanical Engineering and Automation, **1**(1): p. 17-20.

[14] Tagore, G., S.D. Anjikar, and A. Venu Gopal., 2007, *Multi objective optimisation of build orientation for rapid prototyping with fused deposition modeling (FDM).* in *Seventeenth Solid Freeform Fabrication (SFF) Symposium, Austin.*

[15] Chua, C.K., K.F. Leong, and C.S. Lim., 2010 *Rapid prototyping: principles and applications.*: World Scientific.

[16] Abdelrhman, A.M., *Assessment of Geometrical Dimension and Tolereance of FDM Part*, MSc thesis in Faculty of Mechanical Engineering 2008, Universiti Teknologi Malaysia.

IOP Conf. Series: Materials Science and Engineering **694** (2019) 012049 doi:10.1088/1757-899X/694/1/012049

Effect of magnesium fluoride coating on corrosion behaviour of magnesium alloy

O Saliza Azlina[1], Mohd Shafiq Ruba'ai[1], D Kurniawan[2]

[1]Faculty of Mechanical and Manufacturing Engineering, Universiti Tun Hussein Onn Malaysia, Batu Pahat 86400, Malaysia
[2]Faculty of Engineering, Universiti Teknologi Brunei, Gadong BE1410, Brunei Darussalam

E-mail: denni.kurniawan@utb.edu.bn

Abstract. Magnesium and its alloys are explored as potential biomedical materials for being lightweight, bio-absorbable, and having attractive biological properties. A major hindrance for their use is their high corrosion rate, in particular when exposed to body fluids. This study aims at suppressing the corrosion rate of a magnesium alloy (Mg1.0Ca) by coating it with magnesium fluoride (MgF_2). The coating was done by immersion of the workpiece in hydrofluoric acid solution. For comparison, pure magnesium was also coated with MgF_2. The MgF_2 coated magnesium exhibits significantly lower corrosion rate than pure magnesium. The MgF_2 coated magnesium alloy shows even lower corrosion rate. The MgF_2 coating works in inhibiting corrosion on magnesium alloy Mg1.0Ca. The corrosion inhibition was also contributed by other compound formed during reaction between Mg1.0Ca and hydrofluoric acid and the alloy in Mg1.0Ca.

1. Introduction

Metallic biomaterials are usually made of stainless steels, titanium alloys, or cobalt–chromium alloys. They serve well for medical devices that are directly in contact with human body, including implants [1-3]. However, these metallic implants are reported to lead to the problem of stress shielding effect which is caused by the mismatch in mechanical properties between these implant materials and human bone [4]. This mismatch (metallic implant has much higher mechanical properties compared to bone) will cause stress concentration at bone–implant interfaces and consequently reducing stimulation of new bone growth and decreasing implant stability [2, 5-8]. In addition, the implants also remain as permanent fixtures, which is not ideal for bone trauma fixation. The implant needs to be removed with follow up surgery after the bone tissue has healed. This means additional cost and more inconvenience to the patient [1,7].

For this temporary implants purpose and also for tissue engineering scaffolds, biodegradable metals are being developed. Magnesium based alloys are potential candidates because of their physical and mechanical properties, including high specific strength and modulus elasticity close to cartilage and bone, and good effect on bone formation [1,3,5,9]. Magnesium (Mg) is an essential element for human metabolism [2,4,10,11]. When it degrades in body fluids or corrodes in aqueous environments that contain chloride ions, the magnesium can be resorbed by the body [1,9,12].

Magnesium in itself does not have sufficient mechanical properties to be used as medical devices which require biodegradable metals. Common way to increase its mechanical properties is by alloying. Commercially available magnesium alloys intended for structural use can be used for this purpose, but they often contain non-biocompatible alloying elements [13]. This study uses Mg-Ca alloy considering Ca is also an essential element for human bone, it degrades in body fluids with non-toxic degradation product, and it affects the corrosion rate of the alloy [3,14].

Another main problem with magnesium and its alloys is their high corrosion rate. This high corrosion rate leads to loss of mass or structural integrity, which means their mechanical properties become even lower [15]. For the intended application as temporary implants, the biomaterial is expected to degrade after the bone tissue heals, and not before. This expectation cannot be fulfilled by magnesium and its alloys for the time being because their corrosion rate is too high.

Corrosion resistance of magnesium alloys is influenced by multiple factors, including the alloying elements, purity of the alloys, and the presence of protective coatings [8]. In order to enhance the corrosion resistance of magnesium, several techniques have been suggested, including fluoride conversion coatings [5,9,16-18]. This fluoride conversion coating is done by immersion of magnesium workpiece in hydrofluoric acid (HF) at room temperature. On pure magnesium, it was reported that a thin layer of MgF_2 and small amount of $Mg(OH)_2$ was formed on the workpiece surface after this immersion treatment [3,14]. The presence of this thin layer coating increases polarisation resistance and reduces the corrosion current density, resulting an overall lower corrosion rate of treated magnesium compared to pure magnesium when exposed to simulated body fluid [3,14]. Similar fluoride conversion coating was also applied on some magnesium alloys, including AZ31 [19]. The treated magnesium alloy was reported to show higher corrosion resistance compared to untreated magnesium alloy.

There is limited information yet on the effect of fluoride conversion coating on magnesium alloys with calcium as the alloying element. This study aims at suppressing the corrosion rate of a magnesium-calcium alloy (Mg1.0Ca) by fluoride conversion through immersion in hydrofluoric acid solution. Similar treatment was also conducted on pure magnesium for comparison.

2. Materials and methods

The workpiece materials in this study are pure Mg and Mg1.0Ca. The magnesium alloy was obtained through alloying done in house, using pure magnesium ingot (99.99% Mg) and calcium-magnesium master alloy (Mg−32Ca) as starting materials; the process as reported in [14]. The samples were cut into dimension of 10 mm × 10 mm × 10 mm by using saw and then were mounted, ground and polished until the surface are homogeneous, smooth, and free from scratches. Surface roughness test was conducted on the samples' surface using portable surface roughness tester Mitutoyo SJ-400. It was recorded that the samples have a surface roughness (Ra) of 0.04 μm.

Chemicals used were of reagent grade, purchased from Sigma Aldrich. Initial treatment was immersion of the samples in sodium hydroxide solution for two hours. Afterwards, the solution of hydrofluoric acid with concentration of 48wt% was used to perform fluoride conversion coating on the pure Mg and Mg1.0Ca surfaces through immersion technique. The immersion was done at room temperature for 24 hours.

Surface characterization was conducted by using optical microscope and scanning electron microscopy (SEM), and then followed by corrosion testing using potentiodynamic polarisation test. The polarisation test was carried out to study the corrosion behaviour of the Mg and its alloy in simulated body fluid (SBF). The SBF used in this study was physiological saline (0.9% NaCl) solution that was prepared by mixing sodium chloride in distilled water. The electrochemical cell with three electrodes was connected to VersaSTAT Instrument Potentionstat and VersaSTAT software was controlled by computer to measure the polarization curves of samples.

3. Results and discussion

Figure 1 shows the surfaces of pure magnesium (Mg) and magnesium-calcium alloy Mg1.0Ca, untreated and fluoride-coated. The grain size of the magnesium-calcium alloy seems to be larger than pure magnesium's. Based on previous analysis done in house on Mg-Ca alloys, the untreated Mg1.0Ca alloy microstructure has predominantly α-Mg content and some Mg_2Ca phase [14]. The eutectic Mg_2Ca phase are distributed on the grain boundaries and within grains of α-Mg matrix, seen as small precipitates are observed within the microstructure. The width of the grain boundary of the magnesium alloy also was wider compared to pure magnesium, in agreement with previous study [11]. For treated magnesium and magnesium alloy samples, dark regions of magnesium fluoride were formed, with the latter showing more obvious dark regions. The dark regions are likely to be a form of magnesium hydroxide, which might be developed by the magnesium in reaction with sodium hydroxide [20]. In addition to the presence of MgF_2 and $Mg(OH)_2$, there is possibility that CaF_2 and $Ca(OH)_2$ phases were grown on the workpiece surface during the immersion processes. The magnesium alloy showed even more oxidized regions on its surface, indicating more residual magnesium hydroxide, which might be due to the presence of more magnesium or the calcium on its surface.

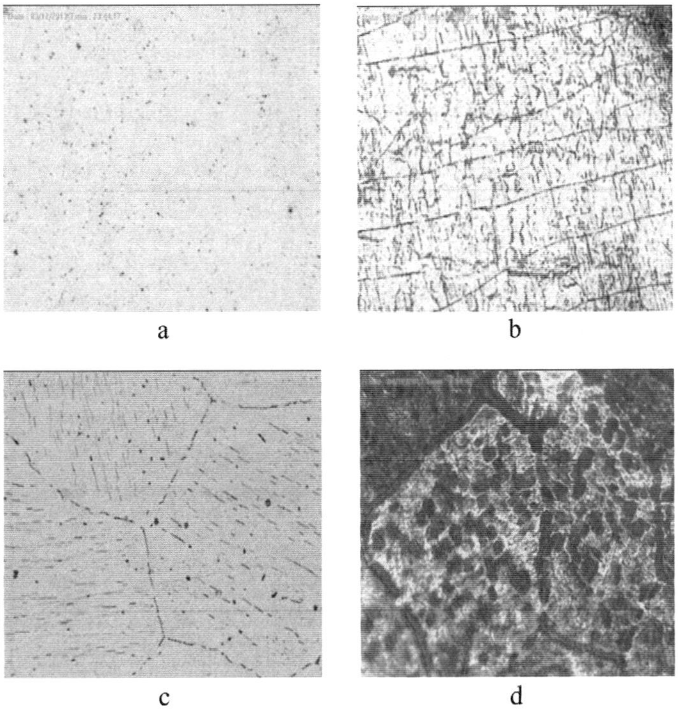

Figure 1. Surface characteristic of (a) uncoated Mg, (b) fluoride-coated Mg, (c) uncoated Mg1.0Ca, and (d) fluoride -coated Mg1.0Ca.

The SEM micrograph of the cross-section of fluoride-coated for both pure Mg and Mg1.0Ca is as depicted in figure 2. Also in figure 2, there is uniform thin layer of MgF_2 coating on the workpiece's surface. The thickness of the MgF_2 coating is about 5±1 μm for both magnesium and magnesium alloy workpieces. The MgF_2 coating appears to be well adhered to the substrate. From the appearance and possible chemical reaction, the MgF_2 coating grew by replacing some of the $Mg(OH)_2$ layer and on top of the residual $Mg(OH)_2$ layer [3,10].

IOP Conf. Series: Materials Science and Engineering **694** (2019) 012049 doi:10.1088/1757-899X/694/1/012049

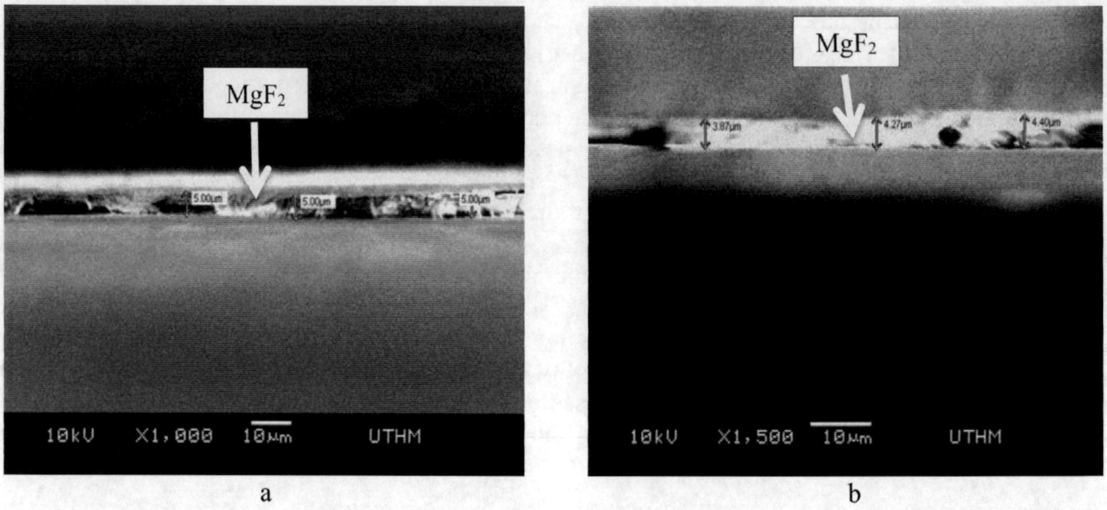

Figure 2. SEM micrograph of the cross section of fluoride coated on (a) pure Mg (b) Mg1.0Ca.

The corrosion rate of the magnesium and magnesium alloy workpieces was determined by using potentiondynamic polarisation method. The effect of current charging was obtained to determine the Tafel slopes and corrosion rate as presented in Fig. 3. For pure magnesium, the corrosion rate was estimated to be about 8.80 mm/year. Meanwhile, fluoride-coated pure magnesium shows a much lower corrosion rate of 0.35 mm/year. This significantly lower corrosion rate of fluoride coated Mg compared is consistent with previous studies [5,9,16-18]. The indicators for higher corrosion resistance of the treated workpiece were the increased voltage and lower current compared to those of the pure Mg (Figs. 3a and 3b). The main contributor is the presence of MgF_2 layer on the treated workpiece's surface. The MgF_2 layer functions as barrier coating, preventing the corrosive action on the Mg substrate. In addition to the presence of MgF_2 layer, small amount of $Mg(OH)_2$ might also present on the workpiece's surface.

It is expected that the corrosion rate of Mg1.0Ca alloy to be slightly lower than pure magnesium. This is related to the slight change in corrosion potential. Addition of 1.0 wt% Ca to pure Mg changed the corrosion potential towards more cathodic values, because Ca has higher electronegativity compared to Mg [14]. Fluoride-coated Mg1.0Ca alloy estimated the corrosion rate to be about 0.25 mm/year. This corrosion rate is higher than that of fluoride coated Mg workpiece. The indicator was that fluoride-coated Mg1.0Ca alloy showed higher voltage and lower current compared to its pure Mg counterpart (Figs. 3b and 3c). These translated to about 40% improvement in corrosion rate. The improvement might be due to the presence of CaF_2 and $Ca(OH)_2$ phases in addition to MgF_2 and $Mg(OH)_2$ which present in both workpieces. Another possible reason for the improvement is the fact that Mg1.0Ca substrate is more cathodic for having Ca which is more electronegative compared to Mg [14].

4. Conclusion

The effect of magnesium fluoride coating on corrosion behavior of magnesium alloy has been investigated. The fluoride coating was done by immersion process. The Mg1.0Ca alloy consisted of α-Mg matrix with Mg_2Ca distributed on the grain boundaries and as small precipitates within the grains. The treated Mg1.0Ca alloy has MgF_2 coating and some amount of $Mg(OH)_2$. After corrosion test in saline solution, it was found that fluoride-coated Mg1.0Ca workpiece showed higher voltage and lower current compared to fluoride coated pure magnesium. In terms of corrosion rate where the lower value is the better, fluoride coated Mg1.0Ca workpiece showed lower corrosion rate (of 0.25 mm/year) compared to fluoride coated pure magnesium (of 0.35 mm/year). This indicates the presence of CaF_2 and $Ca(OH)_2$ phases within the fluoride treated Mg1.0Ca workpiece. When compared to pure Mg, the fluoride coated magnesium and magnesium alloy showed much lower corrosion resistance. This means fluoride coating has the potential to inhibit corrosion to occur in magnesium-calcium alloys.

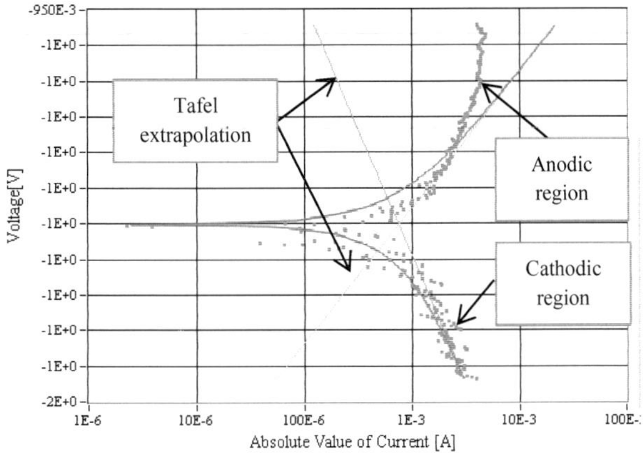

a

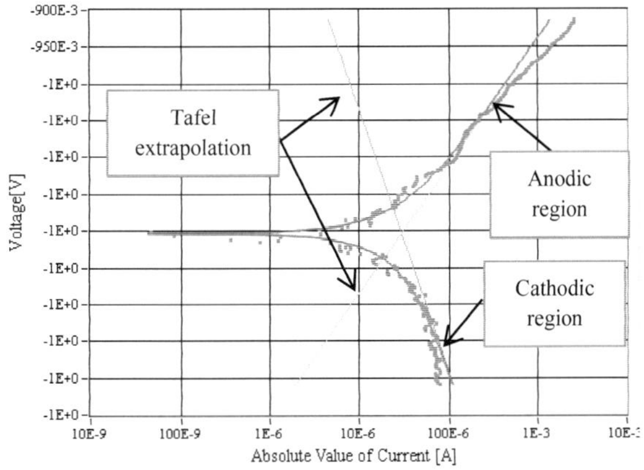

b

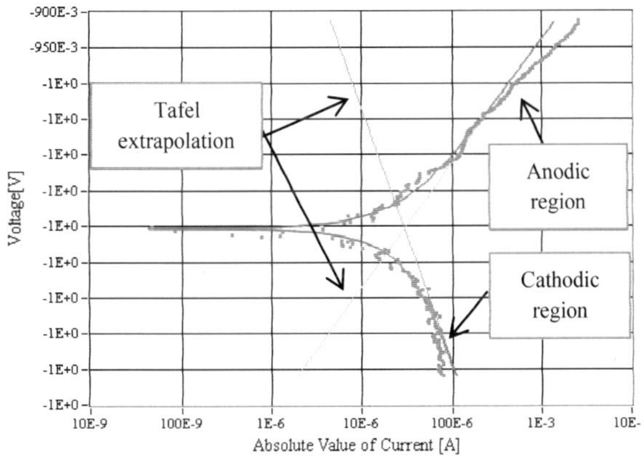

c

Figure 3. Tafel extrapolation at polarisation curves for (a) pure Mg, (b) fluoride-coated Mg, and (c) fluoride-coated Ma1.0Ca.

Acknowledgement
The authors appreciate the facilities provided by Universiti Tun Hussein Onn Malaysia and Universiti Teknologi Malaysia. The authors also thank the assistance from Prof. Dr Izman Sudin of Universiti Teknologi Malaysia and Dr. Murtala Sule Dambatta during the experiments.

References
[1] Gu, X-N., Li, S-S., Li, X-M, Fan, Y-b., "Magnesium Based Degradable Biomaterials: A Review". *Mater. Sci.*, Vol 8, No. 3, pp. 200–218, 2014.
[2] Shadanbaz, S., Dias, G-J., "Calcium Phosphate Coatings on Magnesium Alloys for Biomedical Applications: A Review". *Acta Biomaterialia*, Vol., pp. 20–30, 2012.
[3] Chiu, K.Y., Wong, M.H., Cheng, F.T., Man, H.C., "Characterization and Corrosion Studies of Fluoride Conversion Coating on Degradable Mg Implants". *Surface & Coatings Technology*, Vol. 202 , pp.590–598, 2007.
[4] Sara, K.F.A., Semeth, S., Tan, M.J., "Hydrothermally Deposited Protective and Bioactive Coating for Magnesium Alloys for Implant Application". *Surface & Coatings Technology*, Vol. 258, pp. 931–937, 2014.
[5] Wang, Y-P., He, Y-H., Zhu, Z-J., Jiang, Y., Zhang, J., Niu, J-L. , Mao, L., Yuan, G-Y., "In Vitro Degradation And Biocompatibility of Mg-Nd-Zn-Zr Alloy". *Chinese Science Bulletin*, Vol. 57, No.17, pp. 2163-2170, 2012.
[6] Witte F. "The History of Biodegradable Magnesium Implants: A Review". *Acta Biomater.*, Vol. 6, pp.1680–1692, 2010.
[7] Pietak, A., Mahoney, P., Dias, G. J., Staiger, M. P., "Bone-Like Matrix Formation on Magnesium and Magnesium Alloys". *Journal of Materials Science: Materials in Medicine*, Vol. 19, no. 1, pp. 407–415, 2008.
[8] Song, G., "Control of Biodegradation of Biocompatible Magnesium Alloys". *Corrosion Science*, Vol. 49, pp. 1696–1701, 2007.
[9] Yang, J-X., Cui, F., Lee, I.S., "Surface Modifications of Magnesium Alloys for Biomedical Applications". *Annals of Biomedical Engineering*, Vol. 39, No. 7, pp. 1857–1871, 2011.
[10] Staiger, M.P., Pietak, A.M., Huadmai, J., Dias, G., "Magnesium and Its Alloys as Orthopedic Biomaterials: A Review". *Biomaterials*, Vol. 27, No. 9, pp. 1728-1734, 2006.
[11] Wan, Y., Xiong, G., Luo H., He, F., Huang, Y., Zhou, X.,. "Preparation and Characterization of a New Biomedical Magnesium–Calcium Alloy". *Materials & Design*, Vol. 29, No. 10, pp. 2034–2037, 2008
[12] Hassel, S.P., Bach, T., F-W and Krause, C., "Influence Of The Alloy Composition On The Mechanical And Electrochemical Properties of Binary MgCa Alloys And Its Corrosion Behavior In Solutions At Different Chloride Concentration", *7th International Conference On Magnesium and Their Application*, 06-09 November, in Dresden, 2006.
[13] Dambatta, M.S., Murni, N.S., Izman, S., Kurniawan, D., Froemming, G.R.A., Hermawan, H. (2015) In Vitro Degradation and Cell Viability Assessment of Zn–3Mg Alloy for Biodegradable Bone Implants. *Proceedings of the Institution of Mechanical Engineers, Part H, Journal of Engineering in Medicine*. 229: 335-342.
[14] Harandi, S.E., Mirshahi, M., Koleini, S., Idris, M.H., Jafari, H., Kadir, M.R.A., "Effect of calcium content on the microstructure, hardness and in-vitro corrosion behavior of biodegradable Mg-Ca binary alloy". *Materials Research*, Vol. 16, pp.11-18, 2013.
[15] Dambatta, M.S., Izman, S., Yahaya, B., Lim, J.Y., Kurniawan, D., "Mg-Based Bulk Metallic Glasses for Biodegradable Implant Materials: A Review on Glass Forming Ability, Mechanical Properties, and Biocompatibility". *Journal of Non-Crystalline Solids*, Vol. 426, pp. 110-115, 2015.
[16] Li, T., He, Y., Zhang, H., Wang, X., "Microstructure, Mechanical Property and In Vitro Biocorrosion Behavior of Single-Phase Biodegradable Mg-1.5Zn-0.6Zr Alloy" *Journal of Magnesium and Alloys*, Vol. 2, pp. 181-189, 2014.

[17] Thomann, M., Krause, C., Angrisani, N., Bormann, D., Hassel, T., Windhagen, H., Meyer-Lindenberg, A., "Influence of a Magnesium-Fluoride Coating of Magnesium-Based Implants (MgCa0.8) on Degradation In A Rabbit Model". *Journal of Biomedical Materials Research Part A*, Vol. 93, No. 4, pp. 1609-1619, 2010.

[18] Razavi, M., Fathia, M., Savabi, O., Razavi, S.M., Heidari, F., Manshaei, M., Vashaee, D., Tayebi, L., "In Vivo Study Of Nanostructured Diopside ($CaMgSi_2O_6$) Coating on Magnesium Alloy as Biodegradable Orthopedic Implants". *Applied Surface Science*, Vol. 313, pp.60–66, 2014.

[19] Yan, T., Tan, L. Xiong, D., Liu, X., Zhang, B., Yang, K., "Fluoride treatment and in vitro corrosion behavior of an AZ31B magnesium alloy". *Materials Science and Engineering: C*, Vol. 30, Issue 5, pp. 740-748, 2010.

[20] Bakhsheshi-Rad, H.R., Idris, M.H., Kadir, M.R.A., Daroonparvar, M., "Effect of fluoride treatment on corrosion behavior of Mg–Ca binary alloy for implant application". *Trans. Nonferrous Met. Soc. China*, Vol. 23, pp. 699–710, 2013.

AUTHOR INDEX

AUTHOR INDEX